Methods
of the Classical Theory
of Elastodynamics

Vladimir B. Poruchikov

Methods of the Classical Theory of Elastodynamics

With 74 Figures

Springer-Verlag
Berlin Heidelberg New York
London Paris Tokyo
Hong Kong Barcelona
Budapest

Professor Dr. Vladimir B. Poruchikov
Institute of Mechanics, Moscow State University
1, Michurinsky Prospect, 117192 Moscow, Russia

Translators:
Dr. V.A. Khokhryakov (Chaps. 1–4)
kv. 44, d. 1/10, st. Malaya Ostroumovskaya, Moscow 107014 Russia

Dr. G. P. Groshev (Chaps. 5–7)
kv. 57, d. 4a, st. Serov, Zhukovsky town, Moscow region 140160, Russia

Title of the original Russian edition: *Metody, dinamicheskoĭ teorii uprugosti* Nauka, Moscow 1986

ISBN-13:978-3-642-77101-9 e-ISBN-13:978-3-642-77099-9
DOI: 10.1007/978-3-642-77099-9

Library of Congress Cataloging-in-Publication Data. Poruchikov, V. B. (Vladimir Borisovich) [Metody dinamicheskoĭ teorii uprugosti. English] Methods of the classical theory of elastodynamics / V. B. Poruchikov. p. cm. Translation of: Metody dinamicheskoĭ teorii uprugosti. Includes bibliographical references and index. ISBN-13:978-3-642-77101-9 (U.S.) 1. Elasticity. I. Title. TA418.P6713 1992 531'.382--dc20 92-6511

Softcover reprint of the hardcover 1st edition 1993

Cover Design: Erich Kirchner

Typesetting: Springer Tex in-house system
Production Editor: P. Treiber

54/3140-5 4 3 2 1 0 – Printed on acid-free paper

Preface

Numerous significant problems of present-day theoretical and applied science are closely related to studies of wave propagation in continuous media, determination of stress fields induced within bodies by impingement of waves or impulsive loadings, as well as to calculations of the behavior of structures subjected to sudden shocks. Primarily, this class of problems incorporates studying the effects of blast and seismic waves, hypervelocity impacts of bodies upon obstacles, penetration of solid bodies into compressible media, and crack propagation following impulsive loadings.

To solve these problems, one has to employ methods of continuum mechanics and, in particular, those of elastodynamics, which has recently been most intensively developed. However, until recently, there have been no books available presenting a comprehensive and systematic analysis of rigorous analytical methods of elastodynamics, which, although thoroughly developed during the last 25–30 years, are presented in a variety of hardly to obtain publications.

In 1986 a monograph by the present author, *Methods of the Dynamic Theory of Elasticity* (Nauka, Moscow) which attempted to fill the gap, was published in Russian. It considered various methods of obtaining accurate analytical solutions to dynamic problems of the linear theory of elasticity: both the classical ones which were thoroughly developed during the last decades, and the ones proposed quite recently.

The present book is an English translation of the revised Russian edition. It should be noted that in preparing the English edition the list of references has been thoroughly updated and extended and now adequately reflects recent progress in analytical methods of elastodynamics. References difficult to obtain outside this country were omitted while English language ones were retained. Although many Soviet science journals are presently translated into English, in those cases where the author had no reliable information on the availability of translated editions, references to original Russian publications are given.

The monograph is addressed to research scientists, engineers and specialists working in various branches of mechanical engineering and aerospace technology, structural mechanics, seismology, as well as to students and postgraduates engaged in studying transient processes in continuous media.

The author is sincerely grateful to Dr. V.A. Khokhryakov and Dr. G.P. Groshev whose hard work in translating the book was of great help for publishing it in English.

Moscow, August 1992 *V. B. Poruchikov*

Contents

1. Introduction

Many scientific and technological problems are closely related to studying wave processes in continuous media, the behavior of structures subjected to impulsive loadings, and the phenomena of penetration and destruction. These problems can often be solved by applying the methods of elastodynamics.

Since disturbances arising in both solid and dense liquid media, such as metals, rock materials, and water, may remain small even at large absolute values of applied loadings or pressure drops at the wavefronts, such media may be treated with the linear approach within a wide range of mechanical parameters [1.1]. Therefore, from the mathematical point of view, for such physical problems one has to solve a system of linear equations coupled through linear boundary conditions and satisfying given initial conditions.

Thus, many applied problems may be solved by employing the linear theory of dynamics of a solid deformable body. This has stimulated rapid development of this approach during the last few decades, and has called to life numerous monographs, textbooks and reviews [1.2–26].

It should be noted that analytical methods play an important role in the study of dynamic processes. They are needed for solving model and test problems, carrying out qualitative studies of properties and particularities of solutions, and for obtaining numerical solutions to complicated three-dimensional dynamic problems which require the use of combined numerical and analytical methods. Asymptotic methods have been thoroughly treated by *Babich* and *Buldyrev* [1.27] in their book on short-wavelength asymptotics of diffraction problems of acoustics, and by *Achenbach* et al. [1.20] in the monograph on applying ray methods to elastodynamics.

Until recently there have been no books presenting a systematic discussion of rigorous analytical methods of elastodynamics, and this circumstance has hindered their study.

The present monograph is intended to fill this need. A variety of practical methods of obtaining accurate analytical solutions to dynamic problems in the linear theory of elasticity are considered. The scope of the book is limited to transient problems which play the most important role in dynamic processes. Along with the the classical methods thoroughly developed in recent years, such as the Smirnov–Sobolev method, the integral transform, and the Wiener–Hopf approach, this monograph treats new and highly efficient methods of solving two and three-dimensional problems: transient problems with moving boundaries, wedge problems with mixed boundary conditions, homogeneous solutions for the general anisotropic case (Willis' method).

This material forms the content of the subsequent six chapters of the book (Chaps. 2–7).

Chapter 2 considers the formulation of dynamic problems of the theory of elasticity. It also describes some general results, and in particular, compares various representations of solutions to the Lamé equations.

Chapter 3 is devoted to the Smirnov–Sobolev method. Homogeneous solutions of integer degree in space and time variables are considered in detail. Representations of solutions to three-dimensional problems through two-dimensional ones are given, and numerous examples of solving plane-strain and axisymmetric problems are considered. These are most interesting in application to real problems. In the closing part of the chapter, an alternative derivation of the Smirnov–Sobolev representations is discussed, which makes possible the generalization of the method to anisotropic media.

Chapter 4 considers the application of integral transforms to elastodynamics. Here Cagniard's method is described, and solutions to a number of acoustic and elastic problems are discussed, obtained with the help of various integral representations.

In Chap. 5 a method for solving three-dimensional dynamic problems for wedge-shaped regions with mixed boundary conditions is presented, which has been developed by the author. This method was used to obtain solutions to three-dimensional problems of diffraction of spherical and plane waves on a smooth rigid wedge with an arbitrary vertex angle.

Chapter 6 is on applying the Wiener–Hopf method to dynamics of elastic bodies. Firstly, the method is studied in its traditional form where the entire analysis is performed in the complex plane. This situation is characteristic of problems with regions having different boundary conditions and a boundary that is stationary or moving at a constant speed. Secondly, we consider the case where upon factorization of the functions, the analysis is carried out in the plane of physical variables. This situation occurs in mixed boundary-value problems where the boundary moves at a time-dependent speed. Applications of the Wiener–Hopf method are illustrated by the problem of a punch penetrating acoustic and elastic media and of crack propagation due to tensile and shear stresses.

An efficient method for obtaining homogeneous solutions to dynamic problems for an arbitrary anisotropic continuous medium (i.e., having 21 different elastic constants) is presented in Chap. 7. This method, developed by Willis, until now has been described in only one paper [1.28], and in the author's opinion, is inadequately known. The method allows obtaining homogeneous solutions to complicated two- and three-dimensional problems in anisotropic media. As examples, solutions to Lamb's two and three-dimensional problems and to the two-dimensional self-similar problems of punch indentation and crack propagation are presented as well as to the three-dimensional self-similar problem of elliptic crack expansion.

It should be stressed that these solutions are presented at the level of rigor characteristic of undergraduate continuum mechanics courses. Thus, the problems are solved under the supposition that both the desired and the given functions satisfy the condition which allow application of the mathematical apparatus being employed. Where necessary, the conditions are explicity stated. In all cases a straightforward check may be used to show that the obtained solution satisfies all the formulated conditions.

2. Formulation of Elastodynamic Problems. Some General Results

In this chapter transient dynamic problems are formulated within the linear theory of elasticity. In particular, limitations on the solution in the neighborhood of singular curves and points are considered. The notions of solutions containing weak and strong discontinuities are introduced, and the uniqueness theorem for dynamic problems with strong discontinuities is proven. The Green–Volterra formula and various representations of solutions to the equations of motion of homogeneous isotropic continuous media are presented. These are needed for solving elastodynamic problems. The relationship between the solutions to transient problems and the ones to static, steady-state, and stationary dynamic problems is discussed.

2.1 Fundamental Equations of Elastodynamics

Within the framework of the classical linear theory of elasticity, the equations of motion of an ideally elastic homogeneous anisotropic solid may be presented in a stationary rectilinear Cartesian coordinate system x_1, x_2, x_3 in the form [e.g., Ref. 2.1]:

$$\sigma_{ij,j} + f_i = \varrho \ddot{u}_i \quad (i = 1,2,3) , \tag{2.1.1}$$

where the stress tensor σ_{ij} is related to the strain tensor through Hooke's law

$$\sigma_{ij} = c_{ijkl}\varepsilon_{kl} , \tag{2.1.2}$$

and the components ε_{kl} are defined by the displacement vector components u_1, u_2, u_3 as

$$\varepsilon_{kl} = \tfrac{1}{2}(u_{k,l} + u_{l,k}) \quad (k, l = 1,2,3) . \tag{2.1.3}$$

The elastic potential, or the strain energy density function, (i.e., potential energy of elastic deformations per unit volume) is a positively-definite quadratic form in the components of the strain tensor ε_{ij}:

$$W = \tfrac{1}{2}c_{ijkl}\varepsilon_{ij}\varepsilon_{kl} . \tag{2.1.4}$$

In (2.1.1–4) and hereafter, the following notation is used: $\boldsymbol{f} = (f_1, f_2, f_3)$ is the body force per unit volume, ϱ is the medium density (assumed to be constant), and c_{ijkl} are elastic constants satisfying

$$c_{ijkl} = c_{jikl} = c_{klij} \, . \tag{2.1.5}$$

Partial differentiation with respect to x_i is denoted by the subscript following the comma, $F_{j,i} \equiv \partial F_j/\partial x_i$, while partial differentiation with respect to time is denoted by a point, $\dot{F} \equiv \partial F/\partial t$. In addition, it is convenient to introduce the summation convention whereby a repeated subscript implies summation. For a repeated Roman subscript summation is carried out for 1, 2, 3, and for a repeated Greek subscript for 1, 2 (if not stated otherwise). In this sense,

$$F_{i,i} \equiv \sum_{i=1}^{3} \partial F_i/\partial x_i \, ,$$

$$F_{\alpha,\alpha} \equiv \sum_{\alpha=1}^{2} \partial F_\alpha/\partial x_\alpha \, .$$

By substitution of (2.1.3) into (2.1.2), and then of (2.1.2) into (2.1.1), one arrives at the dynamic equations for homogeneous anisotropic elastic media written in terms of displacements

$$c_{ijkl}u_{k,lj} + f_i = \varrho \ddot{u}_i \quad (i = 1,2,3) \, . \tag{2.1.6}$$

As is known [e.g., Ref. 2.1], the conditions (2.1.5) result in only 21 of the 81 elastic constants c_{ijkl} being independent. However, if a solid is subject to plane deformation, then only six elastic constants remain independent.

If an anisotropic body possesses some symmetric elastic properties, then the number of independent elastic constants c_{ijkl} is less than 21. Some of the most important cases of symmetry of elastic properties have been considered in the classical book by *Love* [2.2].

A solid having the same elastic properties in all directions is said to be fully symmetric and is called isotropic. For it the number of independent elastic constants reduces to two. Then, choosing Lamé constants λ, μ as independent constants, one obtains [2.3]

$$c_{ijkl} = \lambda\delta_{ij}\delta_{kl} + \mu(\delta_{ik}\delta_{jl} + \delta_{il}\delta_{jk}) \, ,$$

and, hence, Hooke's law for an isotropic continuous medium becomes

$$\sigma_{ij} = \lambda\delta_{ij}\varepsilon_{kk} + 2\mu\varepsilon_{ij} \, , \quad \delta_{ij} = \begin{cases} 0 & (i \neq j) \, , \\ 1 & (i = j) \, . \end{cases} \tag{2.1.7}$$

By substitution of (2.1.7) into (2.1.1) and employing (2.1.3) one obtains Lamé equations describing motion of a homogeneous isotropic elastic medium [2.1], written in vector notation as

$$(\lambda + \mu) \, \mathrm{grad \, div} \, \boldsymbol{u} + \mu\Delta\boldsymbol{u} + \boldsymbol{f} = \varrho\ddot{\boldsymbol{u}} \, , \tag{2.1.8}$$

where Δ is Laplace operator.[1]

[1] Often denoted as ∇^2 $(\Delta \equiv \nabla^2)$.

Note that (2.1.8) is presented in vector form and, hence, is valid in any coordinate system, while Hooke's law in the form of (2.1.7) is valid in any orthogonal curvilinear coordinate system.

By putting $\mu = 0$ in (2.1.7, 8), one obtains the linearized equations for an ideal compressible fluid. In this case shear stresses are absent, and the normal components of stresses are independent of the choice of an area element (taken with the opposite sign, they are called the pressure, p):

$$\sigma_{ij} = \lambda\delta_{ij}\varepsilon_{kk} = \begin{cases} 0 & (i \neq j) , \\ \lambda \operatorname{div} \boldsymbol{u} \equiv -p & (i = j) . \end{cases} \tag{2.1.9}$$

Substitution of (2.1.9) into (2.1.1) yields the equations of motion of an ideal compressible fluid (linearized Euler's equations):

$$-\operatorname{grad} p + \boldsymbol{f} = \varrho\dot{\boldsymbol{v}} \quad (\boldsymbol{v} \equiv \dot{\boldsymbol{u}}) . \tag{2.1.10}$$

Then, application of the div operator to both sides of (2.1.10) and exclusion of $\operatorname{div} \dot{\boldsymbol{v}}$ with the help of (2.1.9) twice differentiated with respect to time, result in

$$\Delta p = c_0^{-2}\ddot{p} + \operatorname{div} \boldsymbol{f} \quad \left(c_0^2 = \lambda/\varrho\right) . \tag{2.1.11}$$

Hence, disturbed motion of an ideal compressible fluid is described, in the linear approximation, by the wave equation, and therefore the case $\mu = 0$ is often called acoustic.

In concluding this section, it should be noted that the above linear equations of elastodynamics (2.1.6, 8) are valid if the processes under consideration are characterized by small displacements, strains, velocities, accelerations and density variations of portions of a continuum, while the processes themselves are adiabatic or isothermic. Also, in the case of an adiabatic process the values of elastic constants c_{ijkl} (as well as λ for an isotropic body) will differ with respect to the case of an isothermic process.

2.2 Initial and Boundary Conditions. Interfaces

If an elastic solid is unbounded, then a solution to the equation of motion is fully defined by specifying the displacement vector $\boldsymbol{u}$ and its derivative with respect to time, i.e., the velocity vector $\boldsymbol{v} = \dot{\boldsymbol{u}}$ at the initial moment, for instance, $t = 0$ (the Cauchy problem):

$$\boldsymbol{u} = \boldsymbol{u}_0(\boldsymbol{x}) , \quad \boldsymbol{v} = \boldsymbol{v}_0(\boldsymbol{x}) \quad \boldsymbol{x} = (x_1, x_2, x_3) . \tag{2.2.1}$$

Let us next consider an elastic solid V bounded by the surface S, Fig. 2.1. Let $\boldsymbol{\alpha}$, $\boldsymbol{\beta}$, $\boldsymbol{n}$ be a system of mutually orthogonal unit vectors defined at all points of S where it is smooth, $\boldsymbol{n}$ being a unit outward normal. Besides the initial conditions (2.2.1), one has to specify boundary conditions at the points of the surface S in one of the following four ways:

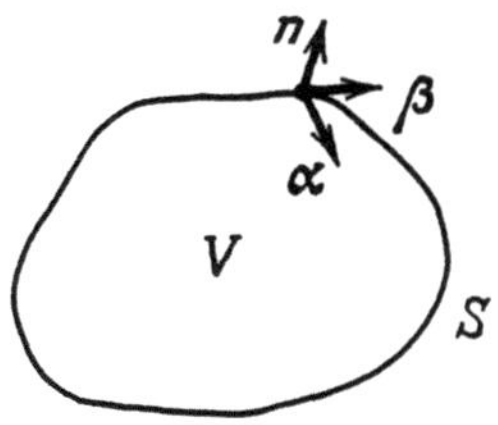

Fig. 2.1. Region V and its boundary S

(1) the stress vector $\boldsymbol{\sigma}_n$ is given on S (the 1st boundary-value problem)

$$\boldsymbol{\sigma}_n = \boldsymbol{\Sigma}_n(\boldsymbol{x}, t) , \quad \boldsymbol{x} \in S ; \tag{2.2.2}$$

(2) the displacement vector $\boldsymbol{u}$ is given on S (the 2nd boundary-value problem)

$$\boldsymbol{u} = \boldsymbol{U}(\boldsymbol{x}, t) , \quad \boldsymbol{x} \in S ; \tag{2.2.3}$$

(3) the normal component of the displacement vector u_n and the tangential components of the stress vector, $\sigma_{n\alpha}$, $\sigma_{n\beta}$ are given on S (the 3rd boundary-value problem)

$$u_n = U_n(\boldsymbol{x}, t) , \quad \sigma_{n\tau} = \Sigma_{n\tau}(\boldsymbol{x}, t) , \quad \boldsymbol{x} \in S \quad (\tau = \alpha, \beta) ; \tag{2.2.4}$$

(4) the normal component of the stress vector σ_{nn} and the tangential components of the displacement vector, u_α, u_β, are given on S (the 4th boundary-value problem)

$$\sigma_{nn} = \Sigma_{nn}(\boldsymbol{x}, t) , \quad u_\tau = U_\tau(\boldsymbol{x}, t) , \quad \boldsymbol{x} \in S \quad (\tau = \alpha, \beta) \tag{2.2.5}$$

(no summation).

Situations may occur when boundary conditions of different types (2.2.2–5) are specified on different parts of the surface S. There exist some other types of boundary conditions, too [2.4], which are not considered in this book.

Hereafter S is assumed to be a piecewise-smooth surface with a continuously rotating outward normal $\boldsymbol{n}$ everywhere except at a finite number of piecewise-smooth lines (the edges of the surface) and points (the vertices of the surface). The corner points of the edges will also be included into the number of vertices under the assumption that the number of the former is finite. The boundary conditions are not defined on the edges and at the vertices of surface S since here the latter does not possess normals. Also, the boundary conditions are not defined on the lines separating the parts of S with boundary conditions of different types. Also, the latter lines are supposed to be piecewise-smooth, and their number to be finite. We shall single out the corner points of these lines, and assume that their number is finite, too. Next, let us call all the lines of surface S, where the boundary conditions are not specified, singular. All the vertices of S as well as the corner points of the lines separating the parts of the surface with different boundary conditions we shall call singular. Conversely, the points of S which do not belong to the singular lines/points will be called regular. Thus, hereafter we shall deal with a finite number of singular points and lines belonging to the surface S.

It will be shown below that in the neighborhood of singular points/lines one has to impose additional constraints onto the behavior of the sought functions, so as to ensure uniqueness of a solution to the problem with specified initial and boundary conditions. We shall also analyze the physical meaning of these additional conditions. It is supposed that the position of the surface S and, hence, those of the singular points/lines may be time-dependent.

Besides the above considered surfaces S of the region V, some problems may involve interfaces S^0 (Fig. 2.2), i.e., surfaces separating elastic bodies with different properties, such as different densities, ϱ^+ and ϱ^-, or elastic constants, c^+_{ijkl} and c^-_{ijkl}, where the signs + and − are indexes for the different bodies. Thus, we do not exclude the possibility of the region V being filled with a number of different elastic bodies.

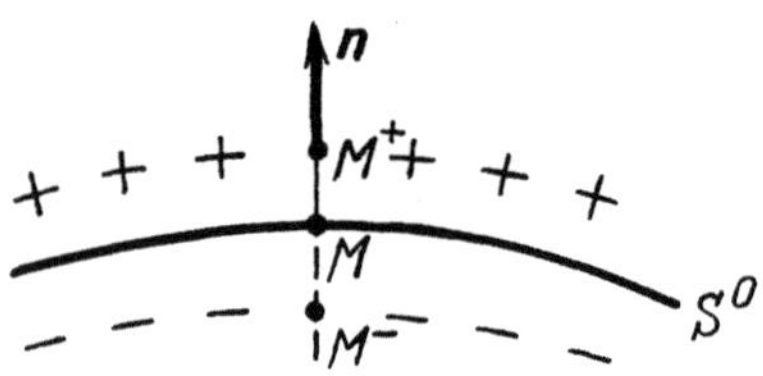

Fig. 2.2. Interface S^0

Usually, the following two types of boundary conditions are considered at regular points of the interface S^0:

(1) welded contact

$$\boldsymbol{\sigma}_n^+ = \boldsymbol{\sigma}_n^- , \quad \boldsymbol{u}^+ = \boldsymbol{u}^- ; \tag{2.2.6}$$

(2) frictionless contact

$$\begin{aligned} &\sigma_{nn}^+ = \sigma_{nn}^- , \qquad u_n^+ = u_n^- , \\ &\sigma_{n\tau}^+ = \sigma_{n\tau}^- = 0 \quad (\tau = \alpha, \beta) . \end{aligned} \tag{2.2.7}$$

Here, f^+, f^- ($f \equiv \boldsymbol{\sigma}_n, \boldsymbol{u}, \sigma_{nn}, u_n, \sigma_{n\tau}$) denote, respectively, finite limiting values of the function f when points M^+, M^- tend to the point $M \in S^0$, moving along the normal $\boldsymbol{n}(M)$ to both sides of S^0, and $\boldsymbol{\alpha}$, $\boldsymbol{\beta}$, $\boldsymbol{n}$ form a system of mutually perpendicular unit vectors defined at every regular point $M \in S^0$. It should be noted that situations may occur when contact conditions of both types coexist on different parts of the interface S^0. (Contact conditions other than (2.2.6, 7) are not considered.)

Supposing that the interfaces S^0 meet the same smoothness conditions as the boundary S, let us single out singular points/curves belonging to the surfaces S^0 in the same way as it was done for the surface S. Thus, below we shall deal with a finite number of singular points/curves belonging to the surfaces S^0, in whose neighborhood one has to impose additional constraints on the behavior of the sought functions, in order to ensure uniqueness of a solution to an elastodynamic problem.

2.3 Constraints Imposed on the Solution Behavior in the Neighborhood of Singular Points/Curves

In ensuring uniqueness of a solution to the dynamic problem, an important part is played by the constraints imposed on the behavior of a sought solution in the neighborhood of singular points/curves which are not considered in the known uniqueness theorems [2.5–8]. However, if one allows for the presence of singular points/curves in the conditions of the uniqueness theorem, then the requirement of the derivatives of displacements being continuous up to the points/curves proves to be too strong and leads to the loss of a broad class of solutions to elastodynamic problems characterized by discontinuity of the first derivatives of displacements at singular points/curves. This class of problems encompasses numerous transient problems related to the motion of a punch along the surface of an elastic continuous medium, propagation of cracks, diffraction of waves on bodies with edges and vertices, etc. However, the latter problems are of the greatest interest for both theory and applications. Therefore, in the discussion below, reasonable physical constraints are imposed on the behavior of the solutions to elastodynamic problems in the neighborhood of singular points/curves, which encompass a broad spectrum of situations and ensure solution uniqueness under specified initial and boundary conditions.

Let us consider a singular curve/point belonging to the surface S or S^0. Let the curve (point) be surrounded with a tubular (spherical) surface S_ε of the radius $r = \varepsilon$. In Fig. 2.3 the cross section of such a surface is shown by a plane perpendicular to the singular curve at a point on that curve (for a singular point it is a cross section by a meridional plane). The singular curve (point) itself and, hence, the associated surface S_ε can move at a velocity $\boldsymbol{\alpha} = \boldsymbol{\alpha}(t)$. By supposing that the external region V is filled with an elastic continuous medium, let us consider the expression for the energy flux I across the surface S_ε

$$I \equiv \lim_{\varepsilon \to 0} \iint_{S_\varepsilon} \left[(\boldsymbol{v} \cdot \boldsymbol{\sigma}_n) + \alpha_n \left(W + \varrho \frac{(\boldsymbol{v})^2}{2} \right) \right] ds \, , \tag{2.3.1}$$

where $\boldsymbol{n} = \boldsymbol{n}(\boldsymbol{x})$ is the outward (with respect to V) normal to S at the point $\boldsymbol{x}$, and $\alpha_n(t, \boldsymbol{x}) = (\boldsymbol{\alpha} \cdot \boldsymbol{n})$. The latter formula specifies the energy flux radiated ($I > 0$) or absorbed ($I < 0$) by a singular curve/point. In the case of a stationary ($\boldsymbol{\alpha} = 0$) singular curve (point) having no internal energy sources, one should assume that it neither absorbs nor radiates energy, i.e., the solution to the elastodynamic problem meets the condition

$$\lim_{\varepsilon \to 0} \iint_{S_\varepsilon} (\boldsymbol{v} \cdot \boldsymbol{\sigma}_n) ds = 0 \, . \tag{2.3.2}$$

Let the displacement vector components be of the order of $\varepsilon^\beta + \text{const}$ ($\varepsilon \to 0$) in the neighborhood of a stationary singular curve/point. Then, the main terms of the

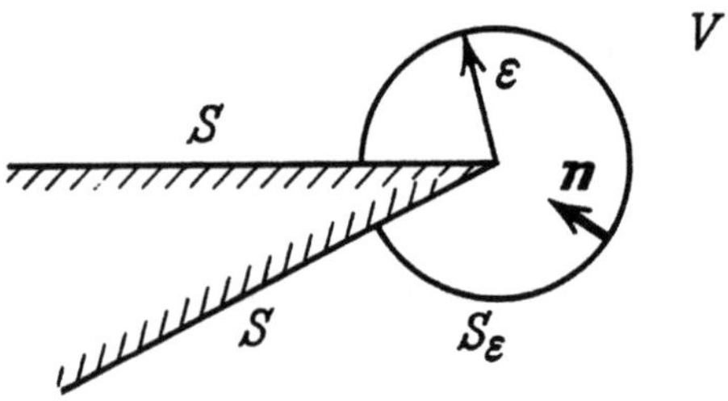

Fig. 2.3. The neighborhood of a singular curve/point

stress components are proportional to $\varepsilon^{\beta-1}$. However, since differentiation with respect to time does not affect the order of magnitude in the vicinity of a stationary curve/point, the velocity vector components are of the order of the displacement vector components (ε^{β}+const). The condition (2.3.2) is met, if the integrand grows slower than ε^{-1} in the neighborhood of a singular curve, or slower than ε^{-2} in the neighborhood of a singular point, since in the former case $ds = \varepsilon d\theta$ (where ε, θ are cylindrical coordinates), while in the latter case $ds = \varepsilon^2 \sin\psi d\psi d\varphi$ (where $\varepsilon, \psi, \varphi$ are spherical coordinates). Hence, in order to meet the condition (2.3.2) it is enough to require that the displacement vector be limited in the neighborhood of a singular curve, i.e., that $\beta > 0$, while in the neighborhood of a singular point the condition $\beta > -1/2$ must be met. In the case of moving ($\alpha \neq 0$) singular curves we shall allow for absorbing/radiating energy in accordance with

$$\lim_{\varepsilon\to 0} \iint\limits_{S_\varepsilon} \left[(\boldsymbol{v}\cdot\boldsymbol{\sigma}_n) + \alpha_n \left(W + \varrho\frac{(\boldsymbol{v})^2}{2}\right)\right] ds = C \quad (C \equiv C(t) \neq \infty) . \quad (2.3.3)$$

This is possible in the case of a crack propagating at a sub-Rayleigh speed when a moving crack edge absorbs energy which is expended in the destruction of the medium.

The condition (2.3.3) is met if the displacement vector satisfies

$$\boldsymbol{u} = \boldsymbol{u}_0 + O\left(\varepsilon^{\beta}\right) , \quad \beta \geq \tfrac{1}{2} , \quad \varepsilon \to 0 , \quad (2.3.4)$$

where $\boldsymbol{u}_0$ is independent of ε (hereafter we denote $O(x^{\beta}) \equiv f(x)$ if $|f(x)| \leq M|x|^{\beta}$ as $x \to 0$ or $x \to \infty$).

In fact, in the neighborhood of a singular curve/point $\boldsymbol{x} = \boldsymbol{x}_0(t)$ moving with the velocity $\boldsymbol{\alpha}(t)$ with respect to the stationary coordinate system x_1, x_2, x_3, one has

$$\boldsymbol{v}(\boldsymbol{x},t) \equiv \dot{\boldsymbol{u}}(\boldsymbol{x},t) = \left[\partial/\partial t - (\boldsymbol{\alpha}\cdot\text{grad})\right] \boldsymbol{u}(\boldsymbol{\xi},t) ,$$

where ξ_1, ξ_2, ξ_3 is a moving local system of coordinates related to the above neighborhood by the formulas

$$\boldsymbol{\xi} = \boldsymbol{x} - \boldsymbol{x}_0(t) , \quad \boldsymbol{\alpha}(t) = \dot{\boldsymbol{x}}_0(t) .$$

Hence, the velocity vector components are of the same order of magnitude as the stress components, namely, $\varepsilon^{\beta-1}$. Thus, we conclude that the integrand in (2.3.3) is of the order of $\varepsilon^{2\beta-2}$, and, hence, (2.3.3) is satisfied if $\beta \geq 1/2$. If $\beta > 1/2$, then $C \equiv 0$ in (2.3.3) or, in other words, a moving singular curve neither radiates, nor absorbs energy.

In the case of moving singular points (which takes place, for instance, when one considers a slender moving body) it will be assumed, without an essential limitation of the wide spectrum of elastodynamic problems, that the points neither radiate, nor absorb energy:

$$\lim_{\varepsilon\to 0} \iint_{S_\varepsilon} \left[(\boldsymbol{v}\cdot\boldsymbol{\sigma}_n) + \alpha_n\left(W + \varrho\frac{(\boldsymbol{v})^2}{2}\right)\right] ds = 0\,. \tag{2.3.5}$$

Since in this case the velocity vector components and the stress tensor components are of the same order of magnitude, $\varepsilon^{\beta-1}$, the condition (2.3.5) is met if $\beta > 0$.

Thus, while solving an elastodynamic problem one imposes the following constraints onto the solution behavior in the neighborhood of singular curves/points belonging to the surfaces S, S^0:

$$\boldsymbol{u} = \boldsymbol{u}_0 + \mathrm{O}\left(\varepsilon^{\beta}\right)\,, \quad \varepsilon \to 0\,, \tag{2.3.6}$$

where $\beta > -1/2$ for a stationary singular point, $\beta > 0$ for both a stationary singular curve and a moving singular point, and $\beta \geq 1/2$ for a moving singular curve; besides, the vector $\boldsymbol{u}_0$ on the right-hand side of (2.3.6) may depend on time and coordinates measured along a singular curve, while the estimates are supposed to be uniform with respect to both space variables and time.

The estimates (2.3.6) were obtained by supposing implicitly that the shear modulus $\mu \neq 0$. Then, for a stationary singularity the stress tensor components possess singularities of a higher order as compared to the velocity components. If, however, $\mu = 0$, then the shear components of the tensor, depending on the derivatives of the displacement vector with respect to coordinates, vanish, and the stress (i.e., pressure p) may prove to be a smoother function than velocity; hence, the case $\mu = 0$ should be considered separately. Then, as it follows from (2.1.10) for a sufficiently smooth density of body forces $\boldsymbol{f}$, if $p = p_0 + \mathrm{O}(\varepsilon^{\beta})$ as $\varepsilon \to 0$, the orders of magnitude allow the following estimates:

(a) for a stationary singularity

$$\dot{\boldsymbol{v}} \sim \boldsymbol{v} \sim \operatorname{grad} p \sim \varepsilon^{\beta-1}\,;$$

(b) for a traveling singularity

$$\boldsymbol{v} \sim \operatorname{grad} p \sim \varepsilon^{\beta-1}\,, \quad \boldsymbol{v} \sim p\,.$$

Then, similarly to (2.3.6) one obtains the following constraints on the behavior of the acoustic solution in the neighborhood of singular points/curves belonging to the boundary S:

$$p = p_0 + \mathrm{O}\left(\varepsilon^{\beta}\right)\,, \quad \varepsilon \to 0 \tag{2.3.7}$$

where $\beta > -1$ for a moving singular point, $\beta \geq -1/2$ for a moving singular curve, $\beta > -1/2$ for a stationary singular point, and $\beta > 0$ for a stationary singular curve, the estimates being supposed uniform with respect to space variables and time. Note that in the case of acoustics, the equality sign in the slack inequality $\beta \geq -1/2$, as well as in the case of an elastic medium for $\beta \geq 1/2$, indicates the possibility of both absorbing and radiating energy by a traveling singular curve.

2.4 Continuous and Discontinuous Solutions

Let a piecewise-homogeneous elastic medium occupy a region V bounded by piecewise-smooth surface S and incorporating a number of piecewise-smooth interfaces S^0. Let also the initial conditions (2.2.1) be specified within V, and the boundary conditions (2.2.2–5) on the boundary S. In addition, the contact conditions of the type (2.2.6,7) are assumed to be met at the interfaces S^0, and constraints (2.3.6) in the vicinity of singular points/curves belonging to S and S^0. We shall be seeking a solution $\boldsymbol{u}$ to the above mixed elastodynamic problem, i.e., a function $\boldsymbol{u}$ satisfying the equations (2.1.6) as well as the above initial and boundary conditions and the contact conditions and constraints in the neighborhood of singular points/curves.

Further, let the smoothness of the specified body forces in the initial and boundary conditions be such that

(1) the sought solution $\boldsymbol{u}$ and its first partial derivatives with respect to coordinates and time are continuous throughout V up to the boundary S with the exclusion of the interfaces S^0 and singular curves/points of S, and
(2) the second derivatives of $\boldsymbol{u}$ with respect to coordinates and time are continuous within V outside the interfaces S^0.

Such a solution will be called a continuous (classical) solution to the above mixed elastodynamic problem.

However, quite often, the specified functions (the body forces, the initial and boundary conditions) do not possess the required degree of smoothness for obtaining continuous second derivatives of $\boldsymbol{u}$ within V (but outside S^0). This is due to the fact that the equations (2.1.6) form a system of linear hyperbolic equations [2.9], and, hence, violation of the smoothness condition for the specified functions results in discontinuities propagating on characteristic surfaces within V. In this case, being limited only by the condition of integrability of the second derivatives of $\boldsymbol{u}$ within V, one obtains a solution containing weak discontinuities.

Furthermore, the specified functions may prove to be inappropriate for ensuring the above continuity of the first partial derivatives of $\boldsymbol{u}$ within V (outside S^0) and, hence, no solution to (2.1.6) exists in the usual sense. However, *Sobolev* [2.6] has shown that a discontinuous (generalized) solution may be constructed containing finite dicontinuities of the first partial derivatives of $\boldsymbol{u}$, on certain moving surfaces. The discontinuities subject to certain conditions were originally introduced by

Sobolev while solving the wave equation and were called by him regular strong discontinuities.

Within the framework of elastodynamics [see (2.5)], this terminology applied to discontinuities on a moving surface

$$F(\boldsymbol{x},t)=0 \quad (\boldsymbol{x}=(x_1,x_2,x_3)) \ , \tag{2.4.1}$$

which involve discontinuities of the first partial derivatives $\dot{u}_i, u_{i,j}$ of the solutions to (2.1.6) and are such that linear combinations of the first derivatives of a certain form remain continuous across the surface (2.4.1); in other words, the following conditions are met:

$$[\boldsymbol{M}_k]=0 \quad (k=1,2,3) \ , \quad \boldsymbol{M}_k \equiv \dot{\boldsymbol{u}} n_k + \alpha_n \frac{\partial \boldsymbol{u}}{\partial x_k} \ ; \tag{2.4.2}$$

$$[\boldsymbol{M}_4]=0 \ , \quad \boldsymbol{M}_4 \equiv \boldsymbol{\sigma}_n + \alpha_n \varrho \dot{\boldsymbol{u}} \ . \tag{2.4.3}$$

Here a symbol $[A]$ denotes a jump in the value of a function A at a point $N(\boldsymbol{x})$ across the surface (2.4.1) (Fig. 2.4), $[A]=A_+ - A_-$, where A_+ and A_- are the limiting values of the function A in approaching the point $N(\boldsymbol{x})$ from below and from above of the surface (2.4.1) respectively, $\boldsymbol{n} \equiv \boldsymbol{n}(N,t) = (n_1,n_2,n_3)$ is the normal to the surface (2.4.1) at the point $N(\boldsymbol{x})$ of the surface at time t, and $\alpha_n = \alpha_n(N,t)$ is the projection of the velocity $\boldsymbol{\alpha}$ [at which surface (2.4.1) travels] at the point $N(\boldsymbol{x})$ at time t on the normal $\boldsymbol{n}$:

$$\boldsymbol{n}=\frac{\operatorname{grad} F}{|\operatorname{grad} F|} \ , \quad \alpha_n = -\frac{\dot{F}}{|\operatorname{grad} F|} = -\frac{\dot{F}}{\sqrt{F_{,i}\,F_{,i}}} \ . \tag{2.4.4}$$

(Note that α_n is equal to the velocity value $|\boldsymbol{\alpha}|$ to within the accuracy of the sign. Besides, if one uses projections onto the coordinate axes $\boldsymbol{M}_k=(M_{k1},M_{k2},M_{k3})$, then the components M_{ki} $(k,i=1,2,3)$ form a second-rank tensor with respect to affine orthogonal coordinates, while the expression $\boldsymbol{M}_4$ remains a vector.) Hereafter it is supposed that the functions $\dot{u}_i$, $u_{i,j}$, σ_{ni} entering (2.4.2, 3) may contain finite discontinuities only, while the surface (2.4.1) is smooth and, hence, $F_{,k}$, $\dot{F}$ are continuous functions of their arguments.

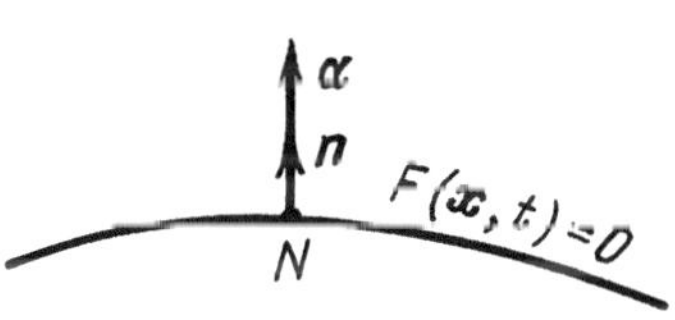

Fig. 2.4. The moving surface $F(\boldsymbol{x},t)=0$

Conditions (2.4.2) are called kinematic conditions of compatibility and have the following meaning: if one considers a four-dimensional space $x_1, x_2, x_3, x_4 = t$ within which the surface (2.4.1) is at rest, then from (2.4.2) it follows that the first partial derivatives of $\boldsymbol{u}$ with respect to tangential directions to the surface (2.4.1) remain continuous across the surface.

Conditions (2.4.3) represent the law of momentum conservation across the surface (2.4.1) and are called dynamic conditions of compatibility.

Thus, regular strong discontinuities on the moving surface (2.4.1) are those discontinuities of the first partial derivatives of the displacement vector $\boldsymbol{u}$ which originate while crossing the surface (2.4.1) and ensure fulfillment of both kinematic and dynamic compatibility conditions (2.4.2, 3).

What is the idea of introducing regular strong discontinuities, i.e., requiring both the kinematic and dynamic compatibility conditions to be fulfilled? The answer is that from the conditions (2.4.2, 3) it follows [2.5, 6, 9] that

(1) surface (2.4.1) must be a characteristic surface,
(2) a solution with regular strong discontinuities thus constructed has an explicit physical meaning since it may be viewed as the limit of a convergent (in a certain sense and within a certain metric) sequence of smooth solutions to the elastodynamic problem under consideration as long as the specified functions (the body forces, initial and boundary conditions) are presented as limits of sequences of the corresponding smooth functions which are convergent in the same sense.

In contrast, if one begins by constructing a discontinuous solution for specified discontinuous functions as the limit of a convergent (in a certain sense) sequence of smooth solutions with corresponding given smooth functions, and supposes that the kinematic compatibility conditions (2.4.2) are met, then it follows immediately that

(1) discontinuities of thus constructed solutions may occur on characteristic surfaces only, and
(2) the discontinuities satisfy the dynamic compatibility conditions (2.4.3), i.e., they are regular strong discontinuities.

Thus, for a solution to an elastodynamic problem with discontinuities of the first derivatives inside V to be physically meaningful (i.e. to be representable as a limit of a convergent sequence of smooth solutions), it is necessary and sufficient that it contains regular strong discontinuities.

Thus, the use of characteristic surfaces allows us to construct solutions containing not only weak discontinuities, but strong discontinuities, too. Furthermore, sometimes specified body forces, as well as initial and boundary conditions, prove to be such that one is unable to construct a generalized solution allowing for only the first derivatives of $\boldsymbol{u}$ inside V being discontinuous. However, in this case, too, one may construct a generalized physically meaningful solution if the specified functions are obtained as a result of a limiting (in a certain sense) passage with respect to a parameter n of some other, smoother, given functions. Then the possibility arises of constructing a sequence of smooth functions $\boldsymbol{u}_n$ converging, in a certain sense, to the function $\boldsymbol{u}$. In such cases one may construct a solution in which the function itself is discontinuous, while according to [2.10] these discontinuities inside V can propagate on chacteristic surfaces only. The construction and justification of generalized solutions of the dynamic theory of elasticity are presented in greater detail in [2.3, 5, 6, 9].

As stated above, both weak and strong discontinuities of derivatives as well as discontinuities of the function itself within V may propagate only on characteristic surfaces in physically meaningful solutions to the linear hyperbolic system (2.16). That is why such characteristic surfaces are called wave fronts.

According to [2.5, 9, 11, 12], in the general case of a homogeneous anisotropic continuous medium there exist three types of characteristic surfaces (wave fronts) propagating through an elastic medium with the speeds $c_1 = c_1(\boldsymbol{n})$, $c_2 = c_2(\boldsymbol{n})$, $c_3 = c_3(\boldsymbol{n})$, respectively, where $c_1 \geq c_2 \geq c_3$, and $c_k(\boldsymbol{n})$ $(k = 1, 2, 3)$ is the phase velocity of a corresponding plane wave propagating in the direction $\boldsymbol{n}$. Hence, the speed of propagation of a characteristic surface c_k depends not only on the elastic constants c_{ijkl} and density ϱ, but on the direction $\boldsymbol{n}$ of propagation of a characteristic surface at a considered point $N(\boldsymbol{x}, t)$, too. (In other words, it depends on the direction of the normal $\boldsymbol{n} = \boldsymbol{n}(N)$ to the characteristic surface at the point N presently located on the surface.) Thus, generally speaking, at a fixed arbitrary moment of time t the speed of propagation of a characteristic surface is different at different points of the latter.

Let us consider briefly wave fronts in isotropic media. In this case, characteristic surfaces of two types coincide $(c_2 = c_3)$, the speeds c_1, c_2 are independent of the direction in which characteristic surfaces propagate, and $c_1 = [(\lambda + 2\mu)/\varrho]^{1/2}$, $c_2 = [\mu/\varrho]^{1/2}$. If the displacement vector $\boldsymbol{u}$ in an isotropic medium is decomposed in the neighborhood of a point on a characteristic surface into the components u_n, u_t, normal and tangential to the surface, respectively, then in the case of a strong discontinuity across the characteristic surface of the first type propagating with the speed c_1, the first partial derivatives of u_n with respect to the normal and time, become discontinuous. However, the first partial derivatives of u_t remain continuous across the considered characteristic surface. Therefore, the first-type characteristic surfaces are called longitudinal wave fronts.

Characteristic surfaces of the other type propagate through an isotropic medium at a speed c_2 such that in the case of a strong discontinuity across these surfaces the first derivatives of the tangential component u_t with respect to the normal and time become discontinuous, while the first derivatives of u_n remain continuous. Characteristic surfaces of this type are called transverse wave fronts.

Thus, longitudinal and transverse waves travel through an isotropic continuous medium with the speeds $c_1 = [(\lambda + 2\mu)/\varrho]^{1/2}$ and $c_2 = [\mu/\varrho]^{1/2}$, respectively.

2.5 Uniqueness Theorem for Solutions to Elastodynamic Problems with Strong Discontinuities

Let us prove the uniqueness theorem for elastodynamic problems within the class of solutions with regular strong discontinuities. Let the region V be occupied with a piecewise-homogeneous elastic medium, and its surface S and interface S^0 be formed of a finite number of singly connected piecewise-smooth surfaces with no more than a finite number of singular points/curves. Let body forces $\boldsymbol{f}$ and initial conditions (2.2.1) be specified inside V, boundary conditions of the type (2.2.2–5) are valid on the surface S, and contact conditions of the type (2.2.6, 7) on the interface S^0, with the solution behavior in the neighborhood of singular points/curves (belonging to S, S^0) being subject to restraints (2.3.6).

Let us suppose that there exist two solutions, $\boldsymbol{u}_1, \boldsymbol{u}_2$ to the considered mixed elastodynamic problem within the class of solutions with regular strong discontinuities, in other words, such functions $\boldsymbol{u}_1, \boldsymbol{u}_2$ which

(1) satisfy the equations (2.1.16) at the points inside V where the functions are continuous together with their first and second partial derivatives with respect to coordinates and time;
(2) satisfy given initial conditions (2.2.1) at the points within V where the specified functions are continuous;
(3) satisfy given boundary conditions (2.2.2–5) at such points on S where the functions specified on S are continuous;
(4) satisfy the conditions (2.3.6) in the neighborhood of singular points/curves belonging to S;
(5) satisfy the contact conditions (2.2.6, 7) at the regular points belonging to the interface S^0;
(6) satisfy the restraints (2.3.6) in the neighborhood of singular points/curves belonging to S^0;
(7) may suffer weak and regular strong discontinuities on the characteristic surfaces F of the equations (2.1.6) while the number of such surfaces is supposed to be finite, and the surfaces themselves are asumed to be smooth.

Consider the difference $\boldsymbol{u} = \boldsymbol{u}_1 - \boldsymbol{u}_2$ which is a solution to the mixed elastodynamic problem in the absence of body forces, satisfies zero initial and boundary conditions and belongs to the class of solutions with regular strong discontinuities. Since due to the condition inside V, one has at any moment of time a finite number of interfaces S^0 and moving characteristic surfaces F where the first and the second partial derivatives of $\boldsymbol{u}$ are discontinuous, the region V is divided by the surfaces into a finite number of subregions V_k within which the solution is continuous together with its first and second partial derivatives with respect to coordinates and time and satisfies (2.1.6) for $f_i \equiv 0$.. Generally speaking, the boundary of a subregion V_k consists of parts of the surfaces S, S^0, F, and each of the subregions V_k is finite at any finite moment of time. This conclusion follows from the homogeneous initial and boundary conditions and the hyperbolicity of the system (2.1.6) which ensures a finite speed of propagation of disturbances within a medium.

Let us next transform the equations for $\boldsymbol{u}$ within the subregion V_k. By multiplying both sides of (2.1.1) (for $f_i \equiv 0$) by $\dot{u}_i$ and summing with respect to i one gets

$$\dot{u}_i \operatorname{div} \boldsymbol{\sigma}_i = \varrho(\dot{\boldsymbol{u}} \cdot \ddot{\boldsymbol{u}}) . \tag{2.5.1}$$

Differentiation of (2.1.4) with respect to t with the use of (2.1.2, 5) results in

$$\dot{W} = c_{ijkl}\dot{\varepsilon}_{ij}\varepsilon_{kl} = \sigma_{ij}\dot{\varepsilon}_{ij} = \operatorname{div}(\boldsymbol{\sigma}_m \dot{u}_m) - \dot{u}_m \operatorname{div} \boldsymbol{\sigma}_m \ . \tag{2.5.2}$$

Then, by expressing the left-hand side of (2.5.1) with the help of (2.5.2), one obtains

$$\frac{\partial}{\partial t}\left[W + \frac{\varrho}{2}(\dot{\boldsymbol{u}})^2\right] = \operatorname{div}(\boldsymbol{\sigma}_m \dot{u}_m)\ . \tag{2.5.3}$$

The latter equality expresses the law of linear elastodynamics energy conservation in the differential form. On the left-hand side of the equation we have the specific energy of the medium, consisting of the potential and kinetic components.

Let us use the latter expression for an integral representation of the energy conservation law for the subregion V_k within which a solution is smooth. Since at the singular points/curves of the boundary of the region V_k the solution may incorporate singularities and, hence, may not be smooth, let us start by singling out the singular curves/points with the help of tubular/spherical surfaces S_k^ε whose radius is equal to ε (Fig. 2.5). (Hereafter, the parts of the surface F as well as of the surfaces S, S^0 outside S_k^ε, which form the boundary of V_k, will be denoted by F_k, S_k S_k^0, respectively.)

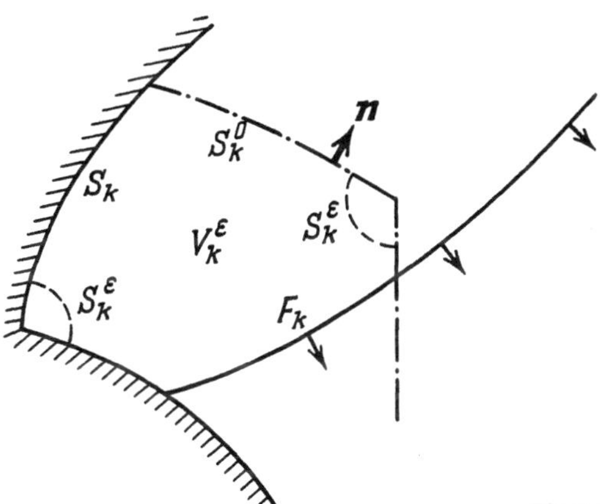

Fig. 2.5. The integration domain V_k^ε

Next we integrate the equality (2.5.3) over the volume V_k^ε obtained from V_k by excluding the neighborhoods of singular curves/points bounded by the surfaces S_k^ε. Since the solution $\boldsymbol{u}$ is smooth within V_k^ε up to the boundary, then one may apply the Gauss-Ostrogradskii formula to the right-hand side of the resultant integral equality, and thus arrive at

$$\iiint\limits_{V_k^\varepsilon} \frac{\partial}{\partial t}\left[W + \varrho\frac{(\dot{\boldsymbol{u}})^2}{2}\right] dV = \iint\limits_{G_k} (\dot{\boldsymbol{u}} \cdot \boldsymbol{\sigma}_n) ds\ , \tag{2.5.4}$$

where $G_k = S_k + S_k^0 + F_k + S_k^\varepsilon$, and the normal $\boldsymbol{n}$ to G_k is outward with respect to the volume V_k^ε.

Let us change the order of integration over V_k^ε and differention with respect to t on the left-hand side of (2.5.4) taking into account that the surface F_k is a moving one and that the surfaces S_k^ε may be moving too. Then,

$$\frac{d}{dt}\iiint\limits_{V_k^\varepsilon}\left[W+\varrho\frac{(\dot{\boldsymbol{u}})^2}{2}\right]dV = \iint\limits_{S_k+S_k^0}(\dot{\boldsymbol{u}}\cdot\boldsymbol{\sigma}_n)ds + \iint\limits_{S_k^\varepsilon}\left\{(\dot{\boldsymbol{u}}\cdot\boldsymbol{\sigma}_n)+\alpha_n\left[W+\varrho\frac{(\dot{\boldsymbol{u}})^2}{2}\right]\right\}ds + \iint\limits_{F_k}\left\{(\dot{\boldsymbol{u}}\cdot\boldsymbol{\sigma}_n)+\alpha_n\left[W+\varrho\frac{(\dot{\boldsymbol{u}})^2}{2}\right]\right\}ds\,, \tag{2.5.5}$$

where $\alpha_n = \alpha_n(\boldsymbol{x},t)$ is the normal component of the velocity of the surface S_k^ε/F_k at the point $\boldsymbol{x}$ at the instant t ($\alpha = 0$ for stationary surfaces S_k^ε).

The latter formula is the integral representation of the energy conservation law for the volume V_k^ε occupied with an elastic medium. On the left we have under the sign of derivative the medium energy composed of the potential and kinetic components. The integrals of $(\dot{\boldsymbol{u}}\cdot\boldsymbol{\sigma}_n)$ on the right-hand side have the dimension of the power (work per unit time). If the surface bounding the volume V_k^ε is moving, then the surface integral of $\alpha_n[W+\varrho(\dot{\boldsymbol{u}})^2/2]$ must be added to the right-hand side, which specifies the energy change due to inflow/outflow of an elastic continuous medium across the moving surface under consideration. Upon summing both the left- and the right-hand sides of (2.5.5) over all the subregions V_k^ε one gets

$$\dot{E}_\varepsilon \equiv \frac{d}{dt}\sum_k\iiint\limits_{V_k^\varepsilon}\left[W+\varrho\frac{(\dot{\boldsymbol{u}})^2}{2}\right]dV = \sum_k\iint\limits_{S_k}(\dot{\boldsymbol{u}}\cdot\boldsymbol{\sigma}_n)ds + \sum_k\iint\limits_{S_k^0}(\dot{\boldsymbol{u}}\cdot\boldsymbol{\sigma}_n)\,ds + \sum_k\iint\limits_{F_k}\left\{(\dot{\boldsymbol{u}}\cdot\boldsymbol{\sigma}_n)+\alpha_n\left[W+\varrho\frac{(\dot{\boldsymbol{u}})^2}{2}\right]\right\} + \sum_k\iint\limits_{S_k^\varepsilon}\left\{(\dot{\boldsymbol{u}}\cdot\boldsymbol{\sigma}_n)+\alpha_n\left[W+\varrho\frac{(\dot{\boldsymbol{u}})^2}{2}\right]\right\}ds\,. \tag{2.5.6}$$

Here the first term (the sum over S_k) on the right-hand side vanishes due to zero boundary conditions for the function $\boldsymbol{u}$.

Since the interfaces S_k^0 are internal with respect to V, i.e., every part of S_k^0 is a boundary between two neighboring sub-regions V_k^ε, then the integral over each of them is taken twice, namely, along the two opposite sides with oppositely directed outward normals. Then, by reversing the normal's direction in one of the integrals, the pair of integrals over the surface S_k^0 may be replaced by a single one

$$\iint\limits_{S_k^0}\left[(\dot{\boldsymbol{u}}\cdot\boldsymbol{\sigma}_n)^+ - (\dot{\boldsymbol{u}}\cdot\boldsymbol{\sigma}_n)^-\right]ds$$

equal to zero due to the contact condition (2.2.6) or (2.2.7). Thus, the total sum over S_k^0 in (2.5.6) becomes zero, too.

Similar manipulations with the sum of F_k ($F = \sum_k F_k$) result in

$$\iint_F \left\{ \left[(\dot{\boldsymbol{u}} \cdot \boldsymbol{\sigma}_n) + \alpha_n \left(W + \varrho \frac{(\dot{\boldsymbol{u}})^2}{2} \right) \right]^+ - \left[(\dot{\boldsymbol{u}} \cdot \boldsymbol{\sigma}_n) + \alpha_n \left(W + \varrho \frac{(\dot{\boldsymbol{u}})^2}{2} \right) \right]^- \right\} ds \,. \tag{2.5.7}$$

If the characteristic surface F is a weak-discontinuity surface, and, hence, the first partial derivatives of the displacement vector remain continuous across F, the integral (2.5.7) becomes zero. Let us consider the case when F is a strong-discontinuity surface. We shall show that in this case the expression

$$L \equiv (\dot{\boldsymbol{u}} \cdot \boldsymbol{\sigma}_n) + \alpha_n \left\{ W + \varrho \frac{(\dot{\boldsymbol{u}})^2}{2} \right\}$$

remains continuous across the surface F . This may be done by transforming L, with the help of (2.1.2–5) to

$$\begin{aligned} L \equiv (\dot{\boldsymbol{u}} \cdot \boldsymbol{\sigma}_n) + \alpha_n \left\{ W + \varrho \frac{(\dot{\boldsymbol{u}})^2}{2} \right\} &= \frac{1}{2} \left\{ (\dot{\boldsymbol{u}} \cdot \boldsymbol{\sigma}_k) n_k + \varrho (\dot{\boldsymbol{u}})^2 \alpha_n \right\} \\ + \frac{1}{2} \left\{ (\dot{\boldsymbol{u}} \cdot \boldsymbol{\sigma}_k) n_k + \alpha_n \left(\boldsymbol{\sigma}_k \cdot \frac{\partial \boldsymbol{u}}{\partial x_k} \right) \right\} &= \frac{1}{2} \left\{ (\dot{\boldsymbol{u}} \cdot \boldsymbol{M}_4) + (\boldsymbol{\sigma}_k \cdot \boldsymbol{M}_k) \right\} , \end{aligned}$$

where $\boldsymbol{M}_k$ ($k = 1, 2, 3$) and $\boldsymbol{M}_4$ are defined by the expressions (2.4.2, 3). Then, taking into account that the functions $\boldsymbol{M}_k$, $\boldsymbol{M}_4$ remain continuous across F, the integrand in (2.5.7) becomes

$$\begin{aligned} [L] &= \frac{1}{2} \left\{ ([\dot{\boldsymbol{u}}] \cdot \boldsymbol{M}_4) + ([\boldsymbol{\sigma}_k] \cdot \boldsymbol{M}_k) \right\} \\ &= \frac{1}{2} \left\{ [\dot{\boldsymbol{u}}] \cdot (\boldsymbol{\sigma}_k n_k + \alpha_n \varrho \dot{\boldsymbol{u}}) + [\boldsymbol{\sigma}_k] \cdot \left(\dot{\boldsymbol{u}} n_k + \alpha_n \frac{\partial \boldsymbol{u}}{\partial x_k} \right) \right\} \\ &= \frac{1}{2} \left\{ \dot{\boldsymbol{u}} \cdot [\boldsymbol{M}_4] + (\boldsymbol{\sigma}_k n_k) \cdot [\dot{\boldsymbol{u}}] + \alpha_n \frac{\partial \boldsymbol{u}}{\partial x_k} \cdot [\boldsymbol{\sigma}_k] \right\} \\ &= \frac{1}{2} \left\{ (\boldsymbol{\sigma}_k n_k) \cdot [\dot{\boldsymbol{u}}] + \alpha_n \frac{\partial \boldsymbol{u}}{\partial x_k} \cdot [\boldsymbol{\sigma}_k] \right\} , \end{aligned} \tag{2.5.8}$$

where the expressions between the square brackets denote jumps in the corresponding variables, while the values of all the other functions are taken simultaneously at one and the same side of F.

However, from (2.4.2) it follows that

$$[\dot{\boldsymbol{u}}] n_k = -\alpha_n \left[\frac{\partial \boldsymbol{u}}{\partial x_k} \right] ,$$

and, hence, (2.5.8) may be reduced to

$$[L] = \frac{\alpha_n}{2}\left\{-\boldsymbol{\sigma}_k \cdot \left[\frac{\partial \boldsymbol{u}}{\partial x_k}\right] + [\boldsymbol{\sigma}_k] \cdot \frac{\partial \boldsymbol{u}}{\partial x_k}\right\} .$$

Since according to (2.1.2, 3), Hooke's law may be written in displacements in the form

$$\sigma_{kl} = c_{klij} u_{i,j} \ ,$$

then, taking into account (2.1.5), one obtains finally

$$[L] = -\frac{\alpha_n}{2}\{c_{klij} u_{i,j} [u_{i,j} [u_{l,k}] - c_{klij} [u_{i,j}] u_{l,k}\} = 0 \ ,$$

and, hence, the integral (2.5.7) becomes zero. Thus, (2.5.6) becomes

$$\dot{E}_\varepsilon = \sum_k \iint_{S_k^\varepsilon} \left\{ (\dot{\boldsymbol{u}} \cdot \boldsymbol{\sigma}_n) + \alpha_n \left(W + \varrho \frac{(\dot{\boldsymbol{u}})^2}{2} \right) \right\} ds \ . \tag{2.5.9}$$

If one passes now to the limit at $\varepsilon \to 0$, then, due to the constraints (2.3.6), all the integrals on the right-hand side vanish, except the ones taken over the tubular surfaces surrounding singular curves on S, S^0, which radiate/absorb energy, i.e., within whose neighborhood the condition

$$\boldsymbol{u} = \boldsymbol{u}_0 + \mathrm{O}\left(\varepsilon^{1/2}\right) \ . \tag{2.5.10}$$

holds.

Thus, (2.5.9) reduces to

$$\dot{E} = \sum_k \lim_{\varepsilon \to 0} \iint_{S_k^\varepsilon} \left\{ (\dot{\boldsymbol{u}} \cdot \boldsymbol{\sigma}_n) + \alpha_n \left(W + \varrho \frac{(\dot{\boldsymbol{u}})^2}{2} \right) \right\} ds \ , \tag{2.5.11}$$

where

$$E = \lim_{\varepsilon \to 0} E_\varepsilon = \lim_{\varepsilon \to 0} \sum_k \iiint_{V_k^\varepsilon} \left\{ W + \varrho \frac{(\dot{\boldsymbol{u}})^2}{2} \right\} dV \tag{2.5.12}$$

is the energy of an elastic medium occupying the region V.

If we next suppose that the right-hand side of (2.5.1) meets the condition

$$\sum_k \lim_{\varepsilon \to 0} \iint_{S_k^\varepsilon} \left\{ (\dot{\boldsymbol{u}} \cdot \boldsymbol{\sigma}_n) + \alpha_n \left(W + \varrho \frac{(\dot{\boldsymbol{u}})^2}{2} \right) \right\} ds \leq 0 \ , \tag{2.5.13}$$

then $\dot{E} \leq 0$, and, hence, $E(t)$ is a decreasing function of time for $t \geq 0$. Then, with the help of the initial condition $E(0) = 0$ we deduce that $E(t) \leq 0$. However, since according to (2.5.12) $E(t) \geq 0$, then we finally get $E(t) \equiv 0$ and hence $\dot{\boldsymbol{u}} \equiv 0$.

Taking into account zero initial conditions, we deduce that $u = u_1 - u_2 \equiv 0$. Thus, the uniqueness theorem is proved.

Note that the theorem remains valid in the case when the characteristic surface F is piecewise-smooth and contains a finite number of curves/points where F is not smooth. In the neighborhood of the points/curves the function σ_n and the derivatives of the function u with respect to coordinates and time may have singularities; however, the latter must be such that the integral of L along the boundary surrounding these curves/points, would tend to zero while the boundary is shrinking to the curves/points.

The above proof implies that the uniqueness theorem is valid, if the solutions u_1, u_2 do not contain moving singular curves or if these curves can neither radiate nor absorb energy (in other words, in the neighborhood of a singular line $u_k = u_k^0 + O(\varepsilon_k^{\beta})$ at $\varepsilon \to 0$, $\beta_k > 1/2$, $k = 1,2$ where u_k^0 are independent of ε), since in these cases the right-hand side of (2.5.11) becomes zero.

Much more complicated is the proof in the case when the moving singular curves are capable of either radiating or absorbing energy (in the neighborhood of such a curve $u_k = u_k^0 + O(\varepsilon^{1/2})$, $\varepsilon \to 0$, $k = 1,2$). In this case the presence of additional conditions ensuring uniqueness of a solution to the elastodynamic problem under consideration, demonstrates a substantial dependence on the speed of motion of a singular curve [2.13].

Thus, in the case of moving singular curves which either absorb or radiate energy, the issue of the solution uniqueness should be considered not in general, but in its connection with each particular elastodynamic problem.

In concluding this section, we should like to note that from the above proof of the uniqueness theorem it follows that the fulfilment of the condition (2.5.13) for the difference $u \equiv u_1 - u_2$ of solutions u_1, u_2 obtained at the same initial and boundary condtions, ensures validity of the uniqueness theorem. Hence, the theorem remains valid also for solutions containing, in the vicinity of singular points/curves, higher-order singularities as compared with the ones considered above; however, the class of these solutions must be such that the difference $u \equiv u_1 - u_2$ of any two solutions belonging to the class (with the same initial and boundary conditions) meet the condition (2.5.13).

2.6 The Green–Volterra Formula

Consider a homogeneous elastic continuous medium occupying volume V, and two solutions u, u^0 to (2.1.1), corresponding to two different distributions of body forces f, f^0. Let us next write down for the solutions the equations of motion (2.1.1) and Hooke's law (2.1.2):

$$\frac{\partial \sigma_k}{\partial x_k} + f = \varrho \ddot{u} , \tag{2.6.1}$$

$$\sigma_{ij} = c_{ijkl} u_{k,l} = c_{ijkl} \varepsilon_{kl} , \tag{2.6.2}$$

$$\frac{\partial \sigma_k^0}{\partial x_k} + f^0 = \varrho \ddot{u}^0 , \tag{2.6.3}$$

$$\sigma_{ij}^0 = c_{ijkl} u_{k,l}^0 = c_{ijkl} \varepsilon_{kl}^0 . \tag{2.6.4}$$

From (2.6.2,4) and the symmetry of elastic constants c_{ijkl} it follows that the following identity

$$\sigma_k \cdot \frac{\partial u^0}{\partial x_k} = \sigma_k^0 \cdot \frac{\partial u}{\partial x_k}$$

holds, which may be rewritten in the form

$$u^0 \cdot \frac{\partial \sigma_k}{\partial x_k} - u \cdot \frac{\partial \sigma_k^0}{\partial x_k} = \frac{\partial}{\partial x_k} \left(\sigma_k \cdot u^0 - \sigma_k^0 \cdot u \right) . \tag{2.6.5}$$

Then, by multiplying (2.6.1) by u^0, and (2.6.3) by u and subtracting the latter from the former, one obtains

$$f \cdot u^0 - f^0 \cdot u = \frac{\partial}{\partial x_k} \left(\sigma_k^0 \cdot u - \sigma_k \cdot u^0 \right) + \varrho \frac{\partial}{\partial t} \left(u^0 \cdot \dot{u} - u \cdot \dot{u}^0 \right) . \tag{2.6.6}$$

The latter relation is called the differential Green-Volterra formula of the theory of elasticity.

Let us now integrate (2.6.6) over the four-dimensional region $\Omega(x_1, x_2, x_3, x_4 = t)$ which is a direct product of the volume V and the time interval $0 \le t \le \tau$. The summands on the right side of (2.6.6) represent a four-dimensional divergence; hence, the Gauss–Ostrogradskii formula may be used to write the corresponding integral in the form of an integral over the boundary Γ of the region Ω:

$$\iiint\limits_{\Gamma} \left[\varrho \left(\dot{u} \cdot u^0 - u \cdot \dot{u}^0 \right) N_4 - \left(\sigma_k \cdot u^0 - \sigma_k^0 \cdot u \right) N_k \right] d\Gamma$$

$$= \iiiint\limits_{\Omega} \left(f \cdot u^0 - f^0 \cdot u \right) d\Omega . \tag{2.6.7}$$

Here $N(N_1, N_2, N_3, N_4)$ is the outward four-dimensional normal to the boundary Γ of the four-dimensional region $\Omega(x_1, x_2, x_3, t)$,

$$\begin{aligned} N_k &= \frac{F_{,k}}{|\mathrm{grad}_t F|} \quad (k = 1, 2, 3) , \\ N_4 &= \frac{\dot{F}}{|\mathrm{grad}_t F|} , \\ |\mathrm{grad}_t F| &= \left(F_{,k} F_{,k} + (\dot{F})^2 \right)^{1/2} , \end{aligned} \tag{2.6.8}$$

and $F \equiv F(x_1, x_2, x_3, t) = 0$ is the equation describing Γ. Equality (2.6.7) is called the Green–Vollterra formula. Upon denoting

$$M_4^0(u) \equiv \sigma_k N_k - \varrho \dot{u} N_4 , \tag{2.6.9}$$

(2.6.7) may be rewritten in the form

$$\iiiint_{\Omega} \left(\boldsymbol{f} \cdot \boldsymbol{u}^0 - \boldsymbol{f}^0 \cdot \boldsymbol{u} \right) d\Omega = \iiint_{\Gamma} \left[\boldsymbol{u} \cdot \boldsymbol{M}_4^0 (\boldsymbol{u}^0) - \boldsymbol{u}^0 \cdot \boldsymbol{M}_4^0(\boldsymbol{u}) \right] d\Gamma \,. \tag{2.6.10}$$

Formulas (2.6.7, 10) have been derived for $\boldsymbol{u}$, $\boldsymbol{u}^0$ which are continuous within Ω together with the first derivatives with respect to x_k ($k = 1,2,3$) and t up to the boundary Γ which is supposed to be piecewise-smooth. However, it follows from (2.6.10) that the Green–Volterra formula remains valid in the case when Ω contains interfaces S^0 (with the same contact conditions for $\boldsymbol{u}$, $\boldsymbol{u}^0$ all over S^0) and regular strong discontinuities of the solutions $\boldsymbol{u}$, $\boldsymbol{u}^0$ (which, as stated above, may occur on the characteristic surfaces only). In fact, if one takes into account (2.4.3, 4) for $\boldsymbol{M}_4$, α_n, and the three-dimensional normal $\boldsymbol{n}$, then the expression $\boldsymbol{M}_4^0$ may be rewritten in the form

$$\boldsymbol{M}_4^0(\boldsymbol{u}) = \boldsymbol{M}_4(\boldsymbol{u}) \left| \frac{\operatorname{grad} F}{\operatorname{grad}_t F} \right| \,. \tag{2.6.11}$$

Hence, since $\boldsymbol{M}_4$ remains continuous across the strong-discontinuity surfaces, the integrand on the right-hand side of (2.6.10) remains continuous across smooth surfaces with strong discontinuities in solutions $\boldsymbol{u}$, $\boldsymbol{u}^0$.

Similarly, since the interface S^0 boundaries are stationary, and the contact conditions for $\boldsymbol{u}$, $\boldsymbol{u}^0$ are the same all over the boundaries, we deduce that across S^0 the integrand on the right-hand side of (2.6.10) remains continuous and equal to

$$\begin{aligned} \left[\boldsymbol{u} \cdot \boldsymbol{M}_4^0 (\boldsymbol{u}^0) - \boldsymbol{u}^0 \cdot \boldsymbol{M}_4^0(\boldsymbol{u}) \right] &= \left(\boldsymbol{u} \cdot \boldsymbol{\sigma}_k^0 N_k - \boldsymbol{u}^0 \cdot \boldsymbol{\sigma}_k N_k \right)^+ \\ &= \left(\boldsymbol{u} \cdot \boldsymbol{\sigma}_k^0 N_k - \boldsymbol{u}^0 \cdot \boldsymbol{\sigma}_k N_k \right)^- , \end{aligned}$$

where the signs $+,-$ correspond to different sides of the surface S^0. Then, application of (2.6.10) to each of the subregions Ω_k formed by dividing the region Ω by the boundaries Γ_k, interfaces S_k^0 and characteristic surfaces F_k of the solutions $\boldsymbol{u}$, $\boldsymbol{u}^0$, and subsequent summation with respect to k, result in the above formula (2.6.10) for the whole of region Ω, since the integrals on the right-hand side of (2.6.10) over the interfaces S_k^0 and characteristic surfaces F_k will enter twice with opposite signs, and hence cancel. Note that the characteristic surfaces F_k, boundaries Γ_k and S_k^0 may be piecewise-smooth, while in the neighborhood of discontinuities the derivatives of solutions $\boldsymbol{u}$, $\boldsymbol{u}^0$ with respect to coordinates and time may possess singularities, such that the integral

$$\iiint_{\Gamma_k + F_k} \left[\boldsymbol{u} \cdot \boldsymbol{M}_4^0 (\boldsymbol{u}^0) - \boldsymbol{u}^0 \cdot \boldsymbol{M}_4^0(\boldsymbol{u}) \right] d\Gamma$$

over a surface surrounding singular curves/points of the integrands tends to zero when Γ_k, S_k^0 and F_k shrink to the curves/points. Furthermore, (2.6.7, 10) remain valid also in the case when one of the solutions $\boldsymbol{u}$, $\boldsymbol{u}^0$ is generalized while the other one is sufficiently smooth.

If the body volume V is time-independent, and the boundaries S are stationary (in the sense that the normal component of the boundary velocity is zero, which is a natural assumption for the considered problems of the linear theory of elasticity, including those of crack propagation and punch penetration with an expanding contact boundary), then (2.6.10) may be presented in the form

$$\int\limits_{-0}^{t_0+0} dt \iiint\limits_V \left(\boldsymbol{f}^0 \cdot \boldsymbol{u} - \boldsymbol{f} \cdot \boldsymbol{u}^0\right) dV = \int\limits_{-0}^{t_0+0} dt \iint\limits_S \left(\boldsymbol{\sigma}_n \cdot \boldsymbol{u}^0 - \boldsymbol{\sigma}_n^0 \cdot \boldsymbol{u}\right) dS + \iiint\limits_V \varrho \left(\dot{\boldsymbol{u}} \cdot \boldsymbol{u}^0 - \boldsymbol{u} \cdot \dot{\boldsymbol{u}}^0\right) dV \Big|_{t=-0}^{t=t_0+0} , \tag{2.6.12}$$

where $\boldsymbol{n}$ is the outward normal to S. Note that the boundary S may expand along itself, i.e., there may exist the singular curves/ points introduced above.

The Green-Volterra formula is a generalization of Betti's reciprocal theorem [2.2, 3] to elastodynamic problems. Note that (2.6.12) remains valid in the case where one of the solutions $\boldsymbol{u}$, $\boldsymbol{u}^0$ is a generalized function, and the other one is sufficiently smooth.

Let us consider the solution $\boldsymbol{u}^0(\boldsymbol{x}, t)$ to the elastodynamic equations (2.1.6) with a concentrated body force $\boldsymbol{f}^0 = (f_1^0, f_2^0, f_3^0)$ applied at a point $\boldsymbol{x}_0$ of volume V, and directed along the x_k-axis

$$f_i^0 = \delta(t)\delta(\boldsymbol{x} - \boldsymbol{x}_0)\delta_{ik}$$
$$\left(\iiint\limits_V f(\boldsymbol{x})\delta(\boldsymbol{x} - \boldsymbol{x}_0) dV = \begin{cases} f(\boldsymbol{x}_0) , & \boldsymbol{x}_0 \in V \\ 0 , & \boldsymbol{x}_0 \notin V \end{cases}\right) . \tag{2.6.13}$$

Let $\boldsymbol{u}^0(\boldsymbol{x}, t)$ be subjected to zero initial conditions

$$\boldsymbol{u}^0 \equiv 0 \quad (t < 0) \tag{2.6.14}$$

and the boundary conditions on S belong to one of the types (2.2.2–5) (in particular, $\boldsymbol{u}^0$ may be a solution for the case of unbounded space).

Since (2.1.6) are invariant with respect to the replacement of t by $-t$ and a shift of the initial time moment, let us consider the solution $\boldsymbol{u}^0(\boldsymbol{x}, t_0, -t)$ which, instead of the condition (2.6.14), meets the condition

$$\boldsymbol{u}^0 \equiv 0 \quad (t > t_0)$$

subject to body forces

$$f_i^0 = \delta(t_0 - t)\delta(\boldsymbol{x} - \boldsymbol{x}_0)\delta_{ik} \quad (i = 1, 2, 3) .$$

By substituting this solution into (2.6.12) and denoting $\boldsymbol{u}^0 \equiv \boldsymbol{u}^{(k)}$, $\boldsymbol{\sigma}_n^0 \equiv \boldsymbol{\sigma}_n^{(k)}$, one obtains ($\boldsymbol{x}_0 \in V$)

$$u_k(\boldsymbol{x}_0, t_0) = \int_0^{t_0} dt \iiint_V f_i(\boldsymbol{x}, t) u_i^{(k)}(\boldsymbol{x}, t_0 - t) dV$$
$$+ \int_0^{t_0} dt \iint_S \Big[\sigma_{in}(\boldsymbol{x}, t) u_i^{(k)}(\boldsymbol{x}, t_0 - t)$$
$$-\sigma_{in}^{(k)}(\boldsymbol{x}, t_0 - t) u_i(\boldsymbol{x}, t)\Big] dS \;, \tag{2.6.15}$$

which is a generalization of Somiliana's formula onto elastodynamic problems [2.3].

According to the latter formula, the solution $\boldsymbol{u}$ to an elastodynamic problem with boundary conditions belonging to one of the four types (2.2.2–5), may be obtained by setting $\boldsymbol{u}^{(k)}$ $(k = 1, 2, 3)$ equal to the solution for concentrated body forces at zero boundary conditions of the same type. Then, the integrand of the integral over S in (2.6.15) depends on both $\boldsymbol{u}^{(k)}$ and the given boundary conditions. Thus, according to (2.6.15), a solution to the elastodynamic problem may be obtained if one knows all the fundamental solutions $\boldsymbol{u}^{(k)}(\boldsymbol{x}, \boldsymbol{x}_0, t)$ $(k = 1, 2, 3)$, i.e., the solutions for the concentrated force (2.6.13) with zero initial and the corresponding zero boundary conditions.

If we denote by $U_{ik} \equiv U_{ik}(\boldsymbol{x}, \boldsymbol{x}_0, t)$ the ith component of the displacement vector in a fundamental solution $\boldsymbol{u}^{(k)}) = \boldsymbol{u}^{(k)}(\boldsymbol{x}, \boldsymbol{x}_0, t)$ (where $\boldsymbol{x}_0$ is the point of application of the instantaneous concentrated force (2.6.13) within an elastic body V, and $\boldsymbol{x}$ is an arbitrary point of V), then it follows from the above that in order to obtain a solution $\boldsymbol{u}$ to the elastodynamic problem at zero initial conditions and given boundary conditions, one has to know three vectors $\boldsymbol{u}^{(k)}(U_{1k}, U_{2k}, U_{3k})$ $(k = 1, 2, 3)$ which comprise Green's displacement tensor. The components U_{ik} of the tensor are called Green's functions of the respective problem. Note that according to Betti's reciprocal theorem

$$U_{ik}(\boldsymbol{x}, \boldsymbol{x}_0, t) = U_{ki}(\boldsymbol{x}_0, \boldsymbol{x}, t) \tag{2.6.16}$$

or, in other words, the ith component of the displacement vector at the point $\boldsymbol{x}$, due to an instantaneous concentrated force applied at the point $\boldsymbol{x}_0$, and directed along the x_k-axis, is equal to the kth component of the displacmement vector at the point $\boldsymbol{x}_0$ the displacement being induced by an instantaneous concentrated force applied at the point $\boldsymbol{x}$ and directed along the x_i-axis.

In fact, if one puts $\boldsymbol{u}, \boldsymbol{u}^0$ in (2.6.12) equal to the fundamental solutions $\boldsymbol{u}^{(k)}(\boldsymbol{x}, \boldsymbol{x}_0, t)$ and $\boldsymbol{u}^{(l)}(\boldsymbol{x}, \boldsymbol{y}_0, t_0 - t)$, respectively, then due to the zero boundary conditions and the conditions

$$\boldsymbol{u}^{(k)} \equiv 0 \quad (t < 0) \;, \qquad \boldsymbol{u}^{(l)} \equiv 0 \quad (t > t_0) \;, \tag{2.6.17}$$

the right-hand side of (2.6.12) vanishes. Hence, the presence of the instantaneous concentrated forces

$$f_i = \delta(t)\delta(\boldsymbol{x} - \boldsymbol{x}_0)\delta_{ik} \;, \quad f_i^0 = \delta(t_0 - t)\delta(\boldsymbol{x} - \boldsymbol{y}_0)\delta_{il}$$

on the left-hand side of (2.6.12) results in the equality

$$U_{lk}(\boldsymbol{y}_0, \boldsymbol{x}_0, t_0) = U_{kl}(\boldsymbol{x}_0, \boldsymbol{y}_0, t_0) .$$

Note that the relation (2.6.16) remains valid in the case of an unbounded elastic space, too.

2.7 Various Representations of Solutions to the Equations of Motion of a Homogeneous Isotropic Medium

In this section we consider the main representations of solutions to elastodynamic equations for a homogeneous isotropic medium. These representations are basic for solving elastodynamic problems and their choice depends on the shape of the elastic body under consideration.

2.7.1 Lamé Representation

As we know [2.1, 14], an arbitrary vector field may be represented in the form

$$\boldsymbol{u} = \operatorname{grad} \varphi + \operatorname{curl} \boldsymbol{\psi} , \quad \operatorname{div} \boldsymbol{\psi} = 0 , \tag{2.7.1}$$

where φ is a scalar potential, and $\boldsymbol{\psi}$ is a vector potential (Lamé representation). Then, (2.1.8) are equivalent to a system of two wave equations for φ and $\boldsymbol{\psi}$ [2.1]:

$$\Delta\varphi = \frac{\partial^2\varphi}{c_1^2\partial t^2} - \frac{\Phi}{\lambda+2\mu} , \quad c_1 = \left(\frac{\lambda+2\mu}{\varrho}\right)^{1/2} , \tag{2.7.2}$$

$$\Delta\boldsymbol{\psi} = \frac{\partial^2\boldsymbol{\psi}}{c_2^2\partial t^2} - \frac{\boldsymbol{\Psi}}{\mu} , \quad c_2 = \left(\frac{\mu}{\varrho}\right)^{1/2} , \tag{2.7.3}$$

where $\Phi, \boldsymbol{\Psi}$ are related to the body force $\boldsymbol{f}$ from (2.1.8) by the formula

$$\boldsymbol{f} = \operatorname{grad} \Phi + \operatorname{curl} \boldsymbol{\Psi} , \quad \operatorname{div} \boldsymbol{\Psi} = 0 . \tag{2.7.4}$$

Let us prove this statement. By employing the identity

$$\Delta \boldsymbol{u} = \operatorname{grad} \operatorname{div} \boldsymbol{u} - \operatorname{curl} \operatorname{curl} \boldsymbol{u} \tag{2.7.5}$$

the Lamé equations (2.1.8) may be rewritten in the form

$$(\lambda + 2\mu) \operatorname{grad} \operatorname{div} \boldsymbol{u} - \mu \operatorname{curl} \operatorname{curl} \boldsymbol{u} + \boldsymbol{f} - \varrho \ddot{\boldsymbol{u}} = 0 . \tag{2.7.6}$$

We shall start by showing that the solution to the equations (2.7.6) may be written in the form (2.7.1) where $\varphi, \boldsymbol{\psi}$ satisfy the system of wave equations (2.7.2, 3). To do this, let us represent the vectors $\boldsymbol{u}, \boldsymbol{f}$ entering (2.7.6) in the form

$$\boldsymbol{u} = \operatorname{grad} \varphi + \operatorname{curl} \boldsymbol{\psi} , \quad \boldsymbol{f} = \operatorname{grad} \Phi + \operatorname{curl} \boldsymbol{\Psi} . \tag{2.7.7}$$

Since the operators Δ, $\partial^2/\partial t^2$, and grad are interchangeable, as well as the operators $\delta\Delta$, $\partial^2/\partial t^2$, curl, and div grad $\equiv \Delta$, div curl $\equiv 0$, curl grad $\equiv 0$, while $\varrho = \text{const}$, one obtains

$$\operatorname{grad}\,[(\lambda+2\mu)\Delta\varphi - \varrho\ddot{\varphi} + \Phi] + \operatorname{curl}\,\left[\mu\Delta\boldsymbol{\psi} - \varrho\ddot{\boldsymbol{\psi}} + \boldsymbol{\Psi}\right] = 0\,. \tag{2.7.8}$$

By application of the divergence operation to the latter equation, we get

$$\Delta\,[(\lambda+2\mu)\Delta\varphi - \varrho\ddot{\varphi} + \Phi] = 0$$

and, hence,

$$(\lambda+2\mu)\Delta\varphi - \varrho\ddot{\varphi} + \Phi = -\varrho A\,, \tag{2.7.9}$$

where A is a harmonic function (hence, $\Delta A = 0$). Then, from (2.7.8) we obtain

$$\operatorname{curl}\,\left[\mu\Delta\boldsymbol{\psi} - \varrho\ddot{\boldsymbol{\psi}} + \boldsymbol{\Psi}\right] = \varrho\operatorname{grad} A\,. \tag{2.7.10}$$

Let

$$\begin{aligned} \varphi &= \varphi_1 + \int\!\!\int A\,dt\,dt\,, \\ \operatorname{curl}\boldsymbol{\psi} &= \operatorname{curl}\boldsymbol{\psi}_1 - \operatorname{grad}\iint A\,dt\,dt\,, \end{aligned} \tag{2.7.11}$$

since grad A is a solenoidal vector (div (grad A) = $\Delta A = 0$). By substitution of (2.7.1) into (2.7.9, 10) one arrives at

$$(\lambda+2\mu)\Delta\varphi_1 - \varrho\ddot{\varphi}_1 + \Phi = 0\,, \quad \operatorname{curl}\,\left(\mu\Delta\boldsymbol{\psi}_1 - \varrho\ddot{\boldsymbol{\psi}}_1 + \boldsymbol{\Psi}\right) = 0\,. \tag{2.7.12}$$

From the latter equation we get

$$\mu\Delta\boldsymbol{\psi}_1 - \varrho\ddot{\boldsymbol{\psi}}_1 + \boldsymbol{\Psi} = -\varrho\operatorname{grad} B\,. \tag{2.7.13}$$

Here we may preserve only the vortex parts of the vector functions $\boldsymbol{\psi}_1, \boldsymbol{\Psi}$ (by putting div$\boldsymbol{\Psi}$ = div$\boldsymbol{\psi}_1$ = 0), because the gradient parts of the vectors $\boldsymbol{\psi}_1, \boldsymbol{\Psi}$ do not affect the values of curl$\boldsymbol{\Psi}$ in (2.7.11, 12) (since curl grad $\equiv 0$). Hence, upon applying the divergence operation to (2.7.13), one deduces that $\Delta B = 0$. Then, by putting

$$\boldsymbol{\psi}_1 = \boldsymbol{\psi}_2 + \operatorname{grad}\iint B\,dt\,dt \tag{2.7.14}$$

we get from (2.7.14)

$$\mu\Delta\boldsymbol{\psi}_2 - \varrho\ddot{\boldsymbol{\psi}}_2 + \boldsymbol{\Psi} = 0\,. \tag{2.7.15}$$

Taking into account that curl grad $\equiv 0$, the displacement vector $\boldsymbol{u}$ may be written in the form

$$\boldsymbol{u} = \operatorname{grad}\varphi + \operatorname{curl}\boldsymbol{\psi} = \operatorname{grad}\varphi_1 + \operatorname{grad}\iint \boldsymbol{A}\,dt\,dt + \operatorname{curl}\boldsymbol{\psi}_2$$
$$- \operatorname{grad}\iint \boldsymbol{A}\,dt\,dt = \operatorname{grad}\varphi_1 + \operatorname{curl}\boldsymbol{\psi}_2 \qquad (2.7.16)$$

besides,

$$\operatorname{div}\boldsymbol{\psi}_2 = 0 \qquad (2.7.17)$$

[due to application of the divergence operation to (2.7.14)].

Hence, omitting the subscripts 1 and 2 of $\varphi_1, \boldsymbol{\psi}_2$ in (2.7.12, 15–17) and taking into account the constraint $\operatorname{div}\boldsymbol{\Psi} = 0$, we arrive finally at (2.7.1–4).

Conversely, it may be shown that the function $\boldsymbol{u}$ defined by (2.7.1–4) is a solution to the Lamé equation (2.1.8). In fact, by substituting $\boldsymbol{u}, \boldsymbol{f}$ from (2.7.1, 4) into the left-hand side of (2.7.6) we arrive at

$$(\lambda + 2\mu)\operatorname{grad}\operatorname{div}\boldsymbol{u} - \mu\operatorname{curl}\operatorname{curl}\boldsymbol{u} + \boldsymbol{f} - \varrho\ddot{\boldsymbol{u}}$$
$$= \operatorname{grad}\left[(\lambda + 2\mu)\Delta\varphi - \varrho\ddot{\varphi} + \Phi\right] + \operatorname{curl}\left[-\mu\operatorname{curl}\operatorname{curl}\boldsymbol{\psi} - \varrho\ddot{\boldsymbol{\psi}} + \boldsymbol{\Psi}\right] . \qquad (2.7.18)$$

However, by using the identity (2.7.5) for the function $\boldsymbol{\psi}$, as well as the identity $\operatorname{curl}\operatorname{grad} \equiv 0$, one may rewrite the latter term on the right-hand side in the form

$$\operatorname{curl}\left[-\mu\operatorname{curl}\operatorname{curl}\boldsymbol{\psi} - \varrho\ddot{\boldsymbol{\psi}} + \boldsymbol{\Psi}\right] = \operatorname{curl}\left[-\mu\Delta\boldsymbol{\psi} - \varrho\ddot{\boldsymbol{\psi}} + \boldsymbol{\Psi}\right] . \qquad (2.7.19)$$

Then, taking into account the wave equations (2.7.2, 3) one finally deduces that the right-hand side of (2.7.18) vanishes.

Note that although we have used the condition $\operatorname{div}\boldsymbol{\psi} = 0$ in the representation of the vector $\boldsymbol{u}$ while proving equivalence of the Lamé equations to (2.7.1–4), this condition is unimportant. In fact, the replacement of the vector $\boldsymbol{\psi}$ in the formula

$$\boldsymbol{u} = \operatorname{grad}\varphi + \operatorname{curl}\boldsymbol{\psi} \qquad (2.7.20)$$

by

$$\boldsymbol{\psi}_0 = \boldsymbol{\psi} + \operatorname{grad}\chi , \qquad (2.7.21)$$

where χ is an arbitrary function of coordinates and time, will not result in altering (2.7.20), since $\operatorname{curl}\operatorname{grad} \equiv 0$. If we also require that χ satisfies the homogeneous wave equation

$$\Delta\chi = c_2^{-2}\ddot{\chi} , \qquad (2.7.22)$$

then (2.7.3) will remain unchanged for $\boldsymbol{\psi}_0$, too, but now

$$\operatorname{div}\boldsymbol{\psi}_0 = \Delta\chi = c_2^{-2}\chi \not\equiv 0 .$$

Of special importance, therefore, is the fact that while solving the system (2.7.1–4) without the condition $\operatorname{div}\boldsymbol{\psi} = 0$, may choose the vector $\boldsymbol{\psi}$ arbitrarily to an accuracy of the gradient of an arbitrary function χ which is a solution to the wave equation (2.7.22). However, since in the general case of a three-dimensional problem the displacement vector $\boldsymbol{u}$ is defined by three components (obtainable with the help of the three scalar Lamé equations with appropriate boundary and initial conditions) while the right-hand side of the representation (2.7.20) depends on four functions (φ and the three components of $\boldsymbol{\psi}$), one has to have an additional condition limiting the number of independent functions defining the vector $\boldsymbol{u}$ to 3. This may be done by employing the condition $\operatorname{div}\boldsymbol{\psi} = 0$. Of course, any other condition may be used while solving the system (2.7.1–4), defined by the constraint on the choice of function χ entering (2.7.21) and satisfying (2.7.22). In particular, if $\boldsymbol{f} \equiv 0$ in (2.7.1–4), then according to [2.15] one may find such a function which, while satisfying (2.7.22), will result in the deletion of a component of the vector $\boldsymbol{\psi}_0 = (\psi_{01}, \psi_{02}, \psi_{03})$ from (2.7.21), say, ψ_{03}. In fact, the condition (2.7.21) at $\psi_{03} \equiv 0$ implies that $\psi_3(x_1, x_2, x_3, t) = -\partial\chi/\partial x_3$, and, hence, integration results in

$$\chi = -\int_c^{x_3} \psi_3(x_1, x_2, x_3, t)\, dx_3 + f_0(x_1, x_2, t)\,.$$

By substitution of this expression for χ into (2.7.22), one arrives at the following equation for determining $f_0(x_1, x_2, t)$:

$$\begin{aligned}\Delta\chi - c_2^{-2}\ddot{\chi} = &-\int_c^{x_3} \left(\Delta - c_2^{-2}\frac{\partial^2}{\partial t^2}\right)\psi_3(x_1, x_2, x_3, t)dx_3 \\ &- \left.\frac{\partial\psi_3(x_1, x_2, x_3, t)}{\partial x_3}\right|_{x_3=c} + \left(\Delta_1 - c_2^{-2}\frac{\partial^2}{\partial t^2}\right) f_0(x_1, x_2, t) = 0 \\ &\left(\Delta_1 \equiv \frac{\partial^2}{\partial x_1^2} + \frac{\partial^2}{\partial x_2^2}\right)\,.\end{aligned}$$

Taking into account the wave equation (2.7.3) (for $\boldsymbol{f} \equiv 0$) the latter expression may be finally reduced to the inhomogeneous wave equation

$$\left(\Delta_1 - c_2^{-2}\frac{\partial^2}{\partial t^2}\right) f_0(x_1, x_2, t) = \left.\frac{\partial\psi_3}{\partial x_3}\right|_{x_3=c}\,,$$

to which the function f_0 entering the expression for χ must satisfy. Thus, for $\boldsymbol{f} \equiv 0$ the vector $\boldsymbol{\psi}$ in the representation (2.7.1–4) may be sought in the form $\boldsymbol{\psi} = (\psi_1, \psi_2, 0)$, and the condition $\operatorname{div}\boldsymbol{\psi} = 0$ must be discarded.

Below, it will be shown that in the plane-strain case the vector $\boldsymbol{\psi}$ depends on one scalar function only and, hence, no additional condition is needed.

Let us now discuss briefly the following case. The above proof of the equivalence of the Lamé equations to the system (2.7.1–4) was based on the solution $\boldsymbol{u}$ and the given body forces $\boldsymbol{f}$ being smooth enough (differentiable with respect to space variables up to the third order and having first partial derivatives with respect to space variables, respectively). However, actually, this equivalence occurs at any degree of smoothness of the solutions under consideration, including any generalized solutions to the Lamé equations and equations (2.7.2, 3) [2.5]. This is due to the fact that any solution to the equations under consideration may be presented as a limit of a sequence of the classical solutions

$$\boldsymbol{u}_n = \operatorname{grad}\varphi_n + \operatorname{curl}\boldsymbol{\psi}_n$$

to our equations, corresponding to sequences $\boldsymbol{f}_n$, Φ_n, $\boldsymbol{\Psi}_n$ of functions of any degree of smoothness which must satisfy the relation

$$\boldsymbol{f}_n = \operatorname{grad}\Phi_n + \operatorname{curl}\boldsymbol{\Psi}_n .$$

In the case under study, an appropriate choice of the functions $\boldsymbol{f}_n$ ensures the field $\boldsymbol{u}_n$ being smooth. The equivalence of the equations under consideration to the Lamé system (2.7.1–4) for $\boldsymbol{u}_n$, φ_n, $\boldsymbol{\psi}_n$ at any n implies their equivalence for the limiting function $\boldsymbol{u}$, φ, $\boldsymbol{\psi}$, too, irrespective of the latter having classical or generalized meaning.

In this connection, it should be emphasized that the equivalence of the equations of motion of an elastic medium to other equations and systems to be considered below, may also be proved in the generalized sense; however, we shall not pay special attention to the proofs maintaining the level of rigour characteristics of undergraduate studies in applied mathematics and mechanics.

Going back to the system (2.7.1–4) once again, it should be pointed out that the scalar potential φ describes propagation of longitudinal waves through an elastic medium. This process results in altering the particles' volume; however, they do not rotate, since

$$\operatorname{div}\boldsymbol{u}_1 \neq 0 , \quad \operatorname{curl}\boldsymbol{u}_1 = 0 \quad (\boldsymbol{u}_1 = \operatorname{grad}\varphi) .$$

Therefore, longitudinal waves propagating with the phase speed c_1 are also called dilatation waves (or irrotational waves, compressional waves, pressure waves, primary waves, P-waves) . The scalar potential φ is called the longitudinal displacement potential.

The vector potential $\boldsymbol{\psi}$ describes propagation of transverse waves through an elastic medium. The latter waves are equivoluminal, but rotational, since

$$\operatorname{div}\boldsymbol{u}_2 = 0 , \quad \operatorname{curl}\boldsymbol{u}_2 \neq 0 \quad (\boldsymbol{u}_2 = \operatorname{curl}\boldsymbol{\psi}) .$$

Therefore, transverse waves propagating with the phase speed c_2 are also called equivoluminal waves (or rotational waves, distortional waves, shear waves, secondary waves, S-waves). The vector potential $\boldsymbol{\psi}$ is called the transverse displacement potential.

Let us consider some special cases of the representations (2.7.1–4). If an elastic body is subject to plane deformation in the plane x_1x_2, then $(\boldsymbol{u}\cdot\boldsymbol{e}_3) = 0$, and

$$\begin{aligned} &\boldsymbol{f} \equiv \boldsymbol{f}(x_1, x_2, t) , \quad \boldsymbol{\Psi} = \Psi\boldsymbol{e}_3 , \quad \Psi \equiv \Psi(x_1, x_2, t) , \\ &\Phi \equiv \Phi(x_1, x_2, t) , \quad \boldsymbol{\psi} = \psi\boldsymbol{e}_3 , \quad \psi \equiv \psi(x_1, x_2, t) , \\ &\varphi \equiv \varphi(x_1, x_2, t) , \end{aligned} \qquad (2.7.23)$$

then, instead of (2.7.2, 3) we have

$$\Delta_1\varphi = c_1^{-2}\ddot{\varphi} - \frac{\Phi}{\lambda+2\mu}\,,\ \Delta_1\psi = c_2^{-2}\ddot{\psi} - \frac{\Psi}{\mu} \tag{2.7.24}$$

and the displacement vector is defined by

$$\boldsymbol{u} = \operatorname{grad}\varphi + \operatorname{curl}(\psi \boldsymbol{e}_3)\,, \tag{2.7.25}$$

while stresses and strains are given by (2.1.3, 7), with $u_3 \equiv 0$, $\sigma_{13} \equiv \sigma_{23} \equiv 0$.

As we know [2.1], transition from the plane-strain equations to those of the generalized plane-stress state is implemented by replacement of the Lamé constant λ by $\lambda' = 2\lambda\mu/(\lambda+2\mu)$. Hence, (2.7.23–25) obtained for the plane-strain case, remain valid for the generalized plane-stress state realized within the plane x_1x_2 of a thin flat plate, if λ in (2.7.23–25) is replaced by λ', and the functions $\boldsymbol{u}$, φ, ψ, $\boldsymbol{f}$, $\boldsymbol{\Phi}$, $\boldsymbol{\Psi}$ are assumed to be averaged across the plate's thickness.

In the case of antiplane deformation when $f_1 \equiv f_2 \equiv 0$, $f_3 \equiv f_3(x_1, x_2, t)$, $u_1 \equiv u_2 \equiv 0$, and $u_3 \equiv u_3(x_1, x_2, t)$, (2.1.8) reduces to

$$\Delta_1 u_3 + f_3 = c_2^{-2}\ddot{u}_3\,, \tag{2.7.26}$$

since $\operatorname{div}\boldsymbol{u} \equiv 0$. However, the stress components σ_{12}, σ_{23} remain nonzero:

$$\sigma_{13} = \mu u_{3,1}\,, \quad \sigma_{23} = \mu u_{3,2}\,. \tag{2.7.27}$$

2.7.2 The Case of a Separable Solution to the Vector Wave Equation

From (2.7.2, 3) it follows that the displacement vector $\boldsymbol{u}$ may always be decomposed into the longitudinal, $\boldsymbol{u}_1$, and transverse, $\boldsymbol{u}_2$, components

$$\boldsymbol{u} = \boldsymbol{u}_1 + \boldsymbol{u}_2\,, \quad \boldsymbol{u}_1 = \operatorname{grad}\varphi\,, \quad \boldsymbol{u}_2 = \operatorname{curl}\boldsymbol{\psi} \tag{2.7.28}$$

satisfying the wave equations for longitudinal and transverse waves, respectively:

$$\Delta\boldsymbol{u}_1 - c_1^{-2}\ddot{\boldsymbol{u}}_1 + \boldsymbol{f}_1/(\lambda+2\mu) = 0\,, \quad \boldsymbol{f}_1 = \operatorname{grad}\Phi\,, \tag{2.7.29}$$

$$\Delta\boldsymbol{u}_2 - c_2^{-2}\ddot{\boldsymbol{u}}_2 + \boldsymbol{f}_2/\mu = 0\,, \quad \boldsymbol{f}_2 = \operatorname{curl}\boldsymbol{\Psi}\,. \tag{2.7.30}$$

Since $\operatorname{curl}\operatorname{grad}\chi \equiv 0$, the vector $\boldsymbol{\psi}$ in the expression for $\boldsymbol{u}_2$ is defined to an accuracy of the gradient of the arbitrary function χ. However, the proof of (2.7.1–4) was based on assuming that the vector $\boldsymbol{\psi}$ satisfies the wave equations, too, and hence we had to require that the arbitrary function χ must satisfy the homogeneous wave equation (2.7.22). As a result, we deduced that $u_2 = \operatorname{curl}\boldsymbol{\psi}$ where the vector $\boldsymbol{\psi}$ satisfying the wave equation (2.7.3), may be written in the form $\boldsymbol{\psi} = \boldsymbol{\psi}' + \operatorname{grad}\chi$ with $\boldsymbol{\psi}'$ and χ satisfying (2.7.3) and (2.7.22), respectively, and $\operatorname{div}\boldsymbol{\psi}' = 0$.

Thus, since the vector potential $\boldsymbol{\psi}$ is always arbitrary due to the possibility of choosing the function χ [which is a solution to (2.7.22)], problems should be solved by subjecting χ to various constraints. Thus, in (2.7.1–4) we put $\chi \equiv 0$ (i.e., $\operatorname{div}\boldsymbol{\psi} = 0$), although χ could have been chosen in a different way, for instance, by putting one of the components of the vector $\boldsymbol{\psi}$ equal to zero in the case when $\boldsymbol{f} \equiv 0$. (Note that the constraint $\boldsymbol{f} \equiv 0$ is unimportant for solving the Lamé equations, since a solution to the nonhomogeneous wave equation

$$\left(\Delta - c_\alpha^{-2}\right) \boldsymbol{w}(\boldsymbol{x}, t) = \boldsymbol{f}_\alpha(\boldsymbol{x}, t)$$

may be expressed by means of the solution $\boldsymbol{w}_0(x, t)$ to the corresponding homogeneous wave equation as

$$\boldsymbol{w}(\boldsymbol{x}, t) = -\frac{1}{4\pi} \iiint\limits_V \frac{\boldsymbol{f}_\alpha(\boldsymbol{\xi}, t - R/c_\alpha)}{R} dV(\boldsymbol{\xi}) + \boldsymbol{w}_0(\boldsymbol{x}, t) ,$$

where $R \equiv R(\boldsymbol{x}, \boldsymbol{\xi})$ is the distance between the points $\boldsymbol{x}, \boldsymbol{\xi}$.)

Formulas (2.7.1–4), as well as their modifications, are difficult to use for solving three-dimensional problems, since in curvilinear coordinates (2.7.3) written in projections on coordinate axes, results, generally, in a system of coupled equations for determining scalar projections of the vector $\boldsymbol{\psi}$, which must be solved simultaneously. An attempt to circumvent the difficulty by using a rectilinear-coordinate representation of $\boldsymbol{\psi}$, runs into similar difficulties while trying to meet boundary conditions at a curvilinear boundary when all the $\boldsymbol{\psi}$-projections prove to be coupled through conditions on a body surface. Hence, while solving three-dimensional problems, one should seek the vector potential $\boldsymbol{\psi}$ (see (2.7.28) in another form which would leave the desired system for potentials free of the above disadvantage.

Bearing this in mind, let us consider the expression $\boldsymbol{u}_2 = \operatorname{curl} \boldsymbol{\psi}$ again, and seek the vector $\boldsymbol{\psi}$ in a form ensuring that $\boldsymbol{u}_2$ satisfies the wave equation (2.7.30) as before, while the vector $\boldsymbol{\psi}$ itself is expressed through two scalar functions, $\psi_1(\boldsymbol{x}, t)$, $\psi_2(\boldsymbol{x}, t)$, only, both satisfying the scalar wave equation for transverse waves. Such a study carried out for the case of $\boldsymbol{f} \equiv 0$ [2.16] has produced the following expression for $\boldsymbol{u}_2$

$$\boldsymbol{u}_2 = \operatorname{curl}\left[w, (\xi_1)\psi_1 \boldsymbol{e}_1\right] + \operatorname{curl}\operatorname{curl}\left[w(\xi_1)\psi_2 \boldsymbol{e}_1\right] , \tag{2.7.31}$$

where $\boldsymbol{e}_1$ is the unit vector directed along the ξ_1-axis of the orthogonal coordinate system ξ_1, ξ_2, ξ_3; $w(\xi_1)$ is a function of the single variable ξ_1, depending on the chosen coordinate system, and ψ_1, ψ_2 are scalar transverse potentials satisfying the wave equation

$$\Delta\psi = c_2^{-2}\ddot{\psi} . \tag{2.7.32}$$

We see that both terms on the right-hand side of (2.7.31) satisfy the vector wave equation for transverse waves:

$$\Delta \boldsymbol{u}_2 = c_2^{-2}\ddot{\boldsymbol{u}}_2 \tag{2.7.33}$$

with the first term, $\operatorname{curl}[w(\xi_1)\psi_1 \boldsymbol{e}_1]$, specifying a vector tangential to the surface $\xi_1 = \text{const}$, and the other term, $\operatorname{curl}\operatorname{curl}[w(\xi_1)\psi_2 \boldsymbol{e}_1]$, giving a vector which does not coincide with the former one (and, generally, is not orthogonal to it), but whose curl is also tangential to the surface $\xi_1 = \text{const}$.

It has been shown in [2.15] that a solution to the vector wave equation may be separated, i.e., decomposed in accordance with (2.7.31) where ψ_1, ψ_2 satisfy (2.7.32), in the following six coordinate systems only: a rectangular one with ξ_1 coinciding with x, y or z; three cylindrical ones (circular, elliptic and parabolic) with ξ_1 coinciding with z; and finally, spherical and conical coordinate systems with ξ_1 coinciding with r $(r = (x^2+y^2+z^2)^{1/2})$. For the former four systems and the latter two systems $w(\xi_1)$ is equal to 1 and r, respectively. Thus, a solution to the vector wave equation may be separated (in the above sense) in six of the eleven coordinate systems in which the variables in the scalar wave equation can be separated.

For the purpose of comparing the vector potential representation described by (2.7.1–4), with analogous expressions given by (2.7.31, 32), the expression (2.7.31) should be rewritten in the form

$$\boldsymbol{u}_2 = \operatorname{curl}\boldsymbol{\psi}\,, \quad \boldsymbol{\psi} = w(\xi_1)\psi_1\boldsymbol{e}_1 + \operatorname{curl}\left[w(\xi_1)\psi_2\boldsymbol{e}_1\right]\,. \tag{2.7.34}$$

Hence, unlike $\boldsymbol{\psi}$ from (2.7.1–4), the vector potential $\boldsymbol{\psi}$ given by (2.7.34) is expressed through two scalar functions ψ_1, ψ_2 only, which are solutions to the scalar wave equation (2.7.32), while the vector potential $\boldsymbol{\psi}$ given by (2.7.34), need not, generally speaking, satisfy the condition $\operatorname{div}\boldsymbol{\psi} = 0$ and the vector wave equation

$$\Delta\boldsymbol{\psi} = c_2^{-2}\ddot{\boldsymbol{\psi}}$$

[however, the vector $\boldsymbol{u}_2 = \operatorname{curl}\boldsymbol{\psi}$ does comply with (2.7.33)].

Furthermore, although in the above six coordinate systems a solution to the vector wave equation may be separated for each of the functions ψ_1, ψ_2 taken alone, nevertheless the difficulties of meeting boundary conditions remain, since the latter contain, as a rule, not only the transverse potentials ψ_1, ψ_2, but also the longitudinal potential φ and, generally speaking, all the potentials are often mixed in such a complicated way that boundary conditions cannot be separated for each of the potentials. As a result, such a problem cannot be solved analytically. From among curvilinear coordinate systems, only the cylindrical (circular) and the spherical ones sometimes permit separation of boundary conditions for the above potentials and solution of the corresponding elastodynamic problem.

Thus, in rectangular, cylindrical (circular), elliptic and parabolic coordinate systems, the Lamé equations

$$(\lambda+\mu)\operatorname{grad}\operatorname{div}\boldsymbol{u} + \mu\Delta\boldsymbol{u} = \varrho\ddot{\boldsymbol{u}} \tag{2.7.35}$$

are equivalent to the following system of equations

$$\Delta\varphi = c_1^{-2}\ddot{\varphi}\,, \quad \Delta\psi_\alpha = c_2^{-2}\ddot{\psi}_\alpha \quad (\alpha = 1,2)\,, \tag{2.7.36}$$

$$\boldsymbol{u} = \operatorname{grad}\varphi + \operatorname{curl}(\psi_1\boldsymbol{e}_z) + \operatorname{curl}\operatorname{curl}(\psi_2\boldsymbol{e}_z) \tag{2.7.37}$$

where the unit vector $\boldsymbol{e}_z$ is always directed along the z-axis, and the coordinates of the respective systems are related to Cartesian coordinates x, y, z as follows:

(1) cylindrical coordinates (r, θ, z):

$$x = r\cos\theta\,,\quad y = r\sin\theta\,,\quad z = z\,; \tag{2.7.38}$$

(2) elliptic coordinates (ξ, θ, z):

$$x = c\cosh\xi\,\cos\theta\,,\quad y = c\sinh\xi\,\sin\theta\,,\quad z = z\,; \tag{2.7.39}$$

(3) parabolic coordinates (ξ, η, z):

$$x = c\xi\eta\,,\quad y = c\left(\eta^2 - \xi^2\right)/2\,,\quad z = z\quad (c = \text{const})\,. \tag{2.7.40}$$

In the cases of spherical and conical coordinates, (2.7.35) is equivalent to the following system of equations

$$\Delta\varphi = c_1^{-2}\ddot{\varphi}\,,\quad \Delta\psi_\alpha = c_2^{-2}\ddot{\psi}_\alpha\quad (\alpha = 1, 2)\,, \tag{2.7.41}$$

$$\boldsymbol{u} = \operatorname{grad}\varphi + \operatorname{curl}(r\psi_1\boldsymbol{e}_r) + \operatorname{curl}\operatorname{curl}(r\psi_2\boldsymbol{e}_r)\,, \tag{2.7.42}$$

where the unit vector $\boldsymbol{e}_r$ is always directed along the radius r, and the coordinates of the respective systems are related to Cartesian coordinates x, y, z by the following relations:

(1) spherical coordinates (r, θ, β):

$$x = r\sin\theta\cos\beta\,,\quad y = r\sin\theta\sin\beta\,,\quad z = r\cos\theta\,; \tag{2.7.43}$$

(2) conical coordinates (r, ξ, η):

$$\begin{aligned} &x = r\mathrm{dn}(\xi, \alpha)\mathrm{sn}(\eta, \gamma)\,,\quad y = r\mathrm{sn}(\xi, \alpha)\mathrm{dn}(\eta, \gamma)\,,\\ &z = r\mathrm{cn}(\xi, \alpha)\mathrm{cn}(\eta, \gamma)\,, \end{aligned} \tag{2.7.44}$$

where cn, dn, sn are elliptic functions [2.14], and α, β are constants.

Note, that in the three-dimensional case, transverse waves are defined by two scalar potentials, ψ_1, ψ_2, and, hence, there exist transverse waves of two types, differing in the polarization vector (the direction of the displacement vector). Thus, in a plane transverse wave of the 1st kind, $\boldsymbol{u} = \operatorname{curl}(\psi_1\boldsymbol{e}_z)$, traveling along the y-axis of Cartesian coordinate system x, y, z, the displacement is parallel to the x-axis, while in the other transverse wave, $\boldsymbol{u} = \operatorname{curl}\operatorname{curl}(\psi_2\boldsymbol{e}_z)$, traveling in the same direction, the displacement is parallel to the z-axis.

Let us discuss in more detail the cylindrical and spherical coordinate systems which we shall use below.

We shall start by recalling that in the cylindrical coordinate system r, θ, z defined by (2.7.38), one has

$$\begin{aligned} &\boldsymbol{u} = \{u_r, u_\theta, u_z\}\,,\quad \Delta = \frac{\partial^2}{\partial r^2} + \frac{1}{r}\frac{\partial}{\partial r} + \frac{\partial^2}{r^2\partial\theta^2} + \frac{\partial^2}{\partial z^2}\,,\\ &\operatorname{div}\boldsymbol{u} = \frac{\partial(ru_r)}{r\partial r} + \frac{\partial u_\theta}{r\partial\theta} + \frac{\partial u_z}{\partial z}\,,\\ &\operatorname{curl}\boldsymbol{u} = \left(\frac{1}{r}\frac{\partial u_z}{\partial\theta} - \frac{\partial u_\theta}{\partial z}\right)\boldsymbol{e}_r + \left(\frac{\partial u_r}{\partial z} - \frac{\partial u_z}{\partial r}\right)\boldsymbol{e}_\theta + \left(\frac{\partial(ru_\theta)}{r\partial r} - \frac{\partial u_r}{r\partial\theta}\right)\boldsymbol{e}_z\,,\\ &\operatorname{grad} f = \frac{\partial f}{\partial r}\boldsymbol{e}_r + \frac{\partial f}{r\partial\theta}\boldsymbol{e}_\theta + \frac{\partial f}{\partial z}\boldsymbol{e}_z\,, \end{aligned} \tag{2.7.45}$$

where e_r, e_θ, e_z are unit vectors directed along r, θ, z, respectively. The stress tensor is related to the strain tensor by (2.1.7) with $i, j = r, \theta, z$, and the strain tensor components are defined by

$$\begin{aligned}
&\varepsilon_{rr} = \frac{\partial u_r}{\partial r}, \quad \varepsilon_{\theta\theta} = \frac{1}{r}\left(\frac{\partial u_\theta}{\partial \theta} + u_r\right), \quad \varepsilon_{zz} = \frac{\partial u_z}{\partial z}, \\
&2\varepsilon_{r\theta} = 2\varepsilon_{\theta r} = r\frac{\partial}{\partial r}\left(\frac{u_\theta}{r}\right) + \frac{1}{r}\frac{\partial u_r}{\partial \theta}, \\
&2\varepsilon_{\theta z} = 2\varepsilon_{z\theta} = \frac{1}{r}\frac{\partial u_z}{\partial \theta} + \frac{\partial u_\theta}{\partial z}, \\
&2\varepsilon_{rz} = 2\varepsilon_{zr} = \frac{\partial u_z}{\partial r} + \frac{\partial u_r}{\partial z}.
\end{aligned} \tag{2.7.46}$$

For an axisymmetrically loaded body of rotation without torsion, one has in (2.7.36, 37):

$$\begin{aligned}
&\varphi = \varphi(r, z, t), \quad \psi_1 = 0, \quad \psi_2 = \psi_2(r, z, t), \\
&\sigma_{\theta r} = \sigma_{\theta z} = 0, \quad u_\theta = 0,
\end{aligned} \tag{2.7.47}$$

where z is the symmetry axis.

Then, (2.7.37) becomes

$$u = \operatorname{grad}\varphi + \operatorname{curl}\left(-\frac{\partial \psi_2}{\partial r} e_\theta\right). \tag{2.7.48}$$

If, however, a body is subject to torsion (with respect to the z-axis), then

$$\varphi = 0, \quad \psi_1 = \psi_1(r, z, t), \quad \psi_2 = 0, \tag{2.7.49}$$

and u_θ, $\sigma_{\theta r}$, $\sigma_{\theta z}$ are nonzero. Hence,

$$u = \operatorname{curl}(\psi_1 e_z) = -\frac{\partial \psi_1}{\partial r} e_\theta. \tag{2.7.50}$$

In the spherical coordinate system r, θ, β defined by (2.7.43), one has

$$\begin{aligned}
&u = \{u_r, u_\theta, u_\beta\}, \\
&\Delta = \frac{\partial^2}{\partial r^2} + \frac{2}{r}\frac{\partial}{\partial r} + \frac{\partial^2}{r^2 \partial \theta^2} + \frac{\cot\theta}{r^2}\frac{\partial}{\partial \theta} + \frac{1}{r^2 \sin^2\theta}\frac{\partial^2}{\partial \beta^2}, \\
&\operatorname{div} u = \frac{1}{r^2}\frac{\partial}{\partial r}(r^2 u_r) + \frac{1}{r\sin\theta}\frac{\partial}{\partial \theta}(u_\theta \sin\theta) + \frac{1}{r\sin\theta}\frac{\partial u_\beta}{\partial \beta}, \\
&\operatorname{curl} u = \frac{1}{r\sin\theta}\left[\frac{\partial}{\partial \theta}\left(u_\beta \sin\theta - \frac{\partial u_\theta}{\partial \beta}\right)\right] e_r + \frac{1}{r}\left(\frac{1}{\sin\theta}\frac{\partial u_r}{\partial \beta} - \frac{\partial (r u_\beta)}{\partial r}\right) e_\theta \\
&\qquad + \frac{1}{r}\left[\frac{\partial (r u_\theta)}{\partial r} - \frac{\partial u_r}{\partial \theta}\right] e_\beta, \\
&\operatorname{grad} f = \frac{\partial f}{\partial r} e_r + \frac{1}{r}\frac{\partial f}{\partial \theta} e_\theta + \frac{1}{r\sin\theta}\frac{\partial f}{\partial \beta} e_\beta,
\end{aligned} \tag{2.7.51}$$

where $\boldsymbol{e}_r, \boldsymbol{e}_\theta, \boldsymbol{e}_\beta$ are unit vectors directed along r, θ, β, respectively. The stress tensor is related to the strain tensor by (2.1.7) where $i, j = r, \theta, \beta$, and the strain tensor components may be written in the form

$$\begin{aligned}
\varepsilon_{rr} &= \frac{\partial u_r}{\partial r}, \quad \varepsilon_{\theta\theta} = \frac{1}{r}\left(\frac{\partial u_\theta}{\partial \theta} + u_r\right), \\
\varepsilon_{\beta\beta} &= \frac{1}{r}\left(\frac{1}{\sin\theta}\frac{\partial u_\beta}{\partial \beta} + u_\theta \cot\theta + u_r\right), \\
2\varepsilon_{r\theta} &= 2\varepsilon_{\theta r} = \frac{1}{r}\left(\frac{\partial u_r}{\partial \theta} - u_\theta\right) + \frac{\partial u_\theta}{\partial r}, \\
2\varepsilon_{\beta\theta} &= 2\varepsilon_{\theta\beta} = \frac{1}{r}\left(\frac{1}{\sin\theta}\frac{\partial u_\theta}{\partial \beta} + \frac{\partial u_\beta}{\partial \theta} + u_\beta \cot\theta\right), \\
2\varepsilon_{r\beta} &= 2\varepsilon_{\beta r} = \frac{1}{r}\left(\frac{1}{\sin\theta}\frac{\partial u_r}{\partial \beta} - u_\beta\right) + \frac{\partial u_\beta}{\partial r}.
\end{aligned} \tag{2.7.52}$$

For an axisymmetrically loaded body of rotation without torsion, one has in (2.7.41, 42):

$$\begin{aligned}
&\varphi = \varphi(r, \theta, t), \quad \psi_1 = 0, \quad \psi_2 = \psi_2(r, \theta, t), \\
&\sigma_{\theta\beta} = \sigma_{r\beta} = 0, \quad u_\beta = 0,
\end{aligned} \tag{2.7.53}$$

if z is the symmetry axis. Then, (2.7.53) becomes

$$\boldsymbol{u} = \operatorname{grad}\varphi + \operatorname{curl}\left(-\frac{\partial\psi_2}{\partial\theta}\boldsymbol{e}_\beta\right). \tag{2.7.54}$$

If, however, a body is subject to torsion, then in (2.7.41, 42)

$$\varphi = 0, \quad \psi_1 = \psi_1(r, \theta, t), \quad \psi_2 = 0, \tag{2.7.55}$$

and $u_\beta, \sigma_{\theta\beta}, \sigma_{r,\beta}$ are nonzero. Hence,

$$\boldsymbol{u} = \operatorname{curl}(r, \psi_1, \boldsymbol{e}_r) = -\frac{\partial\psi_1}{\partial\theta}\boldsymbol{e}_\beta. \tag{2.7.56}$$

The formulas (2.7.48, 50, 54, 56) obtained in cylindrical (r, θ, z) and spherical (r, θ, β) coordinates, can also be used to describe an axisymmetrical motion of an elastic medium in other orthogonal coordinates (q_1, q_2, ω), if ω coincides with β (or with θ, in cylindrical coordinates). In this case it is necessary to replace $\partial\psi_\alpha/\partial\theta$ in (2.7.54, 56) by $(\partial\psi_\alpha/\partial q_1)(\partial q_1/\partial\theta) + (\partial\psi_\alpha/\partial q_2)(\partial q_2/\partial\theta)$ (or $(\partial\psi_\alpha/\partial r)$ in (2.7.48, 50) by $(\partial\psi_\alpha/\partial q_1)(\partial q_1/\partial r) + (\partial\psi_\alpha/\partial q_2)(\partial q_2/\partial r)$).

2.7.3 Iacovache's Representation

Besides representations of the solution $\boldsymbol{u}$ through longitudinal and transverse potentials in the form (2.7.1–4) as well as its modifications (2.7.36, 37), (2.7.41, 42), there exist some alternative representations of solutions to the elastodynamic equations (2.1.8). Let us consider them briefly.

It has been shown by *Iacovache* [2.17] that the function $\boldsymbol{u}$ defined by

$$\boldsymbol{u} = \frac{\lambda+2\mu}{\mu}\Box_1\boldsymbol{\Phi} - \frac{\lambda+\mu}{\mu}\,\mathrm{grad}\,\mathrm{div}\,\boldsymbol{\Phi}\,, \tag{2.7.57}$$

where $\boldsymbol{\Phi}$ satisfies the equation

$$\begin{aligned}&(\lambda+2\mu)\Box_1\Box_2\boldsymbol{\Phi} + \boldsymbol{f} = 0\,,\\ &\Box_\alpha \equiv \Delta - \frac{1}{c_\alpha^2}\frac{\partial^2}{\partial t^2} \quad (\alpha = 1,2)\end{aligned} \tag{2.7.58}$$

is a solution to the Lamé equations. This statement may be proved by straightforward substitution of (2.7.57) into (2.1.8).

Conversely, according to [2.3, 18], any solution $\boldsymbol{u}$ to the Lamé equations may be written in the form (2.7.57, 58).

The latter statement may be proved by substitution of the vector $\boldsymbol{u}$ written in the form

$$\boldsymbol{u} = \mathrm{grad}\,\varphi + \mathrm{curl}\boldsymbol{\psi}\,. \tag{2.7.59}$$

into (2.7.6). Then, one gets

$$\mathrm{grad}\,\Box_1\varphi + \frac{\mu}{\lambda+2\mu}\mathrm{curl}\Box_2\boldsymbol{\psi} + \frac{\boldsymbol{f}}{\lambda+2\mu} = 0\,. \tag{2.7.60}$$

Let us next define the function $\boldsymbol{\Phi}$ with the help of the scalar function θ and the vector function $\boldsymbol{\chi}$ by the formula

$$\boldsymbol{\Phi} = \mathrm{grad}\,\theta + \frac{\mu}{\lambda+2\mu}\mathrm{curl}\boldsymbol{\chi}\,. \tag{2.7.61}$$

Substitution of $\boldsymbol{\Phi}$ into (2.7.58) gives

$$\mathrm{grad}\Box_1\Box_2\theta + \frac{\mu}{\lambda+2\mu}\mathrm{curl}\Box_1\Box_2\boldsymbol{\chi} + \frac{\boldsymbol{f}}{\lambda+2\mu} = 0\,. \tag{2.7.62}$$

Comparing (2.7.60) and (2.7.62) we deduce that (2.7.62) holds if θ and $\boldsymbol{\chi}$ satisfy the equations

$$\Box_2\theta = \varphi\,, \tag{2.7.63}$$

$$\Box_1\boldsymbol{\chi} = \boldsymbol{\psi}\,. \tag{2.7.64}$$

Thus, according to (2.7.63, 64), the functions $\theta, \boldsymbol{\chi}$ are retarded potentials

$$\theta(\boldsymbol{x},t) = -\frac{1}{4\pi}\iiint\limits_V \frac{\varphi(\boldsymbol{\xi},t-R/c_2)}{R(\boldsymbol{x},\boldsymbol{\xi})}dV(\boldsymbol{\xi}) ,$$
$$\chi(\boldsymbol{x},t) = -\frac{1}{4\pi}\iiint\limits_V \frac{\boldsymbol{\psi}(\boldsymbol{\xi},t-R/c_1)}{R(\boldsymbol{x},\boldsymbol{\xi})}dV(\boldsymbol{\xi}) , \tag{2.7.65}$$

where $R(\boldsymbol{x},\boldsymbol{\xi})$ is the distance between points $\boldsymbol{x}$ and $\boldsymbol{\xi}$. Let us apply the operator $\Box_1$ to (2.7.61). Then, taking into account (2.7.64), one gets

$$\Box_1\boldsymbol{\Phi} = \operatorname{grad}\Box_1\theta + \frac{\mu}{\lambda+2\mu}\operatorname{curl}\boldsymbol{\psi} . \tag{2.7.66}$$

By excluding $\operatorname{curl}\boldsymbol{\psi}$ from (2.7.59) with the help of (2.7.66), we obtain

$$\boldsymbol{u} = \operatorname{grad}\left(\varphi - \frac{\lambda+2\mu}{\mu}\Box_1\theta\right) + \frac{\lambda+2\mu}{\mu}\Box_1\boldsymbol{\Phi} . \tag{2.7.67}$$

Upon transforming the expression in parentheses with the help of (2.7.63) and the relation

$$\frac{\lambda+2\mu}{\mu}\Box_1\theta = \Box_2\theta + \frac{\lambda+\mu}{\mu}\Delta\theta , \tag{2.7.68}$$

we arrive at

$$\boldsymbol{u} = \frac{\lambda+2\mu}{\mu}\Box_1\boldsymbol{\Phi} - \frac{\lambda+\mu}{\mu}\operatorname{grad}\Delta\theta . \tag{2.7.69}$$

Taking into consideration $\operatorname{div}\boldsymbol{\Phi} = \Delta\theta$ [which follows from (2.7.61)], we finally transform (2.7.69) into (2.7.57), which was the thing to be shown.

If we now pass to the static problem by allowing ϱ to tend to zero, then c_α in (2.7.58) tends to infinity ($\alpha = 1,2$), and the latter equations transform into the Galerkin equations [2.3]:

$$(\lambda+2\mu)\Delta\Delta\boldsymbol{\Phi} + \boldsymbol{f} = 0 ,$$

where $\boldsymbol{\Phi}$ is the Galerkin vector, and the vector $\boldsymbol{u}$ has the form known from statics

$$\boldsymbol{u} = \frac{\lambda+2\mu}{\mu}\Delta\boldsymbol{\Phi} - \frac{\lambda+\mu}{\mu}\operatorname{grad}\operatorname{div}\boldsymbol{\Phi} .$$

Thus, the vector $\boldsymbol{\Phi}$ introduced by Iacovache is a generalization of the Galerkin vector for the case of dynamics.

It should be emphasized that representation of a solution in the form of (2.7.57, 58) is convenient in the case of body forces acting in an infinite region where one does not have to meet boundary conditions. However, in the case of a finite region one should prefer representations (2.7.1–4), (2.7.36, 37), and (2.7.41, 42) which have been obtained by expressing the displacement vector through longitudinal and transverse potentials. This is due to the fact that in the case of the 2nd boundary-value problem being solved by employing representations (2.7.1–4), boundary conditions contain the first derivatives of the potentials, while in the case of the 1st boundary-value problem they contain the second derivatives. The Iacovache representation involves the second derivatives in the case of the 2nd boundary-value problem, and the third derivatives in the case of the 1st boundary-value problem.

Besides, in curvilinear coordinates, equations (2.7.58) for the components of the vector function $\boldsymbol{\Phi}$ prove to be coupled, thus making the Iacovache representation less convenient compared to the representation via scalar potentials φ, ψ_1, ψ_2 defined by formulas (2.7.36, 37), (2.7.41, 42).

2.7.4 Representation Employing Papkovich–Neuber Functions

Let us now consider the representation of the displacement vector with the help of Papkovich–Neuber functions. The method proposed by the authors is known to be rather useful for solving some boundary-value problems of elastostatics. Let us see whether the Papkovich–Neuber representation

$$\boldsymbol{u} = \operatorname{grad}(\theta + \boldsymbol{R} \cdot \boldsymbol{\chi}) - 4(1-\nu)\boldsymbol{\chi}\,, \tag{2.7.70}$$

(where $\nu = \lambda/2(\lambda+\mu)$ is the Poisson coefficient, $\boldsymbol{R}$ is the radius vector of point $\boldsymbol{x}$) used for analyzing static problems, is convenient for solving elastodynamic problems, too.

We start by substituting (2.7.70) into the Lamé equation (2.1.8). Then, since

$$\operatorname{div}\boldsymbol{u} = \Delta\theta + \boldsymbol{R}\cdot\Delta\boldsymbol{\chi} + (4\nu - 2)\operatorname{div}\boldsymbol{\chi}\,,$$

we get

$$\begin{aligned}&\operatorname{grad}\left[(\lambda+2\mu)(\Delta\theta + \boldsymbol{R}\cdot\Delta\boldsymbol{\chi}) - \varrho\ddot{\theta} - \varrho(\boldsymbol{R}\cdot\ddot{\boldsymbol{\chi}})\right]\\ &\quad - 4(1-\nu)\left[\mu\Delta\boldsymbol{\chi} - \varrho\ddot{\boldsymbol{\chi}}\right] + \boldsymbol{f} = 0\,.\end{aligned} \tag{2.7.71}$$

Let us next choose $\boldsymbol{\chi}$ satisfying the equation

$$-4(1-\nu)\mu\Box_2\boldsymbol{\chi} + \boldsymbol{f} = 0\,. \tag{2.7.72}$$

Then, (2.7.71) becomes

$$\Box_1\theta + \boldsymbol{R}\cdot\Box_1\boldsymbol{\chi} = 0\,. \tag{2.7.73}$$

The zero on the right-hand side of (2.7.73) must be replaced by an arbitrary function $A(t)$:

$$\Box_1\theta + \boldsymbol{R}\cdot\Box_1\boldsymbol{\chi} = A(t)\,.$$

Then, upon substituting

$$\theta' = \theta + c_1^2\iint A(t)\,dt\,dt\,,$$

the equation reduces to the form (2.7.73) for the function θ', with representation (2.7.70) for $\boldsymbol{u}$ preserving its form while passing from θ to θ'. Hence, we can set $A(t) \equiv 0$ from the very beginning.

Equations (2.7.72,73) form a closed system of wave equations for determining θ, χ which enter (2.7.70). In the case of an infinite region, the solution will be sought as follows: First, the function χ is obtained from (2.7.72) and substituted into (2.7.73). Then, (2.7.73) is solved to give θ. For a finite body, the solution of the system of wave equations (2.7.72,73) becomes rather complicated, since the functions θ, χ entering (2.7.73) are coupled.

Note that if one passes to the static problem by tending ϱ to zero, then $c_\alpha \to \infty$ $(\alpha = 1,2)$, and the wave equations (2.7.72,73) become decoupled since $\Box_1\chi = \Box_2\chi = \Delta\chi$ and $\Box_1\chi$ is excluded from (2.7.73):

$$-4(1-\nu)\mu\Delta\chi + \boldsymbol{f} = 0\,, \tag{2.7.74}$$

$$4(1-\nu)\mu\Delta\theta + \boldsymbol{R}\cdot\boldsymbol{f} = 0\,, \tag{2.7.75}$$

and we arrive at the above Papkovich–Neuber representation (2.7.70,72,73) for the static problem where θ, χ are Papkovich–Neuber functions. Thus, the functions θ, χ obtained from (2.7.72,73), generalize the Papkovich–Neuber functions proposed for solving static problems, to the case of elastodynamic problems.

Thus, of all the representations of the solution $\boldsymbol{u}(\boldsymbol{x}, t)$ considered up till now (namely, via the longitudinal and vector transverse potentials, $\varphi, \boldsymbol{\psi}$, the scalar potentials φ, ψ_1, ψ_2, the Iacovache function $\boldsymbol{\Phi}$ and the generalized Papkovich–Neuber functions θ, χ), the most practical is the representation through the potentials $\varphi, \boldsymbol{\psi}$ and, in particular, through the scalar functions φ, ψ_1, ψ_2. The latter representations lead to the simplest wave equations. The Iacovache representation may be conveniently used for determining displacements within an infinite region or half-space. Representations based on the Papkovich–Neuber functions are less convenient because the wave equations used for determining the latter functions are coupled.

2.8 On the Relationships Between Solutions of Transient Dynamic Problems and Those of Static, Steady-State and Stationary Dynamic Problems

If ϱ in (2.1.6) tends to zero, and the accelerations of particles, $\ddot{\boldsymbol{u}}$, are assumed to remain finite, then inertial forces vanish, and the latter equations reduce to those of elastostatics. Correspondingly, the formulas describing solutions to elastodynamic problems transform into the ones of respective static problems. In the case of a bounded solid, this limiting passage may be accomplished if the given boundary stresses are self-balanced at any instant. The static solutions thus obtained contain the time t as a parameter (the so-called quasi-static solutions).

It was noted above that in the limiting case when $\varrho \to 0$, the Iacovache and Papkovich–Neuber representations transform into the respective well-known representations for static problems. However, it must be stressed that in the case of the Lamé representations (2.7.1–4), as well as of (2.7.36, 37) and (2.7.41, 42), the functions $c_1^{-2}\partial^2\varphi/\partial t^2$, $c_2^{-2}\partial^2\boldsymbol{\psi}/\partial t^2$, $c_2^{-2}\partial^2\psi_1/\partial t^2$ and $c_2^{-2}\partial^2\psi_2/\partial t^2$ cannot be set equal to zero while passing to the limit as $\varrho \to 0$ in the wave equations for $\varphi, \boldsymbol{\psi}, \psi_1$, and ψ_2. In fact, since

$$\frac{1}{c_1^2}\frac{\partial^2\varphi}{\partial t^2} = \Delta\varphi + \frac{\Phi}{\lambda+2\mu} = \operatorname{div}\boldsymbol{u} + \frac{\Phi}{\lambda+2\mu}\,, \tag{2.8.1}$$

and the equality

$$\frac{1}{c_2^2}\frac{\partial^2\boldsymbol{\psi}}{\partial t^2} = \Delta\boldsymbol{\psi} + \frac{\boldsymbol{\Psi}}{\mu} = -\operatorname{curl}\boldsymbol{u} + \frac{\boldsymbol{\Psi}}{\mu} \tag{2.8.2}$$

holds taking into account the identity

$$\Delta\boldsymbol{\psi} = \operatorname{grad}\operatorname{div}\boldsymbol{\psi} - \operatorname{curl}\operatorname{curl}\boldsymbol{\psi}\,,$$

by putting $c_1^{-2}\partial^2\varphi/\partial t^2 = c_2^{-2}\partial^2\boldsymbol{\psi}/\partial t^2 = 0$, one obtains in the case of zero body forces ($\Phi = 0$, $\boldsymbol{\Psi} = 0$)

$$\operatorname{div}\boldsymbol{u} = 0\,, \quad \operatorname{curl}\boldsymbol{u} = 0\,.$$

Thus, we arrive at the conclusion that no dilatation and particle rotation take place in the medium, which is generally wrong. However, this is what would result from setting the functions $c_1^{-2}\partial^2\varphi/\partial t^2$ and $c_2^{-2}\partial^2\boldsymbol{\psi}/\partial t^2$ equal to zero.

Usually, by steady-state dynamic problems we mean the problems related to considering harmonic oscillations, in which the displacement vector $\boldsymbol{u}$ is written in the form

$$\boldsymbol{u}(\boldsymbol{x},t,\omega) = \boldsymbol{U}(\boldsymbol{x},\omega)\mathrm{e}^{\mathrm{i}\omega t} \quad (\operatorname{Im}\omega = 0\,, -\infty < t < \infty)\,.$$

Then, the system of equations (2.1.6) for $\boldsymbol{U}(\boldsymbol{x},\omega)$ is elliptic. It can be readily shown that the same system is satisfied by the Laplace transform $\boldsymbol{u}^*(\boldsymbol{x},p)$ of the transient-problem solution $\boldsymbol{u}(\boldsymbol{x},t)$, if $p = \mathrm{i}\omega$. Thus, the solution $\boldsymbol{U}(\boldsymbol{x},\omega)$ to the steady-state dynamic problem may be obtained from the solution of the corresponding transient problem $\boldsymbol{u}(\boldsymbol{x},t)$, if in its transform $\boldsymbol{u}^*(\boldsymbol{x},p)$ one puts $p = \mathrm{i}\omega + \varepsilon$, $\varepsilon > 0$, $\operatorname{Im}\omega = 0$ (taking into account that $\operatorname{Re}p > 0$) and passes to the limit as $\varepsilon \to 0$:

$$\boldsymbol{U}(\boldsymbol{x},\omega) = \lim_{\varepsilon\to 0}\boldsymbol{u}^*(\boldsymbol{x},\mathrm{i}\omega+\varepsilon)\,.$$

Hence, the problems involving harmonic oscillations may often be solved by employing the methods used for solving transient problems.

It should be added that the analysis of steady-state problems by the method of potential and the theory of singular integral equations is discussed in detail in the monographs [2.19, 20]. Application of the separation of variables and integral transforms in the solution of the above problems is treated in [2.21–23].

It must be noted that in elastodynamics the problems of harmonic oscillations play a more modest role than the transient problems. This is due to the fact that wave processes in elastic media are quite often excited by shocks and explosions, and in practice one has to deal, as a rule, with transient problems.

The class of stationary problems incorporates the problems related to steady processes, such as the ones caused by constant-velocity loads, punches, etc. From the physical point of view, these problems may be considered equivalent to the corresponding transient problems as $t \to \infty$. In a frame of axes moving together with a load/punch/crack edge, a solution to the problem is independent of time, and this allows us to use the powerful method based on the theory of functions of the complex variable, developed by *Muskhelishvili* [2.24, 25], for solving boundary-value problems of mathematical physics. The methods used to solve stationary dynamic problems are analyzed in many works, among which of primary significance are *Galin*'s monograph [2.26] and the works [2.27–35].

Taking into account the above remarks as well as the limited scope of the present monograph, we shall only consider transient problems.

3. The Method of Functionally Invariant Solutions (the Smirnov–Sobolev Method)

The method of functionally invariant solutions was developed in the 1930s by Smirnov and Sobolev who called it "the method of complex solutions". It was designed for solving two- and three-dimensional nonstationary problems of isotropic elasticity and was closely related to the refined body of mathematics of theoretical seismology. The most detailed description of the method was presented by *Smirnov* and *Sobolev* in [3.1,2]. The method was further developed and applied to solving elastodynamic problems mostly in the Soviet Union. In the 1930–50s it was used for analyzing and solving a number of problems of nonstationary oscillations of a half-space [3.3,4], elastic wave propagation within a layer [3.5], reflection and refraction of elastic waves, the general theory of Rayleigh waves [3.6–11], and diffraction of plane acoustic waves by a wedge [3.12]. The method was also employed for solving two- [3.13–15] and three-dimensional [3.16] problems of diffraction of plane elastic waves at cuts and welded half-planes.

Yerugin [3.17] made an attempt to generalize the Smirnov–Sobolev method onto the general three-dimensional case. We should also mention the further development of the method for the case of anisotropic media, carried out by *Sveklo* [3.18–21].

Note also that the works dedicated to developing the Smirnov–Sobolev method are closely related to studying the methods of homogeneous solutions, based, in essence, on the Chaplygin transform [3.22]

$$r = \frac{2\varepsilon}{1+\varepsilon^2}$$

originally considered by its author as early as in 1902 in his work on gas jets. *Busemann* [3.23] showed that the Chaplygin transform makes it possible to reduce the wave equation

$$\frac{\partial^2 u}{\partial x^2} + \frac{\partial^2 u}{\partial y^2} = \frac{\partial^2 u}{a^2 \partial t^2}$$

for the self-similar solution

$$u(x,y,t) \equiv u(\xi,\eta) \left(\xi = \frac{x}{at}\,,\ \eta = \frac{y}{at}\right)$$

to the Laplace equation written in the polar coordinate system

$$\varepsilon\frac{\partial}{\partial\varepsilon}\left(\varepsilon\frac{\partial u}{\partial\varepsilon}\right)+\frac{\partial^2 u}{\partial\theta^2}=0\,,$$

if one uses the formulas $\xi = r\cos\theta$, $\eta = r\sin\theta$ to pass in the wave equation from the independent variables (x, y, t) to (ξ, η), and then to (r, θ) and (ε, θ).

The Chaplygin transform was used in fluid dynamics both for studying stationary supersonic flows past bodies (see, e.g., Ward's monograph [3.24]), and solving nonstationary problems [3.25, 26]. *Miles* [3.27] was most successful in developing the method of homogeneous solutions employing the Chaplygin transform, in its application to elastodynamics.

Important contributions to the development of the Smirnov–Sobolev and similar methods of obtaining homogeneous solutions to two- and three-dimensional elastodynamic problems with mixed boundary conditions on a plane boundary were made by *Maue* [3.28], *Kostrov* [3.29–31], *Craggs* [3.32], *Afanas'ev* and *Cherepanov* [3.33–35], *Nikitin* and *Odintsev* [3.36, 37], *Robinson* and *Thompson* [3.38, 39], and *Brock* [3.40–43]. Analogous problems were considered in [3.44–48]. Here *Norwood*'s studies [3.49] should be mentioned, in which the Smirnov–Sobolev representations were obtained with the help of the Laplace transform with respect to time and by using the Cagniard-deHoop method. Also we must mention the fundamental work by *Willis* [3.50] where he developed the method of solving elastodynamic problems for anisotropic media.

Below we discuss in detail the method of functionally invariant solutions, the discussion being based mainly on the aforementioned works by Smirnov and Sobolev, and present the further development of the method for the most interesting case of homogeneous solutions. Firstly homogeneous solutions of integer degree in space variables and time are considered in detail for plane-strain and antiplane problems. Then representations of solutions to three-dimensional problems through two-dimensional ones are given and numerous examples of solving axisymmetric problems are considerd, which are the most interesting from the point of view of applications. At the end of the chapter an alternative derivation of Smirnov–Sobolev representations is presented allowing generalization of the method onto the case of anisotropic media.

3.1 Functionally Invariant Solutions to the Wave Equation

According to [3.1, 2], functionally invariant solutions are constructed in the following way: a function $\Omega(x, y, t)$ (not necessarily a real one) is sought such that any doubly differentiable function[1] $u = f(\Omega)$ is a solution of the equation

$$\frac{\partial^2 u}{\partial x^2}+\frac{\partial^2 u}{\partial y^2}-\frac{\partial^2 u}{a^2\partial t^2}=0\,, \tag{3.1.1}$$

or, in other words,

[1] If $\Omega(x, y, t)$ is a complex function, then $f(\Omega)$ is analytical.

$$f''(\Omega)\left[\left(\frac{\partial\Omega}{\partial x}\right)^2+\left(\frac{\partial\Omega}{\partial y}\right)^2-\frac{1}{a^2}\left(\frac{\partial\Omega}{\partial t}\right)^2\right]$$
$$+f'(\Omega)\left[\frac{\partial^2\Omega}{\partial x^2}+\frac{\partial^2\Omega}{\partial y^2}-\frac{1}{a^2}\frac{\partial^2\Omega}{\partial t^2}\right]=0\,.$$

In view of the arbitrariness of $f(\Omega)$, this gives

$$\frac{\partial^2\Omega}{\partial x^2}+\frac{\partial^2\Omega}{\partial y^2}=\frac{1}{a^2}\frac{\partial^2\Omega}{\partial t^2}\,,\qquad\left(\frac{\partial\Omega}{\partial x}\right)^2=\frac{1}{a^2}\left(\frac{\partial\Omega}{\partial t}\right)^2\,,\tag{3.1.2}$$

and, hence, Ω is simultaneously a solution to the wave equation and the eikonal equation. In order that $u=f(\Omega)$ be a solution to the wave equation, the necessary and sufficient condition is that the system of equations (3.1.2) be satisfied. The validity of the following statement can be proved by direct verification: the general integral of the system (3.1.2) is an expression which is linear in x, y and t:

$$\delta\equiv l(\Omega)t+m(\Omega)x+n(\Omega)y-k(\Omega)=0\,.\tag{3.1.3}$$

The coefficients of the latter equation are related in the following manner:

$$l^2(\Omega)=a^2[m^2(\Omega)+n^2(\Omega)]\,.\tag{3.1.4}$$

In fact, implicit differentiation provides the following expressions for the derivatives with respect to coordinates:

$$\begin{aligned}
&\frac{\partial\Omega}{\partial x}=-\frac{m}{\delta'}\,,\quad\frac{\partial\Omega}{\partial y}=-\frac{n}{\delta'}\,,\quad\frac{\partial\Omega}{\partial t}=-\frac{l}{\delta'}\,,\\
&\frac{\partial^2\Omega}{\partial x^2}=\frac{1}{\delta'}\frac{\partial}{\partial\Omega}\left[\frac{m^2(\Omega)}{\delta'}\right]\,,\\
&\frac{\partial^2\Omega}{\partial y^2}=\frac{1}{\delta'}\frac{\partial}{\partial\Omega}\left[\frac{n^2(\Omega)}{\delta'}\right]\,,\quad\frac{\partial^2\Omega}{\partial t^2}=\frac{1}{\delta'}\frac{\partial}{\partial\Omega}\left[\frac{l^2(\Omega)}{\delta'}\right]\,,\\
&\delta'\equiv\frac{\partial\delta}{\delta\Omega}=l'(\Omega)t+m'(\Omega)x+n'(\Omega)y-k'(\Omega)\,.
\end{aligned}\tag{3.1.5}$$

By substitution of (3.1.5) into (3.1.2) and taking into account (3.1.4), we find that (3.1.3) gives the integral of the system (3.1.2). Thus, we have proved that the wave equation (3.1.1) does possess a class of solutions of the form indicated above. It should be noted that if $f(\Omega)$ is a complex function, the functions $u=\operatorname{Re}f(\Omega)$ and $u=\operatorname{Im}f(\Omega)$ are also solutions to the equation.

Let us write down the second derivatives of the function $f(\Omega)$ with respect to x, y, and t, which shall be required later:

$$
\begin{aligned}
&\frac{\partial^2 f}{\partial x^2} = f''\left(-\frac{m}{\delta'}\right)^2 + \frac{f'}{\delta'}\frac{\partial}{\partial \Omega}\left(\frac{m^2}{\delta'}\right) = \frac{1}{\delta'}\frac{\partial}{\partial \Omega}\left(f'\frac{m^2}{\delta'}\right) ,\\
&\frac{\partial^2 f}{\partial y^2} = \frac{1}{\delta'}\frac{\partial}{\partial \Omega}\left(f'\frac{n^2}{\delta'}\right) , \quad \frac{\partial^2 f}{\partial t^2} = \frac{1}{\delta'}\frac{\partial}{\partial \Omega}\left(f'\frac{l^2}{\delta'}\right) ,\\
&\frac{\partial^2 f}{\partial y \partial t} = \frac{1}{\delta'}\frac{\partial}{\partial \Omega}\left(f'\frac{nl}{\delta'}\right) , \quad \frac{\partial^2 f}{\partial x \partial t} = \frac{1}{\delta'}\frac{\partial}{\partial \Omega}\left(f'\frac{ml}{\delta'}\right) ,\\
&\frac{\partial^2 f}{\partial x \partial y} = \frac{1}{\delta'}\frac{\partial}{\partial \Omega}\left(f'\frac{mn}{\delta'}\right) .
\end{aligned}
\tag{3.1.6}
$$

Here the primes denote differentiation with respect to Ω.

It should be noted that while constructing the solution $u = f(\Omega)$ of the wave equation, it was required that $f(\Omega)$ be differentiable with respect to Ω. Indeed, if for real values of x, y the function $\Omega(x, y, t)$ assumes in the plane of the complex variable $z = x + \mathrm{i}y$ the values that fill a certain domain, it should be assumed that the function $f(\Omega)$ is analytical within the domain. In an implicit form this assumption is known to contain the Laplace equation for the real and imaginary parts of the function f. The solution $u = f(x + \mathrm{i}y)$ is an example of this kind. If, however, the function $f(\Omega)$ assumes a set of values which depend on one real parameter, i.e. lie on a certain line, the condition of analyticity is not necessary. In this case, in order to get the solution $f(\Omega)$ in the classical sense, it is sufficient that the function be twice continuously differentiable along the line. The usual plane waves $u = f(x \pm at)$ are an example of this kind.

It should also be noted that as shown above, the class of functionally invariant solutions of the wave equation is determined by the function $\Omega(x, y, t)$ which satisfies the system (3.1.2) and consequently has the form (3.1.3) if the condition (3.1.4) is satisfied. As indicated above, the functions $f(\Omega)$ themselves in this case may be arbitrary twice differentiable or analytic functions. (The function $f(\Omega)$ and its derivatives may also be considered to be generalized ones, if they do not exist in the usual sense.) It is this property of the solutions which is reflected in the title "method of functionally invariant solutions". The name of the method reflects certain general group properties of the wave equation solutions.

3.2 Plane and Complex Waves

The simplest form of solution is obtained by assuming the coefficients l, m and n to be constant and putting the function $k(\Omega)$ equal to Ω. Then, (3.1.3) becomes

$$
\Omega = lt + mx + ny , \tag{3.2.1}
$$

where $l^2 = a^2(m^2+n^2)$, and the class of solutions of the wave equation is described by functions of the type

$$
u = f(lt + mx + ny) . \tag{3.2.2}
$$

If l, m and n are all real numbers, then one obtains the plane wave which is the simplest solution to the wave equation. Some of the coefficients l, m and n may prove to be complex numbers, e.g. $m = 1$, $n = \pm i$, $l = 0$. Then, we arrive at the general integral of the Laplace equation

$$\frac{\partial^2 u}{\partial x^2} + \frac{\partial^2 u}{\partial y^2} = 0 ,$$

which, naturally, satisfies the wave equation, too. If, however, all the three coefficients are nonzero and complex, we get an entirely new solution called by Smirnov and Sobolev the complex plane wave. (From the physical point of view the latter solution is no longer a plane wave.)

Without loss of generality, the coefficient l in (3.2.2) may be put equal to unity. (In fact, for $l \neq 0$ one has

$$f(lt + mx + ny) = f\left[l\left(t + \frac{m}{l}x + \frac{n}{l}y\right)\right] = f_1(t + m_1x + n_1y) ,$$

and, hence, discarding the subscripts one arrives at the above statement.) Then, denoting m by $-\theta$, we get $n = \pm(a^{-2} - \theta^{-2})^{1/2}$, and hence, the expression (3.2.2) becomes

$$u = f\left(t + \theta x \pm y\sqrt{a^{-2} - \theta^2}\right) . \tag{3.2.3}$$

Since (3.2.3) satisfies the wave equation (3.1.1), the real and imaginary parts

$$u = \operatorname{Re} f\left(t - \theta x \pm y\sqrt{a^{-2} - \theta^2}\right) ,$$
$$u = \operatorname{Im} f\left(t - \theta x \pm y\sqrt{a^{-2} - \theta^2}\right)$$

are also solutions to the equation (3.1.1).

3.2.1 Reflection of Plane Longitudinal and Transverse Waves

By way of an example, let us consider the problem of reflection of a plane wave at the boundary $y = 0$ of an elastic half-plane $y \geq 0$. Let us call the plane longitudinal wave (P-wave) such a solution to the equations

$$\Delta\Phi = \frac{\partial^2\Phi}{c_1^2\partial t^2} , \quad \Delta\Psi = \frac{\partial^2\Psi}{c_2^2\partial t^2} , \tag{3.2.4}$$

for which

$$\Psi(x, y, t) = 0 , \quad \Phi = f\left(t - \theta x \pm y\sqrt{c_1^{-2} - \theta^2}\right) k , \tag{3.2.5}$$

and the coefficients θ and $\sqrt{c_1^{-2} - \theta^{-2}}$ are real. If the problem of elastic vibrations caused by a plane longitudinal wave is considered for the half-plane $y > 0$, then the solution of the type

$$\Phi = f\left(t - \theta x + y\sqrt{c_1^{-2} - \theta^2}\right) , \quad \Psi = 0 \tag{3.2.6}$$

is called the wave propagating towards a boundary, where θ is a real number, $|\theta| < c_1^{-1}$. The wave propagating away from the boundary is described by

$$\Phi = f\left(t - \theta x - y\sqrt{c_1^{-2} - \theta^2}\right) , \quad \Psi = 0 . \tag{3.2.7}$$

Let us call the plane transverse wave (S-wave) a solution to (3.2.4), which can be represented in the form

$$\Phi = 0 , \quad \Psi = f\left(t - \theta x \pm y\sqrt{c_2^{-2} - \theta^2}\right) , \tag{3.2.8}$$

where the coefficients θ and $\sqrt{c_2^{-2} - \theta^{-2}}$ are real too. (It should be noted that the S-waves considered in this section prove to be polarized in the xy-plane, i.e., the displacement vector behind the wave front is perpendicular to the direction of its propagation and lies in the xy-plane. In the world literature such waves are often called SV-waves in order to distinguish them from the SH-waves arising, for instance, due to anti-plane deformation. The latter waves are polarized in the plane perpendicular to the xy-plane, i.e., the displacement vector behind the front of an SH-wave is perpendicular to the xy-plane.) We shall consider two kinds of transverse waves: the ones propagating towards the boundary of the half-space $y = 0$, for which

$$\Phi = 0 , \quad \Psi = f(t - \theta x + y\sqrt{c_2^{-2} - \theta^2}) , \tag{3.2.9}$$

and the ones propagating away from the boundary, for which

$$\Phi = 0 , \quad \Psi = f\left(t - \theta x - y\sqrt{c_2^{-2} - \theta^2}\right) . \tag{3.2.10}$$

The geometrical meaning of the names of the waves is quite obvious. In the waves propagating towards the boundary, the potential retains a constant value on the planes

$$t - \theta x + y\sqrt{c_1^{-2} - \theta^2} = \text{const} \qquad \text{(P-wave)} ,$$
$$t - \theta x + y\sqrt{c_2^{-2} - \theta^2} = \text{const} \qquad \text{(S-wave)} ,$$

which are displaced with increasing t in such a way that the direction $\boldsymbol{N}$ of their motion ($\boldsymbol{N}$ – normal) characterized by the direction cosines

$$\begin{aligned} N_x &= \theta c_1 , \quad N_y = -\sqrt{1 - \theta^2 c_1^2} \qquad \text{(P-wave)} , \\ N_x &= \theta c_2 , \quad N_y = -\sqrt{1 - \theta^2 c_2^2} \qquad \text{(S-wave)} \end{aligned} \tag{3.2.11}$$

forms an obtuse angle with the y-axis. However, for the wave propagating away from the boundary the cosines of the normal N_1 are given by

$$\begin{aligned} N_{1x} &= \theta c_1 , \quad N_{1y} = \sqrt{1-\theta^2 c_1^2} \qquad \text{(P-wave)} , \\ N_{1x} &= \theta c_2 , \quad N_{1y} = \sqrt{1-\theta^2 c_2^2} \qquad \text{(S-wave)} \end{aligned} \tag{3.2.12}$$

and the normal forms an acute angle with the y-axis.

Let us consider a wave incident on the boundary of a half-space free of stresses. Then, at $y = 0$ we have the following boundary conditions:

$$\begin{aligned} \sigma_{xy} &= \varrho c_2^2 \left[2\frac{\partial^2 \Phi}{\partial x \partial y} + \frac{\partial^2 \Psi}{\partial y^2} - \frac{\partial^2 \Psi}{\partial x^2} \right] = 0 , \\ \sigma_{yy} &= \varrho \left[c_1^2 \frac{\partial^2 \Phi}{\partial y^2} + \left(c_1^2 - 2c_2^2\right) \frac{\partial^2 \Phi}{\partial x^2} - 2c_2^2 \frac{\partial^2 \Psi}{\partial x \partial y} \right] = 0 . \end{aligned} \tag{3.2.13}$$

To begin with, let us suppose that a plane longitudinal wave described by the potentials of the type (3.2.6), travels towards the boundary $y = 0$. We shall assign the subscript 1 to the potentials, and shall seek for the waves reflected at the boundary in the form of a sum of longitudinal and transverse waves traveling from the boundary

$$\Phi_2 = Af\left(t - \theta x - y\sqrt{c_1^{-2} - \theta^2}\right) , \quad \Psi_2 = Bf\left(t - \theta x - y\sqrt{c_2^{-2} - \theta^2}\right) . \tag{3.2.14}$$

Here, A and B are unknown constants called the wave amplitudes or the reflection coefficients. In order to determine A and B we shall substitute the expressions for $\Phi = \Phi_1 + \Phi_2$ and $\Psi = \Psi_1 + \Psi_2$ into (3.2.13). As a result of substitution, we get

$$\begin{aligned} &\left[(1 - 2\theta^2 c_2^2)(1 + A) - 2\theta c_2^2 B\sqrt{c_2^{-2} - \theta^2}\right] f''(t - \theta x) = 0 , \\ &\left[-2\theta(1 - A)\sqrt{c_1^{-2} - \theta^2} + \left(c_2^{-2} - 2\theta^2\right) B\right] f''(t - \theta x) = 0 . \end{aligned} \tag{3.2.15}$$

Since f is an arbitrary function, the expressions in the square brackets are equal to zero, and we get from (3.2.15):

$$\begin{aligned} A &= \left[4\theta^2\sqrt{(c_1^{-2} - \theta^2)(c_2^{-2} - \theta^2)} - (2\theta^2 - c_2^{-2})^2\right] / S(\theta^2) , \\ B &= \left[-4\theta(2\theta^2 - c_2^{-2})\sqrt{c_1^{-2} - \theta^2}\right] / S(\theta^2) , \end{aligned} \tag{3.2.16}$$

where

$$S(\theta^2) = (2\theta^2 - c_2^{-2})^2 + 4\theta^2\sqrt{(c_1^{-2} - \theta^2)(c_2^{-2} - \theta^2)} > 0 .$$

Consequently, we obtain the required expressions for the potentials of the reflected waves with the help of (3.2.14, 16). It should be noted that while seeking for the solution to the problem of the reflection of a plane longitudinal wave at the free boundary of a half-space, we have assumed that the reflected waves are described by the same function $f(\Omega)$ as the incident waves. This function describes the profile of the incident wave. It can be seen from the solution (3.2.16) that there exist reflected waves with the same profile.

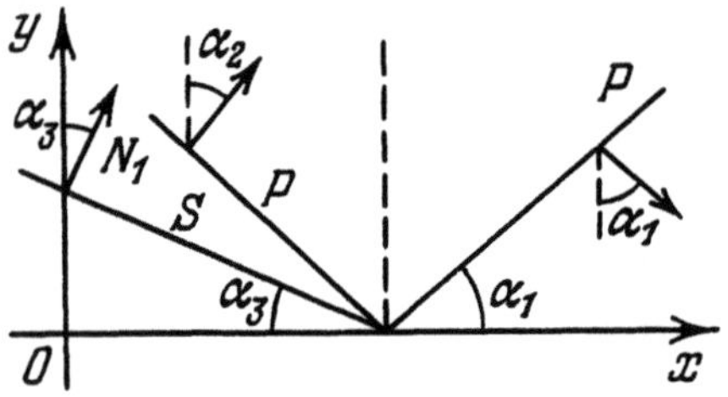

Fig. 3.1. Reflection of an incident longitudinal wave. P and S designate the longitudinal and the transverse waves, respectively

Let us consider some geometrical consequences of the formulas for incident and reflected waves (Fig. 3.1). If we call the angle α_1 between the normal to the surface

$$t - \theta x + y(c_1^{-2} - \theta^2)^{1/2} = \text{const}$$

(at a corresponding choice of the constants on the right-hand side, this surface may be called the wavefront at the moment t) and the direction of the negative semi-axis y, the angle of incidence of the wave, and the angles α_2, α_3 between the normals to the surface

$$t - \theta x - y\sqrt{c_1^{-2} - \theta^2} = \text{const}\,, \quad t - \theta x - y\sqrt{c_2^{-2} - \theta^2} = \text{const}$$

and the direction of the positive semi-axis y, the angles of reflection, then it follows from the formulas for the potentials under study that

(1) $\alpha_1 = \alpha_2$, i.e., the angle of incidence of the longitudinal wave is equal to the angle of reflection.

(2) $\sin\alpha_1 / \sin\alpha_3 = c_1/c_2$, i.e., the ratio of the sines of the angle of incidence of the longitudinal wave and of the angle of reflection of the transverse wave is equal to the ratio of the velocities of propagation of longitudinal and transverse waves. From (3.2.12) it follows that $\theta = c_1^{-1} = \sin\alpha_1$.

Let us now consider a plane transverse wave incident on the boundary $y = 0$, which is described by the potentials

$$\Phi_1 = 0\,, \quad \Psi_1 = f\left(t - \theta x + y\sqrt{c_2^{-2} - \theta^2}\right)\,. \tag{3.2.17}$$

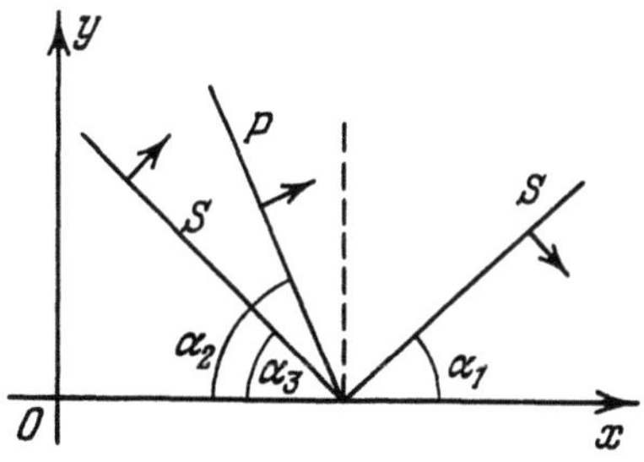

Fig. 3.2. Reflection of an incident transverse wave at $|\theta| < c_1^{-1}$

If $|\theta| < c_1^{-1}$, then the problem is quite analogous to the previous one, and is solved in a similar way, see Fig. 3.2. In this case, if a solution to the problem is sought in the form $\Phi = \Phi_2$, $\Psi = \Psi_1 + \Psi_2$ where

$$\Phi_2 = Cf\left(t - \theta x - y\sqrt{c_1^{-2} - \theta^2}\right), \quad \Psi_2 = Df\left(t - \theta x - y\sqrt{c_2^{-2} - \theta^2}\right), \tag{3.2.18}$$

then one gets for the reflection coefficients (amplitudes of the reflected waves) C, D:

$$\begin{aligned} C &= 4\theta(2\theta^2 - c_2^{-2})\sqrt{c_2^{-2} - \theta^2}/S(\theta^2), \\ D &= \left[4\theta^2\sqrt{(c_1^{-2} - \theta^2)(c_2^{-2} - \theta^2)} - (2\theta^2 - c_2^{-2})^2\right]/S(\theta^2). \end{aligned} \tag{3.2.19}$$

Let us consider the constraint $|\theta| < c_1^{-1}$ imposed while describing the reflection of the transverse waves. Formally, this restriction follows from the real nature of the coefficients entering the expression for $\Omega(x, y, t)$, which is the necessary condition for the existence of a plane longitudinal wave. From the mechanical point of view, this means that the variable $|\theta| = |\theta_0| = c_1^{-1}$ defines the angle of incidence of a plane transverse wave, at which total internal reflection takes place. From (3.2.11) one can find the relation between the angle α_1 of incidence of a plane transverse wave on the boundary $y = 0$ and θ_0:

$$\sin\alpha_1 = \theta_0 c_2 = c_2/c_1. \tag{3.2.20}$$

(For the sake of simplicity, we assume that the direction in which the wave travels froms an acute angle with the positive direction of the x-axis, and hence $\theta > 0$.)

It can easily be seen from Fig. 3.3 that (3.2.20) defines the value of the angle $\alpha_1 = \alpha_0$ ($\alpha_0 \equiv \sin^{-1}(c_2/c_1)$) for which the speed of propagation of transverse waves along the boundary is equal to that of longitudinal waves. The angle α_0 is called the critical angle of total internal reflection. For $\alpha_1 > \alpha_0$, total internal reflection of transverse waves takes place. Longitudinal disturbances arising at the points of the boundary $y = 0$ due to incidence of a transverse wave will outrun the transverse wave. Sometimes it is said that "the sine of the angle of reflection of a longitudinal wave, calculated according to the law of sines, $\sin\alpha_2 = \theta c_1$, exceeds

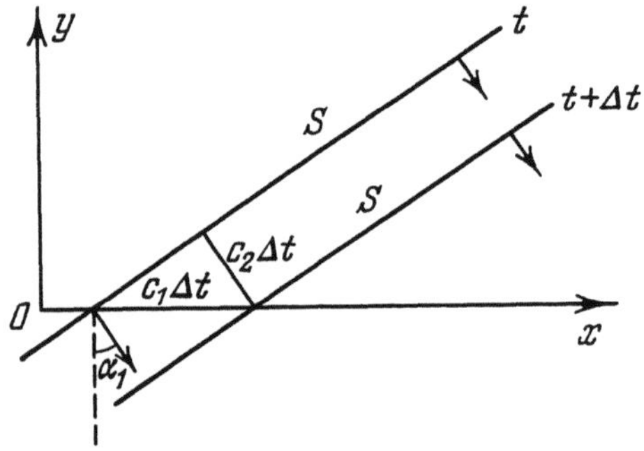

Fig. 3.3. Transition to total internal reflection

unity, and hence, in the usual sense, no real angle of reflection exists for such a longitudinal wave".

Thus, the solution to the problem of reflection, described by the formulas (3.2.18, 19), is valid for $|\theta| < c_1^{-1}$ only (Fig. 3.2), i.e., for the angles of incidence of a transverse wave which are less than the critical angle of total internal reflection: $\alpha_1 < \alpha_0$.

3.2.2 The Case of Total Internal Reflection

Let us further consider the case of total internal reflection $c_2^{-1} > |\theta| > c_1^{-1}$, i.e. $\alpha_1 > \alpha_0$. It is natural to conclude that in this case too one may seek for the solution in the form of a sum of plane waves; however, the requirement of the coefficients in the expression for $\Omega(x, y, t)$ in the solution for the longitudinal potential, being real, must be excluded, since at $c_2^{-1} > |\theta| > c_1^{-1}$ the coefficients of a plane longitudinal wave may become complex. Hence, in accordance with the terminology introduced by Smirnov and Sobolev [3.1,2], we arrive at the expression for longitudinal waves in the form of complex plane waves. It can be shown that the following problem is solvable for $c_2^{-1} > |\theta| > c_1^{-1}$: for a given incident transverse wave (3.2.17), find reflected longitudinal and transverse waves with bounded displacements, such that the resultant disturbances satisfy fixed boundary conditions.

Let us consider two auxiliary cases which will help us in constructing the general solution to the problem of reflection. Suppose that the function $f = f_1$ in (3.2.17) is complex and represents the values of a certain function of a complex variable, $z = x + iy$, on the real axis, which is regular within the domain $y > 0$ and has a bounded derivative there. The solution to the problem is sought in the form

$$\Phi_2 = C_1 f_1\left(t - \theta x + iy\sqrt{\theta^2 - c_1^{-2}}\right), \quad \Psi_2 = D_1 f_1\left(t - \theta x - y\sqrt{c_2^{-2} - \theta^2}\right), \tag{3.2.21}$$

where C_1, D_1 are constants determined from the homogeneous boundary conditions (3.2.13). Comparing (3.2.18) and (3.2.21), it can be easily seen that C_1, D_1

may be obtained from C, D in (3.2.19) by simply replacing the radical $\sqrt{c_1^{-2}-\theta^2}$ by $-\mathrm{i}\sqrt{\theta^2-c_1^{-2}}$.

In an exactly similar manner, we can consider the second auxiliary problem, when $f = f_2$ is an analytical function in the half-plane $y < 0$. In this case, the reflected waves are described with the help of the potentials

$$\begin{aligned} \Phi_2 &= C_2 f_2\left(t-\theta x-\mathrm{i}y\sqrt{\theta^2-c_1^{-2}}\right), \\ \Psi_2 &= D_2 f_2\left(t-\theta x-y\sqrt{c_2^{-2}-\theta^2}\right) \end{aligned} \tag{3.2.22}$$

and C_2, D_2 may be obtained from (3.2.19) by replacing the radical $\sqrt{c_1^{-2}-\theta^2}$ by $\mathrm{i}\sqrt{\theta^2-c_1^{-2}}$.

The auxiliary solutions (3.2.21,22) can be used for solving the problem in the general case. We proceed from the fact that an arbitrary real function of a real variable, $f(x)$, belonging to a sufficiently wide class of functions, may be presented in the form

$$f(x) = \tfrac{1}{2}[f_1(x) + f_2(x)] , \tag{3.2.23}$$

where $f_1(x)$ is the boundary value of a function of a complex variable, which is regular in the half-plane $y > 0$ and having a bounded first derivative there, and $f_2(x)$ is the boundary value of a function regular in the half-plane $y < 0$ and has a bounded first derivative there. This representation is implemented with the help of Cauchy-type integrals, see, e.g. [3.51], and results in the following expressions for the functions $f_1(x)$, $f_2(x)$:

$$f_1(x) = \lim_{y\to+0} f_1(x+\mathrm{i}y) , \quad f_2(x) = \lim_{y\to+0} f_2(x-\mathrm{i}y) , \tag{3.2.24}$$

where

$$\begin{aligned} f_1(x+\mathrm{i}y) &= \frac{1}{\pi\mathrm{i}}\int_{-\infty}^{\infty}\frac{f(\xi)d\xi}{\xi-x-\mathrm{i}y} , \\ f_2(x-\mathrm{i}y) &= -\frac{1}{\pi\mathrm{i}}\int_{-\infty}^{\infty}\frac{f(\xi)d\xi}{\xi-x+\mathrm{i}y} \quad (y>0) . \end{aligned}$$

It follows from the properties of Cauchy-type integrals that the functions $f_1(x)$, $f_2(x)$, as well as the functions $f_1(x+\mathrm{i}y)$ and $f_2(x-\mathrm{i}y)$, are complex conjugate functions, and $\mathrm{Re}\, f_1(x) = \mathrm{Re}\, f_2(x) = f(x)$.

While solving the general problem of reflection of a transverse wave

$$\Phi_1 = 0 , \quad \Psi_1 = f\left(t-\theta x+y\sqrt{c_2^{-2}-\theta^2}\right) ,$$

where f is an arbitrary real function at $c_2^{-1} > |\theta| > c_1^{-1}$, we shall decompose Ψ into two summands using (3.2.23,24):

$$\Psi_1 = \frac{1}{2}\left[f_1\left(t-\theta x+y\sqrt{c_2^{-2}-\theta^2}\right)+f_2\left(t-\theta x+y\sqrt{c_2^{-2}-\theta^2}\right)\right],$$

and shall solve the problem separately for each component

$$\Phi = \frac{1}{2}\left[C_1 f_1\left(t-\theta x+\mathrm{i}y\sqrt{\theta^2-c_1^{-2}}\right)+C_2 f_2\left(t-\theta x-\mathrm{i}y\sqrt{\theta^2-c_1^{-2}}\right)\right],$$

$$\Psi = \frac{1}{2}\left[f_1\left(t-\theta x+y\sqrt{c_2^{-2}-\theta^2}\right)+f_2\left(t-\theta x+y\sqrt{c_2^{-2}-\theta^2}\right)\right]$$
$$+\frac{1}{2}\left[D_1 f_1\left(t-\theta x-y\sqrt{c_2^{-2}-\theta^2}\right)+D_2 f_2\left(t-\theta x-y\sqrt{c_2^{-2}-\theta^2}\right)\right]. \tag{3.2.25}$$

Since the constants D_1, D_2, as well as C_1, C_2, are complex conjugates, the latter formulas may be written in the following final form:

$$\Phi = \mathrm{Re}\left\{C_1 f_1\left(t-\theta x+\mathrm{i}y\sqrt{\theta^2-c_1^{-2}}\right)\right\},$$
$$\Psi = \mathrm{Re}\left\{f_1\left(t-\theta x+y\sqrt{c_2^{-2}-\theta^2}\right)+D_1 f_1\left(t-\theta x-y\sqrt{c_2^{-2}-\theta^2}\right)\right\}. \tag{3.2.26}$$

In order to analyze the case of total internal reflection, let us consider a specific form of the function f_1, which is frequently encountered in applications:

$$f_1\left(t-\theta x+y\sqrt{c_2^{-2}-\theta^2}\right)=A\exp\left[\mathrm{i}\omega\left(t-\theta x+y\sqrt{c_2^{-2}-\theta^2}\right)\right]$$

or

$$f\left(t-\theta x+y\sqrt{c_2^{-2}-\theta^2}\right)=A\cos\left[\omega\left(t-\theta x+y\sqrt{c_2^{-2}-\theta^2}\right)\right].$$

This is the case of the so-called steady-state vibrations, where A is the wave amplitude, and $\omega(t-\theta x+y\sqrt{c_2^{-2}-\theta^2})$ is the wave phase. In this case, the potential of the reflected transverse wave has the form

$$\Psi_2 = A\mathrm{Re}\left\{D_1\exp\left[\mathrm{i}\omega\left(t-\theta x-y\sqrt{c_2^{-2}-\theta^2}\right)\right]\right\}.$$

However, the modulus of the coefficient D_1 of the reflected transverse wave is equal to D_1 if $\sqrt{c_1^{-2}-\theta^2}$ is replaced by $-\mathrm{i}\sqrt{\theta^2-c_1^{-2}}$, so that the amplitude of the reflected transverse wave is equal to that of the incident wave, and hence the energies of the waves are equal.

Seeking for the potential of a reflected longitudinal wave, we find that

$$\Phi = A\exp\left(-\omega y\sqrt{\theta^2-c_1^{-2}}\right)\mathrm{Re}\left\{C_1\exp[\mathrm{i}\omega(t-\theta x)]\right\},$$

and hence longitudinal vibrations attenuate exponentially as y grows, i.e., while they travel from the boundary of the half-space. Thus, a longitudinal wave does not actually carry energy away into the half-space, but accumulates it near the boundary.

Thus, in the case of total internal reflection only the reflected transverse wave carries energy away into the half-space, and its energy is equal to that of the incident transverse wave.

The meaning of the "total internal reflection" term may be understood by considering incidence of a wave onto a plane interface between two media. In this case, both reflected and refracted (i.e., penetrating through the interface) waves arise, which may also be analyzed with the help of the Smirnov–Sobolev method. If the speed of motion along the interface of a wave arriving from the first medium, θ^{-1}, proves to be less than the speed of propagation of a transverse wave through the second medium, then both longitudinal and transverse refracted waves will become (according to the terminology introduced by Smirnov and Sobolev) complex plane waves, and hence they will attenuate exponentially while propagating into the second medium. Thus, both refracted waves will not carry away energy into the second medium, and the entire energy of an incident wave will transform into that of the reflected waves, hence the "total internal reflection" term.

3.2.3 Rayleigh Waves

We shall now seek for a solution for the case when the longitudinal and transverse disturbances are complex waves, and there are no stresses at the boundary of the half-space. Choosing for the potentials such functions of a complex variable, whose derivative vanishes at infinity, we shall get a solution in which the displacement tends to zero at infinity. That is why such waves are called surface waves. Let the longitudinal potential Φ be given by the formula

$$\Phi = f_1\left(t - \theta x + \mathrm{i}\sqrt{\theta^2 - c_1^{-2}}\,y\right) ,$$

where θ is a real number, $|\theta| > c_2^{-1}$, $f_1(x)$ is the boundary value of the function which is analytical in the upper half-plane, and $|f_1'| < M$. The potential Ψ will be sought in the form

$$\Psi = Af_1\left(t - \theta x + \mathrm{i}\sqrt{\theta^2 - c_2^{-2}}\,y\right) .$$

Then, substituting these potentials into the conditions (3.2.13), we get

$$\left[(1 - 2c_2^2\theta^2) + 2\mathrm{i}\,Ac_2^2\theta\sqrt{\theta^2 - c_2^{-2}}\right] f_1''(t - \theta x) = 0 ,$$

$$\left[-2\mathrm{i}\theta\sqrt{\theta^2 - c_1^{-2}} + A\left(c_2^{-2} - 2\theta^2\right)\right] f_1''(t - \theta x) = 0 .$$

We see that this system has a non-zero solution, if the following relation holds:

$$(2\theta^2 - c_2^{-2})^2 - 4\theta^2\sqrt{(\theta^2 - c_1^{-2})(\theta^2 - c_2^{-2})} = 0\,. \tag{3.2.27}$$

This equation is called the Rayleigh equation. It can be shown to have one real positive root, lying in the interval $c_2^{-1} < \theta < \infty$; the other root is equal in magnitude, but has a negative sign. We denote the positive root by c_R^{-1} and substitute $\pm c_R^{-1}$ for θ in all the expressions. Then, we finally get

$$\Phi = f_1\left(t \mp \frac{x}{c_R} + \mathrm{i}\sqrt{c_R^{-2} - c_1^{-2}}\,y\right),$$

$$\Psi = Af_1\left(t \mp \frac{x}{c_R} + \mathrm{i}\sqrt{c_R^{-2} - c_2^{-2}}\,y\right),$$

where

$$A = \left(2c_R^{-2} - c_2^{-2}\right)\left(\pm 2\mathrm{i}c_R^{-1}\sqrt{c_R^{-2} - c_2^{-2}}\right)^{-1},$$

and the upper or lower signs in the expressions are taken simultaneously.

In a similar way, by considering the function f_2 which is the boundary value of the function analytical in the lower half-plane, we arrive at the formulas:

$$\Phi = f_2\left(t \mp \frac{x}{c_R} - \mathrm{i}\sqrt{c_R^{-2} - c_1^{-2}}\,y\right),$$

$$\Psi = Bf_2\left(t \mp \frac{x}{c_R} - \mathrm{i}\sqrt{c_R^{-2} - c_2^{-2}}\,y\right),$$

$$B = \left(2c_R^{-2} - c_2^{-2}\right)\left(\mp 2\mathrm{i}c_R^{-1}\sqrt{c_R^{-2} - c_2^{-2}}\right)^{-1}$$

If the functions $f_1(x)$, $f_2(x)$ are conjugate, then the half-sum of the respective solutions gives the required real function:

$$\begin{aligned}\Phi &= \mathrm{Re}\left\{f_1\left(t \mp \frac{x}{c_R} + \mathrm{i}\sqrt{c_R^{-2} - c_1^{-2}}\,y\right)\right\},\\ \Psi &= \mathrm{Re}\left\{Af_1\left(t \mp \frac{x}{c_R} + \mathrm{i}\sqrt{c_R^{-2} - c_2^{-2}}\,y\right)\right\}\end{aligned} \tag{3.2.28}$$

(in (3.2.28) the upper or lower signs are taken simultaneously).

Such waves are called Rayleigh waves after the man who was the first to establish their existence. It follows from (3.2.28) that the entire pattern of motion is displaced along the x-axis with the velocity c_R, reamining unchanged in the moving system of coordinates. The quantity c_R is called Rayleigh's velocity.

If $f_1(z) = \mathrm{e}^{\mathrm{i}z}$, then it follows from (3.2.28) that

$$\Phi = \cos\left(t \mp \frac{x}{c_R}\right) \exp\left(-\sqrt{c_R^{-2} - c_1^{-2}}\, y\right) ,$$

$$\Psi = A\, \mathrm{i} \sin\left(t \mp \frac{x}{c_R}\right) \exp\left(-\sqrt{c_R^{-2} - c_2^{-2}}\, y\right) ,$$

and hence the amplitude of such a Rayleigh wave decreases exponentially with the growth of the depth.

3.3 Homogeneous Solutions

Here we shall consider in detail an important representation of homogeneous solutions of the wave equation (3.1.1), which appears in a particular case of employing the method of functionally invariant solutions.

3.3.1 Construction of Homogeneous Solutions to the Wave Equation

Let the function $k(\Omega)$ in (3.1.3) be identically equal to zero. Then, if $l(\Omega) \neq 0$, we get $n^2(\Omega)/l^2(\Omega) = a^{-2} - \theta^2$ by dividing both sides of (3.1.3) by $l(\Omega)$ and denoting $m(\Omega)/l(\Omega)$ by $-\theta$. Hence, (3.1.3) may be rewritten in the form

$$t - \theta x + \sqrt{a^{-2} - \theta^2}\, y = 0 . \tag{3.3.1}$$

Thus, we should now write $f(\theta)$ instead of $f(\Omega)$.

Equation (3.3.1) may be rewritten in the form

$$1 - \theta\xi + \sqrt{a^{-1} - \theta^2}\, \eta = 0 \quad (\xi = x/t\,,\ \eta = y/t)\,, \tag{3.3.2}$$

and hence θ is determined from (3.3.2) as a function of only two arguments, ξ and η. In this case, the solutions $f(\theta)$ constructed for (3.1.1) will be functions of the arguments ξ and η. In other words, they will be zero-dimensional homogeneous functions of t, x, y. It is well known [3.52] that s-dimensional homogeneous functions of the variables t, x, y are defined by the relation

$$u(kt, kx, ky) = k^s u(t, x, y)$$

which must be identically true. Such solutions will be called s-dimensional homogeneous solutions. Thus, any twice differentiable (analytical, if complex) function $f(\theta)$, with θ obtained from (3.3.1), is a zero-dimensional solution of the wave equation (3.1.1). Conversely, it may be shown [3.1] that any zero-dimensional homogeneous solution to the equation (3.1.1) may be written in the form $u = f(\theta)$, where θ is a solution to (3.3.1).

Let us analyze (3.3.2) in greater detail. In order to isolate the single-valued branch of the radical $(a^{-2} - \theta^2)^{1/2}$, we make a cut in the complex plane θ along the real axis $(-a^{-1}, a^{-1})$, and fix the branch of the radical $(a^{-2} - \theta^2)^{1/2}$ by imposing the condition $(a^{-2} - \theta^2)^{1/2} > 0$ for $\theta = ib$, where $b > 0$. Solving (3.3.2) for θ, we find, for $\xi^2 + \eta^2 \le a^2$

$$\theta = \left[\xi - i\eta\sqrt{1 - a^{-2}(\xi^2 + \eta^2)}\right] / (\xi^2 + \eta^2) . \tag{3.3.3}$$

In view of the chosen branch of the radical $(a^{-2} - \theta^2)^{1/2}$, the arithmetic value of the root $[1 - a^{-2}(\xi^2 + \eta^2)]^{1/2}$ should be taken in (3.3.1). Under this condition, Eq. (3.3.2) maps the interior of the circle $\xi^2 + \eta^2 \le a^2$ onto the plane θ with the cut $(-a^{-1}, a^{-1})$. In this case, the upper, $\eta > 0$, and the lower, $\eta < 0$, parts of the interior of the circle $\xi^2 + \eta^2 \le a^2$ are mapped onto the lower and the upper half-planes θ, respectively. The point $\xi = \eta = 0$ is mapped onto the infinitely distant point of the plane θ, while the intervals $\eta = 0$, $0 < \xi < a$ and $\eta = 0$, $-a < \xi < 0$ are mapped onto the real semiaxes $\theta > a^{-1}$ and $\theta < -a^{-1}$, respectively. The circle $\xi^2 + \eta^2 = a^2$ is mapped onto the cut $[-a^{-1}, a^{-1}]$. The corresponding points of the circle $\xi + \eta^2 \le a^2$ and the plane θ are shown in Figs. 3.4 and 3.5.

Let $f(\theta)$ be a single-valued analytical function in the plane θ with a cut $(-a^{-1}, a^{-1})$. The solution of (3.1.1) is taken in the form $u = \operatorname{Re} f(\theta)$. It is defined inside the circle $\xi^2 + \eta^2 < a^2$. We shall show that it can be continuously extended

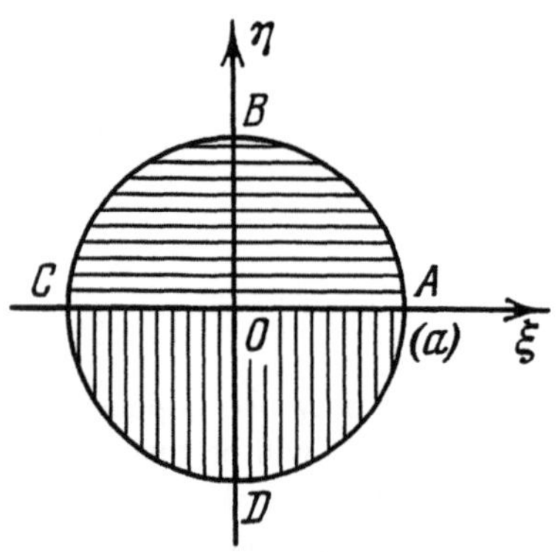

Fig. 3.4. The domain of variation of the self-similar variables ξ, η

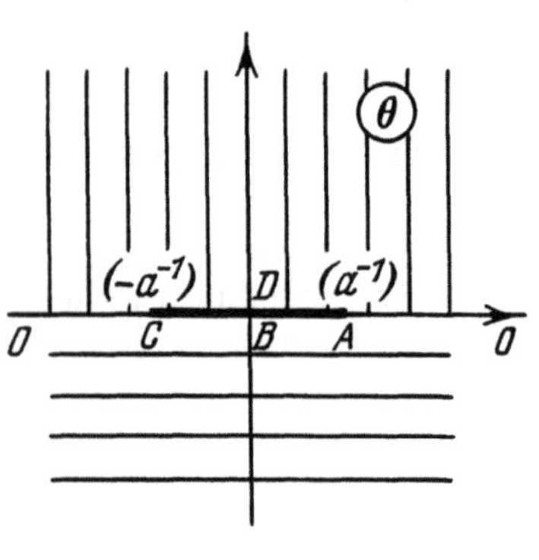

Fig. 3.5. The complex plane θ with a cut

into the exterior of the circle $\xi^2 + \eta^2 > a^2$. Indeed, let us consider (3.3.2) once again. If $\xi^2 + \eta^2 > a^2$, then the solution of the equation may be written in the form

$$\theta = \left[\xi \pm \eta\sqrt{a^{-2}(\xi^2 + \eta^2) - 1}\right] / (\xi^2 + \eta^2)\,, \tag{3.3.4}$$

where we have taken the arithmetic value of the radical. It can be seen from the equation that θ has real values for $\xi^2 + \eta^2 > a^2$. For $\theta = \text{const}$ this equation is the equation of the tangent to the circle $\xi^2 + \eta^2 = a^2$. Consequently, θ remains constant on the tangent to the circle, and the point of tangency divides the tangent into two semitangents: I in the counterclockwise direction from the point of tangency (its equation can be obtained from (3.3.4) by choosing the plus sign for the radical), and II in the clockwise direction (described by (3.3.4) with the minus sign before the radical).

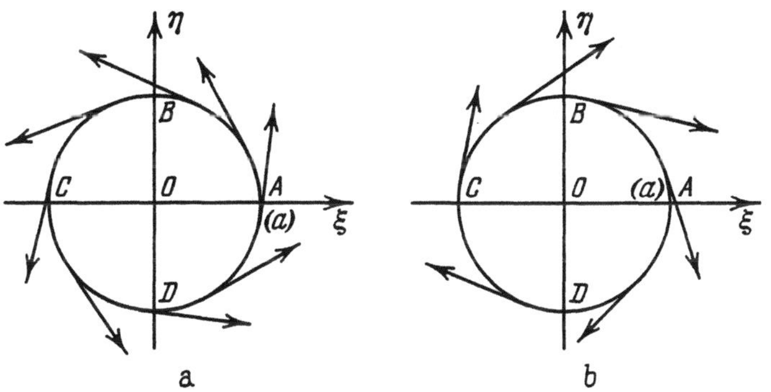

Fig. 3.6a,b. Semi-tangents directed (a) counterblockwise and (b) clockwise

The semitangents are shown in Fig. 3.6a,b. Thus, assuming that θ remains constant on the semitangents (its values being the same at the corresponding points of tangency), one can uniquely map the exterior of the circle $\xi^2 + \eta^2 > a^2$ onto the interval $(-a^{-1}, a^{-1})$ in the plane θ (since each of the semitangents I, II is mapped onto a point of the cut). Thus, we can continuously extend the solution across the boundary $\xi^2 + \eta^2 = a^2$ into the exterior of the circle, $\xi^2 + \eta^2 > a^2$, while the value of the solution $u = \operatorname{Re} f(\theta)$ remains constant along the semitangents I, II. Moreover, if we split the solution $u = \operatorname{Re} f(\theta)$ into two real components $u = u_1(\theta) + u_2(\theta)$ and continue $u_1(\theta)$ into the exterior of the circle, $\xi^2 + \eta^2 > a^2$, along the semitangents I, and $u_2(\theta)$ along the semitangents II, then we shall once again obtain a real solution to the equation (3.4.1), extended continuously into the exterior of the circle across the boundary $\xi^2 + \eta^2 = a^2$. Thus, we have, in essence, an infinite set of various ways of extending solutions, and in all the cases the solution remains continuous while crossing the boundary of the circle $\xi^2 + \eta^2 = a^2$. In specific problems, a method of extending the solution is chosen by considering the motion of the wavefront.

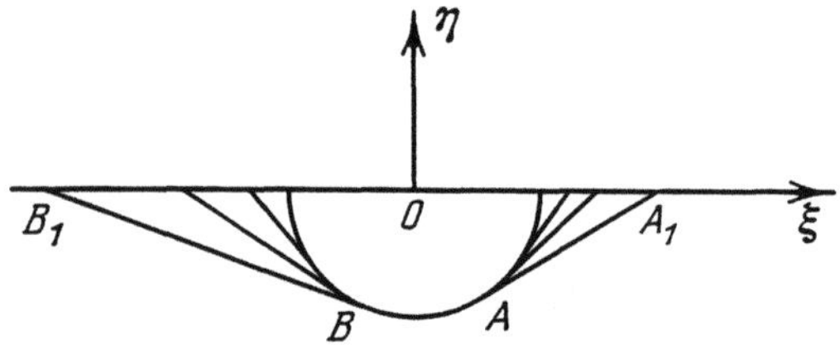

Fig. 3.7. The domain of the disturbed motion in the self-similar variables

The above analysis is valid if the solution is considered in the entire plane. Let us now consider the half-plane $y \leq 0$ (in other words, we shall be considering only the half-plane $\eta \leq 0$ of the plane ξ, η). Suppose that the solution to (3.1.1) inside the semicircle $\xi^2 + \eta^2 \leq a^2$, $\eta \leq 0$ is taken in the form $u = \operatorname{Re} f(\theta)$ and the solution vanishes on a certain arc AB of the semicircle (see Fig. 3.7). Then, it is often more convenient to obtain a single-valued continuation of the solution across the arc of the circle by using the semitangents to the circle I (for $\xi > 0$) and II (for $\xi < 0$), as shown in Fig. 3.7 (hence, one should take the signs plus and minus for $\xi > 0$ and $\xi < 0$ respectively). It follows from the continuation that the solution is identically equal to zero outside the bounded region $A_1ABB_1OA_1$.

It should be noted that the thus constructed solutions, being continuous across the circle $\xi^2 + \eta^2 = a^2$, may have discontinuities of the normal derivatives while crossing the circle. Generally, although the solutions are continuous inside the circle, $\xi^2 + \eta^2 < a^2$, together with all their derivatives (since $f(\theta)$ is an analytical function), outside the circle, i.e., at $\xi^2 + \eta^2 \geq a^2$, both the solutions and their derivatives may contain discontinuities, infinite ones included. This fact may be easily understood if one notes that the circles $(x/t)^2 + (y/t)^2 = a^2$ (i.e., $\xi^2 + \eta^2 = a^2$) as well as their tangents $t - \theta x + \sqrt{a^{-2} - \theta^2}\, y = 0$ (where $\theta = \text{const}$, $-a^{-1} < \theta < a^{-1}$) are cross-sections of the characteristic surfaces by the planes $t = \text{const}$ in the space x, y, t (the characteristic surfaces of the two-dimensional wave equation within the space x, y, t are cones whose axes are parallel to the t-axis and the planes tangent to the cones [3.53].

Returning to the equations (3.3.1,2), we should point out that the complex variable θ may be replaced by another one, z, related to θ by means of an analytical dependence. We shall indicate a fairly convenient functional dependence of this kind, especially suitable for solving elastodynamic problems for corner regions. Let z and θ be related by

$$\theta = \frac{1}{2a}\left(z + \frac{1}{z}\right). \tag{3.3.5}$$

As we know [3.54], in this case the plane θ with the cut $(-a^{-1}, a^{-1})$ is transformed into the unit circle $|z| < 1$. For such a choice of the branch of the radical, the following equality holds:

$$\sqrt{a^{-2} - \theta^2} = \frac{\mathrm{i}}{2a}\left(z - \frac{1}{z}\right). \tag{3.3.6}$$

Then, (3.3.5) may be rewritten in the form

$$1 - \frac{1}{2a}\left(z + \frac{1}{z}\right)\xi + \frac{\mathrm{i}}{2a}\left(z - \frac{1}{z}\right)\eta = 0$$

or

$$1 - \frac{z}{2a}(\xi - \mathrm{i}\eta) - \frac{1}{2az}(\xi + \mathrm{i}\eta) = 0 \,. \tag{3.3.7}$$

Let us introduce for the circle $\xi^2 + \eta^2 \leq a^2$ the polar coordinates $\bar{\varrho}$, φ related to ξ, η by

$$\xi = \bar{\varrho}\cos\varphi \,, \quad \eta = \bar{\varrho}\sin\varphi \quad (0 \leq \bar{\varrho} \leq a \,,\, 0 \leq \varphi \leq 2\pi) \,. \tag{3.3.8}$$

Then, (3.3.7) becomes

$$\bar{\varrho}\,\mathrm{e}^{-\mathrm{i}\varphi}z^2 - 2za + \bar{\varrho}\,\mathrm{e}^{\mathrm{i}\varphi} = 0 \,.$$

An obvious solution to the latter equation is

$$z = R\,\mathrm{e}^{\mathrm{i}\varphi} \,,$$

where R is defined by the quadratic equation

$$\bar{\varrho}R^2 - 2Ra + \bar{\varrho} = 0 \quad (0 \leq R \leq 1) \,.$$

By solving the latter equation, we get

$$R = \bar{\varrho}\left(a + \sqrt{a^2 - \bar{\varrho}^2}\right)^{-1} \,. \tag{3.3.9}$$

Thus, to each point of the circle $\xi^2 + \eta^2 \leq a^2$ a value of the complex variable $z = R\,\mathrm{e}^{\mathrm{i}\varphi}$ corresponds, possessing the same argument, a similar correspondence existing between the points of the circle $\xi^2 + \eta^2 = a^2$ and the unit circle with the same argument. In other words, each radius of the circle $\xi^2 + \eta^2 \leq a^2$ corresponds to a radius of the unit circle $|z| \leq 1$ with the same polar angle.

3.3.2 Diffraction of a Plane Shear Wave by a Wedge

As an example of solving a problem with the help of the transform

$$z = R\,\mathrm{e}^{\mathrm{i}\varphi} \,, \tag{3.3.10}$$

let us consider the problem of diffraction of a plane shear wave (the SH-wave)

$$u_3 = (c_2 t + r\cos(\varphi - \varphi_0))H(c_2 t + r\cos(\varphi - \varphi_0))$$

by a rigid wedge with an arbitrary vertex angle $\alpha (0 < \alpha < \pi)$ immersed frictionlessly in an infinite homogeneous isotropic continuum whose density is ϱ, and the speeds of longitudinal and transverse waves are c_1 and c_2, respectively. [In the above expression, $H(x)$ is the Heaviside function, equal to unity for $x \geq 0$ and to zero for $x < 0$.] The cylindrical system of coordinates r, φ, x_3 ($x_1 = r\cos\varphi$,

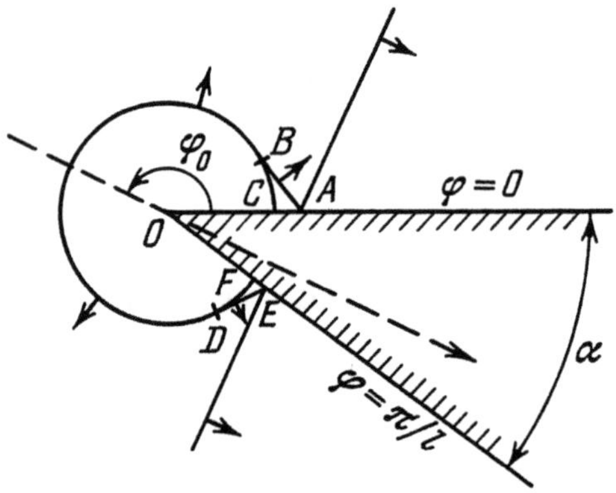

Fig. 3.8. Diffraction of a plane shear wave SH by a wedge

$x_2 = r\sin\varphi$) has been chosen in such a way that its x_3-axis coincides with the wedge's edge (Fig. 3.8) while the elastic continuum outside the wedge fills the region $0 < \varphi < \pi/l$ where $l = \pi/(2\pi - \alpha)$, $1/2 \le l < 1$.

Since the incident wave, as well as both the boundary and initial conditions, are independent of x_3, we have the case of antiplane deformation ($u_r = u_\varphi = 0$, $u_3 \equiv u_3(r,\varphi,t)$), described by (2.7.26). Due to the absence of a characteristic dimension, the problem proves to be self-similar, and hence the velocity, strains and stresses are homogeneous functions of the zeroth dimension with respect to the variables r, t. Then, with the help of the formulas (2.1.7, 3.6) and (2.7.26, 36), we get the following system of equations for determining the dimensionless velocity $v = \dot{u}_3/c_2$

$$\begin{aligned} &\Delta_1 v = c_2^{-2}\ddot{v}\,, \quad \frac{\partial v}{\partial \varphi} = 0(\varphi = 0, \pi/l)\,, \\ &v = H(c_2 t + r\cos(\varphi - \varphi_0)) \quad (t < 0)\,, \quad v = C + \mathrm{O}(r^\beta)\,, \\ &\beta > 0\,, \quad r \to 0\,. \end{aligned} \tag{3.3.11}$$

For the sake of definiteness, we shall consider the case of no geometric shadow, i.e., it will be supposed that $\pi/l - \pi < \varphi_0 < \pi$ (Fig. 3.8). Since in the regions ABC and DEF behind the reflected waves the influence of the edge is absent, the solution here will be the same as in the case of the wave reflection from an infinite wall, and hence the velocity in the regions doubles: $v = 2$. However, within the regions behind an incident wave outside the diffraction region, $v = 1$. Consequently, $v = 2$ on the arcs BC and DF, $v = 1$ on the arc BD, and $\partial v/\partial\varphi = 0$ on the straights OC and OF, and, to solve the problem, one has to determine v within the circular sector $CBDFOC$ (whose radius is $c_2 t$) using the values of v and its normal derivative at the boundary. This can be done by mapping the circular sector $CBDFOC$ shown in Fig. 3.8 onto the circular sector $CBDFOC$ with the same flare angle but of unit radius, lying in the complex plane z, see Fig. 3.9a, using the transform (3.3.10) where R, according to (3.3.8, 9), is given by

$$R = \left(c_2 t/r + \sqrt{(c_2 t/r)^2 - 1} \right)^{-1} . \tag{3.3.12}$$

The solution v is sought as the real part of a certain function $f(z)$ which is analytical within the sector and satisfies the following boundary conditions:

$$\begin{aligned} &\operatorname{Re} f(z) = 2 \quad \text{on the arcs } BC, DF ; \\ &\operatorname{Re} f(z) = 1 \quad \text{on the arc } BD ; \\ &\operatorname{Im} f(z) = 0 \quad \text{on the straights } OC, OF . \end{aligned} \tag{3.3.13}$$

The latter condition follows from the boundary conditions in (3.3.11) and the Cauchy-Riemann relations between the real and imaginary parts of an analytical function. In fact,

$$\frac{\partial v}{r \partial \varphi} = \frac{\partial \operatorname{Re} f}{r \partial \varphi} = - \frac{\partial \operatorname{Im} f}{\partial r} = 0 . \tag{3.3.14}$$

By integrating the latter equality along the straights OC, OF and supposing that the function $f(z)$ is continuous at a point of the boundary $z = 0$, one gets for the straights OC, OF:

$$\operatorname{Im} f(z) = \operatorname{Im} f(0) = C_0 . \tag{3.3.15}$$

Since the required function $f(z)$ is defined to an accuracy of an arbitrary imaginary constant, one can put the constant C_0 equal to zero.

By employing the transform

$$w = \varrho \mathrm{e}^{\mathrm{i}\varphi} = z^l = \left(R \, \mathrm{e}^{\mathrm{i}\varphi} \right)^l ,$$

we map the circular sector $CBDFOC$ in the plane z (Fig. 3.9) onto the upper semicircle $CBDFOC$ in the plane w, Fig. 3.10, preserving the unit radius. Then, due to the boundary condition $\operatorname{Im} f(w) = 0$ on the diameter CF, one can extend the function $f(w)$ analytically onto the lower semicircle $COFD_1B_1C$ by using the formula $f(\bar{w}) = \overline{f(w)}$ where $\bar{F}$ and F are complex conjugates. Thus, we arrive at the problem of defining within the circle $|w| < 1$ an analytical function $f(w)$ by using the known values of its real part on the circle $|w| = 1$:

$$\begin{aligned} &\operatorname{Re} f(w) = 2 , \\ &(0 < \varphi < (\pi - \varphi_0)l , \quad 2\pi - (\pi + \varphi_0)l < \varphi < (\pi + \varphi_0)l , \\ &2\pi - (\pi - \varphi_0)l < \varphi < 2\pi) , \\ &\operatorname{Re} f(w) = 1 , \\ &((\pi - \varphi_0)l < \varphi < 2\pi - (\pi + \varphi_0)l , \quad (\pi + \varphi_0)l < \varphi < 2\pi - (\pi - \varphi_0)l) . \end{aligned}$$

The solution to this problem accurate to an arbitrary imaginary constant is given by the Schwarz integral [3.55]

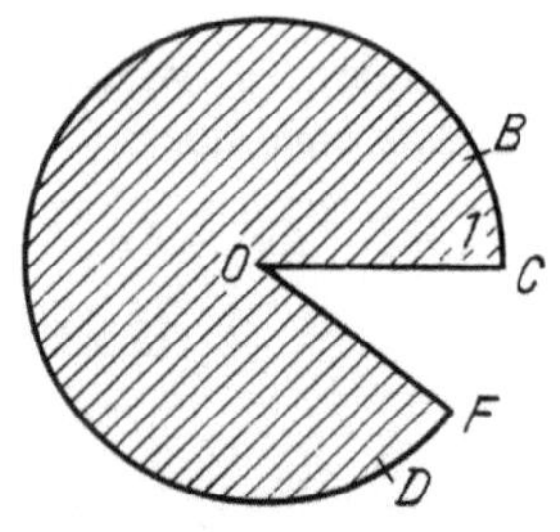

Fig. 3.9. The domain of definition of the analytical function $f(z)$ in the complex plane z

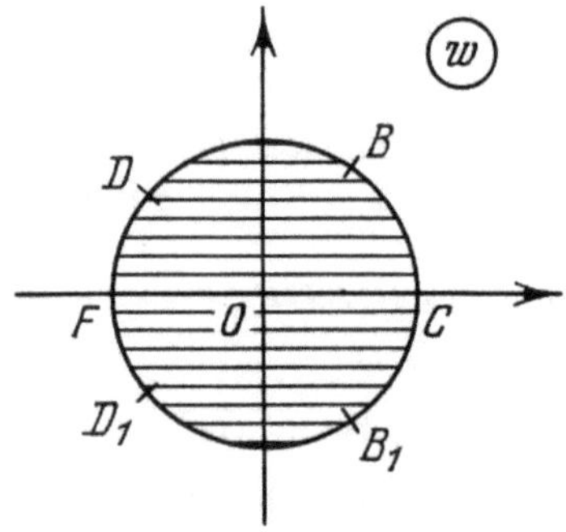

Fig. 3.10. The domain of definition of the analytical function $f(w)$ in the complex plane w

$$f(w) = \frac{1}{2\pi} \int_{|\zeta|=1} \operatorname{Re} f(\zeta) \frac{\zeta + w}{\zeta - w} |d\zeta| + \mathrm{i}C , \tag{3.3.16}$$

where ζ is a point of the circle $|\zeta| = 1$, and $w = R^l \mathrm{e}^{\mathrm{i}\varphi l}$ is a point of the circle $|w| = 1$, $\operatorname{Im} C = 0$.

By performing the quadratures and separating the real part, one gets

$$\begin{aligned} v &= \operatorname{Re} f(w) = 1 + \frac{1}{\pi} \left[\tan^{-1} \Phi(\varphi_0, \varphi) + \tan^{-1} \Phi(-\varphi_0, \varphi)\right] , \\ \Phi(\theta, \varphi) &= \frac{(1 - R^{2l}) \sin[(\pi - \theta)l]}{(1 + R^{2l}) \cos[(\pi - \theta)l] - 2R^l \cos \varphi l} \end{aligned} \tag{3.3.17}$$

[Here the values of arc tangents are taken in the interval $(0, \pi)$.]

If l tends to unity, then (3.3.17) gives the solution to the problem of wave reflection from an infinite rigid wall: $v = 2$. For $r \to 0$, taking into account (3.3.12), one gets from (3.3.17):

$$v \to 2l + \mathrm{O}(r^l) , \tag{3.3.18}$$

and hence the condition on the wedge's edge in (3.3.11), ensuring uniqueness of the problem solution, is met, and according to (3.3.18) strains and stresses in the neighborhood of the wedge's edge prove to be integrable and are of the order of r^{l-1}.

3.4 The Case of an Elastic Half-Plane

In this section, we shall consider homogeneous solutions of plane-strain problems for an elastic half-space.

Note that the problems for a half-space are of interest from both the theoretical and applied points of view. They are interesting also because many elastodynamic problems for a space, in which the boundary conditions are specified on different parts of one and the same plane, say, the $z = 0$ plane, may be reduced, from the mathematical point of view, to the problems for the half-space $z > 0$ (or $z < 0$).

3.4.1 Reduction of the Dynamic Problem to a Superposition of Problems for Symmetric and Antisymmetric Components of the Displacement Vector

The displacement vector $\boldsymbol{u}$ may be always presented as a sum of the symmetric, $\boldsymbol{u}^{(s)}$, and antisymmetric $\boldsymbol{u}^{(a)}$, vectors (with respect to the $z = 0$ plane):

$$\boldsymbol{u} = \boldsymbol{u}^{(s)} + \boldsymbol{u}^{(a)} , \quad \boldsymbol{u} \equiv \boldsymbol{u}(x, y, z, t) , \tag{3.4.1}$$

where x, y, z are rectangular Cartesian coordinates. Here the vectors $\boldsymbol{u}^{(s)}(x, y, z, t)$ and $\boldsymbol{u}^{(a)}(x, y, z, t)$ are expressed through the values of the vector $\boldsymbol{u}$ at the points (x, y, z) and $(x, y, -z)$, which are symmetric with respect to the plane $z = 0$:

$$\begin{aligned} u_x^{(s)}(z) &= \tfrac{1}{2}[u_x(z) + u_x(-z)] , \quad & u_x^{(a)}(z) &= \tfrac{1}{2}[u_x(z) - u_x(-z)] , \\ u_y^{(s)}(z) &= \tfrac{1}{2}[u_y(z) + u_y(-z)] , \quad & u_y^{(a)}(z) &= \tfrac{1}{2}[u_y(z) - u_y(-z)] , \\ u_z^{(s)}(z) &= \tfrac{1}{2}[u_z(z) - u_z(-z)] , \quad & u_z^{(a)}(z) &= \tfrac{1}{2}[u_z(z) + u_z(-z)] . \end{aligned} \tag{3.4.2}$$

We see that $u_x^{(s)}$, $u_y^{(s)}$, $u_z^{(a)}$ are even functions of z, while $u_x^{(a)}$, $u_y^{(a)}$, $u_z^{(s)}$ are odd functions of z. The properties of evenness of the stress tensor components corresponding to the vectors $\boldsymbol{u}^{s}$ and $\boldsymbol{u}^{(a)}$, can be easily established: in fact, $\sigma_{zz}^{(s)}$, $\sigma_{xz}^{(a)}$, $\sigma_{yz}^{(a)}$ and $\sigma_{zz}^{(a)}$, $\sigma_{xz}^{(s)}$, $\sigma_{yz}^{(s)}$ are even and odd functions of z, respectively. Since the displacement and stress components must be continuous on the part of the plane $z = 0$ outside the boundary, hence, the odd components must vanish on this part of the plane, and we arrive at independent boundary conditions for the symmetric and antisymmetric vectors for $z = 0$. The initial conditions (2.2.1) can also be separated for vectors $\boldsymbol{u}^{(s)}$ and $\boldsymbol{u}^{(a)}$ with the help of (3.4.2).

Taking into account these circumstances, as well as the fact that both the symmetric vector $\boldsymbol{u}^{(s)}$ and the antisymmetric vector $\boldsymbol{u}^{(a)}$ must satisfy respectively the Lamé equations for the symmetric and antisymmetric components of bulk forces $\boldsymbol{f}$, we obtain two separate systems for determining each of the vectors $\boldsymbol{u}^{(s)}$ and $\boldsymbol{u}^{(a)}$ within the half-space $z > 0$ (or $z < 0$). It should be noted that, according to (3.4.2), $\boldsymbol{u}^{(s)}$ and $\boldsymbol{u}^{(a)}$ must satisfy the same condition (2.3.6) in the neighborhood of a singular curve (point) as the solution of the original system $\boldsymbol{u}$. Thus, we arrive at the problem of solving two uncoupled systems of equations, boundary and initial conditions, and conditions in the neighborhood of a singular curve (point) for each of the vectors $\boldsymbol{u}^{(s)}$ and $\boldsymbol{u}^{(a)}$.

Let us now refer to plane-strain problems. We shall analyze the elastodynamic problems for an elastic half-plane $y < 0$ with zero initial conditions for the displacement vector and the boundary conditions specified on the $y = 0$ axis.

3.4.2 Homogeneous Solutions

Let the components of the displacement vector $\boldsymbol{u}$ be homogeneous, $(n-1)$-dimensional functions of the variables t, x, y. (Here n is an integer.)

If a body is subjected to plane deformation in the absence of bulk forces, then in accordance with (2.7.23–25), the solution $\boldsymbol{u}(x, y, t)$ to Lamé equations may be presented in the form

$$\boldsymbol{u} = \operatorname{grad}\varphi + \operatorname{curl}(\psi \boldsymbol{e}_z) , \tag{3.4.3}$$

where the potentials φ, ψ satisfy the wave equations

$$\Delta_1\varphi = c_1^{-2}\ddot{\varphi} , \quad \Delta_1\psi = c_2^{-2}\ddot{\psi} \quad \left(\Delta_1 \equiv \frac{\partial^2}{\partial x^2} + \frac{\partial^2}{\partial y^2}\right) . \tag{3.4.4}$$

Let us introduce the operator L_n^α of n-tuple differentiation and integration with respect to α:

$$L_n^\alpha f = \begin{cases} \displaystyle\int_0^{\alpha_0=\alpha} d\alpha_1 \int_0^{\alpha_1} d\alpha_2 \ldots \int_0^{\alpha_{|n|-1}} f(\alpha_{|n|})d\alpha_{|n|} & (n < 0) , \\ \dfrac{\partial f^n}{\partial \alpha^n} & (n \geq 0) . \end{cases} \tag{3.4.5}$$

Then, in the case of homogeneous solutions presently under consideration, the functions $L_n^t\varphi$, $L_n^t\psi$ will be zero-dimensional homogeneous functions satisfying the wave equations (3.4.4) for φ and ψ respectively. (Hereinafter the derivatives with respect to t should be, generally speaking, considered to be generalized ones [3.53].) Consequently, according to the Smirnov–Sobolev method, they could be sought as real parts of the analytical functions $f_1(\theta_1)$ and $f_2(\theta_2)$:

$$L_n^t\varphi = \operatorname{Re} f_1(\theta_1) , \quad L_n^t\psi = \operatorname{Re} f_2(\theta_2) \quad (k = 1, 2) . \tag{3.4.6}$$

Here, according to Sect. 3.2 and the formulas (3.3.2–4), the variables θ_k $(k = 1, 2)$ are defined by

$$\begin{gathered} 1 - \theta_k\xi + \sqrt{c_k^{-2} - \theta_k^2}\,\eta = 0 \\ (\xi = x/t,\ \eta = y/t,\ t > 0,\ k = 1, 2) , \end{gathered} \tag{3.4.7}$$

where the cuts $(-c_k^{-1}, c_k^{-1})$ have been made in the planes θ_k in order to isolate the single-valued branches of the radicals $(c_k^{-2} - \theta_k^2)^{1/2}$, and it is supposed that $(c_k^{-2} - \theta_k^2)^{1/2} > 0$ for $\theta_k = ib$, $b > 0$. They can be also defined explicitly by the formulas:

$$\theta_k = \begin{cases} \dfrac{\xi - i\eta\sqrt{1 - c_k^{-2}(\xi^2 + \eta^2)}}{\xi^2 + \eta^2} & (c_k^2 > \xi^2 + \eta^2) , \\ \dfrac{\xi \pm \eta\sqrt{c_k^{-2}(\xi^2 + \eta^2) - 1}}{\xi^2 + \eta^2} & (c_k^2 \leq \xi^2 + \eta^2) , \quad k = 1, 2 , \end{cases} \tag{3.4.8}$$

where the radicals are assumed to be arithmetic, and the plus and the minus signs in the lower expression are taken for $\xi > 0$ and $\xi < 0$ respectively. Then, for each value of $k(k = 1,2)$, the latter formula maps the lower strip $-c_k < \eta \le 0$ of the real variables ξ, η onto the upper half-plane of the complex variable θ_k in the following way: the interior of the semicircle $\xi^2 + \eta^2 < c_k^2$ is mapped onto the upper half-plane θ_k with the point $\xi = \eta = 0$ being mapped onto the infinitely distant point on the plane θ_k, the intervals $0 < \xi < c_k$ $(\eta = 0)$ and $-c_k < \xi < 0$ $(\eta = 0)$ are mapped onto the real semiaxes $\theta > c_k^{-1}$ and $\theta < -c_k^{-1}$ respectively, and the semicircle $\xi^2 + \eta^2 = c_k^2$ onto the interval $(-c_k^{-1}, c_k^{-1})$. Besides, the points of the region $\xi^2 + \eta^2 > c_k^2$, $-c_k < \eta < 0$ are mapped with the help of the lower formula (3.4.8) onto the points of the interval $(-c_k^{-1}, c_k^{-1})$ in the following way: For a fixed sign and θ_k, the lower expression (3.4.8) may be considered as an equation of a half-line in the plane ξ, η which is tangent to the semicircle at a point ξ_0, η_0 and maps all the points (ξ, η) of the half-line onto the same point of the real interval $(-c_k^{-1}, c_k^{-1})$ belonging to the complex plane θ_k onto which the point of tangency (ξ_0, η_0) is mapped. Correspondence between the points of the region $-c_k < \eta \le 0$ and the half-plane $\operatorname{Im}\theta_k \ge 0$ is shown in Figs. 3.11, 12.

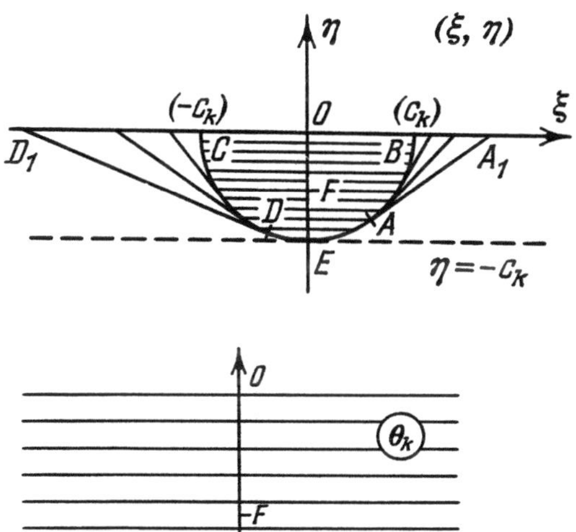

Fig. 3.11. The domain of the disturbed motion in the self-similar variables

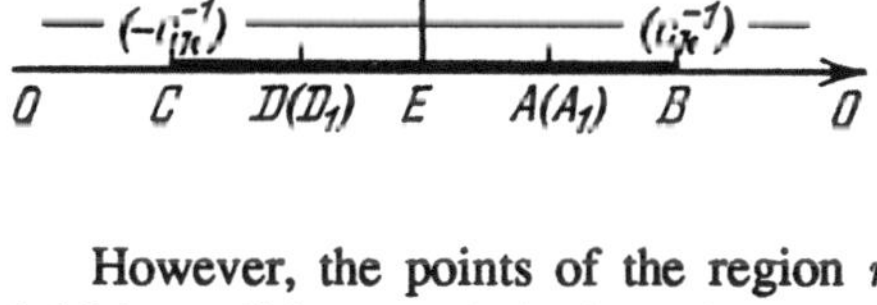

Fig. 3.12. The domain of variation of the complex variable θ_k

However, the points of the region $\eta < -c_k$ are always at rest due to zero initial conditions, and the boundary conditions being specified at the axis $\eta = 0$; hence, we must put $\boldsymbol{u} \equiv 0$ for $\eta < -c_k$.

The following expressions may be obtained for the functions $L_n^t u_x$, $L_n^t u_y$, $L_{n-1}^t\,\sigma_{xx}$, $L_{n-1}^t\,\sigma_{yy}$, $L_{n-1}^t\,\sigma_{xy}$ by using the formulas (2.1.3, 7) (where $u_3 = 0$, $\sigma_{13} = \sigma_{23} = 0$, $x_1 = x$, $x_2 = y$, $x_3 = z$) and (3.4.3–7)

$$
\begin{aligned}
L_n^t u_x &= \operatorname{Re}\left[f_1'(\theta_1)\frac{\partial\theta_1}{\partial x} + f_2'(\theta_2)\frac{\partial\theta_2}{\partial y}\right] ,\\
L_n^t u_y &= \operatorname{Re}\left[f_1'(\theta_1)\frac{\partial\theta_1}{\partial y} - f_2'(\theta_2)\frac{\partial\theta_2}{\partial x}\right] ,\\
&- L_{n-1}^t \sigma_{xx}\\
&= \mu\operatorname{Re}\left[\left(\frac{c_2^{-2} - 2c_1^{-2}}{\theta_1} + 2\theta_1\right) f_1'(\theta_1)\frac{\partial\theta_1}{\partial x} + 2\theta_2 f_2'(\theta_2)\frac{\partial\theta_2}{\partial y}\right] ,\\
- L_{n-1}^t \sigma_{yy} &= \mu\operatorname{Re}\left[\frac{c_2^{-2} - 2\theta_1^2}{\theta_1} f_1'(\theta_1)\frac{\partial\theta_1}{\partial x} - 2\theta_2 f_2'(\theta_2)\frac{\partial\theta_2}{\partial y}\right] ,\\
- L_{n-1}^t \sigma_{xy} &= \mu\operatorname{Re}\left[2\theta_1 f_1'(\theta_1)\frac{\partial\theta_1}{\partial y} + \frac{c_2^{-2} - 2\theta_2^2}{\theta_2} f_2'(\theta_2)\frac{\partial\theta_2}{\partial x}\right] .
\end{aligned}
\tag{3.4.9}
$$

Here the partial derivatives $\partial\theta_k/\partial x$, $\partial\theta_k/\partial y$ have the form

$$
\frac{\partial\theta_k}{\partial x} = \frac{-\theta_k\sqrt{c_k^{-2} - \theta_k^2}}{y\theta_k + x\sqrt{c_k^{-2} - \theta_k^2}} , \quad \frac{\partial\theta_k}{\partial y} = \frac{c_k^{-2} - \theta_k^2}{y\theta_k + x\sqrt{c_k^{-2} - \theta_k^2}} . \tag{3.4.10}
$$

Thus, the solution of the elastodynamic problem may be expressed through the functions $f_1'(\theta_1)$, $f_2'(\theta_2)$ which should be sought from the boundary conditions at $y = 0$. Upon finding the functions, one can obtain the solution of the problem for the displacement and stress components from (3.4.9) by appliyng to the left- and right-hand sides of (3.4.9) the operator $L_{(-n)}^1$ which is an inverse operator with respect to L_n^t due to the zero initial conditions for f:

$$
L_{(-n)}^t L_n^t f = L_n^t L_{(-n)}^t f = f .
$$

In order to express $f_1'(\theta_1)$, $f_2'(\theta_2)$ by means of the problem solution at the boundary, we put $y = 0$ in (3.4.9, 10). Then, since the expressions (3.4.7) give $\theta_1 = \theta_2 = t/x \equiv \theta$ at $y = 0$, (3.4.9) at $y = 0$, i.e., at $\operatorname{Im}\theta = 0$ reduce to

$$
\begin{aligned}
&\operatorname{Re}\left[-\theta^2 F_1(\theta) + \theta\sqrt{c_2^{-2} - \theta^2}F_2(\theta)\right] = xL_n^t u_x ,\\
&\operatorname{Re}\left[\theta\sqrt{c_1^{-2} - \theta^2}F_1(\theta) + \theta^2 F_2(\theta)\right] = xL_n^t u_y ,\\
&\operatorname{Re}\left[\theta(c_2^{-2} - 2c_1^{-2} + 2\theta^2)F_1(\theta) - 2\theta^2\sqrt{c_2^{-2} - \theta^2}F_2(\theta)\right] = \frac{x}{\mu} L_{n-1}^t \sigma_{xx} ,\\
&\operatorname{Re}\left[\theta(c_2^{-2} - 2\theta^2)F_1(\theta) + 2\theta^2\sqrt{c_2^{-2} - \theta^2}F_2(\theta)\right] = \frac{x}{\mu} L_{n-1}^t \sigma_{yy},\\
&\operatorname{Re}\left[-2\theta^2\sqrt{c_1^{-2} - \theta^2}F_1(\theta) + \theta(c_2^{-2} - 2\theta^2)F_2(\theta)\right] = \frac{x}{\mu} L_{n-1}^t \sigma_{xy} ,
\end{aligned}
\tag{3.4.11}
$$

where

$$F_1(\theta) = f_1'(\theta)/\theta\,, \quad F_2(\theta) = f_2'(\theta)/\theta\,. \tag{3.4.12}$$

When obtaining the functions $F_k(\theta)$, which are analytical in the half-plane $\operatorname{Im}\theta > 0$, the boundary conditions for the functions on the axis $\operatorname{Im}\theta = 0$ will be taken from (3.4.11). Consequently, in the framework of elastodynamics, homogeneous solutions (of an arbitrary integer dimension with respect to x, y, t) of the boundary-value problems for a half-plane may be sought as solutions of the corresponding boundary-value problems of the theory of functions of a complex variable for two functions, $F_1(\theta)$, $F_2(\theta)$, which are analytical in the region $\operatorname{Im}\theta > 0$. In the general case of specifying boundary conditions on various parts of the axis $y = 0$, one cannot separate the boundary conditions for the functions $F_1(\theta)$, $F_2(\theta)$, and, as a result, the problem of obtaining an analytical solution becomes much more complicated. Below, we consider only the cases when the boundary conditions for $F_1(\theta)$, $F_2(\theta)$ may be uncoupled, and one arrives at the boundary-value problem for one function of a complex variable only.

Note that while seeking for the functions $F_k(\theta)$ in a number of problems, we shall need expressions obtained by repeated differentiation and integration of (3.4.11) with respect to t. However, since the functions on the left-hand sides of these expressions may contain singularities on the real axis $\operatorname{Im}\theta = 0$ [say, due to differentiating the radicals on the left-hand sides of (3.4.11)], then for a rigorous derivation of the expressions one has, at first, to differentiate/integrate the expressions (3.4.9) with respect to t at $y < 0$, and after that to obtain the needed expressions on the axis $y = 0$ (i.e., $\operatorname{Im}\theta = 0$) by tending y to -0. Thus, by application of the operator L_m^t to the left- and right-hand sides of the first of the formulas (3.4.9), one gets

$$L_{n+m}^t u_x = \operatorname{Re}\left\{M_m\left[f_1'(\theta_1)\frac{\partial\theta_1}{\partial x}\right] + M_m\left[f_2'(\theta_2)\frac{\partial\theta_2}{\partial y}\right]\right\}\,, \tag{3.4.13}$$

where

$$M_m f(\theta_k, x, y) = \begin{cases} \displaystyle\int\limits_0^{\alpha_0=\theta_k} \frac{d\alpha_1}{\beta_1}\int\limits_0^{\alpha_1}\frac{d\alpha_2}{\beta_2}\cdots\int\limits_0^{\alpha_{|m|-1}} f(\alpha_{|m|}, x, y)\frac{d\alpha_{|m|}}{\beta_{|m|}} & (m < 0)\,, \\ \dfrac{\partial f^m}{\partial t^m}\left(\dfrac{\partial f}{\partial t} \equiv \dfrac{\partial f}{\partial\theta_k}\dot\theta_k\right) & (m \geq 0)\,, \end{cases} \tag{3.4.14}$$

$$\beta_n = \frac{\sqrt{c_k^{-2} - \alpha_n^2}}{y\alpha_n + x\sqrt{c_k^{-2} - \alpha_n^2}}\,,$$

and according to (3.4.7)

$$\dot\theta_k \equiv \frac{\partial\theta_k}{\partial t} = \frac{\sqrt{c_k^{-2} - \theta_k^2}}{y\theta_k + x\sqrt{c_k^{-2} - \theta_k^2}} \tag{3.4.15}$$

(Due to zero initial conditions, integration in (3.4.14) with respect to α_n ($n = 1, \ldots, |m|$) is started from $\alpha_n = 0$.) It must be noted here that the independent variables x, y enter the integro-differential operator $M_m f(\theta_k, x, y)$ as parameters, and the contours of integration with respect to α_n belong to the region $\operatorname{Im} \alpha_n \geq 0$ (and hence, circumvent the singular points of the integrands from above). Then, upon applying similar transforms to the other expressions in the formulas (3.4.9), passing to the limit as $y \to -0$, and taking into account (3.4.10, 12, 15), one gets the following relationship for $F_k(\theta)$ between the homogeneous displacements (which are $(n-1)$-dimensional with respect to t, x, y) and the stresses on the axis $y = 0$ ($\operatorname{Im} \theta = 0$):

$$\begin{aligned}
&\operatorname{Re} L_m^\theta \left[-\theta^2 F_1(\theta) + \theta \sqrt{c_2^{-2} - \theta^2} F_2(\theta) \right] = x^{1+m} L_{n+m}^t u_x \,, \\
&\operatorname{Re} L_m^\theta \left[\theta \sqrt{c_1^{-2} - \theta^2} F_1(\theta) + \theta^2 F_2(\theta) \right] = x^{1+m} L_{n+m}^t u_y \,, \\
&\operatorname{Re} L_m^\theta \left[\theta (c_2^{-2} - 2c_1^{-2} + 2\theta^2) F_1(\theta) - 2\theta^2 \sqrt{c_2^{-2} - \theta^2} F_2(\theta) \right] , \\
&\quad = \mu^{-1} x^{1+m} L_{n+m-1}^t \sigma_{xx} \\
&\operatorname{Re} L_m^\theta \left[\theta (c_2^{-2} - 2\theta^2) F_1(\theta) + 2\theta^2 \sqrt{c_2^{-2} - \theta^2} F_2(\theta) \right] = \mu^{-1} x^{1+m} L_{n+m-1}^t \sigma_{yy} \,, \\
&\operatorname{Re} L_m^\theta \left[-2\theta^2 \sqrt{c_1^{-2} - \theta^2} F_1(\theta) + \theta (c_2^{-2} - 2\theta^2) F_2(\theta) \right] = \mu^{-1} x^{1+m} L_{n+m-1}^t \sigma_{xy} \,.
\end{aligned} \tag{3.4.16}$$

Here $\theta = t/x$ ($t > 0$), m is an arbitrary integer, the operator L_m^θ is defined by the formulas (3.4.5) and, as it follows from the derivation of the expresions (3.4.16), at $m < 0$ the contours of $|m|$-uple integration with respect to θ in L_m^θ belong to the region $\operatorname{Im} \theta \geq 0$, and hence circumvent from above the singularities of the integrands on the axis $\operatorname{Im} \theta = 0$, i.e. $\theta = t/x + \mathrm{i}0$. At $m = 0$ (3.4.16) transform into (3.4.11), and hence the relationships (3.4.16) are a generalization of (3.4.11).

In order to define the functions $F_k(\theta)$ ($k = 1, 2$) which are analytical in the upper half-plane θ, one has, in particular, to know their behavior for $\theta \to \infty$. To do this, we shall use (3.4.9) and the asymptotics of the radicals $\sqrt{c_k^{-2} - \theta_k^2} = -\mathrm{i}\theta_k + \mathrm{O}(\theta_k^{-1})$ as $\theta_k \to \infty$, $\operatorname{Im} \theta_k \geq 0$. Then, upon rewriting the upper expression (3.4.8) in the form

$$\begin{aligned}
&\theta_k = \left(t \cos \alpha - \mathrm{i} \sin \alpha \sqrt{t^2 - c_k^{-2} r^2} \right) / r \\
&(x = r \cos \alpha \,, \quad y = r \sin \alpha) \,,
\end{aligned} \tag{3.4.17}$$

we find that

$$\theta_k = \theta + \mathrm{O}(\theta^{-1}) \,, \quad \frac{\partial \theta_k}{\partial x} = -\frac{\theta^2 + \mathrm{O}(1)}{t} \,, \quad \frac{\partial \theta_k}{\partial y} = -\frac{\mathrm{i}\theta^2 + \mathrm{O}(1)}{t} \,, \tag{3.4.18}$$

as $r \to 0$, $t > 0$, $-\pi \leq \alpha \leq 0$, i.e. $\theta \to \infty$, $\operatorname{Im} \theta \geq 0$ ($\theta \equiv t\,\mathrm{e}^{-\mathrm{i}\alpha}/r$).

Then, with the help of the asymptotics (3.4.18) and the notation (3.4.12), we get from (3.4.9) the following estimates as $r \to 0$, $-\pi \le \alpha \le 0$, $t > 0$ (i.e., as $\theta \to \infty$, $\operatorname{Im}\theta \ge 0$):

$$\begin{aligned}
tL_n^t u_x &= -\operatorname{Re}\{F_1(\theta)\theta^3[1+O(\theta^{-2})] + i\theta^3 F_2(\theta)[1+O(\theta^{-2})]\}\,,\\
tL_n^t u_y &= \operatorname{Re}\{[F_1(\theta)\theta^3[1+O(\theta^{-2})] + i\theta^3 F_2(\theta)[1+O(\theta^{-2})]\}\,,\\
\frac{t}{2\mu}\,L_{n-1}^t \sigma_{xx} &= \operatorname{Re}\{F_1(\theta)\theta^4[1+O(\theta^{-2})] + i\theta^4 F_2(\theta)[1+O(\theta^{-2})]\}\,,\\
\frac{t}{2\mu}\,L_{n-1}^t \sigma_{yy} &= -\operatorname{Re}\{F_1(\theta)\theta^4[1+O(\theta^{-2})] + i\theta^4 F_2(\theta)[1+O(\theta^{-2})]\}\,,\\
\frac{t}{2\mu}\,L_{n-1}^t \sigma_{xy} &= -\operatorname{Im}\{F_1(\theta)\theta^4[1+O(\theta^{-2})] + i\theta^4 F_2(\theta)[1+O(\theta^{-2})]\}\,,
\end{aligned} \tag{3.4.19}$$

Thus, the asymptotics of the functions $F_k(\theta)$ $(k=1,2)$ as $\theta \to \infty$ $(\operatorname{Im}\theta \ge 0)$ are related to the asymptotics of the functions $L_n^t u_j$, $L_{n-1}^t \sigma_{lj}$ as $r \to 0$, $\pi \le \alpha \le 0$ by the formulas (3.4.19). These asmyptotic relationships will be used below.

3.4.3 Solution of Some Problems

Let us apply the above results to solving elastodynamic problems. We shall start by considering the case when conditions of the same type are specified at the boundary of a half-plane, i.e. the first or the second boundary-value problem. We shall begin by studying the first boundary-value problem which is more interesting from the point of view of applications (the second boundary-value problem may be considered in a similar way). In this case the functions σ_{xy}, σ_{yy} are specified on the axis $y = 0$ at $t > 0$, and hence, according to (3.4.11), one arrives at the following boundary conditions at the axis $\operatorname{Im}\theta = 0$

$$\operatorname{Re}\Phi_1(\theta) = \Phi_1^0(\theta)\,, \tag{3.4.20}$$

$$\operatorname{Re}\Phi_2(\theta) = \Phi_2^0(\theta)\,, \tag{3.4.21}$$

used for determining the functions $F_1(\theta)$, $F_2(\theta)$ which are analytical in the region $\operatorname{Im}\theta > 0$. The following notation is used in (3.4.20, 21):

$$\begin{aligned}
\Phi_1(\theta) &= \theta\left(c_2^{-2} - 2\theta^2\right)F_1(\theta) + 2\theta^2\sqrt{c_2^{-2} - \theta^2}\,F_2(\theta)\,,\\
\Phi_2(\theta) &= -2\theta^2\sqrt{c_1^{-2} - \theta^2}\,F_1(\theta) + \theta\left(c_2^{-2} - 2\theta^2\right)F_2(\theta)\,,\\
\Phi_1^0(\theta) &= \frac{x}{\mu}L_{n-1}^t\sigma_{yy}(x,0,t)\,,\\
\Phi_2^0(\theta) &= \frac{x}{\mu}L_{n-1}^t\sigma_{xy}(x,0,t)\,.
\end{aligned} \tag{3.4.22}$$

Let the required functions of stresses meet the following conditions

$$\begin{aligned}
&\sigma_{lj}(x,y,t) = O(r^{\varepsilon-1})\,, \quad r \to 0\\
&(\varepsilon > 0, \quad t > 0, \quad r^2 = x^2 + y^2, \quad l,j = x,y)\,.
\end{aligned} \tag{3.4.23}$$

Then, we get for sufficiently smooth (with respect to time) stresses

$$L_{n-1}^{t}\sigma_{lj}(x,y,t) = \mathrm{O}(r^{\varepsilon-1})\,, \quad r \to 0 \quad (\varepsilon > 0, t > 0)\,. \tag{3.4.24}$$

Consequently, if the functions $\sigma_{xy}(x,0,t)$, $\sigma_{yy}(x,0,t)$ specified on the boundary, comply with the estimates (3.4.23), then according to (3.4.22), we have

$$\Phi_k^0(\theta) = \mathrm{O}(\theta^{-\varepsilon})\,, \quad \theta \to \pm\infty \quad (\varepsilon > 0, k = 1,2)\,, \tag{3.4.25}$$

and, taking into account (3.4.20, 21, 25), one should seek for functions $\Phi_k(\theta)$ which decrease as $\theta \to \infty$ ($\operatorname{Im}\theta \geq 0$) as $\mathrm{O}(\theta^{-\varepsilon})$ where $\varepsilon > 0$. Then it can be easily checked that the estimate (3.4.24), and hence (3.4.23), will be satisfied [for this it is necessary to express the functions $F_k(\theta)$ through $\Phi_k(\theta)$ using (3.4.22), and substitute the former into (3.4.19)]. Hence, the estimates (3.4.23) will be correct under the condition

$$\Phi_k(\theta) = \mathrm{O}(\theta^{-\varepsilon})\,, \quad \theta \to \infty\,, \quad \operatorname{Im}\theta \geq 0 \quad (\varepsilon > 0, k = 1,2)\,. \tag{3.4.26}$$

As a result, determination of each of the functions $\Phi_k(\theta)$ with the help of the boundary conditions (3.4.20, 21) where the specified boundary functions $\Phi_k^0(\theta)$ meet the conditions (3.4.25), and the sought functions satisfy the conditions (3.4.26), reduces to solving the Schwarz problem within the region $\operatorname{Im}\theta \geq 0$. The solution is given by the Schwarz integral for the upper half-plane θ [3.51]:

$$\Phi_k(\theta) = \frac{1}{\pi \mathrm{i}} \int_{-\infty}^{\infty} \Phi_k^0(\beta)\frac{d\beta}{\beta - \theta} \quad (k = 1,2,\ \operatorname{Im}\beta = 0)\,. \tag{3.4.27}$$

Although this solution was obtained for a continuous boundary function $\Phi_k^0(\beta)$ possessing no more than a finite number of integrable discontinuities of the form $(\beta-\beta_0)^{\varepsilon-1}$ where $\varepsilon > 0$, it may readily be used for concentrated loadings described by delta-functions. The latter statement may be proved by using, for instance, a specification of concentrated loadings with the help of consequences of smooth functions. In fact, if $\Phi_k^0(\beta) = \delta(\beta - v)$ where $\operatorname{Im} v = 0$, then by formal substitution of $\Phi_k^0(\beta)$ into (3.4.27), one obtains

$$\Phi_k(\theta) = \frac{1}{\pi \mathrm{i}} \int_{-\infty}^{\infty} \frac{\delta(\beta - v)}{\beta - \theta}\, d\beta = \frac{1}{\pi \mathrm{i}}\frac{1}{v - \theta} \quad (\operatorname{Im}\theta > 0)\,. \tag{3.4.28}$$

On the other hand, the same result may be produced by considering, instead of $\delta(\beta - v)$, the function

$$\Phi_k^0(\beta) = \frac{1}{\pi}\frac{\varepsilon}{\varepsilon^2 + (\beta - v)^2} \quad (\varepsilon > 0)\,, \tag{3.4.29}$$

for which it is known [3.53] that

$$\frac{1}{\pi}\frac{\varepsilon}{\varepsilon^2 + (\beta - v)^2} \to \delta(\beta - v)\,,$$

as $\varepsilon \to 0$. In fact, by substituting (3.4.29) into (3.4.27), one finds with the help of the Cauchy theorem

$$\Phi_k(\theta) = -\frac{1}{\pi \mathrm{i}} \frac{1}{(\theta - v) + \mathrm{i}\varepsilon} \quad (\mathrm{Im}\,\theta > 0)\,. \tag{3.4.30}$$

As $\varepsilon \to 0$, the latter expression reduces to (3.4.28).

Having derived $\Phi_k(\theta)$, we can use (3.4.22) to find the functions $F_k(\theta)$, which together with (3.4.9, 10, 12), may be employed for obtaining all the dynamic and kinematic characteristics of disturbed motion of an elastic continuum at any point of the half-plane $y \leq 0$. In particular, many problems are associated with studying the behavior of $\boldsymbol{u}(x, 0, t)$. According to (3.4.16, 22), we find that for $y = 0$ ($\theta = t/x + \mathrm{i}0$):

$$x^{1+m} L^t_{n+m} u_y = \mathrm{Re}\, L^\theta_m \times \left[\frac{\Phi_1(\theta) c_2^{-2} \sqrt{c_1^{-2} - \theta^2} + \theta \Phi_2(\theta) \left(c_2^{-2} - 2\theta^2 - 2\sqrt{c_1^{-2} - \theta^2}\sqrt{c_2^{-2} - \theta^2} \right)}{S(\theta^2)} \right], \tag{3.4.31}$$

$$x^{1+m} L^t_{n+m} u_x = \mathrm{Re}\, L^\theta_m \times \left[\frac{\theta \Phi_1(\theta) \left(2\sqrt{c_1^{-2} - \theta^2}\sqrt{c_2^{-2} - \theta^2} - c_2^{-2} + 2\theta^2 \right) + \Phi_2(\theta) c_2^{-2} \sqrt{c_2^{-2} - \theta^2}}{S(\theta^2)} \right], \tag{3.4.32}$$

where

$$S(\theta^2) = \left(c_2^{-2} + 2\theta^2\right)^2 + 4\theta^2 \sqrt{c_1^{-2} - \theta^2}\sqrt{c_2^{-2} - \theta^2}$$

is the Rayleigh function which is analytical in the complex plane everywhere outside the cuts $(-c_2^{-1}, -c_1^{-1})$, (c_1^{-1}, c_2^{-1}) and takes real values on the axis $\mathrm{Im}\,\theta = 0$ outside the cuts (as pointed out above, the only roots of the Rayleigh equation $S(\theta^2) = 0$ in the complex plane θ are the two real roots $\theta_{1,2} = \pm c_\mathrm{R}^{-1}$ where $c_\mathrm{R} (c_\mathrm{R} < c_2)$ is the speed of the Rayleigh waves).

Let us consider some most interesting examples.

1) At the instant $t = 0$, a constant-value loading starts propagating along an elastic half-plane originally at rest (both velocities, v_1 and v_2, are constant), Fig. 3.13:

$$\begin{aligned} &\sigma_{xy}(x, 0, t) = 0 && (-\infty < x < \infty)\,, \\ &\sigma_{yy}(x, 0, t) = 0 && (x < -v_2 t, \quad x > v_1 t)\,, \\ &\sigma_{yy}(x, 0, t) = -p && (-v_2 t < x < v_1 t)\,. \end{aligned} \tag{3.4.33}$$

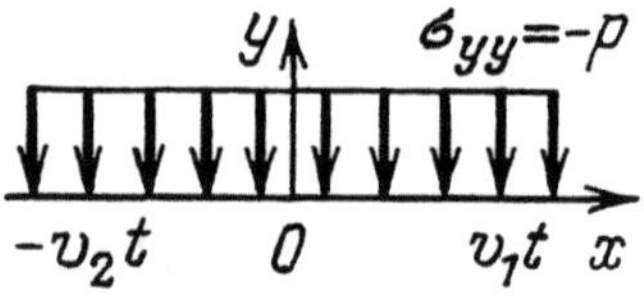

Fig. 3.13. Propagation of a constant loading from the origin of the coordinate system along the boundary of elastic half-plane $y < 0$

Then, $n = 2$, and according to the latter conditions, $\Phi_2^0(\theta) \equiv 0$. In order to derive $\Phi_1^0(\theta)$, let us present the boundary condition $\sigma_{yy}(x, 0, t)$ in the form

$$\sigma_{yy}(x,0,t) = -pH(t - v_1^{-1}x)H(t + v_2^{-1}x) .$$

Then, for $t > 0$ we have

$$\begin{aligned}
\Phi_1^0(\theta) &= \frac{x}{\mu} L_1 \sigma_{yy}(x,0,t) \\
&= -\frac{px}{\mu}[\delta(t - v_1^{-1}x)H(t + v_2^{-1}x) + H(t - v_1^{-1}x)\delta(t + v_2^{-1}x)] \\
&= -\frac{px}{\mu}[\delta(t - v_1^{-1}x) + \delta(t + v_2^{-1}x)] \\
&= -\frac{px}{\mu|x|}\left[\delta\left(\frac{t}{x} - v_1^{-1}\right) + \delta\left(\frac{t}{x} + v_2^{-1}\right)\right] \\
&= -\frac{p|\theta|}{\mu\theta}\left[\delta(\theta - v_1^{-1}) + \delta(\theta + v_2^{-1})\right] \\
&= -\frac{p}{\mu}\left[\delta(\theta - v_1^{-1}) - \delta(\theta + v_2^{-1})\right] .
\end{aligned} \tag{3.4.34}$$

The latter expression was obtained by using the delta-function properties [3.56]:

$$f(x)\delta(x - a) = f(a)\delta(x - a) , \quad \delta(\alpha x) = \frac{1}{|\alpha|}\delta(x) ,$$

where $f(x)$ is continuous at the point $x = a$. As a result, substituting the boundary functions $\Phi_k^0(\theta)$ into (3.4.27), we find (for $\mathrm{Im}\,\theta > 0$):

$$\Phi_1(\theta) = \frac{p}{\mu\pi \mathrm{i}} \frac{v_1^{-1} + v_2^{-1}}{(\theta - v_1^{-1})(\theta + v_2^{-1})} , \quad \Phi_2(\theta) \equiv 0 . \tag{3.4.35}$$

Hence, with the help of (3.4.22) we get $F_k(\theta)$, and then by using (3.4.9, 10) and inversing the operator L_n^t $(n = 1, 2)$ in (3.4.9), we finally find all the components of the displacements and stresses at any point of the half-plane $y \leq 0$. Ignoring the details of the inversion, we forward the interested reader to [3.57].

2) Consider a constant concentrated force p applied (for $t \geq 0$) along the normal to the boundary of the half-plane $y \leq 0$ and moving with a constant speed v along the axis x, the half-plane being originally at rest for $t < 0$, Fig. 3.14.

Then, the boundary conditions are:

$$\sigma_{yy} = -p\delta(x - vt)H(t) , \quad \sigma_{xy} = 0 \quad (y = 0) . \tag{3.4.36}$$

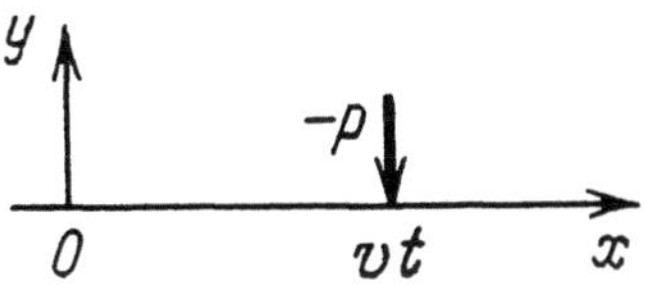

Fig. 3.14. The force application point motion

Here $n = 1$, and for $t > 0$ we have

$$\begin{aligned}
\Phi_2^0(\theta) &= 0\,, \\
\Phi_1^0(\theta) &= \frac{x}{\mu} L_0 \sigma_{yy}(x,0,t) = -\frac{xp}{\mu}\,\delta(x - vt) \\
&= -\frac{p|\theta|}{\mu v \theta}\,\delta(\theta - v^{-1}) = -\frac{p}{\mu v}\,\delta(\theta - v^{-1})\,.
\end{aligned} \tag{3.4.37}$$

Then, the solution to the Schwarz problem for $\operatorname{Im}\theta > 0$ is given by the formulas:

$$\Phi_1(\theta) = \frac{p}{\mu v \pi \mathrm{i}}\,\frac{1}{\theta - v^{-1}}\,, \quad \Phi_2(\theta) = 0\,, \tag{3.4.38}$$

and by putting $n = 1$, $m = -1$ in (3.4.31), we get ($x \neq vt$, $|x| \neq c_R t$):

$$\begin{aligned}
u_y(x,0,t) &= \frac{p}{\pi \mu v c_2^2} \operatorname{Im} \int_0^{t/x} \frac{\sqrt{c_1^{-2} - \theta^2}\,d\theta}{(\theta - v^{-1}) S(\theta^2)}\,, \\
u_x(x,0,t) &= \frac{p}{\pi \mu v} \operatorname{Im} \int_0^{t/x} \frac{2\sqrt{c_1^{-2} - \theta^2}\sqrt{c_2^{-2} - \theta^2} - (c_2^{-2} - 2\theta^2)}{(\theta - v^{-1}) S(\theta^2)}\,\theta\,d\theta\,.
\end{aligned} \tag{3.4.39}$$

As pointed out above, the contours of integration in (3.4.39) circumvent the singular points of the integrands from above (i.e., $\operatorname{Im}\theta > 0$).

Note that the solution of a similar problem by the method of integral transformations is described in detail in [3.58].

If we now let v tend to zero in (3.4.36, 38, 39), then we produce the solution to the problem for a stationary concentrated force $\sigma_{yy}(x,0,t) = -p\delta(x)H(t)$ ($|x| \neq c_R t$):

$$\begin{aligned}
&\Phi_1(\theta) = -\frac{p}{\mu \pi \mathrm{i}}\,, \quad \Phi_2(\theta) = 0\,, \\
&u_y(x,0,t) = -\frac{p}{\pi \mu c_2^2} \operatorname{Im} \int_0^{t/x} \frac{\sqrt{c_1^{-2} - \theta^2}}{S(\theta^2)}\,d\theta\,, \\
&u_x(x,0,t) = -\frac{p}{\pi \mu} \operatorname{Im} \int_0^{t/x} \frac{2\sqrt{c_1^{-2} - \theta^2}\sqrt{c_2^{-2} - \theta^2} - (c_2^{-2} - 2\theta^2)}{S(\theta^2)}\,\theta\,d\theta\,.
\end{aligned} \tag{3.4.40}$$

Finally, by differentiating with respect to t the expressions for the displacements and stresses, given by the formulas (3.4.40), we arrive, in the case of a concentrated loading $\sigma_{yy}(x,0,t) = -p\delta(x)\delta(t)$, at the solution to the Lamb problem [3.59].

In the latter case, the boundary displacements are described by the following formulas ($\theta = t/x + \mathrm{i}0$, $|x| \neq c_{\mathrm{R}}t$):

$$\begin{aligned} u_y(x,0,t) &= -\frac{p}{\pi\mu c_2^2 x}\,\mathrm{Im}\left[\frac{\sqrt{c_1^{-2}-\theta^2}}{S(\theta^2)}\right], \\ u_x(x,0,t) &= -\frac{p\theta}{\pi\mu x}\,\mathrm{Im}\left[\frac{2\sqrt{c_1^{-2}-\theta^2}\sqrt{c_2^{-2}-\theta^2}-c_2^{-2}+2\theta^2}{S(\theta^2)}\right] \end{aligned} \tag{3.4.41}$$

These formulas may also be obtained directly from the formal solution for the load $\sigma_{yy} = -p\delta(x)\delta(t)$, $\sigma_{xy} = 0 (y = 0)$. Then, $n = 0$, and we find for $t > 0$

$$\begin{aligned} &\Phi_2^0(\theta) = 0\,, \quad \Phi_2(\theta) = 0\,, \\ &\Phi_1^0(\theta) = \frac{x}{\mu}L_{-1}\sigma_{yy}(x,0,t) = -\frac{p}{\mu}x\delta(x) = -\frac{p}{\mu}\frac{1}{\theta}\delta\left(\frac{1}{\theta}\right). \end{aligned}$$

Finally, with the help of (3.4.27) we get

$$\begin{aligned} \Phi_1(\theta) &= \frac{1}{\pi\mathrm{i}}\int_{-\infty}^{\infty}\frac{\Phi_1^0(\beta)}{\beta-\theta}\,d\beta = \frac{1}{\pi\mathrm{i}}\int_{-\infty}^{\infty}\frac{\Phi_1^0(1/z)}{z(1-z\theta)}\,dz \\ &= -\frac{p}{\pi\mu\mathrm{i}}\int_{-\infty}^{\infty}\frac{\delta(z)dz}{1-z\theta} = -\frac{p}{\pi\mu\mathrm{i}}\,. \end{aligned}$$

By substituting the thus obtained values of $\Phi_1(\theta)$, $\Phi_2(\theta)$ into (3.4.31) and putting $n = m = 0$, we arrive at the results (3.4.41).

In a similar way, for a tangential concentrated loading $\sigma_{xy}(x,0,t) = -p\delta(x)\delta(t)$, $\sigma_{yy}(x,0,t) = 0$ we find $\Phi_1(\theta) = 0$, $\Phi_2(\theta) = -p/(\pi\mu\mathrm{i})$, and then with the help of (3.4.9, 22) for $n = 0$ we obtain the displacements and stresses in the elastic half-plane $y \leq 0$.

Note: According to the above results for concentrated loadings, an addition of an imaginary constant to the function $\Phi_k(\theta)$ makes a contribution to the e-lastodynamic solution in the form of an additional solution due to a stationary load (normal or tangential for $k = 1$ or 2 respectively) concentrated at the point $x = y = 0$. By generalizing the above result, it may be shown that an addition to the function $\Phi_k(\theta)$ of the first or a higher degree polynomial in θ with pure imaginary coefficients corresponds to an additional contribution to the elastodynamic problem solution, due to concentrated loads with a higher-order singularity at the point $x = y = 0$.

3.5 Mixed Boundary-Value Problems for an Elastic Half-Plane. Crack Propagation

Let us now consider some boundary-value problems for a half-plane, which prove to be more complicated to solve as compared with the first and second boundary-value problems. To be exact, we shall consider elastodynamic problems which are mixed with respect to one of the two boundary conditions, namely, the conditions one of which is of the same type along the entire boundary $y = 0$, and the other is mixed, i.e., it differs in type for different parts of the axis $y = 0$. The class of mixed problems incorporates numerous problems of elastic waves diffraction, punch indentation, and cracks/cuts propagation within elastic media. That is why solving these rather complicated problems is of considerable practical interest.

3.5.1 Representation of Solutions to Mixed Boundary-Value Problems

For the purpose of demonstration of the method for solving the problem, it is enough to consider some of the mixed problems, since all the rest may be solved in a similar way. Therefore, here we shall describe in detail the procedure of solving a problem with the following boundary conditions on the axis $y = 0$:

$$\begin{aligned}
&\sigma_{xy} = 0 && (-\infty < x < \infty)\,, \\
&\sigma_{yy} = \sigma^0_{yy}(x,t) && (-v_2 t < x < v_1 t)\,, \\
&u_y = u^0_y(x,t) && (x < -v_2 t, \quad x > v_1 t) \\
&(0 \le v_j, \quad j = 1,2)\,.
\end{aligned} \tag{3.5.1}$$

Besides, we shall consider a problem with analogous boundary conditions on the axis $y = 0$:

$$\begin{aligned}
&\sigma_{xy} = 0 && (-\infty < x < \infty)\,, \\
&u_y = u^0_y(x,t) && (-v_2 t < x < v_1 t)\,, \\
&\sigma_{yy} = \sigma^0_{yy}(x,t) && (x < -v_2 t, x > v_1 t)\,.
\end{aligned} \tag{3.5.2}$$

As usual, the initial conditions in (3.5.1, 2) will be put in the form:

$$\boldsymbol{u} \equiv 0 \quad \text{for} \quad t < 0\,. \tag{3.5.3}$$

Besides, since we have to account for the points $x = v_1 t$, $x = -v_2 t$ where the boundary conditions are subject to change (according to Sect. 2.3, in the three-dimensional space (x, y, z) these points are treated as singular lines), then, it follows from (2.3.6) that in order to ensure uniqueness of the solution, one should demand that in the neighborhood of the points the displacement vector $\boldsymbol{u}$ satisfy the estimates

$$\boldsymbol{u} = \boldsymbol{u}_0(t) + \mathrm{O}(\varepsilon^\beta)\,, \quad \varepsilon \to 0 \tag{3.5.4}$$

where $\beta > 0$ for a stationary singular curve (i.e. for zero v_j) and $\beta \geq 1/2$ for a moving singular curve (in some cases, the latter condition will be replaced by a stronger one, $\beta > 1/2$).

As stated above, we assume that the displacements are homogeneous functions of the $(n-1)^{\text{th}}$ dimension with respect to x, y, z where n is an arbitrary integer. If the notation (3.4.22) is used, then the first boundary condition in (3.5.1, 2) leads to the Schwarz problem of determining the function $\Phi_2(\theta)$ which is analytical within the region $\operatorname{Im}\theta > 0$ and has a zero real part on the axis $\operatorname{Im}\theta = 0$:

$$\operatorname{Re}\Phi_2(\theta) = 0\,. \tag{3.5.5}$$

Besides, the condition

$$\operatorname{Im}\Phi_2(\theta) \to 0 \quad \text{as} \quad \theta \to \infty\,, \quad \operatorname{Im}\theta \geq 0 \tag{3.5.6}$$

must be met, since otherwise a concentrated load σ_{xy} appears at the point $x = y = 0$, which contradicts the condition $\sigma_{xy} \equiv 0$ for $-\infty < x < \infty$, $y = 0$. Hence, we get from (3.5.5, 6) with the help of (3.4.27):

$$\Phi_2(\theta) \equiv 0 \quad (\operatorname{Im}\theta \geq 0)\,. \tag{3.5.7}$$

Then, taking into account (3.4.22) we get from (3.5.7) for $\operatorname{Im}\theta \geq 0$:

$$F_1(\theta) = \frac{c_2^{-2} - 2\theta^2}{2\theta\sqrt{c_1^{-2} - \theta^2}} F_2(\theta)\,. \tag{3.5.8}$$

Thus, in order to obtain all the kinematic and dynamic characteristics of disturbed motion of an elastic continuum at any point of the half-plane $y \leq 0$ with the help of the representations (3.4.9) and (3.4.12, 5.8), it is enough to define one of the analytical functions $F_1(\theta)$, $F_2(\theta)$ in the region $\operatorname{Im}\theta \geq 0$.

Taking into account (3.5.8), we get from (3.4.16) the following representations on the axis $\operatorname{Im}\theta = 0$:

$$\begin{gathered} \operatorname{Re} L_m^\theta F_2(\theta) = 2c_2^2 x^{1+m} L_{n+m}^t u_y(x,0,t)\,, \\ \operatorname{Re} L_m^\theta \left[\frac{S(\theta^2)F_2(\theta)}{\sqrt{c_1^{-2} - \theta^2}}\right] = \frac{2x^{1+m}}{\mu} L_{n+m-1}^t \sigma_{yy}(x,0,t)\,, \\ \theta = t/x + \mathrm{i}0\,, \end{gathered} \tag{3.5.9}$$

where m is an arbitrary integer.

The latter formulas are the basic ones for solving mixed elastodynamic problems (3.5.1, 2). In particular, for $m = 0$ we find from (3.5.9):

$$\begin{gathered} \operatorname{Re} F_2(\theta) = 2c_2^2 x L_n^t u_y(x,0,t)\,, \\ \operatorname{Re}\left[\frac{S(\theta^2)F_2(\theta)}{\sqrt{c_1^{-2} - \theta^2}}\right] = \frac{2x}{\mu} L_{n-1}^t \sigma_{yy}(x,0,t) \\ (\theta = t/x + \mathrm{i}0)\,. \end{gathered} \tag{3.5.10}$$

Let us start by considering the problem (3.5.1), assuming for the sake of definiteness that $v_j < c_R$, i.e., that the points of the interface move along the axis $y = 0$ at the speeds less than the Rayleigh wave speed for an elastic continuum. Then, we obtain from (3.5.10) ($\theta = t/x + \mathrm{i}0$):

$$\begin{aligned} \operatorname{Re} F_2(\theta) &= 2c_2^2 x L_n^t u_y(x,0,t) \quad (-v_2^{-1} < \theta < v_1^{-1}) , \\ \operatorname{Im} F_2(\theta) &= \frac{2x}{\mu \mathrm{i}} \frac{\sqrt{c_1^{-2} - \theta^2}}{S(\theta^2)} L_{n-1}^t \sigma_{yy}(x,0,t) \quad (\theta < -v_2^{-1},\ \theta > v_1^{-1}) . \end{aligned} \tag{3.5.11}$$

3.5.2 Solution for the $n \geq 1$-Case. Some Examples

The right-hand sides of (3.5.11) prove to be known functions, and we have

$$\begin{aligned} \operatorname{Re} F_2(\theta) &= U(\theta) \quad (-v_2^{-1} < \theta < v_1^{-1}) , \\ \operatorname{Im} F_2(\theta) &= \Sigma(\theta) \quad (\theta < -v_2^{-1},\ \theta > v_1^{-1}) , \end{aligned} \tag{3.5.12}$$

where the following notation is used:

$$\begin{aligned} U(\theta) &= 2c_2^2 x \partial^n u_y^0(x,t)/\partial t^n , \\ \Sigma(\theta) &= \frac{2x}{\mu \mathrm{i}} \frac{\sqrt{c_1^{-2} - \theta^2}}{S(\theta^2)} \frac{\partial^{n-1}\sigma^0(x,t)}{\partial t^{n-1}} \quad (\theta = t/x + \mathrm{i}0) . \end{aligned} \tag{3.5.13}$$

Assuming the validity of the estimates (3.4.24), we deduce from (3.4.26) with the help of (3.4.22, 5.8, 13) that

$$\begin{aligned} F_2(\theta) &= \mathrm{O}(\theta^{-1-\varepsilon}) , \\ \Sigma(\theta) &= \mathrm{O}(\theta^{-1-\varepsilon}) , \quad \varepsilon > 0 \end{aligned} \tag{3.5.14}$$

as $\theta \to \infty$. Here the asymptotic $S(\theta^2) \sim -2\theta^2(c_2^{-2} - c_1^{-2})$ has been used. Besides, from the asymptotic (3.5.4) and (3.5.10) we find:

$$\begin{aligned} F_2(\theta) &= \mathrm{O}\left[(\theta - v_1^{-1})^{-n+1/2}\right] \quad (\theta \to v_1^{-1}) , \\ F_2(\theta) &= \mathrm{O}\left[(\theta + v_2^{-1})^{-n+1/2}\right] \quad (\theta \to -v_2^{-1}) \end{aligned} \tag{3.5.15}$$

and assume that if given boundary functions are sufficiently smooth, then the function $F_2(\theta)$ has no pole at the point $\theta = 0$.

Thus, for $n \geq 1$ we arrive at the mixed boundary-value problem for the function $F_2(\theta)$ which is analytical within the region $\operatorname{Im}\theta > 0$ and satisfies both the conditions (3.5.12) on the axis $\operatorname{Im}\theta = 0$ and the constraints (3.5.14, 15). The solution to the problem is sought in the form:

$$F_2(\theta) = N(\theta) + P(\theta) , \tag{3.5.16}$$

where $N(\theta)$ is a solution satisfying given boundary conditions

$$\begin{aligned} &\operatorname{Re} N(\theta) = U(\theta) \quad (-v_2^{-1} < \theta < v_1^{-1}), \\ &\operatorname{Im} N(\theta) = \Sigma(\theta) \quad (\theta < -v_2^{-1},\ \theta > v_1^{-1}) \end{aligned} \tag{3.5.17}$$

and bounded both at infinity and the points $\theta = v_1^{-1}$, $\theta = -v_2^{-1}$. The solution of the mixed problem for $N(\theta)$ within the region $\operatorname{Im}\theta > 0$ is given by the Keldysh-Sedov formula [3.51]:

$$N(\theta) = \frac{1}{\pi \mathrm{i} G(\theta)} \left[\int_{-v_2^{-1}}^{v_1^{-1}} \frac{U(\alpha)G(\alpha)}{\alpha - \theta} d\alpha + \mathrm{i} \int_{-\infty}^{v_2^{-1}} \frac{\Sigma(\alpha)G(\alpha)}{\alpha - \theta} d\alpha \right.$$

$$\left. + \mathrm{i} \int_{v_1^{-1}}^{\infty} \frac{\Sigma(\alpha)G(\alpha)}{\alpha - \theta} d\alpha \right] + \frac{C_0}{G(\theta)} . \tag{3.5.18}$$

Here $G(\theta) = \sqrt{\theta - v_1^{-1}}/\sqrt{\theta + v_2^{-1}}$, and the branch of the radical $G(\theta)$ has been chosen in such a way that $G(\theta) \to 1$ as $\theta \to \infty$, $\operatorname{Im}\theta \geq 0$. In accordance with (3.5.18) and the specified boundary conditions, $\operatorname{Im} N(\theta) = \mathrm{O}(\theta^{-1-\varepsilon})$, $\varepsilon > 0$, $\operatorname{Re} N(\theta) \to C_0$ as $\theta \to \infty$, and the function $N(\theta)$ is bounded at the point $\theta = -v_2^{-1}$.

However, at the point $\theta = v_1^{-1}$ the function $N(\theta)$ is, generally speaking, unbounded (for an arbitrary C_0). The condition of $N(\theta)$ being bounded for $\theta = v_1^{-1}$, allows us to obtain the following expression for the constant C_0:

$$C_0 = -\frac{1}{\pi \mathrm{i}} \left[\int_{-v_2^{-1}}^{v_1^{-1}} \frac{U(\alpha)}{Q(\alpha)} d\alpha + \mathrm{i} \int_{-\infty}^{-v_2^{-1}} \frac{\Sigma(\alpha)}{Q(\alpha)} d\alpha + \mathrm{i} \int_{v_1^{-1}}^{\infty} \frac{\Sigma(\alpha)}{Q(\alpha)} d\alpha \right] , \tag{3.5.19}$$

where $Q(\theta) = [(\theta - v_1^{-1})(\theta + v_2^{-1})]^{1/2}$ and the branch of the radical is chosen in such a way that $Q(\theta) \sim \theta$ as $\theta \to \infty$ within the region $\operatorname{Im}\theta \geq 0$. Hence, taking into account (3.5.19), the desired function $N(\theta)$ may be written in the form ($\operatorname{Im}\theta > 0$):

$$N(\theta) = \frac{Q(\theta)}{\pi \mathrm{i}} \left[\int_{-v_2^{-1}}^{v_1^{-1}} \frac{U(\alpha) d\alpha}{(\alpha - \theta)Q(\alpha)} + \mathrm{i} \int_{-\infty}^{-v_2^{-1}} \frac{\Sigma(\alpha) d\alpha}{(\alpha - \theta)Q(\alpha)} + \mathrm{i} \int_{v_1^{-1}}^{\infty} \frac{\Sigma(\alpha) d\alpha}{(\alpha - \theta)Q(\alpha)} \right] . \tag{3.5.20}$$

In order to define the function $P(\theta)$, let us consider an auxiliary function $P_0(\theta) = P(\theta)[Q(\theta)]^{2n-1}$. It can be easily seen that the function $P_0(\theta)$ being analytical within the region $\operatorname{Im}\theta > 0$ and bounded at the points of the axis $\theta = -v_2^{-1}$, $\theta = v_1^{-1}$, meets at the axis $\operatorname{Im}\theta = 0$ the homogeneous boundary condition

$$\operatorname{Im} P_0(\theta) = 0 \quad (-\infty < \theta < \infty) . \tag{3.5.21}$$

Besides, it increases at infinity no faster than $M|\theta|^{2n-1}$ where M is a constant. In order to find such an analytical function $P_0(\theta)$ which is continuous up to the boundary, the principle of symmetry [3.51] will be used, according to which $P_0(\theta)$ can be analytically extended onto the lower half-plane by using the formula $P_0(\bar{\theta}) = \overline{P_0(\theta)}$. Then, we arrive at the function $P_0(\theta)$ which is analytical throughout the complex plane θ and increases at infinity no faster than $M|\theta|^{2n-1}$. In accordance with the generalized Liouville theorem [3.60], the function is a polynomial whose degree is no higher than $2n-1$:

$$P_0(\theta) = \sum_{k=0}^{2n-1} A_k\theta^k \,. \tag{3.5.22}$$

At the same time it follows from the condition (3.5.21) that the coefficients of the polynomial, A_k, must be real numbers. Hence,

$$P(\theta) = [Q(\theta)]^{1-2n} \sum_{k=0}^{2n-1} A_k\theta^k \,,$$

and the desired function $F_2(\theta)$ must be finally written in the form:

$$F_2(\theta) = N(\theta) + [Q(\theta)]^{1-2n} \sum_{k=0}^{2n-1} A_k\theta^k \quad (\operatorname{Im}\theta > 0) \,, \tag{3.5.23}$$

where $N(\theta)$ is given by (3.5.20), $Q(\theta) = [(\theta - v_1^{-1})(\theta + v_2^{-1})]^{1/2}$ (the branch $Q(\theta)$ is chosen in such a way that $Q(\theta) \sim \theta$ as $\theta \to \infty$, $\operatorname{Im}\theta \geq 0$), and A_k are real constants.

For determining the $2n$ constants $A_k (k = 0, 1, \ldots, 2n-1)$ one must use the first of the estimates (3.5.14) which provides the following two conditions

$$A_{2n-1} = -C_0 \,, \tag{3.5.24}$$

$$\lim_{\theta\to\infty} [\theta F_2(\theta)] = 0 \tag{3.5.25}$$

and the boundary conditions obtained from the lower expression (3.5.9) at $m = -1, -2, \ldots, -(n-1)$ for each of the cases $\theta < -v_2^{-1}$ and $\theta > v_1^{-1}$:

$$\operatorname{Re} L_m^\theta \left[\frac{S(\theta^2)F_2(\theta)}{\sqrt{c_1^{-2} - \theta^2}}\right] = \frac{2x^{1+m}}{\mu} L_{n+m-1}^t \sigma_{yy}^0(x,t) \quad (\theta = t/x + i0) \,. \tag{3.5.26}$$

[Note, that C_0 entering (3.5.24) is given by the formula (3.5.19).] It should be pointed out that the $2n-2$ conditions (3.5.26) are obtained, in essence, by applying the m-uple operator of integrating with respect to t, to the lower expression (3.5.10) for each of the cases $-v_2t < x < 0$ and $0 < x < v_1t$.

Thus, the solution to the problem (3.5.1), satisfying the conditions (3.4.24, 5.4) ($v_j < c_R$) is given for $n \geq 1$ by the formulas (3.5.23–26).

Let us now use the above results for solving some elastodynamic problems. Consider an elastic continuum occupying the plane $x0y$ and originally at rest. Let a crack (cut) start propagating in both directions at the instant $t = 0$ from the point $x = 0$, $y = 0$. The edges of the cut move at constant speeds v_1, v_2 ($v_1 < c_R$, $v_2 < c_R$), and the sides of the cut $-v_2 t < x < v_1 t$ are loaded by the normal stress $\sigma_{yy} = \sigma^0_{yy}(x,t)$ (Fig. 3.15), where $\sigma^0_{yy}(x,t)$ is a homogeneous function of the $(n-2)^{\text{th}}$ dimension with respect to x, t. Since the problem is symmetric with respect to the $0x$-axis, the displacement and stress components u_y, σ_{xy} should be sought in the form of functions which are odd with respect to y. Hence, due to continuity of the functions outside the cut, we deduce that $u_y = 0$, $\sigma_{xy} = 0$ at the $0x$-axis outside the cut, i.e. for $x < -v_2 t$, $x > v_1 t$. Thus, solution of this elastodynamic problem reduces to solving a mixed problem for the half-plane $y < 0$ subject to the boundary condition (3.5.1). If we suppose that the stresses are integrable in the neighborhood of the point $x = y = 0$ [i.e., that the conditions (3.4.25) hold], if the boundary conditions are smooth enough with respect to time, and the stresses in the vicinity of moving edges of the cut may contain a singularity which is no stronger than $\varepsilon^{-1/2}$ [i.e., that the condition (3.5.4) is met for u], then the solution to the above problem for $n \geq 1$ is given by the formulas (3.5.23–26) with $u^0_y(x,t)$ everywhere identically equal to zero.

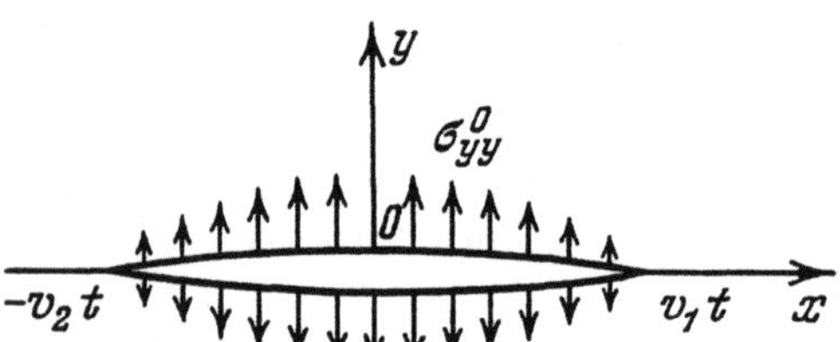

Fig. 3.15. Extension of a cut from the point $x = y = 0$

Let us consider some of the most interesting special cases of the above problem.

1) Let

$$
\begin{aligned}
&\sigma_{yy}(x,0,t) = -p \quad && (p = \text{const},\ -v_2 t < x < v_1 t)\,, \\
&u_y(x,0,t) = 0 && (xy - v_2 t,\ x > v_1 t)\,, \\
&\sigma_{xy}(x,0,t) = 0 && (-\infty < x < \infty)\,.
\end{aligned}
\tag{3.5.27}
$$

Then $n = 2$, and we get from (3.5.13, 20, 23–26):

$$
U(\theta) = \Sigma(\theta) = N(\theta) = 0\,, \quad A_2 = A_3 = 0\,,
$$

hence

$$
F_2(\theta) = \frac{A_1\theta + A_0}{[Q(\theta)]^3} \quad \left(\operatorname{Im}\theta > 0\,,\ Q(\theta) \equiv \sqrt{(\theta + v_2^{-1})(\theta - v_1^{-1})}\right)\,, \tag{3.5.28}
$$

where according to (3.5.26), A_0, A_1 are defined by the expressions

$$\operatorname{Re}\int_0^{\theta}\frac{(A_1\tau+A_0)S(\tau^2)d\tau}{\sqrt{c_1^{-2}-\tau^2}[Q(\tau)]^3}=-\frac{2p}{\mu}\quad(\theta>v_1^{-1})\,,$$
$$\operatorname{Re}\int_{\theta}^{0}\frac{(A_1\tau+A_0)S(\tau^2)d\tau}{\sqrt{c_1^{-2}-\tau^2}[Q(\tau)]^3}=\frac{2p}{\mu}\quad(\theta<-v_2^{-1})\,. \tag{3.5.29}$$

(It should be remembered that the integration contours in the latter formulas circumvent the singular points of the integrands from above, i.e., $\theta=t/x+\mathrm{i}0$.)

From (3.5.29) we find the following expressions for the coefficients A_0, A_1:

$$A_0=\frac{2p}{\mu}\frac{b_1-a_1}{a_1b_0-a_0b_1}\,,\qquad A_1=\frac{2p}{\mu}\frac{a_0-b_0}{a_1b_0-a_0b_1}\,, \tag{3.5.30}$$

where

$$a_k=\operatorname{Re}\int_{c_1^{-1}}^{M_0}\frac{\theta^kS(\theta^2)d\theta}{\sqrt{c_1^{-2}-\theta^2}[Q(\theta)]^3}\quad(v_1^{-1}<M_0<\infty,\ k=0,1)\,,$$
$$b_k=-\operatorname{Re}\int_{M_1}^{-c_1^{-1}}\frac{\theta^kS(\theta^2)d\theta}{\sqrt{c_1^{-2}-\theta^2}[Q(\theta)]^3}\quad(-\infty<M_1<-v_2^{-1},\ k=0,1)\,. \tag{3.5.31}$$

Note that a_k, b_k are independent of the constants M_0, M_1, since a variation of the constants in the above ranges results in changing the imaginary parts of the integrals only.

Let us prove that $a_0=b_0$. In fact, by replacing the limits of integration (c_1^{-1}, M_0), $(M_1, -c_1^{-1})$ by $(0,+\infty)$, $(-\infty,0)$ respectively, and deforming the contour $(-\infty,0)$ of the integral b_0 into the counter $(0,\infty)$, we arrive at the equality $b_0=a_0$. Then, the expressions (3.5.30) reduce, finally, to

$$A_1=0\,,$$
$$A_0=-\frac{2p}{\mu a_0}=-\frac{2p}{\mu}\left[\operatorname{Re}\int_{c_1^{-1}}^{M}\frac{S(\theta^2)d\theta}{\sqrt{c_1^{-2}-\theta^2}[(\theta+v_2^{-1})(\theta-v_1^{-1})]^{3/2}}\right]^{-1} \tag{3.5.32}$$
$$(v_1^{-1}<M\le\infty)\,.$$

Thus,

$$F_2(\theta)=\frac{A_0}{[(\theta+v_2^{-1})(\theta-v_1^{-1})]^{3/2}}\quad(\operatorname{Im}\theta>0)\,, \tag{3.5.33}$$

where A_0 is given by (3.5.32) (where the integration contour circumvents the singular point of the integrand $\theta=v_1^{-1}$ from above).

From the formulas (3.5.9) and with the help of (3.5.33) we get the following expressions for the displacement u_y at the lower side of the cut and the stresses σ_{yy} at the cut's extention:

$$
\begin{aligned}
u_y(x,0,t) &= \frac{x}{2c_2^2}\mathrm{Re}\left[\int_0^\theta d\tau_1 \int_0^{\tau_1} F_2(\tau)d\tau\right] \\
&= \frac{-2A_0 x}{c_2^2(v_2^{-1}+v_1^{-1})^2}[(\theta+v_2^{-1})(\theta-v_1^{-1})]^{1/2} \\
&\quad (\theta < -v_2^{-1}, \theta > v_1^{-1}) ,
\end{aligned}
\tag{3.5.34}
$$

$$
\begin{aligned}
\sigma_{yy}(x,0,t) &= \frac{\mu}{2}\mathrm{Re}\left[\int_0^\theta F_2(\tau)\frac{S(\tau^2)}{\sqrt{c_1^{-2}-\tau^2}}d\tau\right] \\
&= \begin{cases} \frac{\mu A_0}{2}\mathrm{Re}\int_{c_1^{-1}}^{\theta} \frac{S(\tau^2)d\tau}{\sqrt{c_1^{-2}-\tau^2}[(\tau+v_2^{-1})(\tau-v_1^{-1})]^{3/2}} & (0<\theta<v_1^{-1}) , \\ -\frac{\mu A_0}{2}\mathrm{Re}\int_{\theta}^{-c_1^{-1}} \frac{S(\tau^2)d\tau}{\sqrt{c_1^{-2}-\tau^2}[(\tau+v_2^{-1})(\tau-v_1^{-1})]^{3/2}} & (-v_2^{-1}<\theta<0) , \end{cases} \\
&\quad (\theta = t/x + \mathrm{i}0) .
\end{aligned}
\tag{3.5.35}
$$

In the special case of a symmetric cut, when $v_2 = v_1 = v$, we get from (3.5.34, 35):

$$
u_y(x,0,t) = -\frac{B_0 v}{2c_2^2}\sqrt{v^2t^2-x^2} \quad (|x|<vt) , \tag{3.5.36}
$$

$$
\sigma_{yy}(x,0,t) = \frac{B_0\mu}{2}\mathrm{Re}\int_{c_1^{-1}}^{t/|x|}\frac{S(\theta^2)d\theta}{\sqrt{c_1^{-2}-\theta^2}(\theta-v^{-2})^{3/2}} \quad (|x|>vt) , \tag{3.5.37}
$$

$$
B_0 = -\frac{2p}{\mu}\left[\mathrm{Re}\int_{c_1^{-1}}^{\infty}\frac{S(\theta^2)d\theta}{\sqrt{c_1^{-2}-\theta^2}(\theta^2-v^{-2})^{3/2}}\right]^{-1} . \tag{3.5.38}
$$

These formulas coincide with the known ones [3.61], if we take into account that the radical in (3.5.36) takes on arithmetical value.

In particular, for the vanishing density ϱ of a continuum (which enters the expressions for the velocities c_1, c_2) we get from (3.5.36–38):

$$
u_y(x,0,t) = -\frac{p(\lambda+2\mu)}{2\mu(\lambda+\mu)}\sqrt{v^2t^2-x^2} \quad (|x|<vt) , \tag{3.5.39}
$$

$$
\sigma_{yy}(x,0,t) = -p\left[1-\frac{|x|}{\sqrt{x^2-v^2t^2}}\right] \quad (|x|>vt) . \tag{3.5.40}
$$

The latter formulas coincide with the analogous ones for the static problem, if the width of the crack is put equal to $2vt$ [3.62].

If we assume now that one of the cut's edges is stationary (i.e., $v_2 \to 0$, $v_1 \equiv v$), then

$$u_y(x,0,t) = -c_2^{-2} B_1 x \sqrt{\theta - v^{-1}} \quad (v^{-1} < \theta) , \tag{3.5.41}$$

$$\sigma_{yy}(x,0,t) = \begin{cases} \frac{\mu}{2} B_1 \operatorname{Re} \int\limits_{c_1^{-1}}^{\theta} \frac{S(\tau^2)d\tau}{\sqrt{c_1^{-2}-\tau^2}(\tau-v^{-1})^{3/2}} & (0 < \theta < v^{-1}) , \\ -\frac{\mu}{2} B_1 \operatorname{Re} \int\limits_{\theta}^{-c_1^{-1}} \frac{S(\tau^2)d\tau}{\sqrt{c_1^{-2}-\tau^2}(\tau-v^{-1})^{3/2}} & (\theta < 0) , \end{cases} \tag{3.5.42}$$

$$B_1 = -\frac{2p}{\mu} \left[\operatorname{Re} \int\limits_{c_1^{-1}}^{M} \frac{S(\theta^2)d\theta}{\sqrt{c_1^{-2} - \theta^2}(\theta - v^{-1})^{3/2}} \right]^{-1} \quad (v^{-1} < M < \infty) . \tag{3.5.43}$$

The branches of the roots $(\theta - v^{-1})^\alpha$ $(\alpha = 1/2, 3/2)$ in (3.5.41–43) are defined by the asymptotic $(\theta - v^{-1})^\alpha \sim |\theta^\alpha|$ as $\theta \to +\infty$ and the integration path circumvents the point $\theta = v^{-1}$ from above [the cut for singling out the branches is drawn along the real semi-axis $(-\infty, v^{-1})$].

From (3.5.41,42) we see, in particular, that the stress and the velocity have singularities of the same order of magnitude in the neighborhood of the moving edge ($x = vt$), while in the neighborhood of the stationary edge ($x = 0$) the velocity remains bounded, while the stress preserves the singularity of the same order ($\varepsilon^{-1/2}$).

Note that upon superposing onto the solution of the above problem a tensile stress $\sigma_{yy} = p$, we shall get a solution to the problem of propagation of a cut with unloaded sides through a bounded elastic continuum, originally in the homogeneous stressed state caused by a tensile loading $\sigma_{yy} = p$ applied at infinity.

2) Next let a concentrated force be applied to the sides of a propagating cut at the instant $t = 0$, with the point of application moving at a speed v_0 along the positive semi-axis $0x$. Then, the boundary conditions may be presented in the form:

$$\begin{aligned} &\sigma_{yy}(x,0,t) = -p\delta(x - v_0 t)H(t) \quad (-v_2 t < x < v_1 t,\ 0 < v_0 < v_1) , \\ &u_y(x,0,t) = 0 \quad (x < -v_2 t,\ x > v_1 t) , \\ &\sigma_{xy}(x,0,t) = 0 \quad (-\infty < x < \infty) . \end{aligned} \tag{3.5.44}$$

In this case, $n = 1$, and by using (3.5.13,20,23–25), we get

$$U(\theta) = 0\,,$$

$$\sum(\theta) = -\frac{2|\theta|p}{\mu i v_0}\frac{\sqrt{c_1^{-2}-\theta^2}}{\theta S(\theta^2)}\delta(\theta - v_0^{-1}) = \frac{2p\sqrt{v_0^{-2}-c_1^{-2}}}{v_0\mu S(v_0^{-2})}\delta(\theta - v_0^{-1})\,,$$

$$N(\theta) = \frac{A_1}{v_0^{-1}-\theta}\sqrt{(\theta + v_2^{-1})(\theta - v_1^{-1})}\,, \qquad (3.5.45)$$

$$A_1 = \frac{2p\sqrt{v_0^{-2}-c_1^{-2}}}{\pi v_0\mu S(v_0^{-2})\sqrt{(v_0^{-1}+v_2^{-1})(v_0^{-1}-v_1^{-1})}}\,, \qquad (3.5.46)$$

$$A_0 = A_1(v_2^{-1} + v_0^{-1} - v_1^{-1})$$

and hence

$$F_2(\theta) = \frac{A_1(v_0^{-1}-v_1^{-1})(v_0^{-1}+v_2^{-1})}{(v_0^{-1}-\theta)\sqrt{(\theta+v_2^{-1})(\theta-v_1^{-1})}} \quad (\operatorname{Im}\theta > 0)\,. \qquad (3.5.47)$$

[In (3.5.45, 46) the radicals containing no variable θ are positive.] Then, by putting $n = 1$, $m = -1$ in the first of the formulas (3.5.9), and $n = 1$, $m = 0$ in the second one, we arrive at the following expressions for the displacement of the lower side of the cut and stress at the cut's extension:

$$u_y(x,0,t) = \frac{A_1(v_0^{-1}-v_1^{-1})(v_0^{-1}+v_2^{-1})}{2c_2^2}\operatorname{Re}\int_0^\theta \frac{d\tau}{(v_0^{-1}-\tau)\sqrt{(\tau+v_2^{-1})(\tau-v_1^{-1})}}$$

$$(\theta < -v_2^{-1},\ \theta > v_1^{-1})\,, \qquad (3.5.48)$$

$$\sigma_{yy}(x,0,t) = \frac{\mu A_1}{2x}(v_0^{-1}-v_1^{-1})(v_0^{-1}+v_2^{-1}) \times \operatorname{Re}\frac{S(\theta^2)}{\sqrt{c_1^{-2}-\theta^2}(v_0^{-1}-\theta)\sqrt{(\theta+v_2^{-1})(\theta-v_1^{-1})}}$$

$$(-v_2^{-1} < \theta < v_1^{-1})\,. \qquad (3.5.49)$$

If $\theta > v_0^{-1}$, then the integration path in (3.5.47) circumvents the singular point $\tau = v_0^{-1}$ of the integrand from above, since $\theta = t/x + \mathrm{i}0$. However, if $v_0 \to 0$, then for the concentrated force $\sigma_{yy}(x,0,t) = -p\delta(x)H(t)$, (3.5.47–49) become

$$F_2(\theta) = -\frac{p}{\pi\mu(c_2^{-2} - c_1^{-2})} \frac{1}{\sqrt{(\theta + v_2^{-1})(\theta - v_1^{-1})}} \quad (\operatorname{Im}\theta > 0)\,,$$

$$u_y(x,0,t) = -\frac{pc_1^2}{2\pi\mu(c_1^2 - c_2^2)} \times \ln\left[\frac{2|\theta + (v_2^{-1} - v_1^{-1})/2 + \sqrt{(\theta + v_2^{-1})(\theta - v_1^{-1})}|}{v_1^{-1} + v_2^{-1}}\right] \qquad (3.5.50)$$

$$(\theta < -v_2^{-1},\ \theta > v_1^{-1},\ \theta = t/x + \mathrm{i}0)\,,$$

$$\sigma_{yy}(x,0,t) = -\frac{p}{2\pi x(c_2^{-2} - c_1^{-2})} \operatorname{Re} \frac{S(\theta^2)}{\sqrt{c_1^{-2} - \theta^2}\sqrt{(\theta + v_2^{-1})(\theta - v_1^{-1})}}$$

$$(\theta = t/x + \mathrm{i}0), -v_2^{-1} < \theta < v_1^{-1})\,.$$

In particular, at $v_1 = v_2 = v$ (3.5.50) reduce to:

$$F(\theta) = -\frac{p}{\pi\mu(c_2^{-2} - c_1^{-2})} \frac{1}{\sqrt{\theta^2 - v^{-2}}} \quad (\operatorname{Im}\theta > 0)\,,$$

$$u_y(x,0,t) = -\frac{pc_1^2 \ln|v(\theta + \sqrt{\theta^2 - v^{-2}})|}{2\pi\mu(c_1^2 - c_2^2)}$$

$$(|\theta| > v^{-1},\ \theta = t/x + \mathrm{i}0)\,, \qquad (3.5.51)$$

$$\sigma_{yy}(x,0,t) = -\frac{p}{2\pi x(c_2^{-2} - c_1^{-2})} \operatorname{Re}\left[\frac{S(\theta^2)}{\sqrt{c_1^{-2} - \theta^2}\sqrt{\theta^2 - v^{-2}}}\right]$$

$$(|\theta| < v^{-1},\ \theta = t/x + \mathrm{i}0)\,.$$

3.5.3 Solution for the $n \le 0$-Case

For $t > 0$ ($v_j < c_R$), the right-hand sides of (3.5.10) (denoted by $U_0(\theta)$ and $\Sigma_0(\theta)$ where $\theta = t/x + \mathrm{i}0$) become for the mixed problem (3.5.1):

$$U_0(\theta) \equiv 2c_2^2 x L_n^t u_y(x,0,t) = \frac{2c_2^2 x}{|n|!} \int_0^t (t-\tau)^{-n} \dot{u}_y^0(x,\tau)d\tau$$

$$= \frac{2c_2^2}{|n|!} \int_0^\theta (\theta - \eta)^{-n} \dot{u}_y^0(1,\eta)d\eta \quad (-v_2^{-1} < \theta < v_1^{-1})\,, \qquad (3.5.52)$$

$$\Sigma_0(\theta) \equiv \frac{2x}{\mu\mathrm{i}} \frac{\sqrt{c_1^{-2} - \theta^2}}{S(\theta^2)} L^t_{n-1}\sigma_{yy}(x,0,t)$$

$$= \frac{2x\sqrt{c_1^{-2} - \theta^2}}{|n|!\mathrm{i}\mu S(\theta^2)} \int_0^t (t-\tau)^{-n}\sigma_{yy}(x,0,\tau)d\tau$$

$$= \begin{cases} \frac{2x\sqrt{c_1^{-2}-\theta^2}}{|n|!\mathrm{i}\mu S(\theta^2)} \left[\int\limits_0^{x/v_1} (t-\tau)^{-n}\sigma_{yy}(x,0,\tau)d\tau \right. \\ \left. \quad + \int\limits_{x/v_1}^t (t-\tau)^{-n}\sigma^0_{yy}(x,\tau)d\tau \right] \quad (0 < x < v_1 t)\,, \\ \frac{2x\sqrt{c_1^{-2}-\theta^2}}{|n|!\mathrm{i}\mu S(\theta^2)} \left[\int\limits_0^{-x/v_2} (t-\tau)^{-n}\sigma_{yy}(x,0,\tau)d\tau \right. \\ \left. \quad + \int\limits_{-x/v_2}^t (t-\tau)^{-n}\sigma^0_{yy}(x,\tau)d\tau \right] \quad (-v_2 t < x < 0)\,, \\ \frac{2\sqrt{c_1^{-2}-\theta^2}}{|n|!\mathrm{i}\mu S(\theta^2)} \left[\sum\limits_{m=0}^{|n|} A_m\theta^m + \int\limits_{v_1^{-1}}^{\theta} (\theta-\eta)^{-n}\sigma^0_{yy}(1,\eta)d\eta \right] \quad (\theta > v_1^{-1})\,, \\ \frac{2\sqrt{c_1^{-2}-\theta^2}}{|n|!\mathrm{i}\mu S(\theta^2)} \left[\sum\limits_{m=0}^{|n|} B_m\theta^m - \int\limits_{\theta}^{-v_2^{-1}} (\theta-\eta)^{-n}\sigma^0_{yy}(1,\eta)d\eta \right] \quad (\theta < -v_2^{-1})\,, \end{cases} \tag{3.5.53}$$

Here

$$A_m = \frac{|n|!}{m!(-n-m)!} \int_0^{v_1^{-1}} (-\eta)^{-n-m}\sigma_{yy}(1,0,\eta)d\eta\,,$$

$$B_m = \frac{|n|!}{m!(-n-m)!} \int_0^{-v_2^{-1}} (-\eta)^{-n-m}\sigma_{yy}(1,0,\eta)d\eta\,, \tag{3.5.54}$$

$$\dot{u}^0_y(1,\eta) \equiv du^0_y(1,\eta)/d\eta\,, \quad 0! = 1\,.$$

Since the values of $\sigma_{yy}(x,0,t)$ are unknown for $x < -v_2 t$ and $x > v_1 t$, the real constants A_m, B_m entering the expression for $\Sigma_0(\theta)$ in (3.5.53), are unknown too. Thus, we have obtained the following mixed boundary conditions

$$\begin{aligned} &\mathrm{Re}\, F_2(\theta) = U_0(\theta) \quad (-v_2^{-1} < \theta < v_1^{-1})\,, \\ &\mathrm{Im}\, F_2(\theta) = \Sigma_0 \quad (\theta < -v_2^{-1},\ \theta > v_1^{-1}) \end{aligned} \tag{3.5.55}$$

for determining the function $F_2(\theta)$ which is analytical in the upper half-plane. Here $U_0(\theta)$, $\Sigma_0(\theta)$ are defined by the formulas (3.5.52–54), and the expression for $\Sigma_0(\theta)$ contains $2|n| + 2$ unknown constants.

Let us suppose that the given functions $\sigma^0_{yy}(x,t)$, $u^0_y(x,t)$ are integrable with respect to $t(x \neq 0)$, and $\sigma^0_{yy}(x,y) = \mathrm{O}(x^{-1+\varepsilon})$ as $x \to 0$, $\varepsilon > 0$ $(t > 0)$. Then, we get as $\theta \to \infty$

$$\sigma^0_{yy}(1,\theta) = x^{-n+2}\sigma^0_{yy}(x,t) = \mathrm{O}(\theta^{n-1-\varepsilon}) \,. \tag{3.5.56}$$

In this case the expression for $\Sigma_0(\theta)$ from (3.5.53) behaves as $\mathrm{O}(\theta^{-n-1})$ when $\theta \to \infty$. On the other hand, according to (3.5.52) the expression $\theta^n U_0(\theta)$ is integrable at $\theta = 0$. Thus, in the case of integrable stresses one has to construct a function $\theta^n F_2(\theta)$ analytical within the region $\mathrm{Im}\,\theta > 0$ and decreasing as $\mathrm{O}(\theta^{-1})$ when $\theta \to \infty$, by using the values of its real part $\theta^n U_0(\theta)$ in the interval $-v_2^{-1} < \theta < v_1^{-1}$ and the values of its imaginary part $\theta^n \Sigma_0(\theta)$ in the intervals $\theta < -v_2^{-1}$, $\theta > v_1^{-1}$. The function $\theta^n F_2(\theta)$ is assumed to be integrable at the point $\theta = 0$. The solution to this boundary-value problem for $\theta^n F_2(\theta)$ is given by ($\mathrm{Im}\,\theta > 0$):

$$\theta^n F_2(\theta) = \frac{Q(\theta)}{\pi \mathrm{i}} \left[\int\limits_{-v_2^{-1}}^{v_1^{-1}} \frac{U_0(\alpha)\alpha^n d\alpha}{Q(\alpha)(\alpha-\theta)} + \mathrm{i} \int\limits_{-\infty}^{-v_2^{-1}} \frac{\Sigma_0(\alpha)\alpha^n d\alpha}{Q(\alpha)(\alpha-\theta)} + \mathrm{i} \int\limits_{v_1^{-1}}^{\infty} \frac{\Sigma_0(\alpha)\alpha^n d\alpha}{Q(\alpha)(\alpha-\theta)} \right], \tag{3.5.57}$$

where $Q(\theta)$ denotes the same function as in (3.5.19).

The boundedness of the function $\theta^{n+1}F_2(\theta)$ when $\theta \to \infty$ leads to the following two conditions for defining the constants:

$$\int\limits_{-v_2^{-1}}^{v_1^{-1}} \frac{U_0(\alpha)\alpha^n}{Q(\alpha)} d\alpha + \mathrm{i} \int\limits_{-\infty}^{-v_2^{-1}} \frac{\Sigma_0(\alpha)\alpha^n d\alpha}{Q(\alpha)} + \mathrm{i} \int\limits_{v_1^{-1}}^{\infty} \frac{\Sigma_0(\alpha)\alpha^n d\alpha}{Q(\alpha)} = 0 \,, \tag{3.5.58}$$

$$\lim_{\theta \to +\infty} \theta^{n+1}\Sigma_0(\theta) = \lim_{\theta \to -\infty} \theta^{n+1}\Sigma_0(\theta) \,. \tag{3.5.59}$$

[Note, in particular, that if we do not fulfill the latter condition, then according to (3.5.57), this will result in the appearance of the logarithmic singularity in $\theta^{n+1}F_2(\theta)$ as $\theta \to \infty$.] Besides, according to the estimate (3.5.4) the solution to the problem for $n < 0$ must have at the points $\theta = -v_2^{-1}$, $\theta = v_1^{-1}$ bounded derivatives of the function $\theta^n F_2(\theta)$ up to the $|n|^{\text{th}}$ order inclusively:

$$\lim_{\theta \to v_1^{-1}} \frac{d^k}{d\theta^k}(\theta^n F_2(\theta)) = \mathrm{O}(1) \,, \qquad \lim_{\theta \to -v_2^{-1}} \frac{d^k}{d\theta^k}(\theta^n F_2(\theta)) = \mathrm{O}(1) \tag{3.5.60}$$

$(k = 1, 2, \ldots, |n|)$.

Thus, the solution to the problem (3.5.1) for the case $n \le 0$ is described by the formulas (3.5.52–54, 57) in which the $2|n|+2$ arbitrary real constants are defined from $2|n|+2$ conditions (3.5.58–60).

In the case $n = 0$, in the presence of symmetry, $v_1 = v_2 = v$, $\sigma^0_{yy}(-x,t) = \sigma^0_{yy}(x,t)$, $u^0_y(-x,t) = u^0_y(x,t)$, we have only one constant $A_0 = -B_0$ which, according to (3.5.53,59) is given by the expression

$$A_0 = -\int_{v^{-1}}^{\infty} \sigma^0_{yy}(1,\eta)d\eta = -x\int_{x/v}^{\infty} \sigma^0_{yy}(x,t)dt\ . \tag{3.5.61}$$

Finally, the solution to the problem for $F_2(\theta)$ is defined explicitly by (3.5.57) where one must put $v_1 = v_2 = v$, $n = 0$:

$$\begin{aligned} U_0(\theta) &= 2c_2^2 x u^0_y(x,t)\ , \\ \Sigma_0(\theta) &= \frac{2\mathrm{i}\sqrt{c_1^{-2}-\theta^2}}{\mu S(\theta^2)}\, x\int_t^{\infty} \sigma^0_{yy}(x,\tau)d\tau \quad (\theta = t/x + \mathrm{i}0)\ . \end{aligned} \tag{3.5.62}$$

For example if a cut $|x| < vt$ at whose sides the stress $\sigma^0_{yy}(x,t) = -pH(t)/(t^2+x^2)$ is given ($u^0_y(x,t) = 0$ for $|x| > vt$) starts propagating from the state of rest, then (3.5.62) give

$$\begin{aligned} U_0(\theta) &\equiv 0\ , \\ \Sigma_0 &= \frac{2p\sqrt{c_1^{-2}-\theta^2}\cot^{-1}\theta}{\mathrm{i}\mu S(\theta^2)} \qquad (0 < \cot^{-1}\theta < \pi)\ , \end{aligned} \tag{3.5.63}$$

and hence, the solution to the boundary-value problem is written in the form

$$F_2(\theta) = \frac{4p\sqrt{\theta^2 - v^{-2}}}{\mu\pi\mathrm{i}}\int_{v^{-1}}^{\infty} \frac{\alpha\sqrt{c_1^{-2}-\alpha^2}\cot^{-1}\alpha\, d\alpha}{S(\alpha^2)\sqrt{\alpha^2 - v^{-2}}(\alpha^2-\theta^2)} \quad (\mathrm{Im}\,\theta > 0)\ . \tag{3.5.64}$$

In particular, for the displacement at the lower side of the cut we get from (3.5.64) the following expression ($\theta = t/x + \mathrm{i}0$):

$$\begin{aligned} u_y(x,0,t) = \mathrm{Re}\, F_2(\theta) &= \frac{2p\sqrt{\theta^2 - v^{-2}}}{\mu\pi\mathrm{i}xc_2^2}\int_{v^{-1}}^{\infty} \frac{\alpha\sqrt{c_1^{-2}-\alpha^2}\cot^{-1}\alpha\, d\alpha}{S(\alpha^2)\sqrt{\alpha^2 - v^{-2}}(\alpha^2-\theta^2)}\ , \\ &(|\theta| > v^{-1})\ , \end{aligned} \tag{3.5.65}$$

where the integral is taken in the sense of the principal value, due to the singularity in the point $\alpha = \theta$. As we see from (3.5.65), the displacement for $t > 0$ is limited at the point $x = 0$:

$$u_y(0,0,t) = \frac{2p\mathrm{i}}{\pi\mu t c_2^2}\int_{v^{-1}}^{\infty} \frac{\alpha\sqrt{c_1^{-2}-\alpha^2}\cot^{-1}\alpha}{S(\alpha^2)\sqrt{\alpha^2 - v^{-2}}}\, d\alpha\ . \tag{3.5.66}$$

(Note that due to the chosen branches of the radical in the formulas (3.5.64–66), we have: $\sqrt{c_j^{-2}-\theta^2} \sim -\theta\mathrm{i}$, $\sqrt{\theta^2-v^{-2}} \sim \theta$ as $|\theta| \to \infty$.)

3.6 Solution of Analogous Mixed Boundary-Value Problems. Wedge-Shaped Punch

Let us now consider the analogous mixed problem (3.5.2) associated with the problem of a punch indentation into an elastic half-space and a solid body impact onto an obstacle,both problems being interesting from the point of view of mechanics.

3.6.1 Solution for the $n \geq 1$-Case

In principle, the solution to the above problem may be obtained similarly to that of (3.5.1), if the function $F_2(\theta)$ entering (3.5.10) is substituted by $F_2^*(\theta) \equiv S(\theta^2)F_2(\theta)/\sqrt{c_1^{-2}-\theta^2}$. Then, assuming for the sake of definiteness that $v_j < c_\mathrm{R}$ ($j = 1,2$) and limiting our analysis by the most interesting case $n \geq 1$, we get from (3.5.10) the following boundary conditions for the function $F_2^*(\theta)$ on the axis $\mathrm{Im}\,\theta = 0$ ($\theta = t/x + \mathrm{i}0$):

$$\begin{aligned}
&\mathrm{Re}\,F_2^*(\theta) = \Sigma_1(\theta) \quad (-v_2^{-1} < \theta < v_1^{-1})\,,\\
&\mathrm{Im}\,F_2^*(\theta) = U_1(\theta) \quad (\theta < v_2^{-1},\ \theta > v_1^{-1})\,,\\
&\Sigma_1(\theta) = \frac{2x}{\mu}\,\frac{\partial^{n-1}\sigma_{yy}^0(x,t)}{\partial t^{n-1}}\,, \quad U_1(\theta) = \frac{2S(\theta^2)xc_2^2}{\mathrm{i}\sqrt{c_1^{-2}-\theta^2}}\,\frac{\partial^n u_y^0(x,t)}{\partial t^n}\,.
\end{aligned} \tag{3.6.1}$$

Let the given function u_y^0 satisfy the condition

$$u_y^0(x,t) = C(t) + \mathrm{O}(x^\varepsilon)\,, \quad \varepsilon > 0,\ x \to 0\,, \tag{3.6.2}$$

and the required functions of stresses – the estimate (3.4.24). Then, if the given functions are smooth enough with respect to time, the expressions $F_2^*(\theta)$, $\Sigma_1(\theta)$, $U_1(\theta)$ are integrable, and the estimates

$$\begin{aligned}
&F_2^*(\theta) = \mathrm{O}(\theta^{-\varepsilon})\,,\\
&U_1(\theta) = \mathrm{O}(\theta^{-\varepsilon})\,, \quad \varepsilon > 0,\ \theta \to \infty
\end{aligned} \tag{3.6.3}$$

are valid. (The estimate (3.6.3) for $U_1(\theta)$ has been obtained under the condition that $u_y^0(x,t)$ is a homogeneous function of the $(n-1)^{\text{th}}$ dimension with respect to t and x, and hence the condition (3.6.2) becomes $u_y^0(x,t) = C_0t^{n-1} + \mathrm{O}(x^\varepsilon)$ as $x \to 0$.)

Furthermore, taking into account the physical considerations for the problems of punches, we shall demand that the displacements in the neighborhood of $x = v_1 t$, $x = -v_2 t$ satisfy the condition (3.5.4) (for $\beta > 1/2$), i.e.,

$$\boldsymbol{u} = \boldsymbol{u}_0(t) + \mathrm{O}(\varepsilon^\beta)\,, \quad \beta > 1/2\,, \quad \varepsilon \to 0 \tag{3.6.4}$$

due to which these lines neither absorb, nor radiate energy. The latter condition results in the requirement that the function $F_2^*(\theta)$ should have singularities no stronger than $\varepsilon^{\beta-n}$ in the neighborhood of the points $\theta = -v_2^{-1}$, $\theta = v_1^{-1} (v_2 \neq 0, v_1 \neq 0)$.

Note also that, similarly to the problems of continuous-profile punches, the constants v_2, v_1 will be assumed unknown and will be sought in the process of solving the problem.

Let us now construct the solution $F_2^*(\theta)$ in a way similar to that of constructing the function $F_2(\theta)$ from (3.5.23). Then, $F_2^*(\theta)$ will be represented in the form

$$F_2^*(\theta) = M(\theta) + \frac{f(\theta)}{[Q(\theta)]^{2n-3}}\,, \tag{3.6.5}$$

where the function $Q(\theta)$ is the same as in (3.5.23), and

$$M(\theta) = \frac{Q(\theta)}{\pi \mathrm{i}} \left[\int\limits_{-v_2^{-1}}^{v_1^{-1}} \frac{\Sigma_1(\alpha)d\alpha}{Q(\alpha)(\alpha-\theta)} + \mathrm{i} \int\limits_{-\infty}^{-v_2^{-1}} \frac{U_1(\alpha)d\alpha}{Q(\alpha)(\alpha-\theta)} + \mathrm{i} \int\limits_{v_1^{-1}}^{\infty} \frac{U_1(\alpha)d\alpha}{Q(\alpha)(\alpha-\theta)} \right] \quad (\mathrm{Im}\,\theta > 0)\,, \tag{3.6.6}$$

where the branches of the radicals were chosen in the same way as in (3.5.20). Then the function $f(\theta)$ which is analytical within the region $\mathrm{Im}\,(\theta) > 0$, must increase as $\theta \to \infty$ no faster than θ^{2n-3}, with $\mathrm{Im}\, f(\theta) = 0$ on the real axis, and the function $f(\theta)$ being integrable at the points $\theta = -v_2^{-1}$, $\theta = v_1^{-1}$.

Then, similarly to the way of obtaining (3.5.22), we get

$$f(\theta) = \sum_{k=0}^{2n-3} A_k \theta^k\,, \tag{3.6.7}$$

where A_k are real constants. For defining the $2n$ constants A_k ($k = 0, 1, \dots, 2n-3$) and $v_j (j = 1, 2)$ we shall use the condition (3.6.3) and the boundary condition-s from the representation (3.5.9) at $m = -1, -2, \dots, -n$, which results in the following equations:

$$A_{2n-3} = \frac{1}{\pi \mathrm{i}} \left[\int\limits_{-v_2^{-1}}^{v_1^{-1}} \frac{\Sigma_1(\alpha)d\alpha}{Q(\alpha)} + \mathrm{i} \int\limits_{-\infty}^{-v_2^{-1}} \frac{U_1(\alpha)d\alpha}{Q(\alpha)} + \mathrm{i} \int\limits_{v_1^{-1}}^{\infty} \frac{U_1(\alpha)d\alpha}{Q(\alpha)} \right]\,, \tag{3.6.8}$$

$$\mathrm{Re}\, L_m^\theta F_2(\theta) = 2c_2^2 x^{1+m} L_{n+m}^t u_y^0(x,t)\,,$$

$$m = -1, -2, \dots, -n \quad (\theta = t/x + \mathrm{i}0,\ \theta < -v_2^{-1},\ \theta > v_1^{-1})\,.$$

Since the latter equation is applicable for both $\theta < -v_2^{-1}$ and $\theta > v_1^{-1}$, it produces $2n$ equations for determining constants, and hence in (3.6.8) we have $2n+1$ conditions.

However, at $m = -1$ one of the two equations (3.6.8) is a consequence of the other one, and there are $2n$ equations left for determining the $2n$ constants. In fact, at $m = -1$ (3.6.8) give a pair of equations

$$\mathrm{Re} \int_0^\theta F_2(\eta)d\eta = 2c_2^2 \frac{\partial^{n-1} u_y^0(x,t)}{\partial t^{n-1}} \quad (\theta > v_1^{-1}),$$

$$-\mathrm{Re} \int_\theta^0 F_2(\eta)d\eta = 2c_2^2 \frac{\partial^{n-1} u_y^0(x,t)}{\partial t^{n-1}} \quad (\theta < -v_2^{-1}),$$

$$\theta = t/x + \mathrm{i}0,$$

one of which is a consequence of the other. This may be shown by applying the Cauchy theorem to the integral of $F_2(\theta)$ along a closed contour within the region $\mathrm{Im}\,\theta > 0$, which passes along the real axis from $-\infty$ to $+\infty$ and along a semicircle of an infinite radius, and using the estimate: $F_2(\theta) = O(\theta^{-1-\varepsilon})$ $\varepsilon > 0$, $\theta \to \infty$. Then, we arrive at the equality

$$\int_{-\infty}^{\infty} F_2(\eta)d\eta = 0.$$

If we use now the equations

$$\mathrm{Re} \int_\theta^\infty F_2(\eta)d\eta = 2c_2^2(n-1)!C_0 - 2c_2^2 \frac{\partial^{n-1} u_y^0(x,t)}{\partial t^{n-1}} \quad (\theta > v_1^{-1}),$$

$$\mathrm{Re} \int_{-\infty}^\theta F_2(\eta)d\eta = -2c_2^2(n-1)!C_0 + 2c_2^2 \frac{\partial^{n-1} u_y^0(x,t)}{\partial t^{n-1}} \quad (\theta < -v_2^{-1}),$$

obtained by integrating the boundary conditions (3.6.1), then we arrive at the desired statement.

As a result, the solution to the problem (3.5.2), satisfying the conditions (3.4.24, 6.4) and $v_j < c_R$, is given, for $n \geq 1$, by the formula ($\mathrm{Im}\,\theta > 0$)

$$F_2^*(\theta) \equiv \frac{F_2(\theta)S(\theta^2)}{\sqrt{c_1^{-2} - \theta^2}} = M(\theta) + \frac{\sum_{k=0}^{2n-3} A_k \theta^k}{[Q(\theta)]^{2n-3}}, \tag{3.6.9}$$

where $Q(\theta) = [(\theta + v_2^{-1})(\theta - v_1^{-1})]^{1/2}$, the function $M(\theta)$ is represented by the expression (3.6.6), and the unknown real constants A_k, v_1, v_2 are defined from the $2n$ conditions (3.6.8), since at $m = -1$ it is enough to take the condition for $\theta > v_1^{-1}$ only.

3.6.2 Indentation of a Wedge with a Sub-Rayleigh Contact Speed

Let us apply the above results to considering the contact problem for a wedge. Let a wedge-shaped punch with the vertex angle 2α be indented into the half-plane $y \leq 0$ at a constant speed v_0 (Fig. 3.16). The wedge is assumed to be blunt, i.e., the vertex angle $\pi - 2\alpha$ is small, and, generally speaking, asymmetric with respect to the $0y$-axis. The angle between the wedge bisectrix and the $0y$-axis will be denoted by β. At the instant $t = 0$ the vertex of the wedge coincides with the origin of coordinates. Friction between the wedge and the elastic half-plane is absent, and zero initial conditions are assumed. Then, the problem is characterized by the following boundary conditions ($y = 0$):

$$
\begin{aligned}
&u_y = x\cot(\alpha - \beta) - v_0 t \quad (0 < x < v_1 t)\,,\\
&u_y = -x\cot(\alpha + \beta) - v_0 t \quad (-v_2 t < x < 0)\,,\\
&\sigma_{yy} = 0 \quad (x < -v_2 t,\ x > v_1 t)\,,\\
&\sigma_{xy} = 0 \quad (-\infty < x < \infty)\,.
\end{aligned}
\tag{3.6.10}
$$

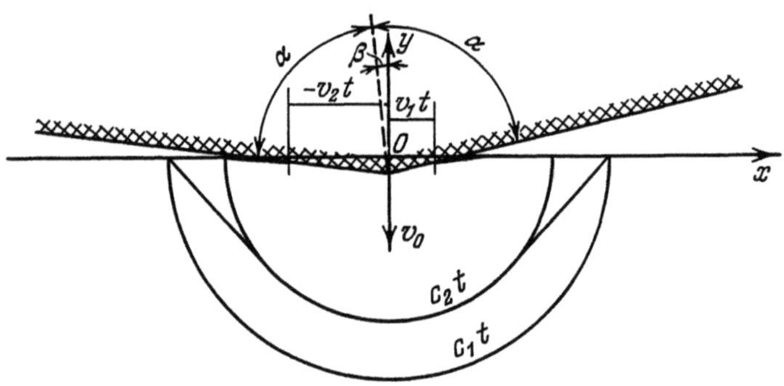

Fig. 3.16. Indentation of a wedge-shaped punch

In the case under consideration $n = 2$, and we get from (3.6.1, 9)

$$
\begin{aligned}
&\Sigma_1(\theta) = U_1(\theta) = 0\,,\\
&\frac{F_2(\theta)S(\theta^2)}{\sqrt{c_1^{-2} - \theta^2}} = \frac{A_0 + A_1\theta}{\sqrt{(\theta + v_2^{-1})(\theta - v_1^{-1})}} \quad (\operatorname{Im}\theta > 0)\,.
\end{aligned}
\tag{3.6.11}
$$

According to (3.6.8) the real constants A_0, A_1, v_1, v_2 are defined from the following conditions

$$A_1 = 0\,, \quad \mathrm{Re}\int_0^\theta F_2(\eta)d\eta = -2c_2^2 v_0 \quad (\theta > v_1^{-1})\,,$$

$$\mathrm{Re}\int_0^\theta (\theta-\eta)F_2(\eta)d\eta = \begin{cases} 2c_2^2[\cot(\alpha-\beta) - v_0\theta] & (\theta > v_1^{-1})\,, \\ -2c_2^2[\cot(\alpha+\beta) + v_0\theta] & (\theta < -v_2^{-1})\,. \end{cases} \tag{3.6.12}$$

Taking into account analyticity of the integrands within the region $\mathrm{Im}\,\eta > 0$, as well as the fact that the contours of integration with respect to η circumvent the singular points of the integrands from above, we get from (3.6.12) the following relationships for determining the constants A_0, v_1, v_2:

$$A_0 = -4c_2^2(c_2^{-2} - c_1^{-2})[\cot(\alpha+\beta) + \cot(\alpha-\beta)]/\pi\,,$$

$$A_0\mathrm{Re}\int_{c_1^{-1}}^{v_1^{-1}} \varphi(\eta)d\eta = -2c_2^2 v_0\,,$$

$$A_0\mathrm{Re}\int_{c_1^{-1}}^{v_1^{-1}} \varphi(\eta)\eta d\eta = -2c_2^2\cot(\alpha-\beta)\,, \tag{3.6.13}$$

$$\varphi(\eta) = \frac{\sqrt{c_1^{-2} - \eta^2}}{S(\eta^2)\sqrt{(\eta + v_2^{-1})(\eta - v_1^{-1})}}\,,$$

where the equality

$$\mathrm{Re}\int_{c_1^{-1}}^{v_1^{-1}} \varphi(\eta)\eta d\eta + \mathrm{Re}\int_{-v_2^{-1}}^{-c_1^{-1}} \varphi(\eta)\eta d\eta$$

$$= \mathrm{Re}\int_{-\infty}^{\infty} \varphi(\eta)\eta d\eta = \frac{\pi}{2(c_2^{-2} - c_1^{-2})} \tag{3.6.14}$$

has been used.

According to (3.6.13), A_0 is defined independently of v_1, v_2.

By employing (3.5.9, 6.11), we obtain the stress distribution under the punch, the force P acting upon the punch, the moment M of the force with respect to the point $x = y = 0$, and the coordinates x_P, 0 of the point of application:

$$\sigma_{yy}(x,0,t) = \frac{A_0\mu}{2}\mathrm{Re}\int_0^\theta \frac{d\theta}{\sqrt{(\theta + v_2^{-1})(\theta - v_1^{-1})}}$$

$$= \frac{A_0\mu}{2}\ln\left|\frac{2tv_1v_2 + x(v_1 - v_2) + 2\sqrt{(v_1t - x)(v_2t + x)v_1v_2}}{(v_1+v_2)x}\right|$$

$$(-v_2t \le x \le v_1t)\,, \tag{3.6.15}$$

$$P = -\int_{-v_2 t}^{v_1 t} \sigma_{yy}(x,0,t)dx = -\frac{\pi t A_0}{2}\mu\sqrt{v_1 v_2}\,,$$

$$M = -\int_{-v_2 t}^{v_1 t} \sigma_{yy}(x,0,t)dx = \frac{\pi}{8}t^2 A_0 \mu\sqrt{v_1 v_2}(v_2 - v_1)\,,$$

$$x_P = \frac{t}{4}(v_1 - v_2)\,,$$

where the radical is assumed to be arithmetic.

In the special symmetric case when $\beta = 0$, $v_1 = v_2 = v$, (3.6.15) become

$$\begin{aligned}\sigma_{yy}(x,0,t) &= \frac{A_0\mu}{2}\ln\left|\frac{vt + \sqrt{(vt)^2 - x^2}}{x}\right| \quad (|x| \leq vt)\,,\\ P &= -\frac{\pi}{2}vA_0\mu t\,, \quad M = 0\,, \quad x_P = 0\,,\end{aligned} \tag{3.6.16}$$

where the constants A_0, v are defined by the equalitites

$$A_0 = -\frac{8}{\pi}\cot\alpha(1 - c_2^2/c_1^2)\,, \tag{3.6.17}$$

$$\int_0^\infty \frac{\sqrt{c_1^{-2} + x^2}\,dx}{\sqrt{x^2 + v^{-2}}[(c_2^{-2} + 2x^2)^2 - 4x^2\sqrt{(c_1^{-2} + x^2)(c_2^{-2} + x^2)}]} = \frac{\pi v_0 \tan\alpha}{4(c_2^{-2} - c_1^{-2})}\,, \tag{3.6.18}$$

which follow from (3.6.13) at $v_1 = v_2 = v$ and $\beta = 0$, with the radicals entering (3.6.18) assumed to be arithmetical. The final expression in (3.6.18) may be easily obtained by deforming the contour of integration in the second of the equations (3.6.13) into the imaginary axis and noting that the third equation is equivalent to the first one, because, according to (3.6.14) the equality

$$\mathrm{Re}\int_{c_1^{-1}}^{v_1^{-1}} \varphi(\eta)\eta d\eta = \frac{\pi}{4(c_2^{-2} - c_1^{-2})}$$

holds at $v_1 = v_2$. The results (3.6.16–18) coincide with those of [3.33, 34, 38, 63]. Also, according to (3.6.16), the stress distribution under a wedge-shaped punch coincides, within the framework of the elastodynamic problem under consideration, with the static one, if in the case of the static problem the contact area width is assumed to be equal to $2vt$ [3.62].

It should be pointed out that the problem for a wedge in the special symmetric case was originally solved by Kostrov [3.29] who used the Smirnov–Sobolev method. However, Kostrov assumed that the contact boundary speed v was equal to $v_0 \tan\alpha$, although v must be found by solving the problem. Later, Kostrov's approach was used by Robinson and Thompson [3.38] for solving the symmetric problem with v being sought within the entire range $0 < v < \infty$.

Note that if we assume that the above problem involves the presence of singularities of the order of $\varepsilon^{-1/2}$ in the neighborhood of the contact zone boundaries, $x = v_1 t, x = -v_2 t$ (i.e. the existence of finite energy fluxes at the points $x = v_1 t, x = -v_2 t$), then an additional term will appear in the solution [3.29, 38], and the solution will become ambiguous. In order to single out a unique solution, one will have to introduce an additional condition. In the above solution, it was the absence of the sources of energy absorbtion/radiation at the points of the contact area between the wedge and the continuum. Such was Kostrov's motivation for his approach in solving the wedge problem. However, if we allow for the existence of finite energy-flux sources at the points $x = -v_2 t, x = v_1 t$, then a different additional condition [3.38] may be employed for ensuring the uniqueness of the problems solution, which has a clear geometrical meaning and guarantees fulfillment of boundary conditions at the free surface outside the wedge. This condition may be formulated in the following way: everywhere outside the contact region, the deformed surface of a continuum is supposed to lie below the wedge's surface. The additional stress terms containing a singularity of the order of $\varepsilon^{-1/2}$ do not comply with the condition (for $v_j < c_R, j = 1, 2$). In fact, since at the points of the contact zone boundary, $x = -v_2 t, z = v_1 t$, the velocities and stresses possess the same singularity, and the problem is self-similar, the tangent to the free surface at the contact boundary becomes vertical, and in the case under consideration, $v_j < c_R$, the continuum's surface looks like shown in Fig. 3.17a (while in the absence of the terms containing the $\varepsilon^{-1/2}$-singularity the surface of the continuum at the contact boundary at $v_j < c_R$ looks like shown in Fig. 3.17b). Hence, the surface of the deformed continuum lies above the wedge's surface in the neighborhood of the points $x = -v_2 t, x = v_1 t$, and we must discard the additional terms responsible for the $\varepsilon^{-1/2}$ singularity in the expressions for stresses.

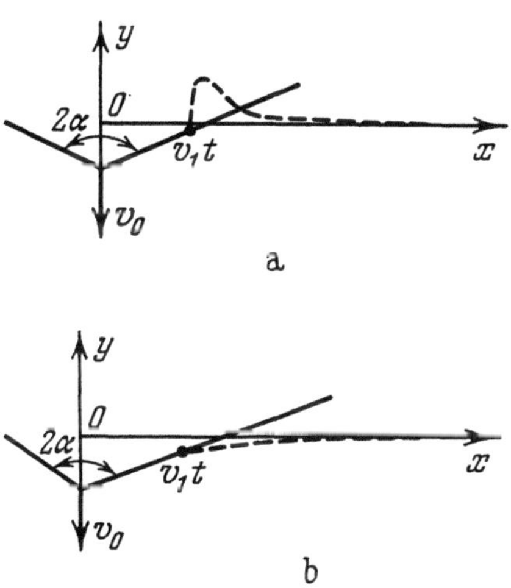

Fig. 3.17a,b. The free surface in the vicinity of a wedge: (**a**) in the case $v_1 < c_R$, taking into accunt the term with the $\varepsilon^{-1/2}$-singularity; (**b**) in the case $v_1 < c_R$, ignoring the term with the $\varepsilon^{-1/2}$-singularity

Note that the solutions to the above mixed problems were obtained for the $v_j < c_R$ velocity range ($j = 1, 2$) which is most interesting from the viewpoint of mechanics. As shown above, the considered problems for this velocity range reduce to the special case of the Riemann–Hilbert problem subject to certain restraints onto the solution's behavior in the vicinity of the points $\theta = -v_2^{-1}$, $\theta = v_1^{-1}$. The required solution to the problem was obtained using the Keldysh–Sedov formula.

3.6.3 Indentation of a Wedge with Super-Rayleigh Contact Speed

The mixed problems may be considered for the case of $v_j > c_R$ too. However, within this velocity range they generally reduce to the boundary-value Riemann-Hilbert problem of defining an analytical function in the upper half-plane, while the solution to the problem must satisfy certain conditions in the neighborhood of the points $\theta = -v_2^{-1}$, $\theta = v_1^{-1}$. Such a problem may be solved in a manner similar to that of the case $v_j < c_R$ by substituting the Keldysh–Sedov formula by the corresponding formula describing the general solution to the Riemann–Hilbert problem. Let us demonstrate it by solving the symmetric problem for a wedge-shaped punch:

$$\begin{aligned} &u_y = |x| \cot\alpha - v_0 t \quad (|x| < vt,\ y = 0,\ v > c_R)\,, \\ &\sigma_{yy} = 0 \quad (|x| > vt,\ y = 0)\,, \\ &\sigma_{xy} = 0 \quad (-\infty < x < \infty,\ y = 0)\,. \end{aligned} \tag{3.6.19}$$

In this case, $n = 2$, and the following boundary conditions are obtained from (3.5.10) for the function $F_2(\theta)$ (which is analytical within the region $\operatorname{Im}\theta > 0$) on the axis $\operatorname{Im}\theta = 0$:

$$\begin{aligned} &\operatorname{Re} F_2(\theta) = 0 \quad (|\theta| > v^{-1})\,, \\ &\operatorname{Re}\left[\frac{S(\theta^2)F_2(\theta)}{\sqrt{c_1^{-2} - \theta^2}}\right] = 0 \quad (|\theta| < v^{-1})\,. \end{aligned} \tag{3.6.20}$$

The function $F_2(\theta)$ is bounded on the axis $\operatorname{Im}\theta = 0$ outside the points $|\theta| = v^{-1}$, and according to (3.6.3, 4) we have

$$\begin{aligned} &F_2(\theta) = O(\theta^{-1-\varepsilon})\,, \quad \theta \to \infty \quad (\operatorname{Im}\theta \geq 0,\ \varepsilon > 0)\,, \\ &F_2(\theta) = O[(|\theta| - v^{-1})^{\beta-2}]\,, \quad \theta \to \pm v^{-1} \quad (\beta > 1/2)\,. \end{aligned} \tag{3.6.21}$$

The latter condition means that neither radiation, nor absorption of energy takes place at the points $x = \pm vt$.

Let us divide the range $v > c_R$ into three parts:

$$(1)\ c_R < v < c_2\,, \quad (2)\ c_2 < v < c_1\,, \quad (3)\ v > c_1\,.$$

Within the first one, $c_R < v < c_2$, the solution will be sought in the same way as in the case $v < c_R$, if one follows the conditions (3.6.21). Then, taking into account the problem's symmetry, we get from (3.6.11)

$$\frac{F_2(\theta)S(\theta^2)}{\sqrt{c_1^{-2}-\theta^2}} = \frac{A}{\sqrt{\theta^2 - v^{-2}}} \quad (\mathrm{Im}\, A = 0) \, . \tag{3.6.22}$$

However, if we now try to satisfy the boundary conditions by choosing the constant A, it probes to be impossible. In fact, from (3.5.9, 6.10) and with the help of (3.6.22) we get for $\theta > v^{-1}$ $(\theta = t/x + \mathrm{i}0)$:

$$\mathrm{Re} \int_0^\theta F_2(\eta)d\eta = A\,\mathrm{Re} \int_0^\theta \frac{\sqrt{c_1^{-2}-\eta^2}}{S(\eta^2)} \frac{d\eta}{\sqrt{\eta^2 - v^{-2}}} = -2c_2^2 v_0 \, .$$

It is obvious that the real part of the latter integral is not constant for $\theta > v^{-1}$, since its values at $\theta < c_\mathrm{R}^{-1}$ and $\theta > c_\mathrm{R}^{-1}$ differ by the semiresidue at the point $\theta = c_\mathrm{R}^{-1}$.

Consequently, the problem has no solution in the interval $c_\mathrm{R} < v < c_2$ subject to (3.6.21). In order to obtain a solution in the interval, we shall assume that finite energy fluxes do exist at the points $x = \pm vt$. In other words we shall substitute the second of the conditions (3.6.21) by the condition

$$F_2(\theta) = \mathrm{O}[(|\theta| - v^{-1})^{-3/2}] \, , \quad \theta \to \pm v^{-1} \, . \tag{3.6.23}$$

Then, taking into account the problem's symmetry, we get, instead of (3.6.22),

$$\frac{F_2(\theta)S(\theta^2)}{\sqrt{c_1^{-2}-\theta^2}} = \frac{A_1 + A_0\theta^2}{(\theta^2 - v^{-2})^{3/2}} \quad (\mathrm{Im}\, A_j = 0, \ j = 0, 1) \, . \tag{3.6.24}$$

For determining the constants from the boundary conditions we get the following expressions

$$\mathrm{Re} \int_0^\theta F_2(\eta)d\eta = -2c_2^2 v_0 \quad (\theta > v^{-1}) \, , \tag{3.6.25}$$

$$\mathrm{Re} \int_0^\theta (\theta - \eta)F_2(\eta)d\eta = 2c_2^2[\cot\alpha - v_0\theta] \quad (\theta > v^{-1}) \tag{3.6.26}$$

$(\theta = t/x + \mathrm{i}0)$,

whose left-hand sides, generally speaking, suffer discontinuities of the first kind while passing through integrand's pole $\theta = c_\mathrm{R}^{-1}$. However, the equalitites (3.6.25, 26) cannot be satisfied without eliminating the discontinuity; in other words, the function $F_2(\theta)$ must have no pole at the point $\theta = c_\mathrm{R}^{-1}$. This generates the additional equality $A_1 = -A_0 c_\mathrm{R}^{-2}$ which, together with the two expressions (3.6.25, 26), allows us to calculate all the three unknown constants A_0, A_1, v.

Hence, we deduce that A_0 is described by the formula (3.6.17) as before, and v may be obtained from the equality

$$\int_0^\infty \frac{(x^2 + c_R^{-2}\sqrt{c_1^{-2} + x^2}\,dx}{(x^2 + v^{-2})^{3/2}[(c_2^{-2} + 2x^2)^2 - 4x^2\sqrt{(c_1^{-2} + x^2)(c_2^{-2} - x^2)}]}$$
$$= \frac{\pi v_0 \tan\alpha}{4(c_2^{-2} - c_1^{-2})}, \tag{3.6.27}$$

where all the radicals are assumed to be arithmetical.

As a result, we get

$$\frac{F_2(\theta)S(\theta^2)}{\sqrt{c_1^{-2} - \theta^2}} = \frac{A_0(\theta^2 - c_R^{-2})}{(\theta^2 - v^{-2})^{3/2}} \tag{3.6.28}$$

and the stress distribution under the punch is given by

$$\sigma_{yy}(x, 0, t) = \frac{A_0}{2}\mu \operatorname{Re} \int_0^{t/x} \frac{(\eta^2 - c_R^{-2})d\eta}{(\eta^2 - v^{-2})^{3/2}}$$
$$= \frac{A_0}{2}\mu \left[\ln \frac{vt + \sqrt{v^2t^2 - x^2}}{|x|} + \frac{(v^2 - c_R^2)vt}{c_R^2\sqrt{v^2t^2 - x^2}}\right]$$
$$(|x| < vt,\ \sqrt{v^2t^2 - x^2} > 0)\,. \tag{3.6.29}$$

From the above solution it follows that in the interval $c_R < v < c_2$ the stresses and the velocities at the points $x = \pm vt$ possess singularities of the order of $-1/2$.

Below we shall try to explain the physical meaning of the singularity; however, presently we note that according to the solution $\partial u_y/\partial x \to -\infty$ as $x \to vt + 0$, and hence the free surface near the contact boundary looks like shown in Fig. 3.18 (the tangent to the free surface becomes vertical as $x \to vt + 0$).

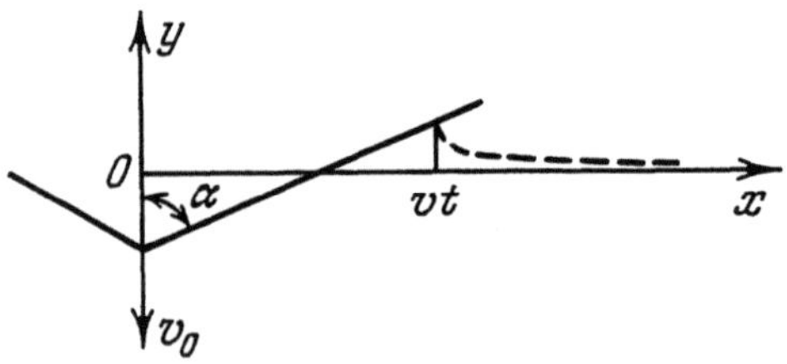

Fig. 3.18. The free surface in the vicinity of a wedge in the case $c_R < v < c_2$

In the second interval, $c_2 < v < c_1$, the boundary conditions (3.6.20) may be conveniently presented in the form ($\operatorname{Im}\theta = 0$):

$$\operatorname{Im} F_3(\theta) = 0 \quad (|\theta| > v^{-1},\ |\theta| < c_1^{-1}),$$
$$\operatorname{Im}[F_3(\theta)S(\theta^2)] = 0 \quad (c_1^{-1} < |\theta| < v^{-1}), \tag{3.6.30}$$

where

$$F_3(\theta) = \frac{F_2(\theta)\sqrt{\theta^2 - v^{-2}}}{\sqrt{c_1^{-2} - \theta^2}} , \tag{3.6.31}$$

and this function, being analytical in the region $\operatorname{Im}\theta > 0$ and bounded on the real axis outside the points $\theta = \pm v^{-1}$, must satisfy, according to (3.6.21), the conditions

$$\begin{aligned} &F_3(\theta) = O(\theta^{-1-\varepsilon}) , \quad \theta \to \infty \quad (\operatorname{Im}\theta \geq 0,\ \varepsilon > 0) , \\ &F_3(\theta) = O[(|\theta| - v^{-1})^{-1+\delta}] , \quad \theta \to \pm v^{-1} \quad (\delta > 0) . \end{aligned} \tag{3.6.32}$$

By representing $S(\theta^2)$ in the form of a sum with real and imaginary parts, the boundary-value problem subject to the conditions (3.6.30) may be reduced to the homogeneous Riemann–Hilbert problem for the upper half-plane. However, in our opinion, it is more convenient to solve the problem by reducing it to the Hilbert problem [3.51]. In order to realize this idea, let us use the upper condition (3.6.30) to extend $F_3(\theta)$ analytically into the lower half-plane across the intervals $|\theta| > v^{-1}$, $|\theta| < c_1^{-1}$ by employing the formula $F_3(\bar{\theta}) = \overline{F_3(\theta)}$. Then, we arrive at the following problem of conjugation: find a function $F_3(\theta)$, analytical throughout the complex plane θ outside the real axis intervals $(-v^{-1}, -c_1^{-1})$ and (c_1^{-1}, v^{-1}) at whose upper (+) and lower (−) sides the limiting values of the function $F_3(\theta)$ are related by

$$F_3^+(\theta) = G(\theta)F_3^-(\theta) , \tag{3.6.33}$$

where

$$G(\theta) = \frac{\overline{S(\theta^2)}}{S(\theta^2)} = \mathrm{e}^{2\mathrm{i}\arg\overline{S(\theta^2)}} \quad (\theta = t/x + \mathrm{i}0) .$$

The solution to this homogeneous problem subject to the conditions (3.6.32) may be written by using the well-known formulas for the case of unclosed contours [3.51]:

$$\begin{aligned} &F_3(\theta) = \frac{A\theta + B}{\theta^2 - v^{-2}}\,\mathrm{e}^{\Phi(\theta)} , \\ &\Phi(\theta) = \frac{1}{2\pi\mathrm{i}}\left[\int\limits_{-v^{-1}}^{-c_1^{-1}} \frac{\ln G(t)dt}{t-\theta} + \int\limits_{c_1^{-1}}^{v^{-1}} \frac{\ln G(t)dt}{t-\theta}\right] , \end{aligned} \tag{3.6.34}$$

where the following branches of the function $\ln G(t)$ are employed

$$\begin{aligned} &-\pi < -\mathrm{i}\ln G(t) < 0 \quad (-v^{-1} < t < c_1^{-1}) , \\ &0 < \mathrm{i}\ln C(t) < \pi \quad (c_1^{-1} < t < v^{-1}) . \end{aligned}$$

By transforming the integrals entering the power of the exponent in (3.6.34) in the complex plane, putting $A = 0$ (due to the symmetry) and taking into account (3.6.31), we finally arrive at the following expression for $F_2(\theta)$:

$$F_2(\theta) = \frac{2B(c_1^{-2} - c_2^{-2})(\theta^2 - c_R^{-2})\sqrt{c_1^{-2} - \theta^2}}{(\theta^2 - v^{-2})^{3/2} S(\theta^2)} e^{N(\theta)} ,$$

$$N(\theta) = \frac{1}{\pi} \int_{v^{-2}}^{c_2^{-2}} \frac{\operatorname{Im} \ln S(t) dt}{t - \theta^2} , \tag{3.6.35}$$

where, as it was done above, the branches should be chosen subject to the condition that $\operatorname{Im} \theta \geq 0$ (and hence, $\operatorname{Im} \ln S(t) < 0$).

In order to find the unknown constants B and v from the boundary conditions, (3.6.34) must be substituted into the expressions (3.6.25,26). Then, we get $B = A_0/2(c_1^{-2} - c_2^{-2})$ where A_0 is still described by (3.6.17), and v is defined by

$$\int_0^\infty \frac{(x^2 + c_R^{-2})\sqrt{c_1^{-2} + x^2} \exp\left\{\frac{1}{\pi} \int_{v^{-2}}^{c_2^{-2}} \frac{\operatorname{Im} \ln S(t) dt}{t + x^2}\right\} dx}{(x^2 + v^{-2})^{3/2}[(c_2^{-2} + 2x^2)^2 - 4x^2\sqrt{(c_1^{-2} + x^2)(c_2^{-2} + x^2)}]}$$
$$= \frac{\pi v_0 \tan \alpha}{4(c_2^{-2} - c_1^{-2})} . \tag{3.6.36}$$

Note that as needed in the interval under consideration $c_2 < v < c_1$ the stresses and velocities near the edges of the wedge-shaped punch, possess singularities which are weaker than $\varepsilon^{-1/2}$, i.e., no energy absorption/radiation takes place at the points $x = \pm vt$. In fact, evaluation of the integral in the power of the exponent in (3.6.35) as $\theta \to \pm v^{-1}$, gives

$$F_2(\theta) = \mathrm{O}[(|\theta| - v^{-1})^{-3/2+\alpha}] ,$$
$$\alpha = -\frac{1}{\pi} \arg S(v^{-2}) \quad (0 < \alpha < 1/2) \tag{3.6.37}$$

and hence, when $\theta \to \pm v^{-1}$ the stresses and velocities possess singularities of the order of $(-1/2+\alpha)$. The expression for $\partial u_y/\partial x$ when $x \to \pm vt$, has a singularity of the same order; note also that $\partial u_y/\partial x \to -\infty$ as $x \to vt + 0$, and hence the shape of the free surface at the contact edge is similar to that of the free surface in the case of $c_R < v < c_2$ shown in Fig. 3.18.

Finally, in the third case when $v > c_1$, the formulas (3.6.20) may be written in the form ($\operatorname{Im} \theta = 0$)

$$\operatorname{Re} F_2(\theta) = 0 \quad (|\theta| > v^{-1},\ |\theta| < v^{-1}) , \tag{3.6.38}$$

and hence, we are dealing with the Schwarz problem. The function $F_2(\theta)$ $(\theta^2 - v^{-2})$ satisfies also the Schwarz problem's boundary conditions, and, taking into account the conditions (3.6.21), is integrable at the points $\theta = \pm v^{-1}$; it may grow slower than θ as $|\theta| \to \infty$. Consequently,

$$F_2(\theta)(\theta^2 - v^{-2}) = A\,\mathrm{i} \quad (\operatorname{Im} A = 0) . \tag{3.6.39}$$

From the conditions (3.6.25, 26) we find with the help of (3.6.38):

$$A = -\frac{4c_2^2 v_0}{\pi v}, \quad v = v_0 \tan\alpha , \tag{3.6.40}$$

and (3.6.39) becomes

$$F_2(\theta) = \frac{4c_2^2 \cot\alpha}{\pi \mathrm{i}(\theta^2 - v_0^{-2}\cot^2\alpha)} . \tag{3.6.41}$$

In this case the free surface proves to be undisturbed ahead of the punch. The displacements, velocities and stresses are given by (3.5.9), and, according to (3.6.41), the velocities and stresses are bounded near the edges of the punch.

Note that the above formulas (3.6.28, 35, 41) for the function $F_2(\theta)$ coincide with the known ones, as well as the corresponding expressions for the stresses [3.38].

For the force $P(t)$ exerted by the continuum on a wedge-shaped punch, we get (for $0 < v < \infty$)

$$\begin{aligned} P(t) &= -\int_{-vt}^{vt} \sigma_{yy}(x,0,t)dx \\ &= -t\left[\int_{-\infty}^{-v^{-1}} \sigma_{yy}(1,0,\theta)\frac{d\theta}{\theta^2} + \int_{v^{-1}}^{\infty} \sigma_{yy}(1,0,\theta)\frac{d\theta}{\theta^2}\right] . \end{aligned}$$

However, due to (3.5.10)

$$\sigma_{yy}(1,0,\theta) = \frac{\mu}{2}\,\mathrm{Re}\int_0^\theta \frac{S(\eta^2)F_2(\eta)}{\sqrt{c_1^{-2} - \eta^2}}\,d\eta \quad (\theta = t/x + \mathrm{i}0,\ \mathrm{Im}\,\eta \geq 0) .$$

Finally, we get

$$\begin{aligned} P(t) &= -\frac{t\mu}{2}\,\mathrm{Re}\left[\int_{-\infty+0\mathrm{i}}^{\infty+0\mathrm{i}} \frac{d\theta}{\theta^2}\int_0^\theta \frac{S(\eta^2)F_2(\eta)}{\sqrt{c_1^{-2} - \eta^2}}\,d\eta - \int_{-v^{-1}+0\mathrm{i}}^{v^{-1}+0\mathrm{i}} \frac{d\theta}{\theta^2}\int_0^\theta \frac{S(\eta^2)F_2(\eta)}{\sqrt{c_1^{-2} - \eta^2}}\,d\eta\right] \\ &= -\frac{\pi t\mu \mathrm{i}}{2}\left(\frac{F_2(\theta)S(\theta^2)}{\sqrt{c_1^{-2} - \theta^2}}\right)\Bigg|_{\theta=+0\mathrm{i}} \end{aligned} \tag{3.6.42}$$

Here the integral with the limits $-\infty + 0\mathrm{i}$, $+\infty + 0\mathrm{i}$ vanishes according to the Cauchy theorem, the value of the expression between the parentheses is computed at $\theta = -0\mathrm{i}$ from (3.6.11, 28, 35, 41) depending on the interval of v. Figure 3.19 shows the results of using the above formulas for calculating the nondimensional force, $P^* = 2P/(|A_0|\mu c_2 \pi t)$, and the nondimensional velocity, v/c_2, as functions of the parameter $(2v_0 c_2^{-1}/|A_0|)$ where A_0 is given by (3.6.17), and $c_2^2/c_1^2 = 0.3$.

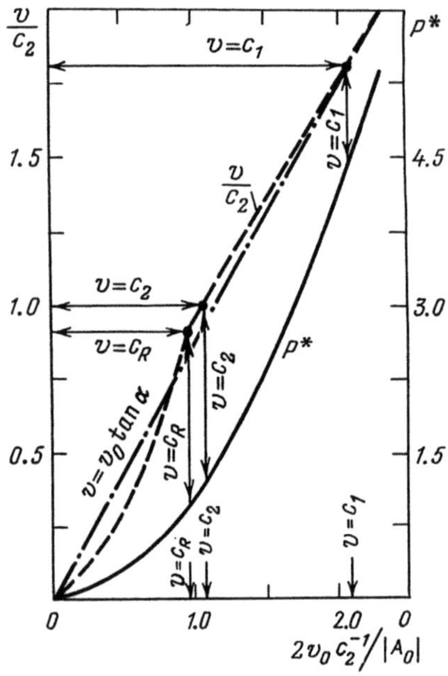

Fig. 3.19. The force acting upon a wedge, and the contact boundary velocity versus the speed of indentation of a wedge

3.6.4 Solution Singularities at the Edges of the Contact Region

Let us, finally, consider the stress and velocity singularities at the edges of a punch, $x = \pm vt$, in the interval $c_R < v < c_2$. Since according to formulas (3.6.27,28), these singularities are of the order of $-1/2$, then finite-energy fluxes must be present at the points $x = \pm vt$. It may be shown that the energy is absorbed in the singular lines $x = \pm vt$.

In fact, in accordance with the derivation of the formula (3.13) from the work [3.64], it is easily seen that in the case under consideration, (2.3.1) furnishes the following expression for the energy flux I (per unit length, since the problem is plane-strain) at the edge of a punch, $x = \pm vt$,

$$I = -\frac{\pi k_1^2 \sqrt{v^{-2} - c_1^{-2}}}{2\mu c_2^2 S(v^{-2})} < 0 ,$$

where k_1 is taken from the relationship $\sigma_{yy} \sim k_1[2(vt - x)]^{-1/2}$ as $x \to vt - 0$, and it is supposed that $\sqrt{v^{-2} - c_1^{-2}} > 0$, $c_R < v < v_2$.

The question arises: What is the energy near the punch's edges converted to? To answer this question, let us, first, consider the case of penetration of a wedge-shaped punch into an incompressible fluid. (Note that the solution for the case may be obtained from the above solution for an elastic continuum for $c_2 < v < c_1$ by means of two sequential limiting passages: first, $\mu \to 0$ (an acoustic medium), and then $c_1 \to \infty$.) As known, see, e.g. [3.65], in the framework of the linear formulation of the problem of a wedge penetration into incompressible fluid, the pressure and the velocity at the points $x = \pm vt$ also possess singularities of the order of $-1/2$, and the shape of the free surface near the punch's edges proves to be analogous to that of an elastic medium at $c_R < v < c_2$ (Fig. 3.18). In order to compute the energy flux I within an incompressible fluid at the points $x = \pm vt$, let us use the known velocity potential φ, see, e.g. [3.65], which has the following asymptotics in the vicinity of $x = vt$:

$$\varphi \sim v_0\sqrt{2vtr}\,\sin\frac{\theta}{2}\,, \quad \frac{\partial\varphi}{\partial t} \sim vv_0\sqrt{\frac{vt}{2r}}\,\sin\frac{\theta}{2}\,, \quad r \to 0\,, \tag{3.6.43}$$

where $x - vt = r\cos(\theta)$, $y = r\sin\theta$, $-\pi \le \theta \le 0$, $\dot{\boldsymbol{u}} = \operatorname{grad}\varphi$. Then, according to (2.3.1), we get for the energy flux (per unit length) at the punch's edge, $x = vt$,

$$\begin{aligned} I &= -\lim_{r\to 0}\int_{-\pi}^{0}\left\{\varrho\frac{\partial\varphi}{\partial r}\frac{\partial\varphi}{\partial t} + \frac{v\varrho}{2}\left[\left(\frac{\partial\varphi}{\partial r}\right)^2 + \left(\frac{\partial\varphi}{r\partial\theta}\right)^2\right]\cos\theta\right\} r\,d\theta \\ &= -\frac{\pi}{4}v_0^2v^2t\varrho\,. \end{aligned} \tag{3.6.44}$$

Consequently, in the case of incompressible fluid, energy is absorbed at the punch's edges, $x = \pm vt$, just as in the case of an elastic medium at $c_R < v < c_2$.

However, if we next use the results of studying the problem of penetration of a symmetric wedge with the vertex angle 2α into incompressible fluid, then according to [3.65], the energy T of a spray jet AB (Fig. 3.20) tends to $\pi\varrho v^2v_0^2t^2/8$ as $\alpha \to \pi/2$ and hence the energy flux I carried away by the spray jet, is equal to

$$I = -\frac{dT}{dt} = -\frac{\pi}{4}\,\varrho v^2 v_0^2 t\,. \tag{3.6.45}$$

Thus, since the energy flux (3.6.44) calculated in the framework of the linear problem coincides with that for the nonlinear problem, (3.6.45), it may be assumed that the energy spent near the edges, $x = \pm vt$, in the case of the linear problem, is equal to the energy carried away by the spray jets generated during penetration of the wedge into a fluid.

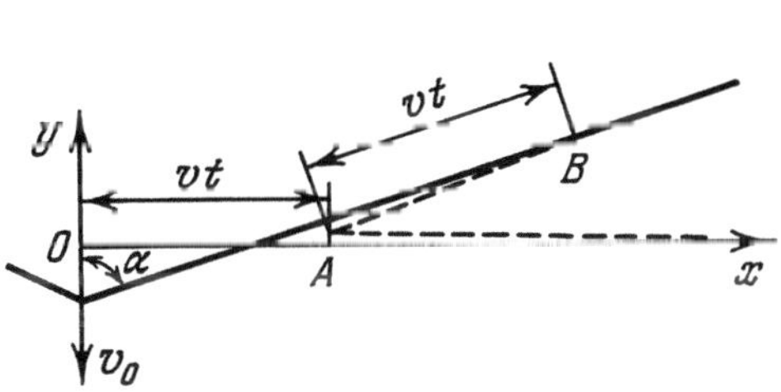

Fig. 3.20. The spray jet, AB, due to a wedge penetration into incompressible fluid

An analogous explanation may be given to the phenomenon of energy absorption in the singular lines $x - \pm vt$ in the framework of the elastic problem at $c_R < v < c_2$; namely, in the case of penetration of a wedge-shaped punch into an elastic half-plane, its edges moving with a super-Rayleigh speed, "cut off" a thin layer of the medium which carries away the energy. (Here we deal with a zero mass flux; however, the energy flux at the edges, $x = \pm vt$, is nonzero and finite.) Thus, for the elastic problem under consideration, the energy spent near the punch's edges, $x = \pm vt$, is the very energy which is carried away by the medium's particles cut off by the moving edges of a punch.

3.7 Interrelation Between Three- and Two-Dimensional Problems

As shown by Smirnov and Sobolev in [3.66], the Principle of Superposition may be used for composing new solutions of two-dimensional solutions of the wave equation, and thus for analyzing wave equations in the general three-dimensional case. The Smirnov–Sobolev formula relating the solution to the three-dimensional wave equation, $U(x, y, z, t)$, to that of the two-dimensional one, $u(x, z, t)$, is based on the Principle of Superposition too.

Let the function $u(X, z, t, \lambda)$ satisfy the equation

$$\frac{\partial^2 u}{\partial X^2} + \frac{\partial^2 u}{\partial z^2} = \frac{\partial^2 u}{c^2 \partial t^2} . \tag{3.7.1}$$

Also, let $X = x \cos \lambda + y \sin \lambda$. Then, the function $u(x \cos \lambda + y \sin \lambda, z, t, \lambda)$ may be readily shown to satisfy the three-dimensional wave equation

$$\frac{\partial^2 u}{\partial x^2} + \frac{\partial^2 u}{\partial y^2} + \frac{\partial^2 u}{\partial z^2} = \frac{\partial^2 u}{c^2 \partial t^2} . \tag{3.7.2}$$

The function

$$U(x, y, z, t) = \int_{-\pi}^{\pi} u(x \cos \lambda + y \sin \lambda, z, t, \lambda) d\lambda \tag{3.7.3}$$

also satisfies (3.7.2). The parameter λ is the angle of rotation of the coordinate system (X, Y, z) about the z-axis with respect to the coordinate system (x, y, z) (Fig. 3.21) and the function $u(X, z, t, \lambda)$ is a solution to the two-dimensional problem in the coordinate system (X, z). It may be shown that any solution to the equation (3.7.2) may be represented in the form (3.7.3) within a broad class of functions. In fact, if in accordance with [3.67], we introduce cylindrical coordinates $x = r \cos \alpha$, $y = r \sin \alpha$, then the function $U(r, \alpha, z, t)$ may be represented by the Fourier series in α and the Hankel transform with respect to r, in the form

$$U(r, \alpha, z, t) = \frac{1}{2\pi} \sum_{n=-\infty}^{\infty} e^{in\alpha} \int_0^{\infty} f_n(q, z, t) q J_n(qr) dq , \tag{3.7.4}$$

where $J_n(qr)$ is the Bessel function of the nth order,

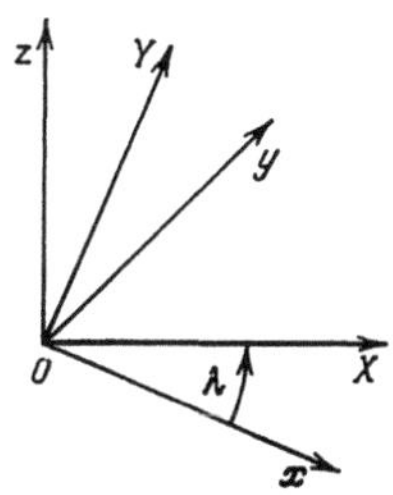

Fig. 3.21. Coordinate systems (x, y, z) and (X, Y, z)

$$f_n(q, z, t) = \int_{-\pi}^{\pi} \int_{0}^{\infty} U(r, \alpha, z, t) e^{-in\alpha} r J_n(qr) dr\, d\alpha .$$

Since $U(r, \alpha, z, t)$ satisfies the three-dimensional wave equation, the function $f_n(q, z, t)$ must satisfy the equation

$$\frac{\partial^2 f}{\partial z^2} - \frac{\partial^2 f}{c^2 \partial t^2} - q^2 f = 0 .$$

Then, if we put

$$u(X, z, t, \lambda) = \frac{1}{4\pi^2} \sum_{n=-\infty}^{\infty} e^{in(\lambda+\pi/2)} \int_{0}^{\infty} f_n(q, z, t) q e^{-iqX} dq , \tag{3.7.5}$$

where $X = x \cos\lambda + y \sin\lambda = r \cos(\lambda - \alpha)$, then the function $u(X, z, t, \lambda)$ will satisfy the wave equation (3.7.1). By substituting (3.7.5) into (3.7.3), changing the sequence of integration, and taking into account that [3.68]

$$\frac{1}{2\pi} \int_{-\pi}^{\pi} e^{-ni\alpha + iqr \sin\alpha} d\alpha = J_n(qr) ,$$

we arrive at the expression (3.7.4).

Thus, the function (3.7.3) allows us to represent the scalar solution to the three-dimensional wave equation (3.7.2) as a combination of solutions to the two-dimensional problems, defined in the planes rotated through various angles λ with respect to the plane (x, z).

The axisymmetric case, $U \equiv U(r, z, t)$, may be obtained from (3.7.3) by supposing that u is independent of λ (i.e. $u \equiv u(X, z, t)$), $x = r\cos\alpha$, $y = \sin\alpha$

$$U(r, z, t) = 2 \int_{0}^{\pi} u(r, \cos\omega, z, t) d\omega . \tag{3.7.6}$$

Here we have taken account of the fact that the integrands are periodic in λ. The latter formula may be shown to establish a one-to-one correspondence between the solutions to the wave equations, $U(r, z, t)$ and $u(X, z, t)$, the latter one being an even function of X. In fact, for an arbitrary function $u(X, z, t)$ the formula (3.7.6) may be rewritten in the form

$$U(r, z, t) = 2 \int_{-r}^{r} \frac{u(X, z, t)}{\sqrt{r^2 - X^2}} \, dX \; . \tag{3.7.7}$$

Hence, the solution $U(r, z, t)$ depends only on the even (with respect to X) part of the function $u(X, z, t)$. To prove this we can decompose $u(X, z, t)$ into the even, u_1, and odd, u_2, functions:

$$u(X, z, t) = u_1(X, z, t) + u_2(X, z, t) \; ,$$

where

$$u_1(X, z, t) = [u(X, z, t) + u(-X, z, , t)]/2 \; ,$$
$$u_2(X, z, t) = [u(X, z, t) - u(-X, z, t)]/2 \; .$$

Here u_1, u_2 are also solutions to two-dimensional wave equations. Then, by substitution of u into (3.7.7) we deduce that the integral of u_2 vanishes, and for the remaining part $u = u_1$, the formula (3.7.7) becomes

$$U(r, z, t) = 4 \int_{0}^{r} \frac{u_1(X, z, t) dX}{\sqrt{r^2 - X^2}} \; . \tag{3.7.8}$$

By inverting this equation, we get

$$u_1(X, z, t) = \frac{1}{2\pi} \frac{d}{dX} \int_{0}^{X} \frac{sU(s, z, t) ds}{\sqrt{X^2 - s^2}} \; . \tag{3.7.9}$$

The latter two equations describe a mutually unique dependence between the two solutions of wave equations, $U(r, z, t)$ and $u_1(X, z, t)$.

An analogous representation may be obtained for the general solution of the equations of elastodynamics (2.7.1–4) in the absence of bulk forces ($\boldsymbol{f} \equiv 0$), since in this case the potentials $\varphi, \boldsymbol{\psi}$ satisfy the three-dimensional wave equations

$$\Delta\varphi = c_1^{-2}\ddot{\varphi} \; , \quad \Delta\boldsymbol{\psi} = c_2^{-2}\ddot{\boldsymbol{\psi}} \left(\Delta \equiv \frac{\partial^2}{\partial x^2} + \frac{\partial^2}{\partial y^2} + \frac{\partial^2}{\partial z^2} \right) \; . \tag{3.7.10}$$

The x- and y-projections of the vector potential $\boldsymbol{\psi}_0 = \psi_Y(X, z, t, \lambda)\boldsymbol{e}_Y$ corresponding to the two-dimensional problem in the (X, z)-coordinate system have the form:

$$\psi_{0x} = -\psi_Y \sin\lambda \; , \quad \psi_{0y} = \psi_Y \cos\lambda \; . \tag{3.7.11}$$

(As pointed out above, the coordinate system X, Y, z was rotated about the z-axis through the angle λ with respect to the coordinate system x, y, z.)

Then the potentials ψ_x, ψ_y defined by (3.7.3) where u is substituted by ψ_{0x}, ψ_{0y}, respectively, satisfy the three-dimensional wave equation. The potentials φ and ψ_z are defined by (3.7.3) directly from the solutions of two-dimensional problems for $\varphi_0(X,t,z,\lambda)$, $\psi_{0z}(X,t,z,\lambda)$. As a result, we have

$$\begin{aligned}
\varphi(x,y,t,z) &= \int\limits_{-\pi}^{\pi} \varphi_0(x\cos\lambda + y\sin\lambda, z, t, \lambda)d\lambda\ ,\\
\psi_x(x,y,z,t) &= -\int\limits_{-\pi}^{\pi} \psi_Y(x\cos\lambda + y\sin\lambda, z, t, \lambda)\sin\lambda d\lambda\ ,\\
\psi_y(x,y,z,t) &= \int\limits_{-\pi}^{\pi} \psi_Y(x\cos\lambda + y\sin\lambda, z, t, \lambda)\cos\lambda d\lambda\ ,\\
\psi_z(x,y,z,t) &= \int\limits_{-\pi}^{\pi} \psi_{0z}(x\cos\lambda + y\sin\lambda, z, t, \lambda)\lambda d\lambda\ ,
\end{aligned} \tag{3.7.12}$$

where $\varphi_0(X,z,t,\lambda)$, $\psi_Y(X,z,t,\lambda)$, $\psi_{0z}(X,z,t,\lambda)$ satisfy the two-dimensional equations

$$\begin{aligned}
&\Delta_1\varphi_0 = c_1^{-2}\ddot{\varphi}_0\ , \quad \Delta_1\psi_Y = c_2^{-2}\ddot{\psi}_Y\ , \quad \Delta_1\psi_{0z} = c_2^{-2}\ddot{\psi}_{0z}\\
&(\Delta_1 \equiv \partial^2/\partial X^2 + \partial^2/\partial z^2)\ .
\end{aligned} \tag{3.7.13}$$

According to (3.7.12), the potentials ψ_x, ψ_y are related by the equality

$$\frac{\partial\psi_x}{\partial x} + \frac{\partial\psi_y}{\partial y} = 0\ , \tag{3.7.14}$$

which replaces the condition $\operatorname{div}\boldsymbol{\psi} = 0$ in removing ambiguity in determining the vector potential $\boldsymbol{\psi}$ entering (2.7.1–4) (for $\boldsymbol{f} \equiv 0$). Since the coordinate axes may be designated arbitrarily, it follows from (3.7.14) that the condition $\operatorname{div}\boldsymbol{\psi} = 0$ in (2.7.1–4) (for $\boldsymbol{f} \equiv 0$) may be replaced by any of the following three conditions:

$$\frac{\partial\psi_x}{\partial x} + \frac{\partial\psi_y}{\partial y} = 0\ , \quad \frac{\partial\psi_y}{\partial y} + \frac{\partial\psi_z}{\partial z} = 0\ , \quad \frac{\partial\psi_z}{\partial z} + \frac{\partial\psi_x}{\partial x} = 0\ . \tag{3.7.15}$$

However, a solution to the three-dimensional problem of elastodynamics may be also expressed through the solution of the two-dimensional problem without employing potentials. Let us consider in the coordinate system X, Y, z (where $X = x\cos\lambda + y\sin\lambda = r\cos(\alpha-\lambda)$, $Y = -x\sin\lambda + y\cos\lambda = r\sin(\alpha-\lambda)$) a two-dimensional solution to the equations of motion of an elastic medium, for which all the variables are independent of Y, and the displacement vector lies in the plane Xz. Let the displacement vector, the strain tensor, and the stress tensor of the solution be designated by $\boldsymbol{u}_1(X,z,t,\lambda)$, $\boldsymbol{\varepsilon}_1(X,z,t,\lambda)$, $\boldsymbol{\sigma}_1(X,z,t,\lambda)$, respectively. Then, the expressions $\boldsymbol{u}(x,y,z,t)$, $\boldsymbol{\varepsilon}(x,y,z,t)$, $\boldsymbol{\sigma}(x,y,z,t)$ defined by

$$\boldsymbol{u} = \int\limits_{-\pi}^{\pi} \boldsymbol{u}_1(X, z, t, \lambda)d\lambda \,,$$

$$\boldsymbol{\varepsilon} = \int\limits_{-\pi}^{\pi} \boldsymbol{\varepsilon}_1(X, z, t, \lambda)d\lambda \,, \quad \boldsymbol{\sigma} = \int\limits_{-\pi}^{\pi} \boldsymbol{\sigma}_1(X, z, t, \lambda)d\lambda \tag{3.7.16}$$

form the solution of a certain three-dimensional problem of elastodynamics. In fact, they are a superposition of solutions of equations of motion, and hence, themselves form a solution. (Since the equations are tensor ones, their form does not depend on the choice of the coordinate system.) In particular, if the functions $\boldsymbol{u}_1$, $\boldsymbol{\varepsilon}_1$, $\boldsymbol{\sigma}_1$ are independent of λ, then the formulas

$$\boldsymbol{u} = \int\limits_{-\pi}^{\pi} \boldsymbol{u}_1(X, z, t)d\lambda \,, \quad \boldsymbol{\varepsilon} = \int\limits_{-\pi}^{\pi} \boldsymbol{\varepsilon}_1(X, z, t)d\lambda \,, \quad \boldsymbol{\sigma} = \int\limits_{-\pi}^{\pi} \boldsymbol{\sigma}_1(X, z, t)d\lambda \tag{3.7.17}$$

give a solution of a certain axisymmetric problem (without torsion), since in this case we have in the projection on the directions r, α, z:

$$u_r = 2\int\limits_0^{\pi} u_{1X}(r\cos\omega, z, t)\cos\omega d\omega \,, \quad u_\alpha = 0 \,,$$

$$u_z = 2\int\limits_0^{\pi} u_{1z}(r\cos\omega, z, t)d\omega \,, \quad \sigma_{\alpha z} = 0 \,,$$

$$\sigma_{rz} = 2\int\limits_0^{\pi} \sigma_{1Xz}(r\cos\omega, z, t)\cos\omega d\omega \,,$$

$$\sigma_{zz} = 2\int\limits_0^{\pi} \sigma_{1zz}(r\cos\omega, z, t)d\omega \,. \tag{3.7.18}$$

Just as in the case of (3.7.6), the relationships (3.7.18) may be shown to establish a one-to-one correspondence between the solutions of plane-strain and axisymmetric problems, and the solution to the plane-strain problem is symmetric with respect to $X = 0$.

Note that the above relationships (3.7.12) employing potentials may be written in the form of (3.7.16), if $\boldsymbol{u}_1$, $\boldsymbol{\varepsilon}_1$, $\boldsymbol{\sigma}_1$ are represented by the following solution to an elastodynamic problem, defined by the displacement vector (in the axes X, Y, z):

$$\boldsymbol{u}_1(X, z, t, \lambda) = \boldsymbol{u}_{10}(X, z, t, \lambda) + \boldsymbol{u}_{11}(X, z, t, \lambda) \,. \tag{3.7.19}$$

Here

$$
\begin{aligned}
&\boldsymbol{u}_{10}(X,z,t,\lambda) = \operatorname{grad}\varphi_0 + \operatorname{curl}[\psi_Y \boldsymbol{e}_Y] ,\\
&\boldsymbol{u}_{11}(X,z,t,\lambda) = \operatorname{curl}[\psi_{0z}\boldsymbol{e}_z] ,\\
&\varphi_0 \equiv \varphi_0(X,z,t,\lambda),\ \psi_{0z} \equiv \psi_{0z}(X,z,t,\lambda),\ \psi_Y \equiv \psi_Y(X,z,t,\lambda) ,
\end{aligned}
$$

and the vector $\boldsymbol{u}_{10}$ is, obviously, polarized in the plane Xz, while the vector $\boldsymbol{u}_{11}$ is orthogonal to the plane.

In conclusion, it should be pointed out that if in (3.7.16) the solution $\boldsymbol{u}_1$, $\boldsymbol{\varepsilon}_1$, $\boldsymbol{\sigma}_1$ to a elastodynamic problem is defined in the axes X, Y, z by the displacement vector $\boldsymbol{u}_1(X,z,t,\lambda)$ in the form

$$
\begin{aligned}
&\boldsymbol{u}_1 = \boldsymbol{u}_{10} + \boldsymbol{u}_{11} ,\\
&\boldsymbol{u}_{10} = \operatorname{grad}\varphi_0 + \operatorname{curl}[\psi_Y \cos\lambda\, \boldsymbol{e}_Y] ,\\
&\boldsymbol{u}_{11} = \operatorname{curl}[\psi_Y \sin\lambda\, \boldsymbol{e}_X + \psi_{0z}\boldsymbol{e}_z] ,\\
&\varphi_0 \equiv \varphi_0(X,z,t,\lambda),\ \psi_Y \equiv \psi_Y(X,z,t,\lambda),\ \psi_{0z} \equiv \psi_{0z}(X,z,t,\lambda)
\end{aligned}
\tag{3.7.20}
$$

(where the vector $\boldsymbol{u}_{10}$ is polarized in the plane Xz, and $\boldsymbol{u}_{11}$ is orthogonal to the plane), then the solution to the three-dimensional problem, $\boldsymbol{u}(x,y,z,t)$, defined by the formulas (3.7.16) may be easily shown to be representable in the form:

$$
\begin{aligned}
&\varphi(x,y,z,t) = \int_{-\pi}^{\pi} \varphi_0(X,z,t,\lambda)d\lambda , \quad \psi_x = 0 ,\\
&\psi_y(x,y,z,t) = \int_{-\pi}^{\pi} \psi_Y(X,z,t,\lambda)d\lambda ,\\
&\psi_z(x,y,z,t) = \int_{-\pi}^{\pi} \psi_{0z}(X,z,t,\lambda)d\lambda
\end{aligned}
\tag{3.7.21}
$$

$$(X = x\cos\lambda + y\sin\lambda,\ \boldsymbol{u} = \operatorname{grad}\varphi + \operatorname{curl}\boldsymbol{\psi},\ \boldsymbol{\psi} = \{\psi_x, \psi_y, \psi_z\}) .$$

Here the scalar potentials φ_0, ψ_y, ψ_{0z} satisfy the two-dimensional wave equations

$$\Delta_1\varphi_0 = c_1^{-2}\ddot{\varphi}_0 , \quad \Delta_1\psi = c_2^{-2}\ddot{\psi} \quad (\psi = \psi_Y, \psi_{0z}) \tag{3.7.22}$$

as before.

Thus, (3.7.21) give the known representation for a solution of the three-dimensional problem of elastodynamics (Sect. 2.7.1), in which the ambiguity of choosing the vector potential $\boldsymbol{\psi}$ is removed with the help of the condition of one of the components ψ_x, ψ_y, ψ_z becoming identical zero. Note also that by changing the designation of the coordinate axes, one may substitute the condition $\psi_x = 0$ by $\psi_y = 0$ or $\psi_z = 0$.

3.8 Application of the Smirnov–Sobolev Method to Solving Axisymmetric Elastodynamic Problems

The formulas derived in the previous section allow to seek for the solutions to three-dimensional problems by using those to two-dimensional problems, and hence make it possible to utilize the method of functionally invariant solutions for obtaining homogeneous solutions to three-dimensional problems. Below we demonstrate the application of this method to solving some axisymmetric problems which are most interesting from the practical point of view. First, upon deriving the ncessary relationships, we shall consider the elastodynamic problem with zero initial conditions and the following boundary conditions:

$$\begin{aligned} &\sigma_{zr}(r,z,t)=0 && (z=0,\ 0\le r<\infty)\,,\\ &\sigma_{zz}(r,z,t)=\sigma^0_{zz}(r,t) && (z=0,\ 0\le r<\infty)\,. \end{aligned} \tag{3.8.1}$$

Later (in Sect. 3.9), we shall consider in detail the solution of the problem with zero initial conditions and mixed boundary conditions

$$\begin{aligned} &\sigma_{zr}(r,z,t)=0 && (z=0,\ 0\le r<\infty)\,,\\ &\sigma_{zz}(r,z,t)=\sigma^0_{zz}(r,t) && (z=0,\ 0\le r<vt)\,,\\ &u_z(r,z,t)=u^0_z(r,t) && (z=0,\ vt<r<\infty) \end{aligned} \tag{3.8.2}$$

as well as an analogous elastodynamic problem

$$\begin{aligned} &\sigma_{zr}(r,z,t)=0 && (z=0,\ 0\le r<\infty)\,,\\ &\sigma_{zz}(r,z,t)=\sigma^0_{zz}(r,t) && (z=0,\ vt<r<\infty)\,,\\ &u_z(r,z,t)=u^0_z(r,t) && (z=0,\ 0\le r<vt)\,. \end{aligned} \tag{3.8.3}$$

Here r, z, α is a cylindrical coordinate system ($x = r\cos\alpha$, $y = r\sin\alpha$, $z = z$) and the given functions $u^0_z(r,t)$, $\sigma_{zz}(r,t)$ are supposed to be homogeneous functions of the $(n-1)^{\text{th}}$ and $(n-2)^{\text{th}}$ degree, respectively in t and r (note that the homgeneity condition $\sigma_{zr}(r,0,t) = 0$ in the problems (3.8.1–3) is used for the sake of definiteness only, and by substituting it by the nonhomogeneous one, $\sigma_{zr}(r,0,t) = \sigma^0_{zr}(r,t)$, we do not impose any substantial constraints on the solutions to the problems under consideration).

3.8.1 Representation of Axisymmetric Solutions by Employing Analytical Functions

First, with the help of (3.7.18) and the global condition $\sigma_{zr} = 0$ $(z = 0)$, let us give representations for the functions $\sigma_{zz}(r,z,t)$ and $u_z(r,z,t)$ from the solution to the axisymmetric dynamic problem. The condition $\sigma_{zr} = 0$ at $z = 0$ and the one-to-one relationship between the solutions to the plane-strain and the axisymmetric problems in the formulas (3.7.18), result in the condition $\sigma_{1Xz} = 0$ at $z = 0$. Since the displacement vector $\boldsymbol{u}(r,z,t)$ entering (3.7.18) is a homogeneous function of the $(n-1)^{\text{th}}$ dimension with respect to t,r,z, the corresponding solution to the plane-strain problem, $\boldsymbol{u}_1(X,z,t)$, considered in the (X,Y,z)-frame is also a homogeneous function of the $(n-1)^{\text{th}}$ dimension with respect to t,X,z. Then, upon introducing the scalar displacement potentials, φ, ψ, as functions of t,X,z:

$$\boldsymbol{u}_1 = \operatorname{grad}\varphi + \operatorname{curl}[-\psi \boldsymbol{e}_Y] , \tag{3.8.4}$$

we deduce that the functions $L_n^t\varphi$, $L_n^t\psi$ where L_n^t is defined by (3.4.5), are homogeneous functions of the zeroth dimension with respect to t,X,z. [Since the coordinate system X,Y,z is oriented in the same way as x,y,z in (3.4.3), then in order to make the axes x,y,z correspondent to $X,z,-Y$ one has to put the minus sign before ψ in (3.8.4).] Then, according to (3.4.6), the functions $L_n^t\varphi$, $L_n^t\psi$ may be sought in the form of the real parts of the complex functions $f_1(\theta_1)$, $f_2(\theta_2)$ which are analytical in the upper half-planes $\operatorname{Im}\theta_1 > 0$ and $\operatorname{Im}\theta_2 > 0$ respectively. Thus, in accordance with the results obtained in Sects. 3.4, 5, one may represent the displacements, $u_{1j}(X,z,t)$, and stresses, $\sigma_{1ij}(X,z,t)$, of the plane-strain solution by (3.4.3, 9–12, 16), having substituted everywhere u_j by u_{1j}, σ_{ij} by σ_{1ij}, x by X, y by z, and z by $-Y$. Then, the condition $\sigma_{1Xz} = 0$ (at $z=0$) leads to the relationship (3.5.8) between the functions $F_1(\theta)$, $F_2(\theta)$ entering (3.4.11, 16), and hence, to (3.5.9) which in our notation becomes

$$\begin{aligned}
&\operatorname{Re} L_m^\theta F_2(\theta) = 2c_2^2 X^{1+m} L_{n+m}^t u_{1z}(X,0,t) ,\\
&\operatorname{Re} L_m^\theta \left[\frac{S(\theta^2)F_2(\theta)}{\sqrt{c_1^{-2}-\theta^2}}\right] = \frac{2}{\mu} X^{1+m} L_{n+m-1}^t \sigma_{1zz}(X,0,t) ,
\end{aligned} \tag{3.8.5}$$

where m is an arbitrary integer, $\theta = t/X + \mathrm{i}0$, and the operator L_m^θ is defined as in (3.4.16).

By applying the integro-differential operators L_{n+m}^t, L_{n+m-1}^t to (3.7.18) relating u_z to u_{1z} and σ_{zz} to σ_{1zz}, and using (3.8.5), we get the following representations for the required functions $u_z(r,z,t)$ and $\sigma_{zz}(r,z,t)$ through the function $F_2(\theta)$ which is analytical in the upper half-plane θ:

$$\begin{aligned}
&\operatorname{Re}\int_0^\pi L_m^\theta F_2(\theta)\frac{d\omega}{X^{1+m}} = c_2^2 L_{n+m}^t u_z(r,0,t) ,\\
&\operatorname{Re}\int_0^\pi L_m^\theta \left[\frac{F_2(\theta)S(\theta^2)}{\sqrt{c_1^{-2}-\theta^2}}\right]\frac{d\omega}{X^{1+m}} = \frac{1}{\mu} L_{n+m-1}^t \sigma_{zz}(r,0,t)\\
&(X = r\cos\omega,\ \theta = t/r\cos\omega + \mathrm{i}0) .
\end{aligned} \tag{3.8.6}$$

In order to remove the nonintegrable singularity in the integrands on the left-hand sides of (3.8.6) at $\omega = \pi/2$ and $m \geq 0$, the path of integration $(0,\pi)$ should be assumed to circumvent from above the point $\omega = \pi/2$ in the complex plane ω. In fact, strictly speaking, the expressions (3.8.6) must be viewed as the limits (as $z \to -0$) of the corresponding expressions for u_z, σ_{zz} given by (3.7.18), upon applying the integro-differential operators L^t_{n+m}, L^t_{n+m-1} and using the representation of the plane-strain solution $u_1(X, z, t)$ through the functions $F_1(\theta_1)$ and $F_2(\theta_2)$ related by (3.5.8). Then, considering, for example, the expression for the displacement u_z, we get (for $z < 0$):

$$L^t_{n+m}u_z(r,z,t) = 2\int_0^\pi L^t_m \mathrm{Re}\left[F_1(\theta_1)\theta_1\frac{\partial\theta_1}{\partial z} - F_2(\theta_2)\theta_2\frac{\partial\theta_2}{\partial X}\right]d\omega$$

$$= 2\,\mathrm{Re}\left[\int_0^\pi \Phi_1(\theta_1,z)d\omega + \int_0^\pi \Phi_2(\theta_2,z)d\omega\right] \qquad (X = r\cos\omega)\,, \tag{3.8.7}$$

where θ_1, θ_2 are given by (3.4.8) (upon substituting ξ by X/t, and η by z/t), and the integrands Φ_1, Φ_2 are analytical functions in the upper half-planes θ_1, θ_2 respectively (and also depend on z as a parameter).

The functions θ_j and their derivatives with respect to t, X, z depend on X, and hence, on ω, and are analytical functions of the parameter ω in the region $\mathrm{Im}\,\omega > 0$ (at least, in the neighborhood of the real axis $\mathrm{Im}\,\omega > 0$).

This can be shown by expressing the functions θ_j and their derivatives with respect to t, X, z through ω explicitly with the help of (3.4.9, 10, 15) where y is put equal to z, and $x = X = r\cos\omega$. Then, having deformed the paths of integration in (3.8.7) in such a way that they circumvent the point $\omega = \pi/2$ from above, and passing in (3.8.7) to the limit $z \to -0$, we arrive at the upper expression (3.8.6) in which the path of integration with respect to ω circumvents the point $\omega = \pi/2$ from above. The second of the formulas (3.8.6) may be obtained in a similar way.

Going in (3.8.6) to the integration variable θ we reduce them to the form

$$\begin{aligned} &\mathrm{Re}\int_l \Phi^0_m(\theta)\frac{d\theta}{\sqrt{\theta^2-\theta_0^2}} = U_m(\theta_0)\,,\\ &\mathrm{Re}\int_l \Phi^1_m(\theta)\frac{d\theta}{\sqrt{\theta^2-\theta_0^2}} = \Sigma_m(\theta_0)\,, \end{aligned} \tag{3.8.8}$$

where $\theta_0 = t/r > 0$, m is an arbitrary integer,

$$\begin{aligned} &U_m(\theta_0) = c_2^2 r t^m L^t_{n+m}u_z(r,0,t) = c_2^2 L^{\theta_0}_{n+m}u_z(1,0,\theta_0)\,,\\ &\Sigma_m(\theta_0) = \frac{1}{\mu} r t^m L^t_{n+m-1}\sigma_{zz}(r,0,t) = \frac{\theta_0^m}{\mu} L^{\theta_0}_{n+m-1}(1,0,\theta_0)\,,\\ &\Phi^0_m(\theta) = \theta^m L^\theta_m F_2(\theta)\,, \quad \Phi^1_m(\theta) = \theta^m L^\theta_m\left[F_2(\theta)\frac{S(\theta^2)}{\sqrt{c_1^{-2}-\theta^2}}\right]\,, \end{aligned} \tag{3.8.9}$$

and the path of integration l, Fig. 3.22, passes within the region $\operatorname{Im}\theta \geq 0$ from the point $\theta = \theta_0$ to the point $\theta = -\theta_0$ in an arbitrary manner [due to analyticity of the integrands in the region $\operatorname{Im}\theta > 0$ in (3.8.8)]. The branches of the radical $\sqrt{\theta^2 - \theta_0^2}$ are singled out by drawing the cuts $(-\infty, -\theta_0)$, $(\theta_0, +\infty)$ in the plane θ along the real axis, and it is assumed that $\sqrt{\theta^2 - \theta_0^2} \sim \theta$ as $\theta \to \infty$, while the branches of the remaining integrals are taken as in the case of the plane-strain problem: $\sqrt{c_j^{-2} - \theta^2} = c_j^{-1}$ for $\theta = 0$.

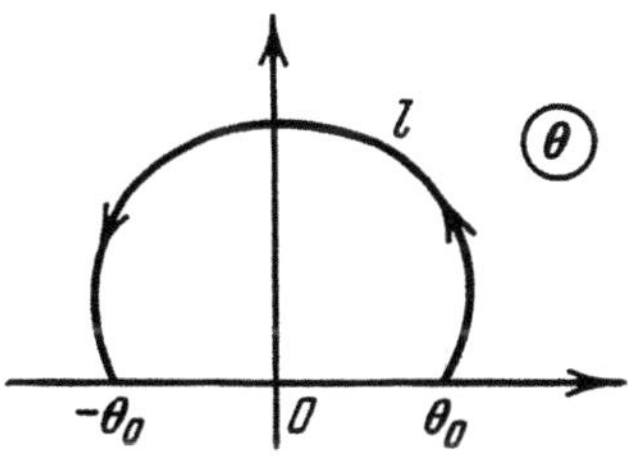

Fig. 3.22. Contour l

Referring now to (3.8.8, 9), we note that the function $F_2(\theta)$ should be assumed decreasing as $\theta \to 0$, so as one could close the contour l for $\theta_0 \to 0$ and to get the required initial conditions: $U_m(0) = \Sigma_m(0) = 0$. On the other hand, the conditions of the displacement vector continuity and the stress integrability at the point $r = 0$, $z = 0$ imply that (for $t > 0$)

$$
\begin{aligned}
&u_z(r, 0, t) = Ct^{n-1} + O(r^{\varepsilon}) , \\
&\sigma_{zz}(r, 0, t) = O(r^{-1+\varepsilon}) , \\
&C = \text{const}, \ r \to 0, \ \varepsilon > 0 ,
\end{aligned}
\tag{3.8.10}
$$

and hence we get (for sufficiently smooth functions) the following conditions:

$$
U_m(\theta_0) = O(\theta_0^{-1}) , \quad \Sigma_m(\theta_0) = O(\theta_0^{-\varepsilon}) , \quad \theta_0 \to \infty , \quad \varepsilon > 0 . \tag{3.8.11}
$$

Then, according to (3.8.8), the conditions (3.8.11), and hence (3.8.10), will be met if we require that

$$
F_2(\theta) = O(\theta^{-1-\varepsilon}) , \quad \theta \to \infty \quad (\operatorname{Im}\theta \geq 0, \ \varepsilon > 0) . \tag{3.8.12}
$$

Note also that, as indicated while deriving the formulas (3.7.18), the plane-strain solution $\boldsymbol{u}_1(X, z, t)$ corresponding to the axisymmetric solution $\boldsymbol{u}(r, z, t)$ under consideration, is symmetric with respect to the axis $X = 0$, and hence $u_z(X, z, t)$ is an even function of X. Then, from (3.5.9), we get

$$
\operatorname{Re} F_2(-\theta) = -\operatorname{Re} F_2(\theta) \quad (\operatorname{Im}\theta = 0) , \tag{3.8.13}
$$

and from the Cauchy-Riemann relationship for analytical functions, the latter expression and the condition (3.8.12), we find

$$\operatorname{Im} F_2(-\theta) = \operatorname{Im} F_2(\theta) \quad (\operatorname{Im}\theta = 0)\,. \tag{3.8.14}$$

As a result, we deduce that the functions $\Phi^j_m(\theta)$, $j = 0, 1$ from (3.8.8) are analytical within the region $\operatorname{Im}\theta > 0$ and meet the conditions

$$\begin{aligned}&\operatorname{Re}\Phi^j_m(-\theta) = -\operatorname{Re}\Phi^j_m(\theta)\,, \quad \operatorname{Im}\Phi^j_m(-\theta) = \operatorname{Im}\Phi^j_m(\theta)\\ &(\operatorname{Im}\theta = 0,\ j = 0, 1)\,;\end{aligned} \tag{3.8.15}$$

$$\begin{aligned}&\Phi^0_m(\theta) = \mathrm{O}(\theta^{-1-\varepsilon})\,, \quad \Phi^1_m(\theta) = \mathrm{O}(\theta^{-\varepsilon})\,, \quad \theta \to \infty\\ &(\operatorname{Im}\theta \geq 0,\ \varepsilon > 0)\,.\end{aligned} \tag{3.8.16}$$

Taking into account the above estimates and the relationships (3.8.11, 15, 16), the inversion formulas for each of the equations (3.8.8) reduce to determining the function $\Phi(\theta)$ from the equation

$$\operatorname{Re}\int_l \Phi(\theta)\frac{d\theta}{\sqrt{\theta^2 - \theta_0^2}} = N(\theta_0) \quad (\theta_0 > 0)\,, \tag{3.8.17}$$

where $N(\theta_0) = \mathrm{O}(\theta_0^{-\varepsilon})$ as $\theta_0 \to \infty$, $\varepsilon > 0$; $N(\theta_0) \to 0$ as $\theta_0 \to 0$; $\Phi(\theta)$ is an analytical function in the region $\operatorname{Im}\theta > 0$, integrable at $\theta = 0$; and $\Phi(\theta) = \mathrm{O}(\theta^{-\varepsilon})$, $\varepsilon > 0$ as $\theta \to \infty$, with the relationships $\operatorname{Re}\Phi(-\theta) = -\operatorname{Re}\Phi(\theta)$, $\operatorname{Im}\Phi(-\theta) = \operatorname{Im}\Phi(\theta)$ being valid on the axis $\operatorname{Im}\theta = 0$.

Thus, thanks to the given conditions, the equation (3.8.17) has two different representations for determining the real and the imaginary parts of the function $\Phi(\theta)$ on the axis $\operatorname{Im}\theta = 0$. The first one can be easily obtained by deforming the contour l into the segment $[-\theta_0, \theta_0]$. As a result, we get

$$\operatorname{Re}\int_l \Phi(\theta)\frac{d\theta}{\sqrt{\theta^2 - \theta_0^2}} = -2\int_0^{\theta_0} \operatorname{Im}\Phi(\theta)\frac{d\theta}{\sqrt{\theta_0^2 - \theta^2}} = N(\theta_0)$$
$$\left(\sqrt{\theta_0^2 - \theta^2} > 0\right)\,.$$

Inverting the latter equation one obtains for $\theta > 0$

$$\operatorname{Im}\Phi(\theta) = -\frac{1}{\pi}\frac{d}{d\theta}\int_0^{\theta}\frac{\theta_0 N(\theta_0)d\theta_0}{\sqrt{\theta^2 - \theta_0^2}} = -\frac{\theta}{\pi}\int_0^{\theta}\frac{N'(\theta_0)d\theta_0}{\sqrt{\theta^2 - \theta_0^2}}\,, \tag{3.8.18}$$

where the second integral was derived by using the condition $N(\theta_0) \to 0$ as $\theta_0 \to 0$. [Note that the solution in the form (3.8.18) may be obtained directly by applying the inversion formula (3.7.9) to (3.7.8).]

On the other hand, making use of analyticity of the integrand which decreases as $\theta \to \infty$, we can transform the contour l in (3.7.9) and obtain, taking into account the branches of the radical $\sqrt{\theta^2 - \theta_0^2}$, another equation

$$\mathrm{Re} \int\limits_{l} \Phi(\theta) \frac{d\theta}{\sqrt{\theta^2 - \theta_0^2}} = 2 \int\limits_{\theta_0}^{\infty} \mathrm{Re}\, \Phi(\theta) \frac{d\theta}{\sqrt{\theta^2 - \theta_0^2}} = N(\theta_0) \,.$$

By inversion of the latter (reducing it to the Abel equation by substituting θ^2 by ν, and θ_0^2 by ν_0), we get for $\theta > 0$

$$\mathrm{Re}\, \Phi(\theta) = -\frac{\theta}{\pi} \int\limits_{\theta}^{\infty} \frac{N'(\theta_0) d\theta_0}{\sqrt{\theta_0^2 - \theta^2}} \quad \left(\sqrt{\theta_0^2 - \theta^2} > 0 \right) . \tag{3.8.19}$$

(Note that the derivatives $N'(\theta_0)$ entering (3.8.18, 19) should be treated, generally speaking, as generalized ones, and hence delta-functions will appear under the integral, if the function $N(\theta_0)$ suffers a discontinuity of the first kind.)

3.8.2 Solutions to Axisymmetric Problems

Let us use the above results for solving the axisymmetric problems (3.8.1–3).

As far as the problem (3.8.1) is concerned, it can be solved quite easily. To do this, we shall apply any of the inversion formulas (3.8.18, 19) to the second of the equations (3.8.8) (at $m = 0$). Then, carrying out the inversion with the help of, say, (3.8.19), we get

$$\mathrm{Re}\, \Phi_0^1(\theta) = -\frac{\theta}{\pi} \int\limits_{|\theta|}^{\infty} \frac{\Sigma_0'(\theta_0) d\theta_0}{\sqrt{\theta_0^2 - \theta^2}} \quad (-\infty < \theta < \infty) \,, \tag{3.8.20}$$

and hence determination of the function

$$\Phi_0^1(\theta) \equiv F_2(\theta) S(\theta^2) / \sqrt{c_1^{-2} - \theta^2} \,,$$

analytical in the region $\mathrm{Im}\, \theta > 0$ and decreasing as $|\theta| \to 0$ reduces to solving the Schwarz problem. With the help of the Schwarz integral, the solution may be written in the form

$$\frac{F_2(\theta) S(\theta^2)}{\sqrt{c_1^{-2} - \theta^2}} = \frac{1}{\pi \mathrm{i}} \int\limits_{-\infty}^{\infty} \frac{\Phi_0^1(\beta) d\beta}{\beta - \theta} \quad (\mathrm{Im}\, \theta > 0,\ \mathrm{Im}\, \beta > 0) \,, \tag{3.8.21}$$

where the value of the integrand is defined by the right-hand side of the expression (3.8.20).

As an example, let us consider the half-space $z < 0$ subjected to an expanding loading of the form $\sigma_{zz}^0(r,t) = -pH(\alpha t - r)$, $\alpha = \mathrm{const}$, Fig. 3.23. In this case $n = 2$,

$$\Sigma_0(\theta_0) = \frac{r}{\mu} L_1^t \sigma_{zz}^0(r,t) = -\frac{p}{\mu} \delta(\theta_0 - \alpha^{-1}) \,,$$

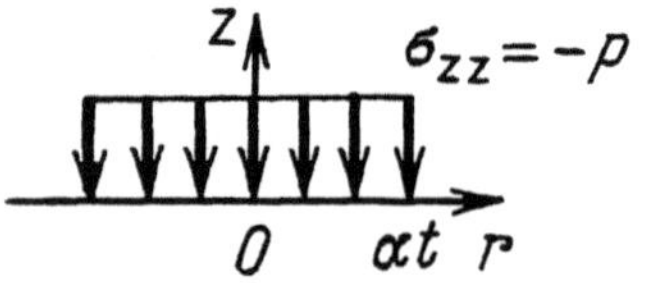

Fig. 3.23. Propagation of a constant load from the origin of coordinates along the boundary of an elastic half-space $z < 0$

$$\operatorname{Re}\Phi_0^1(\theta) = \frac{\theta p}{\mu}\int\limits_{|\theta|}^{\infty}\frac{\delta'(\theta_0-\alpha^{-1})}{\sqrt{\theta_0^2-\theta^2}}d\theta_0$$

$$= \frac{p}{\pi\mu}\frac{d}{d\theta}\left[\int\limits_{|\theta|}^{\infty}\frac{\theta_0\delta(\theta_0-\alpha^{-1})d\theta_0}{\sqrt{\theta_0^2-\theta^2}}\right] = \frac{p}{\pi\mu\alpha}\frac{d}{d\theta}\left[\frac{H(\alpha^{-1}-|\theta|)}{\sqrt{\alpha^{-2}-\theta^2}}\right] . \tag{3.8.22}$$

Then, with the help of (3.8.21) we find for $\operatorname{Im}\theta > 0$:

$$\frac{F_2(\theta)S(\theta^2)}{\sqrt{c_1^{-2}-\theta^2}} = \frac{p}{\pi\mathrm{i}}\int\limits_{-\infty}^{\infty}\frac{d}{d\beta}\left[\frac{H(\alpha^{-1}-|\beta|)}{\sqrt{\alpha^{-2}-\beta^2}}\right]\frac{d\beta}{\pi\mu\alpha(\beta-\theta)}$$

$$= \frac{p}{\pi^2\mu\alpha\mathrm{i}}\int\limits_{-\alpha^{-1}}^{\alpha^{-1}}\frac{d\beta}{\sqrt{\alpha^{-2}-\beta^2}(\beta-\theta)^2} = \frac{p\theta}{\pi\mu\alpha(\alpha^{-2}-\theta^2)^{3/2}} , \tag{3.8.23}$$

where the branch of the radical was chosen so as $(\alpha^{-2}-\theta^2)^{3/2} > 0$ for $\theta = 0$. Hence, the displacement $u_z(r,0,t)$ may be obtained from the upper formula (3.8.8) at $n = 2$; $m = -2$, upon substituting $F_2(\theta)$ from (3.8.23).

Let now the loading be a concentrated force, p, applied at the instant $t = 0$ at the point $r = z = 0$, Fig. 3.24. In this case, $\sigma_{zz}^0(r,t) = -pH(t)\delta(r)/2\pi r$ where $\delta(r)$ is a one-dimensional Dirac delta-function satisfying the conditions $\delta(r) = 0 (r > 0)$ and

$$\int\limits_0^{\infty} f(r)\delta(r)dr = f(0) ,$$

hence

$$2\pi\int\limits_0^{\infty}\sigma_{zz}^0(r,t)r\,dr = -pH(t) .$$

For the problem under consideration, $n = 0$, $\Sigma_0(\theta_0) = -p\delta(1/\theta_0)/2\pi\mu$, and the formulas (3.8.18, 19) give for $-\infty < \theta < \infty$:

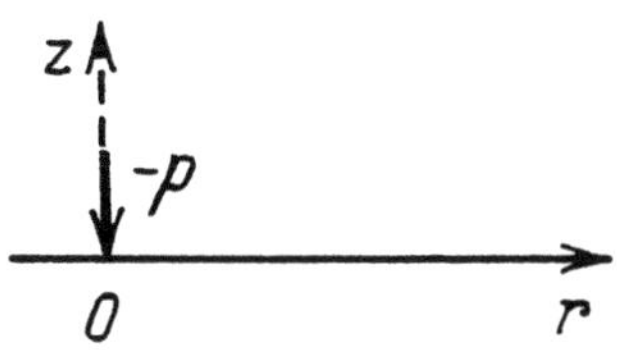

Fig. 3.24. A suddenly applied concentrated force

$$\operatorname{Im}\Phi_0^1(\theta) = \frac{|\theta| p}{2\pi^2\mu}\int_0^{|\theta|}\frac{\delta'_{\theta_0}(1/\theta_0)d\theta_0}{\sqrt{\theta^2-\theta_0^2}} = -\frac{|\theta| p}{2\pi^2\mu}\int_{1/|\theta|}^{\infty}\frac{z\delta'(z)dz}{\sqrt{\theta^2 z^2-1}} = 0\,,$$
$$\operatorname{Re}\Phi_0^1(\theta) = \frac{\theta p}{2\pi^2\mu}\int_{|\theta|}^{\infty}\frac{\delta'_{\theta_0}(1/\theta_0)d\theta_0}{\sqrt{\theta_0^2-\theta^2}} = -\frac{\theta p}{2\pi^2\mu}\int_0^{1/|\theta|}\frac{z\delta'(z)dz}{\sqrt{1-\theta^2 z^2}} = \frac{\theta p}{2\pi^2\mu}\,. \tag{3.8.24}$$

Thus, we get

$$\Phi_0^1(\theta) \equiv F_2(\theta)\frac{S(\theta^2)}{\sqrt{c_1^{-2}-\theta^2}} = \frac{\theta p}{2\pi^2\mu} \quad (\operatorname{Im}\theta \geq 0)\,. \tag{3.8.25}$$

With the help of (3.8.25) one can find from (3.8.8) the following expression for the vertical displacement at the boundary $z = 0$

$$u_z(r,0,t) = \frac{1}{rc_2^2}\operatorname{Re}\int_l F_2(\theta)\frac{d\theta}{\sqrt{\theta^2-\theta_0^2}} \quad \left(\theta_0 = \frac{t}{r}\right)\,, \tag{3.8.26}$$

which may be reduced to the form

$$u_z(r,0,t) = -\left.\frac{p\sqrt{\theta^2-c_1^{-2}}H(\theta-\theta_0)\theta}{[dS(\theta^2)/d\theta]\mu\pi c_2^2 r\sqrt{\theta^2-\theta_0^2}}\right|_{\theta=c_R^{-1}}$$
$$-\frac{p(\lambda+2\mu)}{4\pi\mu r(\lambda+\mu)} \quad (\theta_0 > c_2^{-1})\,,$$
$$u_z(r,0,t) = \frac{pH(\theta_0-c_1^{-1})}{\pi^2\mu c_2^2 r} \tag{3.8.27}$$
$$\times\int_{c_1^{-2}}^{\theta_0^2}\frac{\sqrt{z-c_1^{-2}}(c_2^{-2}-2z)^2dz}{\sqrt{\theta_0^2-z}[(c_2^{-2}-2z)^4+16z^2(z-c_1^{-2})(c_2^{-2}-z)]} \quad (\theta_0 < c_2^{-1})$$

(Here all the radicals are assumed to be arithmetic.) By substitution of the integration variable $\sqrt{z-c_1^{-2}}/\sqrt{\theta_0^2-z}$ by y the integral in (3.8.27) is reduced to that of a rational function, and hence $u_z(r,0,t)$ is expressed through elementary functions over the entire range $0 < \theta_0 < \infty$. A detailed analysis of this and other similar cases is presented in [3.69].

If in (3.8.27) the medium's density ϱ tends to zero, then $c_i \to \infty$ ($i = 1, 2, \mathrm{R}$), and we arrive at the expression

$$u_z(r, 0) = -\frac{p(\lambda + 2\mu)}{4\pi\mu(\lambda + \mu)r}\,, \tag{3.8.28}$$

coincident with the known static solution to the Boussinesq problem, see, e.g. [3.70].

Note that the solution (3.8.25) was derived by using the formula (3.8.19) which, as shown above, may be used, if the required function $\Phi_0^1(\theta)$ decreases at infinity. However, the expression (3.8.25) does not satisfy this condition. Nevertheless, the above solution for a concentrated force is valid. This statement may be proved by replacing the singular loading $\Sigma_0(\theta_0) = -p(2\pi\mu)^{-1}\delta(1/\theta_0)$ by a smoothed one

$$\Sigma_0(\theta_0) = -p\varepsilon/2\pi\mu\theta_0(\varepsilon^2 + \theta_0^{-2})^{3/2}$$

for which the formula (3.8.19) is valid. Then, if in the solution obtained for the smoothed loading, ε tends to zero, then, taking into account that $\varepsilon/(\varepsilon^2 + r^2)^{3/2} \to \delta(r)/r$ (see, e.g., [3.62]), we arrive at the solution (3.8.25).

Upon differentiating (3.8.27) with respect to time, one gets the known solution to the Lamb problem [3.59]. The three-dimensional problem for the case of applying a tangential load at the half-space boundary, was considered in [3.71].

3.9 Solutions to Some Axisymmetric Problems with Mixed Boundary Conditions

In order to solve the mixed problem (3.8.2), let us invert the first of the equations (3.8.8) with the help of (3.8.18), and the second one with the help of (3.8.19); in order to solve the analogous problem (3.8.3), we shall do vice versa, namely, we shall invert the first of the equations (3.8.8) with the help of (3.8.19), and the second one with the help of (3.8.18). As a result, taking into account the evenness properties of (3.8.15), we get from (3.8.8) for the problem (3.8.2):

$$\begin{aligned}
\operatorname{Im}\Phi_m^0(\theta) &= -\frac{|\theta|}{\pi}\int_0^{|\theta|}\frac{U_m'(\theta_0)d\theta_0}{\sqrt{\theta^2-\theta_0^2}} \qquad (-v^{-1} < \theta < v^{-1})\,,\\
\operatorname{Re}\Phi_m^1(\theta) &= -\frac{\theta}{\pi}\int_{|\theta|}^{\infty}\frac{\Sigma_m'(\theta_0)d\theta_0}{\sqrt{\theta_0^2-\theta^2}} \qquad (\theta < -v^{-1},\ \theta > v^{-1})\,,
\end{aligned} \tag{3.9.1}$$

and for the problem (3.8.3)

$$\operatorname{Im} \Phi^1_m(\theta) = -\frac{|\theta|}{\pi} \int_0^{|\theta|} \frac{\Sigma'_m(\theta_0)d\theta_0}{\sqrt{\theta^2-\theta_0^2}} \quad (-v^{-1} < \theta < v^{-1}) ,$$
$$\operatorname{Re} \Phi^0_m(\theta) = -\frac{\theta}{\pi} \int_{|\theta|}^{\infty} \frac{U'_m(\theta_0)d\theta_0}{\sqrt{\theta_0^2-\theta^2}} \quad (\theta < -v^{-1},\ \theta > v^{-1}) . \tag{3.9.2}$$

As a result, taking into account the notation (3.8.9), one can easily deduce that employing the representations (3.9.1, 2) at $m = 0$, one arrives at the Riemann–Hilbert boundary-value problem of determining the function $F_2(\theta)$, analytical in the region $\operatorname{Im}\theta > 0$, with the help of the boundary conditions on the axis $\operatorname{Im}\theta = 0$. (Note, that in the case $n \le 0$ unknown constants will appear on the right-hand sides of formulas (3.9.1, 2). Upon solving the Riemann–Hilbert problem, the constants must be found with the help of additional conditions using, for example, the solution behavior at the point $\theta_0 = v^{-1}$.)

3.9.1 Solution for the $n \geq 1$-Case

For the sake of definiteness, let us consider the most interesting case, when $v < c_R$, $n \ge 1$. Then, with the help of the representations (3.9.1) at $m = 0$, the problem (3.8.2) reduces to the Schwarz problem for the function $F_2(\theta)/\mathrm{i}\theta$ in the region $\operatorname{Im}\theta > 0$:

$$\operatorname{Re}\left[\frac{F_2(\theta)}{\mathrm{i}\theta}\right] = \alpha(\theta) . \tag{3.9.3}$$

Here

$$\alpha(\theta) = \begin{cases} -\dfrac{\operatorname{sgn}\theta}{\pi} \displaystyle\int_0^{|\theta|} \frac{U'_0(\theta_0)d\theta_0}{\sqrt{\theta^2-\theta_0^2}} & (-v^{-1} < \theta < v^{-1}) , \\[2ex] -\dfrac{\sqrt{c_1^{-2}-\theta^2}}{\pi \mathrm{i} S(\theta^2)} \displaystyle\int_{|\theta|}^{\infty} \frac{\Sigma'_0(\theta_0)d\theta_0}{\sqrt{\theta_0^2-\theta^2}} & (\theta < -v^{-1},\ \theta > v^{-1}) , \end{cases} \tag{3.9.4}$$

where $\operatorname{sgn}\theta = 1$ for $\theta > 0$ and $\operatorname{sgn}\theta = -1$ for $\theta < 0$, and according to (3.8.11) we have

$$\alpha(\theta) = \mathrm{O}(\theta^{-2-\varepsilon}) , \quad \theta \to \pm\infty , \quad \varepsilon > 0 . \tag{3.9.5}$$

With the help of the representations (3.9.2) at $m = 0$ the problem (3.8.3) may also be reduced to the Schwarz problem for the function $F_2(\theta)S(\theta^2)/\mathrm{i}\theta\sqrt{c_1^{-2}-\theta^2}$:

$$\operatorname{Re}\left[\frac{F_2(\theta)S(\theta^2)}{\mathrm{i}\theta\sqrt{c_1^{-2}-\theta^2}}\right] = \beta(\theta) . \tag{3.9.6}$$

Here

$$\beta(\theta) = \begin{cases} -\dfrac{\operatorname{sgn}\theta}{\pi} \displaystyle\int_0^{|\theta|} \dfrac{\Sigma_0'(\theta_0)d\theta_0}{\sqrt{\theta^2-\theta_0^2}} & (-v^{-1} < \theta < v^{-1}) , \\ -\dfrac{S(\theta^2)}{\pi i\sqrt{c_1^{-2}-\theta^2}} \displaystyle\int_{|\theta|}^{\infty} \dfrac{U_0'(\theta_0)d\theta_0}{\sqrt{\theta^2-\theta^2}} & (\theta < -v^{-1},\ \theta > v^{-1}) , \end{cases} \tag{3.9.7}$$

and according to the estimate $U_0(\theta) = O(\theta^{-1-\varepsilon})$ as $\theta \to \pm\infty$, $\varepsilon > 0$, $n \geq 1$ [which is more refined than (3.8.11), and follows from (3.8.10) for $n \geq 1$, $m = 0$], we get

$$\beta(\theta) = O(\theta^{-1-\varepsilon}) , \quad \theta \to \pm\infty , \quad \varepsilon > 0 . \tag{3.9.8}$$

Using the Schwarz integral (3.4.27) for solving the Schwarz problem and taking into account the evenness conditions (3.8.13, 14) we get the following expressions for the boundary conditions (3.9.3, 6) (at $\operatorname{Im}\theta > 0$):

$$\frac{F_2(\theta)}{i\theta} = \frac{2}{\pi i}\int_0^{\infty} \frac{\alpha(\tau)\tau d\tau}{\tau^2-\theta^2} + \frac{1}{i}\sum_{k=1}^{p} \frac{A_k}{(\theta^2-v^{-2})^k} \quad (\operatorname{Im} A_k = 0) , \tag{3.9.9}$$

$$\frac{F_2(\theta)S(\theta^2)}{i\theta\sqrt{c_1^{-2}-\theta^2}} = \frac{2}{\pi i}\int_0^{\infty} \frac{\beta(\tau)\tau d\tau}{\tau^2-\theta^2} + \frac{1}{i}\sum_{k=1}^{m} \frac{B_k}{(\theta^2-v^{-2})^k} \quad (\operatorname{Im} B_k = 0) . \tag{3.9.10}$$

Since the boundary functions $\alpha(\theta)$, $\beta(\theta)$ are not defined at the points $\theta = \pm v$ [which were excluded from consideration while deriving the formulas (3.9.1, 2)], we must assume the presence of poles at the points, which lead to the appearance of the sums on the right-hand sides of (3.9.9, 10). Both the unknown real constants A_k, B_k and the orders of the poles p, m are to be found by analyzing the behavior of the required functions at infinity, the boundary conditions and the order of the stress singularity in the neighborhood of the point $r = vt$.

As far as the problem (3.8.2) follows from the elastodynamic problem of the constant-speed expansion of disk-shaped cracks/cuts at $u_z^0 \equiv 0$, let us assume that the stresses and velocities in the framework of the problem (3.8.2) have, at $r \to vt$, $z = 0$, singularities of the order of $-1/2$. Thus, it follows from (3.8.8) (at $m = 0$, $n \geq 1$) that, if $\theta_0 \to v^{-1}$, then $\Sigma_0(\theta_0) = O(|\theta_0 - v^{-1}|^{-n+1/2})$, and hence, as $\theta \to \pm v^{-1}$, the multiplicity of the poles of the function $F_2(\theta)$ must not exceed n. Hence, $p = n$ in (3.9.9). On the other hand, the condition (3.8.12) leads to

$$A_1 = \frac{2}{\pi}\int_0^{\infty} \alpha(\tau)\tau d\tau , \tag{3.9.11}$$

and the remaining $n-1$ constants $A_k (k = 2, \ldots, n)$ are defined from the boundary conditions, namely, from the equations following from (3.8.8)

$$\operatorname{Re} \int_l L_m^\theta \left[\frac{F_2(\theta) S(\theta^2)}{\sqrt{c_1^{-2} - \theta^2}} \right] \frac{\theta^m d\theta}{\sqrt{\theta^2 - \theta_0^2}} = \Sigma_m(\theta_0) , \tag{3.9.12}$$
$$\theta_0 > v^{-1}, \ m = -1, -2, \ldots, -(n-1) .$$

Thus, the solution to the problem (3.8.2) at $n \geq 1$, $v < c_R$ is given by (3.9.9) where the n unknown real constants A_k are defined from the conditions (3.9.11, 12).

It should be noted that the solution to the problem (3.8.2) at $n = 2$ (the self-similar problem) and $u_z^0 = 0$, has been obtained in [3.37].

3.9.2 Circular Crack Expansion and a Conical Punch Indentation

As an example, let us consider a constant-speed expansion of a disk-shaped cut whose sides are subjected to a constant stress $\sigma_{zz} = -P$, see Fig. 3.25, at zero initial conditions. In this case $\sigma_{zz}^0 = -P$, $u_z^0 = 0$ in (3.8.2). The solution to the problem is given by (3.9.9, 11, 12) where $n = 2$, $\alpha(\theta) = 0$, $A_1 = 0$. Thus, we have

$$F_2(\theta) = \frac{A_2 \theta}{(\theta^2 - v^{-2})^2} , \tag{3.9.13}$$

where A_2 is defined by the equation

$$A_2 \operatorname{Re} \int_l \frac{d\theta}{\theta \sqrt{\theta^2 - \theta_0^2}} \int_0^\theta \frac{S(\eta^2)\eta d\eta}{(\eta^2 - v^{-2})^2 \sqrt{c_1^{-2} - \eta^2}}$$
$$= -\frac{P}{\mu \theta_0} \quad \left(\theta_0 = \frac{t}{r} > v^{-1} \right) . \tag{3.9.14}$$

(As pointed out above, the path of integration with respect to η circumvents the singular points $\eta = \pm v^{-1}$ of the integrand, from above.)

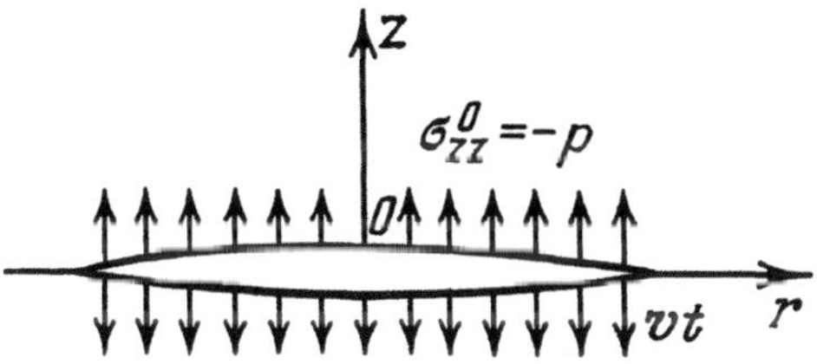

Fig. 3.25. Expansion of a disk-shaped cut

The contour l of the outer integral in (3.9.14) may be deformed into the semiaxes $(-\infty, -\theta_0)$, $(\theta_0, +\infty)$. Since we are interested in the real part of the left-hand expression only, the path of integration of the inner integral may be extended to $\pm\infty$, and hence the multiple integral may be presented in the form

$$\int_l f(\theta)d\theta \int_0^\theta F(\eta)d\eta = \int_{-\infty}^{-\theta_0} f(\theta)d\theta \int_0^{-\infty} F(\eta)d\eta + \int_{\theta_0}^{\infty} f(\theta)d\theta \int_0^{+\infty} F(\eta)d\eta \,.$$

Since the outer integrals are taken independently of the inner ones by deforming the paths of integration with respect to η into the imaginary semiaxis, we get

$$A_2 = \frac{2P}{\mu\pi}\left[\int_0^\infty \frac{S(-x)dx}{(x+v^{-2})^2\sqrt{c_1^{-2}+x}}\right]^{-1}, \tag{3.9.15}$$

where $S(-x) = (c_2^{-2}+2x)^2 - 4x\sqrt{(c_1^{-2}+x)(c_2^{-2}+x)}$, and all the integrals are assumed to be arithmetic.

By substitution of the expression (3.9.13) into the first of the formulas (3.8.8) at $n = 2$, $m = -1$ we find the velocity $\dot{u}_z(r, z, t)$ at the points of the lower side of the cut ($z = -0$, $r < vt$):

$$\begin{aligned}\dot{u}_z(r,-0,t) &= \frac{A_2\theta_0}{c_2^2}\,\mathrm{Re}\int_l \frac{d\theta}{\theta\sqrt{\theta^2-\theta_0^2}}\int_0^\theta \frac{\eta d\eta}{(\eta^2-v^{-2})^2} \\ &= -\frac{\pi A_2 t v^3}{2c_2^2\sqrt{v^2t^2-r^2}} \,.\end{aligned} \tag{3.9.16}$$

Then, integration with respect to t gives the displacements of the sides:

$$u_z(r,-0,t) = -\frac{\pi A_2 v}{2c_2^2}\sqrt{v^2t^2-r^2} \quad (0 \le r < vt) \,. \tag{3.9.17}$$

It should be noted that the above problem about a cut was originally solved in [3.30], and our results (3.9.13, 15–17) differ from those of [3.30] in notation only.

Note, that by tending to zero the density ϱ (entering the expressions for the velocities c_1, c_2) in the formulas (3.9.15, 17), one arrives at the formulas which coincide with the similar ones for the static problem, if the crack radius is put equal to vt [3.62].

As for the problem (3.8.3) whose solution at $n \ge 1$, $v < c_R$ was written in the form (3.9.10), it becomes, at $\sigma_{zz}^0 \equiv 0$, the dynamic problem about a punch indentation. Therefore, we shall assume that the velocities and stresses at the punch edge, $r = vt$, $z = 0$, have singularities which are weaker than $\varepsilon^{-1/2}$; besides, the constant speed of the contact boundary, v, will be assumed to be an unknown quantity to be found. Then, in the formula (3.8.8) for $m = 0$, $n \ge 1$, the function $\Sigma_0(\theta_0)$ has, as $\theta_0 \to v^{-1}$, a singularity which is weaker than $\varepsilon^{-n+1/2}$, and hence $F_2(\theta)$ at $\theta \to \pm v^{-1}$ must have poles whose multiplicity is smaller than n, i.e., one must put $m = n - 1$ in the formula (3.9.10).

Note that since condition (3.8.12) is fulfilled for any values of B_k and v in (3.9.10), there are n boundary conditions left for determining the B_k constants and v, the boundary conditions being given by the equations (3.8.8):

$$\operatorname{Re} \int_l L_m^\theta F_2(\theta) \frac{\theta^m d\theta}{\sqrt{\theta^2-\theta_0^2}} = U_m(\theta_0)\,, \quad \theta_0 = \frac{t}{r} > v^{-1}\,, \tag{3.9.18}$$
$$m = -1, -2, \ldots, -n\,.$$

Thus, the solution to (3.8.3) at $n \geq 1$, $v < c_R$ is given by (3.9.10) where the n unknown real constants B_k and v are defined from the conditions (3.9.18).

Let us now consider the elastodynamic problem of indenting a conical punch with the vertex angle α into the half-space $z < 0$ with a constant speed v_0, Fig. 3.26. Zero initial conditions are assumed, and the boundary conditions are defined by

$$\begin{aligned} &\sigma_{zr}(r,0,t) = 0 \quad (0 \leq r < \infty)\,, \\ &\sigma_{zz}(r,0,t) = 0 \quad (vt < r < \infty)\,, \\ &u_z(r,0,t) = -v_0 t + r\cot\alpha \quad (0 \leq r < vt)\,. \end{aligned} \tag{3.9.19}$$

Thus, the solution to the problem is obtained from (3.9.10, 18) by putting $u_z^0 = -v_0 t + r\cot\alpha$, $\sigma_{zz}^0 = 0$, $n = 2$. Then,

$$\frac{F_2(\theta)S(\theta^2)}{\theta\sqrt{c_1^{-2}-\theta^2}} = \frac{B_1}{\theta^2 - v^{-2}}\,, \tag{3.9.20}$$

and the constants B_1, v are defined by

$$\begin{aligned} &B_1 \operatorname{Re} \int_l \frac{d\theta}{\theta\sqrt{\theta^2-\theta_0^2}} \int_0^\theta \frac{\eta\sqrt{c_1^{-2}-\eta^2}\,d\eta}{(\eta^2-v^{-2})S(\eta^2)} = -\frac{c_2^2 v_0}{\theta_0}\,, \\ &B_1 \operatorname{Re} \int_l \frac{d\theta}{\theta^2\sqrt{\theta^2-\theta_0^2}} \int_0^\theta d\eta \int_0^\eta \frac{\tau\sqrt{c_1^{-2}-\tau^2}\,d\tau}{(\tau^2-v^{-2})S(\tau^2)} = -c_2^2\left(\frac{v_0}{\theta_0} - \frac{\cot\alpha}{\theta_0^2}\right)\,, \end{aligned} \tag{3.9.21}$$

where $\theta_0 > v^{-1}$, and the paths of integration with respect to η and τ, as always, circumvent the integrand's singularities from above.

By transforming the left-hand side of the upper expression (3.9.21) in the same way as in (3.9.14), we get the following expression relating B_1 to v:

$$B_1 = -\frac{2c_2^2 v_0}{\pi}\left[\int_0^\infty \frac{\sqrt{c_1^{-2}+x}\,dx}{(x+v^{-2})S(-x)}\right]^{-1}\,, \tag{3.9.22}$$

where the radicals are assumed to be arithmetic. Taking into account that the paths of integration with respect to τ and η in the lower expression (3.9.21) circumvent the singular points from above, we can change the sequence of integration and reduce the lower expression (with the help of the upper one) to

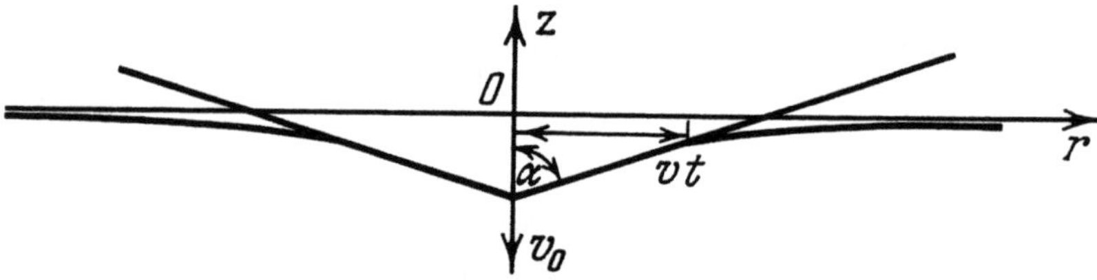

Fig. 3.26. Conical punch indentation

$$B_1 \mathrm{Re} \int\limits_{l} \frac{d\theta}{\theta^2\sqrt{\theta^2-\theta_0^2}} \int_0^{\theta} \frac{\eta^2\sqrt{c_1^{-2}-\eta^2}\,d\eta}{(\eta^2-v^{-2})S(\eta^2)}$$
$$= -\frac{c_2^2}{\theta_0^2}\cot\alpha \quad (\theta_0 > v^{-1}) , \tag{3.9.23}$$

where the path of integration of the inner integral circumvents the integrand's singular points from above. Next, we transform the double integral entering the latter expression in the following way (taking into account that we are interested in the real part only):

$$\mathrm{Re} \int\limits_{l} f(\theta)d\theta \int\limits_{0}^{\theta} F(\eta)d\eta = \mathrm{Re}\left[\int\limits_{-\infty}^{-\theta_0} f(\theta)d\theta \int\limits_{0}^{-\theta_0} F(\eta)d\eta + \int\limits_{\theta_0}^{\infty} f(\theta)d\theta \int\limits_{0}^{\theta_0} F(\eta)d\eta\right]$$
$$= \frac{1}{\theta_0^2}\mathrm{Re}\left[\int\limits_{0}^{\theta_0} F(\eta)d\eta - \int\limits_{0}^{-\theta_0} F(\eta)d\eta\right]$$
$$= \frac{1}{\theta_0^2}\mathrm{Re}\int\limits_{-\infty}^{\infty} F(\eta)d\eta = \frac{\pi}{2\theta_0^2(c_2^{-2}-c_1^{-2})} ,$$
$$\left(\int\limits_{-\infty}^{\infty} F(\eta)d\eta \equiv \lim_{k\to\infty}\int\limits_{-k}^{k} F(\eta)d\eta\right) . \tag{3.9.24}$$

Then, we get from (3.9.23):

$$B_1 = -\frac{2}{\pi}\cot\alpha(1-c_2^2/c_1^2) . \tag{3.9.25}$$

Equating this expression for B_1 to that of (3.9.22), we arrive at the following transcendental equation for v

$$\int\limits_{0}^{\infty} \frac{\sqrt{c_1^{-2}+x}\,dx}{(x+v^{-2})S(-x)} = \frac{v_0\tan\alpha}{c_2^{-2}-c_1^{-2}} . \tag{3.9.26}$$

From the second formula (3.8.8) we get, at $n = 2$, $m = -1$, the stress distribution under the punch ($\theta_0 > v^{-1}$)

$$\sigma_{zz}(r,0,t) = \mu\theta_0 B_1 \mathrm{Re} \int_l \frac{d\theta}{\theta\sqrt{\theta^2-\theta_0^2}} \int_0^\theta \frac{\eta d\eta}{\eta^2 - v^{-2}}$$

$$= \mu\theta_0 \mathrm{Re}\left[-\int_{\theta_0}^{v^{-1}} \frac{\pi i}{2} \frac{d\theta}{\theta\sqrt{\theta^2-\theta_0^2}} + \int_{-v^{-1}}^{-\theta_0} \frac{\pi i}{2} \frac{d\theta}{\theta\sqrt{\theta^2-\theta_0^2}}\right]$$

$$= B_1\mu\pi \ln\left[\frac{vt+\sqrt{v^2t^2-r^2}}{r}\right] \quad \left(\sqrt{v^2t^2-r^2}>0\right). \tag{3.9.27}$$

Note that the latter expression for σ_{zz} coincides with the analogous one obtained for the static problem, if the contact area radius is put equal to vt [3.72].

According to (3.9.27), the stress distribution under the punch has a logarithmic singularity at $r \to 0$.

Integration of (3.9.27) yields the force acting upon the punch

$$P = -2\pi \int_0^{vt} \sigma_{zz}(r,0,t) r\, dr = -B_1\mu\pi^2 v^2 t^2 . \tag{3.9.28}$$

The above results, (3.9.20, 25–28), coincide with those presented in [3.39, 63]. Note that *Kostrov* [3.29] was the first to consider the axisymmetric elastodynamic problem for a conical punch; however, he supposed that $v = v_0 \tan\alpha$, though v must be determined from (3.9.26). Later, Kostrov's approach was used by *Robinson* and *Thompson* [3.39] for solving the problem within the entire range $0 < v < \infty$.

It should be noted that (3.9.1, 2) were derived subject to the asymptotic condition (3.8.12), whose fulfillment ensured validity of the estimates (3.8.10, 11), and, thus, the stresses proved to be integrable as $r \to 0$. If the given boundary conditions imply a stress singularity as $r \to 0$, then the condition (3.8.12) does not hold. Then, the lower formulas in (3.9.1, 2) should be, generally speaking, derived taking into account the contribution due to the integrals (3.8.8) over a semicircle of infinitely large radius while deforming the contour l into the semiaxes $(-\infty, -\theta_0)$, $(\theta_0, +\infty)$. Thus, the lower formulas in (3.9.1, 2) may become invalid, and one has to employ some other method of solution. Nevertheless, in this case too, the problem may be reduced to the previous one for which the required function has the asymptotic (3.8.12). Let us show this for the problem (3.8.2). [A similar procedure is applicable to (3.8.3).] In fact, according to the representation (3.8.8) $F_2(\theta)$ must satisfy the following conditions:

$$\operatorname{Re} \int_l L_m^\theta F_2(\theta) \frac{\theta^m \, d\theta}{\sqrt{\theta^2 - \theta_0^2}} = U_m(\theta_0) \quad (\theta_0 < v^{-1}) \,,$$

$$\operatorname{Re} \int_l L_m^\theta \left[\frac{F_2(\theta) S(\theta^2)}{\sqrt{c_1^{-2} - \theta^2}} \right] \frac{\theta^m \, d\theta}{\sqrt{\theta^2 - \theta_0^2}} = \Sigma_m(\theta_0) \quad (\theta_0 > v^{-1}) \,, \tag{3.9.29}$$

where the function $\Sigma_m(\theta_0)$ is not, generally speaking, bounded at infinity. Let us represent it in the form

$$\Sigma_m(\theta_0) = \Sigma_m^0(\theta_0) + \Sigma_m^1(\theta_0) \,, \tag{3.9.30}$$

where $\Sigma_m^0(\theta_0)$ is defined over the entire range $0 < \theta_0 < \infty$, and

$$\Sigma_m^1(\theta_0) = O(\theta_0^{-\varepsilon}) \quad \text{as} \quad \theta_0 \to \infty \,, \quad \varepsilon > 0 \,.$$

The required function $F_2(\theta)$ will be sought in the form

$$F_2(\theta) = F_2^0(\theta) + F_2^1(\theta) \,, \tag{3.9.31}$$

where the function $F_2^0(\theta)$ is supposed to be analytical within the region $\operatorname{Im} \theta > 0$ and to meet the condition

$$\operatorname{Re} \int_l L_m^\theta \left[\frac{F_2^0(\theta) S(\theta^2)}{\sqrt{c_1^{-2} - \theta^2}} \right] \frac{\theta^m \, d\theta}{\sqrt{\theta^2 - \theta_0^2}} = \Sigma_m^0(\theta_0) \tag{3.9.32}$$

over the entire range $0 < \theta_0 < \infty$.

By applying (3.8.18) for inverting (3.9.32) and the first of the relationships (3.9.29), we get with the help of $\operatorname{Im} \Phi(\theta) = \operatorname{Im} \Phi(-\theta)$

$$\operatorname{Im} \left\{ \theta^m L_m^\theta \left[\frac{F_2^0(\theta) S(\theta^2)}{\sqrt{c_1^{-2} - \theta^2}} \right] \right\} = -\frac{|\theta|}{\pi} \int_0^{|\theta|} \frac{[\Sigma_m^0)(\theta_0)]' \, d\theta_0}{\sqrt{\theta^2 - \theta_0^2}} \tag{3.9.33}$$

$$(-\infty < \theta < \infty) \,,$$

$$\operatorname{Im} \{\theta^m L_m^\theta F_2^1(\theta)\} = -\frac{|\theta|}{\pi} \int_0^{|\theta|} \frac{U_m'(\theta_0) \, d\theta_0}{\sqrt{\theta^2 - \theta_0^2}} - \operatorname{Im} \{\theta^m L_m^\theta F_2^0(\theta)\} \tag{3.9.34}$$

$$(-v^{-1} < \theta < v^{-1}) \,.$$

On the other hand, subtracting (3.9.32) from the second relationship in (3.9.29), we deduce that the function $F_2^1(\theta)$ meets the second relationship in (3.9.29), with, however, the right-hand side equal to $\Sigma_m^1(\theta_0)$ and decreasing at infinity as $O(\theta_0^{-\varepsilon})$, $\varepsilon > 0$. Hence, the function $F_2^1(\theta)$ must be sought as decreasing at infinity as $O(\theta^{-1-\varepsilon})$. Then, inverting the latter relationship for $F_2^1(\theta)$, we get with the help of $\operatorname{Re} \Phi(-\theta) = -\operatorname{Re} \Phi(\theta)$

$$\mathrm{Re}\left\{\theta^m L^\theta_m\left[\frac{F^1_2(\theta)S(\theta^2)}{\sqrt{c_1^{-2}-\theta^2}}\right]\right\} = -\frac{\theta}{\pi}\int\limits_{|\theta|}^{\infty}\frac{[\Sigma^1_m)(\theta_0)]'\,d\theta_0}{\sqrt{\theta_0^2-\theta^2}} \tag{3.9.35}$$

$$(\theta < -v^{-1},\ \theta > v^{-1})\,.$$

Supposing now that the function $F^0_2(\theta)$ on the right-hand side of (3.9.34) is a known function [to be found from the condition (3.9.33)], we come to the conclusion that the formulas (3.9.34, 35) derived for determining the decreasing-at-infinity function $F^1_2(\theta)$ coincide with the analogous formulas (3.9.1) derived earlier for $F_2(\theta)$.

In a similar way, the representations (3.9.33–35) may be used for solving the problem (3.8.2) for $n \le 0$. However, unlike the $n \ge 1$ case, the right-hand sides of the expressions (3.9.33, 34) will incorporate unknown constants to be found from the additional conditions imposed on the solution's behavior as $\theta_0 \to \infty$, $\theta_0 \to v^{-1}$.

3.9.3 Expansion of a Circular Crack Due to a Concentrated Load

As an example, let us consider the problem of a constant-speed ($v < c_R$) expansion of a circular cut whose sides starting from the instant $t = 0$ at the point $r = 0$ are loaded with a concentrated compressive force p, see Fig. 3.27. Then, in (3.8.2) $u^0_z(r,t) = 0$, $\sigma^0_{zz}(r,t) = -pH(t)\delta(r)/2\pi r$, and hence for the problem under consideration $n = 0$:

$$U_0(\theta_0) = 0 \quad (\theta_0 < v^{-1})\,,$$

$$\Sigma_0(\theta_0) = \frac{r}{\mu}\int\limits_0^t \sigma_{zz}(r,0,\tau)\,d\tau = \frac{r}{\mu}\int\limits_0^{r/v}\sigma_{zz}(r,0,\tau)d\tau$$

$$-\frac{pr}{2\pi\mu}\int\limits_{r/v}^{t}\frac{\delta(r)}{r}H(\tau)d\tau = A - \frac{p}{2\pi\mu}\left(1-\frac{1}{v\theta_0}\right)\delta\left(\frac{1}{\theta_0}\right)$$

$$(\theta_0 > v^{-1},\ \mathrm{Im}\,A = 0)\,.$$

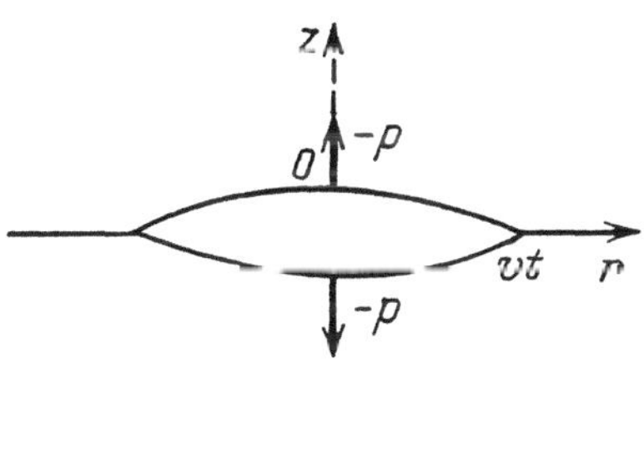

Fig. 3.27. Expansion of a circular cut due to a concentrated force suddenly applied at the origin of coordinates

Then, $F_2(\theta)$ may be found from the following equations [obtained from (3.9.29)]:

$$\operatorname{Re} \int_l F_2(\theta) \frac{d\theta}{\sqrt{\theta^2 - \theta_0^2}} = 0 \quad (\theta_0 < v^{-1}) ,$$

$$\operatorname{Re} \int_l \frac{F_2(\theta) S(\theta^2)}{\sqrt{c_1^{-2} - \theta^2}} \frac{d\theta}{\sqrt{\theta^2 - \theta_0^2}} = A - \frac{p}{2\pi\mu}\left(1 - \frac{1}{v\theta_0}\right)\delta\left(\frac{1}{\theta_0}\right) \tag{3.9.36}$$

$(\theta_0 > v^{-1})$.

By inverting this first of the equations (3.9.36), we get

$$\operatorname{Im} F_2(\theta) = 0 \quad (-v^{-1} < \theta < v^{-1}) . \tag{3.9.37}$$

In order to solve the second equation, let us write $F_2(\theta)$ in the form (3.9.31) where $F_2^0(\theta)$ may be obtained from

$$\operatorname{Re} \int_l \frac{F_2^0(\theta) S(\theta^2)}{\sqrt{c_1^{-2} - \theta^2}} \frac{d\theta}{\sqrt{\theta^2 - \theta_0^2}}$$
$$= A - \left(1 - \frac{1}{v\theta_0}\right)\delta\left(\frac{1}{\theta_0}\right)\frac{p}{2\pi\mu} \quad (0 < \theta_0 < \infty) . \tag{3.9.38}$$

The decreasing-at-infinity function $F_2^1(\theta)$ is defined by the equation

$$\operatorname{Re} \int_l \frac{F_2^1(\theta) S(\theta^2)}{\sqrt{c_1^{-2} - \theta^2}} \frac{d\theta}{\sqrt{\theta^2 - \theta_0^2}} = 0 \quad (v^{-1} < \theta_0)$$

whose solution gives

$$\operatorname{Re} \left[\frac{F_2^1(\theta) S(\theta^2)}{\sqrt{c_1^{-2} - \theta^2}}\right] = 0 \quad (\theta < -v^{-1},\ \theta > v^{-1}) . \tag{3.9.39}$$

Taking into account the property of the delta-function $[f(r + r_0)\delta(r) = f(r_0)\delta(r)]$, one can rewrite the right-hand side of (3.9.38) in the form

$$A - \frac{p}{2\pi\mu}\left(1 - \frac{1}{v\theta_0}\right)\delta\left(\frac{1}{\theta_0}\right) = A - \frac{p}{2\pi\mu}\delta\left(\frac{1}{\theta_0}\right) , \tag{3.9.40}$$

and the component of the solution $F_{20}^0(\theta)$ for the term incorporating the delta-function in (3.8.38) has, according to (3.8.25), the form $(F_2^0(\theta) \equiv F_{20}^0(\theta) + F_{21}^0(\theta))$

$$\frac{F_{20}^0(\theta) S(\theta^2)}{\sqrt{c_1^{-2} - \theta^2}} = \frac{\theta p}{2\pi^2\mu} \quad (\operatorname{Im}\theta \geq 0) . \tag{3.9.41}$$

For the term A on the right-hand side of (3.9.38), the equation (3.9.38) leads to the solution

$$\operatorname{Im}\Phi(\theta) = -\frac{A}{\pi} \left(-\infty < \theta < \infty,\ \Phi(\theta) = \frac{F_{21}^{0}(\theta)S(\theta^2)}{\sqrt{c_1^{-2} - \theta^2}} \right) . \tag{3.9.42}$$

Hence, if $\Phi(\theta)$ is bounded, then it must be constant, i.e.

$$\Phi(\theta) = \frac{A}{\pi \mathrm{i}} + B \quad (\operatorname{Im} B == 0,\ \operatorname{Im}\theta \geq 0) . \tag{3.9.43}$$

However, from the condition (3.8.15) we have

$$\operatorname{Re}\Phi(-\theta) = -\operatorname{Re}\Phi(\theta) \tag{3.9.44}$$

and hence,

$$B = 0 , \tag{3.9.45}$$

and

$$\Phi(\theta) = \frac{a}{\pi \mathrm{i}} \quad (\operatorname{Im}\theta \geq 0) . \tag{3.9.46}$$

Taking into account (3.9.41, 46), we arrive at the solution to (3.9.38):

$$\frac{F_2^0(\theta)S(\theta^2)}{\sqrt{c_1^{-2} - \theta^2}} = \frac{A}{\pi \mathrm{i}} + \frac{\theta p}{2\pi^2\mu} \quad (\operatorname{Im}\theta \geq 0) , \tag{3.9.47}$$

and, with the help of (3.9.39, 47), the following boundary condition is obtained for determining the function $F_2(\theta)$

$$\operatorname{Re}\left[\frac{F_2(\theta)S(\theta^2)}{\sqrt{c_1^{-2} - \theta^2}}\right] = \frac{\theta p}{2\pi^2\mu} \quad (\theta < -v^{-1},\ \theta > v^{-1}) . \tag{3.9.48}$$

The latter condition may be presented in the form

$$\operatorname{Im} F_2(\theta) = \frac{p\theta\sqrt{c_1^{-2} - \theta^2}}{2\pi^2\mu \mathrm{i} S(\theta^2)} \quad (\theta < -v^{-1},\ \theta > v^{-1}) \tag{3.9.49}$$

(note the absence of the constant A).

Finally, by solving the boundary-value problem subject to the boundary conditions (3.9.37, 49), we find, with the help of the Schwarz integral, the following expression for the required function $F_2(\theta)$ bounded at infinity:

$$F_2(\theta) - \mathrm{i}\,\mathrm{Im}\,F_2(\infty) = \frac{1}{\pi}\int\limits_{v^{-1}}^{\infty}\left[\frac{p\alpha\sqrt{c_1^{-2}-\alpha^2}}{2\pi^2\mu\mathrm{i}S(\alpha^2)} - \mathrm{Im}\,F_2(\infty)\right]\frac{d\alpha}{\alpha-\theta}$$

$$-\frac{1}{\pi}\int\limits_{-v^{-1}}^{v^{-1}}\mathrm{Im}\,F_2(\infty)\frac{d\alpha}{\alpha-\theta}$$

$$+\frac{1}{\pi}\int\limits_{-\infty}^{-v^{-1}}\left[\frac{p\alpha\sqrt{c_1^{-2}-\alpha^2}}{2\pi^2\mu\mathrm{i}S(\alpha^2)} - \mathrm{Im}\,F_2(\infty)\right]\frac{d\alpha}{\alpha-\theta} + D$$

$$(\mathrm{Im}\,D = 0,\ \mathrm{Im}\,\theta > 0)\,. \tag{3.9.50}$$

But, according to (3.9.47) as $|\theta| \to \infty$, $D = 0$, and the expression (3.9.50) becomes

$$F_2(\theta) = -\frac{\theta p\mathrm{i}}{\pi^3\mu}\int\limits_{v^{-1}}^{\infty}\frac{\sqrt{c_1^{-2}-\alpha^2}\,\alpha\,d\alpha}{S(\alpha^2)(\alpha^2-\theta^2)} \quad (\mathrm{Im}\,\theta > 0)\,. \tag{3.9.51}$$

Substituting the latter expression for $F_2(\theta)$ into the first of the formulas (3.8.8) at $n = m = 0$, we get the following relationship for the displacement at the points of the lower side of the cut ($\theta_0 > v^{-1}$):

$$u_z(r,-0,t) = -\frac{1}{2c_2^2 r}\,\mathrm{Re}\int\limits_{l}\frac{\theta p\mathrm{i}}{\pi^3\mu}\frac{d\theta}{\sqrt{\theta^2-\theta_0^2}}\int\limits_{v^{-2}}^{\infty}\frac{\sqrt{c_1^{-2}-z}\,dz}{S(z)(z-\theta^2)}$$

$$= \frac{p}{2\pi^3 c_2^2\mu r}\,\mathrm{Re}\int\limits_{l}\frac{\theta d\theta}{\mathrm{i}\sqrt{\theta^2-\theta_0^2}}\int\limits_{v^{-2}}^{\theta_0^2}\frac{\sqrt{c_1^{-2}-z}\,dz}{S(z)(z-\theta^2)}$$

$$= \frac{p}{4\pi^3 c_2^2\mu r}\,\mathrm{Re}\int\limits_{v^{-2}}^{\theta_0^2}\frac{\sqrt{c_1^{-2}-z}\,dz}{S(z)}\int\limits_{l_1}\frac{d\nu}{\mathrm{i}(z-\nu)\sqrt{\nu-\theta_0^2}}\,. \tag{3.9.52}$$

Figure 3.28 shows the closed contour l_1 in the plane $\nu = \theta^2$, enveloping the pole at the point $\nu = z$. Note that the cut for the root $\sqrt{\nu-\theta_0^2}$ passes along the real axis from the point $\nu = \theta_0^2$ and into infinity (this cut appeared due to the cuts $(-\infty,-\theta_0)$ and $(\theta_0,+\infty)$ in the plane for the root $\sqrt{\theta^2-\theta_0^2}$).

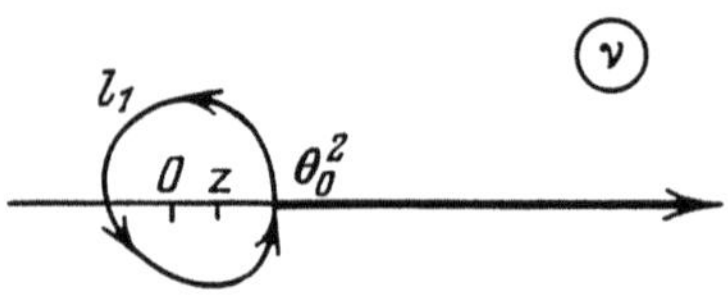

Fig. 3.28. Closed contour l_1

Since the inner integral in (3.9.52) is equal to $2\pi i/\sqrt{\theta_0^2 - z}$, we get finally the expression

$$u_z(r,-0,t) = \frac{p}{2\pi^2 c_2^2 \mu r} \int\limits_{v^{-2}}^{\theta_0^2} \frac{\sqrt{z - c_1^{-2}}\,dz}{S(z)\sqrt{\theta_0^2 - z}} \quad (\theta_0 > v^{-1}) , \tag{3.9.53}$$

where all the radicals are assumed to be arithmetic.

In particular, for a vanishing density ϱ (entering the expression for c_1, c_2) we get

$$S(z) \to -2z(c_2^{-2} - c_1^{-2}) + o(\varrho) ,$$

and hence (3.9.53) reduces to the expression

$$u_z(r,-0,t) = -\frac{p(\lambda + 2\mu)}{2\pi^2\mu(\lambda + \mu)r}\left[\frac{\pi}{2} - \sin^{-1}\frac{r}{vt}\right] , \tag{3.9.54}$$

which coincides with the analogous expression from the solution to the static problem, if the crack's radius is put equal to vt [3.62].

Note that the same results, (3.9.51,53), may also be obtained without employing the delta-function $\delta(1/\theta_0)$. This can be done by replacing $\delta(1/\theta_0)$ in (3.9.36) by a smooth function $\varepsilon\theta_0^{-1}(\varepsilon^2 + \theta_0^{-2})^{-3/2}(\varepsilon > 0)$ and passing in the thus obtained solution to the limits as $\varepsilon \to 0$, taking into account that $\varepsilon(\varepsilon^2 + r^2)^{-3/2} \to \delta(r)/r$.

3.10 An Alternative Derivation of the Smirnov-Sobolev Representations

Here we consider an alterntive derivation of the Smirnov–Sobolev relationships (3.4.7–10) for a half-plane, which allows, in particular, to generalize the above results onto the case of anisotropic elastic media.

Let the displacement vector $\boldsymbol{u}(t,x,y)$, being a homogeneous function of the $(n-1)^{\text{th}}$ dimension (where n is an integer) with respect to the variables t, x, y satisfy the Lamé equations (2.1.8) in the case of plane deformation in the absence of bulk forces.

Consider the function $L_n^t \boldsymbol{u}$ [see (3.4.5)] which is also a solution to the plane-strain Lamè equations.

The derivation is based on representing the function $L_n^t \boldsymbol{u}$ by superposing plane waves. The possibility of such a representation is indicated, in particular, by the form of the integral representing the function $L_n^t \boldsymbol{u}$ on the axis $y = 0$.

In fact, let $U(\theta)$ be an analytical function in the upper half-plane of the complex variable θ. By the Cauchy theorem,

$$U(\theta) = \frac{i}{2\pi}\int\limits_{\Gamma} \frac{U(\tau)d\tau}{\theta - \tau} , \tag{3.10.1}$$

where Γ is a closed contour presented in Fig. 3.29. The virtual singular points on the real axis of the function $U(\theta)$ are circumvented from above. Let $F(\theta)$ be a boundary value of the real part of the function $U(\theta)$ on the real axis, where $F(\theta) = \operatorname{Re} U(\theta)$ at $\operatorname{Im}\theta = 0$. Then, taking into account (3.10.1), we get

$$F(\theta) = \lim_{\varepsilon \to +0} \operatorname{Re} U(\theta + \mathrm{i}\varepsilon) = \operatorname{Re} \frac{\mathrm{i}}{2\pi} \int_\Gamma \frac{U(\tau)d\tau}{\theta + \mathrm{i}0 - \tau}, \tag{3.10.2}$$

where the point θ in the integral is circumvented from below (Fig. 3.29).

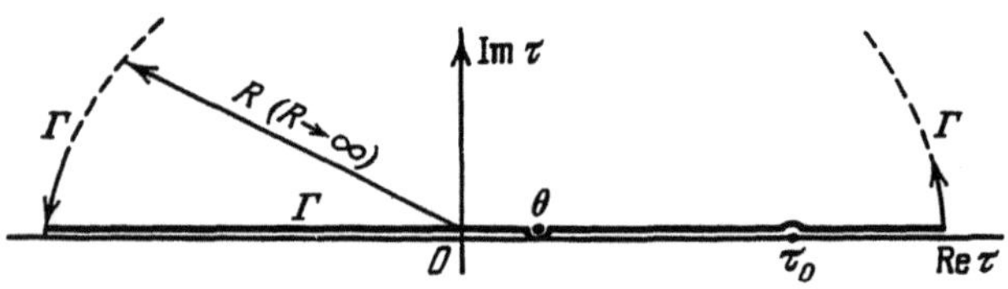

Fig. 3.29. Closed contours Γ [τ_0 is the singular point of the function $U(\tau)$]

In the framework of the problem under consideration, the function $L_n^t \boldsymbol{u}$ is a homogeneous function of the dimension -1; hence, according to (3.10.2) we have on the axis $y = 0$:

$$\begin{aligned} L_n^t u_\alpha(t,x,0) &= \frac{1}{x} F_\alpha\left(\frac{t}{x}\right) = \frac{1}{x}\operatorname{Re}\frac{\mathrm{i}}{2\pi}\int_\Gamma \frac{U_\alpha(\tau)d\tau}{t/x + \mathrm{i}0 - \tau} \\ &= \operatorname{Re}\frac{\mathrm{i}}{2\pi}\int_\Gamma \frac{U_\alpha(\tau)d\tau}{t - (\tau - \mathrm{i}0)x} \quad (\alpha = x, y) . \end{aligned} \tag{3.10.3}$$

In the latter integral integration is carried out taking into account the remark made above (Fig. 3.29) where $\theta = t/x + \mathrm{i}0$). Thus, according to (3.10.3), the function $L_n^t \boldsymbol{u}$ may be presented on the axis $y = 0$ by superposing the functions $1/[t - (\tau - \mathrm{i}0)x]$ which are the boundary values of the plane-wave functions $1/[t - (\tau - \mathrm{i}0)x - \omega y]$ at $y = 0$.

The structure of the expression (3.10.3) suggests that the functions $L_n^t u_\alpha(t,x,y)$ may be represented by superposing the plane waves of the type

$$L_n^t u_\alpha(t,x,y) = \frac{U_\alpha(\tau)}{t - \tau x - \omega y} . \tag{3.10.4}$$

By applying the operator L_n^t to the plane-strain Lamé equations and replacing the functions $L_n^t u_\alpha$ with the help of (3.10.4), we arrive at the following system:

$$\begin{aligned} &[(\lambda + 2\mu)\tau^2 + \mu\omega^2 - \varrho]U_x + (\lambda + \mu)\tau\omega U_y = 0 , \\ &(\lambda + \mu)\tau\omega U_x + [(\lambda + 2\mu)\omega^2 + \mu\tau^2 - \varrho]U_y = 0 . \end{aligned} \tag{3.10.5}$$

The existence of the non-zero solution (U_x, U_y) requires that the determinant of the system (3.10.5) be equal to zero; hence, we get the following four roots:

$$\omega_1 = -\sqrt{c_1^{-2} - \tau^2}\,, \quad \omega_2 = -\sqrt{c_2^{-2} - \tau^2}\,,$$
$$\omega_3 = \sqrt{c_1^{-2} - \tau^2}\,, \qquad \omega_4 = \sqrt{c_2^{-2} - \tau^2}\,.$$

For the sake of definiteness, the branches of the radicals $\sqrt{c_\beta^{-2} - \tau^2}$ ($\beta = 1, 2$) are fixed in such a way that $\sqrt{c_\beta^{-2} - \tau^2} > 0$ for $\tau = ib$ ($b > 0$) (the cuts being drawn along the segments $(-c_\beta^{-1}, c_\beta^{-1})$.

As a result, the representation of the function $L_n^t u_\alpha$ in the region $y < 0$ by superposing the solutions (3.10.4) leads to the expression

$$L_n^t u_\alpha(t,x,y) = \sum_{\beta=1,2} \operatorname{Re} \frac{\mathrm{i}}{2\pi} \int\limits_{\Gamma} \frac{U_\alpha^{(\beta)}(\tau)d\tau}{t - \tau x - \omega_\beta y} \quad (y < 0)\,, \tag{3.10.6}$$

where through $U_\alpha^{(\beta)}$ we denote the solution U_α corresponding to the chosen root ω_β, and the contour Γ is taken as shown in Fig. 3.29. Note, that the sum does not include the terms due to the roots ω_3, ω_4, since the terms are equal to zero. In fact, if $x^2 + y^2 < c_\beta^2 t^2$, $t > 0$, $y < 0$, then the only root of the equation $t - \tau x - y\sqrt{c_\beta^{-2} - \tau^2} = 0$ belongs (taking into account the chosen branch of the radical) to the lower half-plane τ:

$$\tau = \frac{\xi + \mathrm{i}\eta\sqrt{1 - c_\beta^{-2}(\xi^2 + \eta^2)}}{\xi^2 + \eta^2} \left(\xi = \frac{t}{x},\ \eta = \frac{t}{y},\ \sqrt{1 - c_\beta^{-2}(\xi^2 + \eta^2)} > 0\right).$$

Hence, as far as the closed contour Γ belongs to the upper half-plane where the integrands are analytical, the corresponding integrals in (3.10.6) vanish.

Since in the upper half-plane τ all the integrands of the integrals comprising the sum (3.10.6) have, at $x^2+y^2 < c_\beta^2 t^2$, $y < 0$, $t > 0$, one simple pole $\tau = \theta_\beta$ (the root of the equation $t - \tau x + y\sqrt{c_\beta^{-2} - \tau^2} = 0$), we get, by applying the theorem of residues to (3.10.6),

$$L_n^t u_\alpha(t,x,y) = \sum_{\beta=1,2} \operatorname{Re} \frac{U_\alpha^{(\beta)}(\theta_\beta)\sqrt{c_\beta^{-2} - \theta_\beta^2}}{\theta_\beta + x\sqrt{c_\beta^{-2} - \theta_\beta^2}}\,, \tag{3.10.7}$$

where

$$\theta_\beta = \frac{\xi - \mathrm{i}\eta\sqrt{1 - c_\beta^{-2}(\xi^2 + \eta^2)}}{\xi^2 + \eta^2}\,, \quad \xi = \frac{x}{t}\,, \quad \eta = \frac{y}{t}\,;$$
$$\alpha = x, y\,; \quad \beta = 1, 2\,; \quad \sqrt{1 - c_\beta^{-2}(\xi^2 + \eta^2)} > 0\,. \tag{3.10.8}$$

At $t < 0$, $y < 0$, $x^2 + y^2 < c_\beta^2 t^2$ we have $\operatorname{Im}\theta_\beta < 0$, and hence,

$$L_n^t u_\alpha(t,x,y) \equiv 0\,. \tag{3.10.9}$$

Note that the results (3.10.7–9) obtained at $x^2+y^2<c_\beta^2t^2$ remain valid at $x^2+y^2\geq c_\beta^{-2}t^2$ too, if we take into account the continuity of the passage across the circle $x^2+y^2=c_\beta^2t^2$. According to (3.10.8), the root θ_β becomes real in this case only (namely, $\operatorname{Im}\theta_\beta=+0$ at $t>0$, and $\operatorname{Im}\theta_\beta=-0$ at $t<0$).

Since from the first equation of the system (3.10.5) it follows that

$$U_y^{(1)}(\theta)=-\frac{\sqrt{c_1^{-2}-\theta^2}}{\theta}U_x^{(1)}(\theta)\,,\quad U_y^{(2)}(\theta)=\frac{\theta U_x^{(2)}(\theta)}{\sqrt{c_2^{-2}-\theta^2}}\,,$$

and differentiation of the identity $t-\theta_\beta x+y\sqrt{c_\beta^{-2}-\theta_\beta^2}=0$ with respect to x and y results in the relationships

$$\frac{\sqrt{c_\beta^{-2}-\theta_\beta^2}}{y\theta_\beta+x\sqrt{c_\beta^{-2}-\theta_\beta^2}}=-\frac{1}{\theta_\beta}\frac{\partial\theta_\beta}{\partial x}=\frac{1}{\sqrt{c_\beta^{-2}-\theta_\beta^2}}\frac{\partial\theta_\beta}{\partial y}\quad(\beta=1,2)\,,$$

(3.10.7) may be finally rewritten in the form

$$\begin{aligned}L_n^tu_x(t,x,y)&=\operatorname{Re}\left[-\frac{U_x^{(1)}(\theta_1)}{\theta_1}\frac{\partial\theta_1}{\partial x}+\frac{U_x^{(2)}(\theta_2)}{\sqrt{c_2^{-2}-\theta_2^2}}\frac{\partial\theta_2}{\partial y}\right],\\ L_n^tu_y(t,x,y)&=\operatorname{Re}\left[-\frac{U_x^{(1)}(\theta_1)}{\theta_1}\frac{\partial\theta_1}{\partial y}-\frac{U_x^{(2)}(\theta_2)}{\sqrt{c_2^{-2}-\theta_2^2}}\frac{\partial\theta_2}{\partial x}\right].\end{aligned}\tag{3.10.10}$$

Upon denoting

$$f_1'(\theta_1)=-U_x^{(1)}(\theta_1)/\theta_1\,,\qquad f_2'(\theta_2)=U_x^{(2)}(\theta_2)/\sqrt{c_2^{-2}-\theta_2^2}\,,$$

the latter expressions coincide with the Smirnov–Sobolev representations (3.4.9) derived above.

As compared to the traditional one, the advantage of the present approach to deriving the Smirnov–Sobolev representations is due to the fact that it is not in need of a preliminary representation of the displacement vector $\boldsymbol{u}$ through the potentials of the longitudinal and transverse waves, but deals directly with the equations of motion of elastic media. This approach allows to generalize the Smirnov–Sobolev representations for obtaining solutions to two-dimensional nonstationary problems involving fully anisotropic media.

It should be mentioned that the idea of the present representation of homogeneous solutions to elastodynamic problems in the form of superposition of plane waves, was originally proposed by Burridge who used it for obtaining the solution to Lamb's problem in a new form [3.73], as well as for solving the problems of a crack expanding in an anisotropic medium[2].

[2] The solution was obtained by Burridge in his work "The General Two-Dimensional Self-Similar Crack in an Anisotropic Elastic Medium" (1968) which is referred to by him and *Willis* [3.74] as an unpublished one.

4. Integral Transforms in Elastodynamics

Originally, the method of integral transforms in elastodynamics was used by Lamb [4.1] for solving the problem of a concentrated load acting onto an elastic half-space. The method was widely used starting from the 1950s in connection with an intense development of studies in the field of elastodynamics and the necessity of solving a number of applied problems. Important contributions were made by Petrashen' with his colleagues [4.2] and Pekeris [4.3, 4] as well as by numerous researchers cited in the review [4.5]. Studies in the field of elastodynamics were further promoted by Cagniard [4.6] and de Hoop [4.7] who pioneered the use of integral transforms and succeeded in creating an original method of inversing the Laplace transform while using simultaneously integral Fourier and Laplace transforms (the Cagniard-de Hoop method). Further intense developments of s-tudies in the field of elastodynamics are closely associated with the names of Maue [4.8], Flitman [4.9, 10], Kostrov [4.11, 12] who employed the Wiener–Hopf method permitting the solution of a number of complicated problems with mixed boundary conditions. A great contribution to the solution of complicated axisymmetric problems was made by Skalak [4.13], Atkinson [4.14] and the authors of [4.15–17]. Recent achievements due to Freund [4.18], Kostrov [4.19, 20], Slepyan [4.21] have made it possible to use the method of integral transforms for solving a number of complicated elastodynamic problems with mixed boundary conditions involving unsteady motion of their boundaries. Studies in the field of elastodynamics employing integral transforms were further developed by Willis [4.22] who succeeded in solving complicated homogeneous problems for an anisotropic space/half-space, as well as by the works [4.23, 24] dedicated to solving two- and three-dimensional problems for wedge-shaped and conic regions.

Application of integral transforms to elastodynamics was carried out by many authors whose works are reviewed in [4.25–31].

4.1 Application of Integral Transforms to Solving Elastodynamic Problems

As is known (see, e.g. [4.32, 33]), a linear partial differential equation can be reduced to an ordinary equation by multiplying all the terms by the kernel of a certain integral transform and integrating the resultant equality between appropriate limits. Let us start by noting that an application of the Laplace transform (with respect to time t) to the dyamic equations (2.1.6) or the wave equations for longitudinal and transverse potentials makes it possible to reduce the above hyperbolic systems to elliptic ones for transforms of the desired functions. Hence, the images must be analytic functions of their spatial variables everywhere within the physical domain under consideration (excluding the wave sources). This often facilitates considerably obtaining a formal solution to the problem.

In the case of Cartesian coordinates x_1, x_2, x_3 the derivatives with respect to a spatial variable changing between infinite limits may be excluded by subjecting the equations of elastodynamics to the Fourier (or Laplace) transform defined over an infinite interval with respect to the corresponding variable. If the equations are written in the cylindrical system of coordinates r, θ, z ($x_1 = r\cos\theta, x_2 = r\sin\theta, x_3 = z$) it is often advantageous to use the Hankel transform with respect to the radial coordinate r, and the Fourier transform between finite limits with respect to the angular coordinate θ, i.e. to use an expansion into the trigonometric Fourier series in θ.

In the spherical system of coordinates r, θ, ϕ ($x_1 = r\cos\phi\sin\theta$, $x_2 = r\sin\theta\sin\phi$, $x_3 = r\cos\theta$), it is sometimes convenient to employ the Fourier transform over a finite interval with respect to θ (an expansion in the Legendre functions); also the Kontorovich–Lebedev transform in r may be used.

Note that the Laplace transform is usually implemented with respect to a variable which denotes time and is defined over a semi-infinite interval, since in the case of nonstationary problems processes are considered for $t > 0$. However, as the Fourier transform is, in essence, equivalent to the Laplace one, both can be used for carrying out transformations with respect to both spatial and time variables. In fact, the Fourier transform in the complex domain may be shown to be equivalent to the Laplace transform.

Let the function $f(x)$ be integrable over any finite interval and satisfying the relationships

$$f(x) = \begin{cases} \mathrm{O}\left(e^{\tau_- x}\right) & (x \to \infty), \\ \mathrm{O}\left(e^{\tau_+ x}\right) & (x \to -\infty, \ \tau_- < \tau_+). \end{cases} \tag{4.1.1}$$

Let us next consider the Fourier transform of the function $f(x)$:

$$f^*(\alpha) = \int_{-\infty}^{\infty} f(x) e^{i\alpha x}\, dx, \quad f(x) = \frac{1}{2\pi}\int_{i\tau-\infty}^{i\tau+\infty} f^*(\alpha) e^{-i\alpha x}\, d\alpha \tag{4.1.2}$$

$(\tau_- < \tau < \tau_+)$.

The transform $f^*(\alpha)$ is a function which is analytic within the horizontal strip $\tau_- < \mathrm{Im}\,\alpha < \tau_+$ where the integrand of the first integral in (4.1.2) decreases exponentially as $x \to \pm\infty$.

By substituting α by $\mathrm{i}p$ one gets from (4.1.2)

$$F(p) = \int_{-\infty}^{\infty} f(x)\,\mathrm{e}^{-px}\,dx\,,$$
$$f(x) = \frac{1}{2\pi\mathrm{i}} \int_{c-\mathrm{i}\infty}^{c+\mathrm{i}\infty} F(p)\,\mathrm{e}^{px}\,dp \quad (\tau_- < c < \tau_+)\,, \tag{4.1.3}$$

where $F(p) \equiv f^*(\mathrm{i}p)$, and $F(p)$ is analytic in the vertical strip $\tau_- < \mathrm{Re}\,p < \tau_+$.

Formulas (4.1.3) represent the so-called double-sided Laplace transform [4.32]. In particular, if $f(x) \equiv 0$ for $x < 0$, then the formulas (4.1.3) give the conventional Laplace transform

$$F(p) = \int_{0}^{\infty} f(x)\,\mathrm{e}^{-px}\,dx\,,$$
$$f(x) = \frac{1}{2\pi\mathrm{i}} \int_{c-\mathrm{i}\infty}^{c+\mathrm{i}\infty} F(p)\,\mathrm{e}^{px}\,dp \quad (\tau_- < c)\,. \tag{4.1.4}$$

In this case the analyticity strip, $\tau_- < \mathrm{Re}\,p < \tau_+$, is replaced by the right-hand half-plane, $\tau_- < \mathrm{Re}\,p$, where the function $F(p)$ defined by the first integral in (4.1.4), is analytic. Thus, there is practically no difference between the Fourier transform in the complex domain and the Laplace transform.

Concluding this section, let us note that the choice of an appropriate integral transform for solving a concrete problem, depends not only on the geometry of the boundary surface under consideration, but on given boundary conditions too. Therefore, an application of integral transforms to solving elastodynamic problems will be demonstrated by considering some particular examples of great practical importance.

4.2 Lamb's Problem for a Half-Plane

Let the boundary of an elastic isotropic half-space $y \geq 0$ at rest for $t < 0$ be subjected to an impulsive force

$$\sigma_{yy}(t, x, y) = -I\delta(x)\delta(t)\,, \quad y = 0 \quad (I = \mathrm{const}) \tag{4.2.1}$$

concentrated on the axis z and applied at the instant $t = 0$ (Fig. 4.1). Since neither boundary nor initial conditions depend on z, the problem is plane-strain and hence,

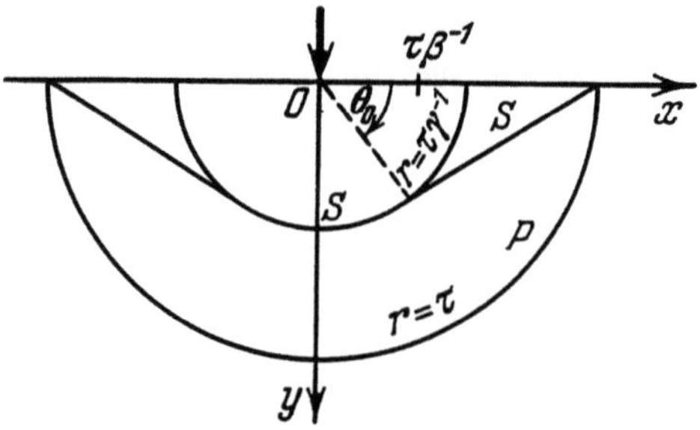

Fig. 4.1. Wave front pattern

the system of the equations of motion including boundary and initial conditions, has the following form (if written in the potentials φ, ψ, i.e. when $u = \text{grad}\varphi + \text{curl}(\psi e_3)$):

$$\Delta\varphi = \frac{\partial^2\varphi}{\partial\tau^2}\,, \quad \Delta\psi = \gamma^2\frac{\partial^2\psi}{\partial\tau^2}$$
$$\left(\tau = c_1 t\,, \quad \gamma = \frac{c_1}{c_2}\,, \quad \Delta \equiv \frac{\partial^2}{\partial x^2} + \frac{\partial^2}{\partial y^2}\right), \tag{4.2.2}$$

$$\sigma_{yy} = \left(c_1^2 - 2c_2^2\right)\varrho\Delta\varphi + 2c_2^2\varrho\left(\frac{\partial^2\varphi}{\partial y^2} - \frac{\partial^2\psi}{\partial x\partial y}\right)$$
$$= -I\delta(x)\delta(\tau)/c_1 \quad (y = 0)\,, \tag{4.2.3}$$
$$\sigma_{xy} = c_2^2\varrho\left(2\frac{\partial^2\varphi}{\partial x\partial y} + \frac{\partial^2\psi}{\partial y^2} - \frac{\partial^2\psi}{\partial x^2}\right) = 0 \quad (y = 0)\,,$$

$$\varphi = \psi = \frac{\partial\varphi}{\partial\tau} = \frac{\partial\psi}{\partial\tau} = 0 \quad (\tau = 0)\,. \tag{4.2.4}$$

4.2.1 Solution to the Problem

By subjecting the system (4.2.2–4) first to the Laplace transform with respect to τ, and then to the Fourier transform with respect to x, one obtains

$$\frac{d^2\bar{\varphi}^*}{dy^2} = (p^2 + q^2)\bar{\varphi}^*\,, \quad \frac{d^2\bar{\psi}^*}{dy^2} = (\gamma^2p^2 + q^2)\bar{\psi}^*\,, \tag{4.2.5}$$

$$(\gamma^2 - 2)p^2\bar{\varphi}^* + 2\left(\frac{d^2\bar{\varphi}^*}{dy^2} + \mathrm{i}q\frac{d\bar{\psi}^*}{dy}\right) = -\frac{I}{\mu c_2} \quad (y = 0)\,,$$
$$\frac{d^2\bar{\psi}^*}{dy^2} + q^2\bar{\psi}^* - 2q\mathrm{i}\frac{d\bar{\varphi}^*}{dy} = 0 \quad (y = 0)\,, \tag{4.2.6}$$

where

$$\bar{f}(p,x,y) = \int_{-0}^{\infty} f(\tau,x,y)\mathrm{e}^{-p\tau}d\tau\ , \quad \operatorname{Re} p > 0\ ,$$
$$\bar{f}^*(p,q,y) = \int_{-\infty}^{\infty} \bar{f}(p,x,y)\mathrm{e}^{\mathrm{i}xq}dx\ , \quad \operatorname{Im} q = 0 \quad (f = \varphi, \psi)\ . \tag{4.2.7}$$

Solving equations (4.2.5) for $\bar{\varphi}^*, \bar{\psi}^*$, one gets

$$\bar{\varphi}^* = A\mathrm{e}^{y\sqrt{p^2+q^2}} + B\mathrm{e}^{-y\sqrt{p^2+q^2}}\ ,$$
$$\bar{\psi}^* = C\mathrm{e}^{y\sqrt{\gamma^2p^2+q^2}} + D\mathrm{e}^{-y\sqrt{\gamma^2p^2+q^2}}\ . \tag{4.2.8}$$

In order to isolate the single-valued branches of the radicals $\sqrt{p^2+q^2}$ and $\sqrt{\gamma^2p^2+q^2}$, we have drawn cuts in the plane q , which extend from the points $\pm pi, \pm\gamma pi$ and into infinity and are directed along the rays $\arg q = \arg p \pm \pi/2$, see Fig. 4.2; besides, it is assumed that the equalities $\sqrt{p^2+q^2} = p$, $\sqrt{\gamma^2p^2+q^2} = \gamma p$ are valid for $q = 0$. Then it may be shown that $\operatorname{Re}\sqrt{p^2+q^2} > 0$, $\operatorname{Re}\sqrt{\gamma^2p^2+q^2} > 0$ for $\operatorname{Im} q = 0$, $\operatorname{Re} p > 0$ (this can be easily done by observing the variations of the angles $\alpha_0, \alpha_0', \alpha_1, \alpha_1'$ while point q travels along the real axis).

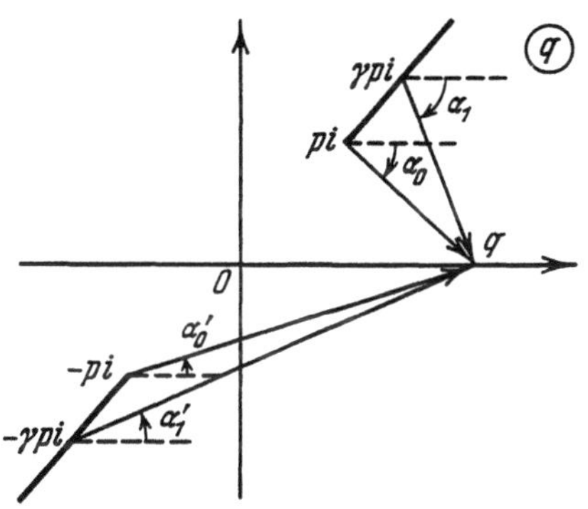

Fig. 4.2. Cuts in the q-plane

Since the medium is at rest for $\tau < 0$, no disturbances φ, ψ are present at $\sqrt{x^2+y^2} > \tau > 0$ and hence the Laplace transforms must decrease as $\sqrt{x^2+y^2} \to \infty$. However, this means that $\bar{\varphi}^*, \bar{\psi}^*$ must also be supposed to be decreasing as $\sqrt{x^2+y^2} \to \infty$. Hence, one must put $A = C = 0$ in (4.2.8). Then, substituting $\bar{\varphi}^*, \bar{\psi}^*$ from (4.2.8) into the boundary conditions (4.2.6), one gets the following relationships for B, D:

$$B = -\frac{(\gamma^2p^2+2q^2)I}{R(p,q)c_1\mu}\ ,$$
$$D = \frac{2q\mathrm{i}\sqrt{(p^2+q^2)}I}{R(p,q)c_1\mu}\ , \tag{4.2.9}$$

where

$$R(p,q) = (\gamma^2 p^2 + 2q^2)^2 - 4q^2\sqrt{(\gamma^2 p^2 + q^2)(p^2 + q^2)}\,.$$

Since, according to (2.7.25), the displacement vector projections u_x, u_y are related to the potentials φ, ψ by the formulas

$$u_x = \frac{\partial\varphi}{\partial x} + \frac{\partial\psi}{\partial y}\,, \qquad u_y = \frac{\partial\varphi}{\partial y} - \frac{\partial\psi}{\partial x}\,,$$

then by applying to the latter the Laplace transform with respect to τ and the Fourier transform with respect to x one gets

$$\bar{u}_x^* = -\mathrm{i}q\bar{\varphi}^* + \frac{d\bar{\psi}^*}{dy}\,, \qquad \bar{u}_y^* = \frac{d\bar{\varphi}^*}{dy} + \mathrm{i}q\bar{\psi}^*. \tag{4.2.10}$$

By substituting the expressions (4.2.8) into (4.2.10) (at the values of B, D given by (4.2.9), and $A = C = 0$), and subjecting the images $\bar{u}_x^*, \bar{u}_y^*$ to the inverse Fourier transform with respect to q, one gets

$$\bar{u}_j(p,x,y) = \frac{1}{2\pi}\int\limits_{-\infty}^{\infty} \mathrm{e}^{-\mathrm{i}qx}\bar{u}_j^* dq \quad (j = x, y)\,,$$

$$\bar{u}_x^* = \frac{\mathrm{i}qI}{\mu c_1 R(p,q)} \times \left[(\gamma^2 p^2 + 2q^2)\mathrm{e}^{-y\sqrt{p^2+q^2}} - 2\sqrt{(p^2+q^2)(\gamma^2p^2+q^2)}\,\mathrm{e}^{-y\sqrt{\gamma^2p^2+q^2}}\right]\,,$$

$$\bar{u}_y^* = \frac{I\sqrt{p^2+q^2}}{uc_1 R(p,q)}\left[(\gamma^2p^2 + 2q^2)\mathrm{e}^{-y\sqrt{p^2+q^2}} - 2q^2\mathrm{e}^{-y\sqrt{\gamma^2p^2+q^2}}\right]\,. \tag{4.2.11}$$

4.2.2 Cagniard-de Hoop Method

Formulas (4.2.11) define $\bar{u}_x, \bar{u}_y$, i.e. the Laplace transforms of the required solution to the problem for the components of the displacement vector at an arbitrary point of the elastic half-plane $y \geq 0$. In order to express the originals themselves, it is necessary to apply the inverse Laplace transform with respect to p to the expressions (4.2.11), and to try to simplify the formal solution obtained in the form of double integrals by employing contour integration. This may be done in a variety of ways [4.34]. In particular, here we use an original method developed by Cagniard [4.6], and modified by de Hoop [4.7]. Essentially, the method is based on transforming the integrals of the inverse Fourier transform with respect to q (which represent the expressions for $\bar{u}_x, \bar{u}_y$ in (4.2.11)) into integrals of the form of the Laplace transform with respect to τ, i.e. into integrals of the form

$$\int_0^{\infty} f(\tau,x,y)\mathrm{e}^{-p\tau}d\tau\,. \tag{4.2.12}$$

This leads directly to the original of the expression (4.2.12) which must have the form of the integrand $f(\tau, x, y)$. In order to reduce u_x, u_y given by (4.2.11) to the form (4.2.12) let us represent them in the form:

$$\begin{aligned}
\bar{u}_x &= \bar{u}_{x1} + \bar{u}_{x2}\,, \quad \bar{u}_y = \bar{u}_{y1} + \bar{u}_{y2}\,,\\
\bar{u}_{x1} &= \frac{\mathrm{i}I}{2\pi c_1 \mu} \int\limits_{-\infty}^{\infty} \frac{q(\gamma^2 p^2 + 2q^2)}{R(p,q)} \mathrm{e}^{-y\sqrt{p^2+q^2}-\mathrm{i}qx} dq\,,\\
\bar{u}_{x2} &= \frac{-\mathrm{i}I}{\pi \mu c_1} \int\limits_{-\infty}^{\infty} \frac{q\sqrt{(p^2+q^2)(\gamma^2p^2+q^2)}}{R(p,q)} \mathrm{e}^{-y\sqrt{\gamma^2p^2+q^2}-\mathrm{i}qx} dq\,,\\
\bar{u}_{y1} &= \frac{I}{2\pi c_1 \mu} \int\limits_{-\infty}^{\infty} \frac{\sqrt{p^2+q^2}(\gamma^2p^2+q^2)}{R(p,q)} \mathrm{e}^{-y\sqrt{p^2+q^2}-\mathrm{i}qx} dq\,,\\
\bar{u}_{y2} &= \frac{-I}{\pi c_1 \mu} \int\limits_{-\infty}^{\infty} \frac{q^2\sqrt{p^2+q^2}}{R(p,q)} \mathrm{e}^{-y\sqrt{\gamma^2p^2+q^2}-\mathrm{i}qx} dq\,.
\end{aligned} \tag{4.2.13}$$

Then, by substituting the integration variable q by $-\mathrm{i}ps$, (4.2.13) may be reduced to

$$\begin{aligned}
\bar{u}_{x1} &= -\frac{\mathrm{i}I}{2\pi c_1 \mu} \int\limits_{L} \frac{s(\gamma^2 - 2s^2)}{R(s)} \mathrm{e}^{-p[sx+y\sqrt{1-s^2}]} ds\,,\\
\bar{u}_{x2} &= \frac{\mathrm{i}I}{\pi c_1 \mu} \int\limits_{L} \frac{s\sqrt{(1-s^2)(\gamma^2-s^2)}}{R(s)} \mathrm{e}^{-p[sx+y\sqrt{\gamma^2-s^2}]} ds\,,\\
\bar{u}_{y1} &= \frac{-\mathrm{i}I}{2\pi c_1 \mu} \int\limits_{L} \frac{\sqrt{1-s^2}(\gamma^2-2s^2)}{R(s)} \mathrm{e}^{-p[sx+y\sqrt{1-s^2}]} ds\,,\\
u_{y2} &= \frac{-\mathrm{i}I}{\pi c_1 \mu} \int\limits_{L} \frac{s^2\sqrt{1-s^2}}{R(s)} \mathrm{e}^{-p[sx+y\sqrt{\gamma^2-s^2}]} ds\,,
\end{aligned} \tag{4.2.14}$$

where the contour L is presented in Fig. 4.3.

Then, by putting $\sqrt{p^2+q^2} = p\sqrt{1-s^2}$, $\sqrt{\gamma^2p^2+q^2} = p\sqrt{\gamma^2 - s^2}$, and taking into account the above chosen branches of the radicals $\sqrt{p^2+q^2}$, $\sqrt{\gamma^2p^2+q^2}$, one deduces that the branches of the radicals $\sqrt{1-s^2}$, $\sqrt{\gamma^2-s^2}$ are defined as $\sqrt{1-s^2}=1$, $\sqrt{\gamma^2-s^2}=\gamma$ for $s=0$, while the cuts for singling out these unique branches coincide with the real semi-axes $(-\infty,-1), (1,+\infty), (-\infty,-\gamma), (\gamma,+\infty)$. The complex function

$$R(s) = (\gamma^2 - 2s^2)^2 + 4s^2\sqrt{(\gamma^2-s^2)(1-s^2)} \tag{4.2.15}$$

is the left-hand side of the Rayleigh equation $R(s)=0$ which has two real roots $s_{1,2} = \pm\beta$ in the complex plane s, where $\beta > \gamma$ (in particular, for $\lambda = \mu, \beta =$

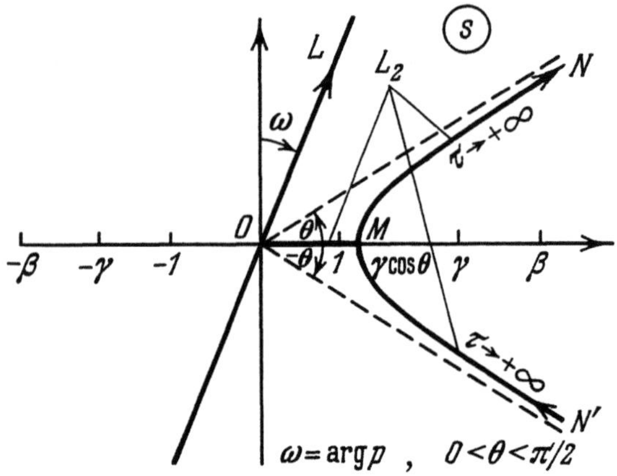

Fig. 4.3. Paths L and L_2

$\gamma[2-2/\sqrt{3}]^{-1/2} \approx \gamma/0.9194$). Let us next deform the integration path L in the formulas for $\bar{u}_{x_1}, \bar{u}_{y_1}$ into the path L_1 along which the expression $sx+y\sqrt{1-s^2}$ is real. Similarly, the contour L in the expressions for $\bar{u}_{x_2}, \bar{u}_{y_2}$ must be deformed into L_2 along which the expression $sx+y\sqrt{\gamma^2-s^2}$ is real. Let us start by determining the contour L_2. To do this, we put

$$sx + y\sqrt{\gamma^2 - s^2} = \tau \ , \tag{4.2.16}$$

where the real quantity τ should be viewed as a parameter varying along the integration path L_2 in the complex plane s. Later, we shall substitute the integration variable s by τ which will be assumed to denote time. That is why we have denoted the quantity $sx + y\sqrt{\gamma^2 - s^2}$ by τ.

By solving (4.2.16) for s, one obtains

$$s = \frac{\tau x \pm y\sqrt{\gamma^2(x^2+y^2) - \tau^2}}{x^2+y^2} \quad (|\tau| < \gamma\sqrt{x^2+y^2}) \ , \tag{4.2.17}$$

$$s = \frac{\tau x \pm \mathrm{i}y\sqrt{\tau^2 - \gamma^2(x^2+y^2)}}{x^2+y^2} \quad (|\tau| > \gamma\sqrt{x^2+y^2}) \ , \tag{4.2.18}$$

where the radicals are assumed to be arithmetic.

For a varying τ (at fixed values of x, y), the expressions (4.2.17, 18) describe some contours $s = s(\tau)$ lying in the plane s . According to the chosen branch of the radical $\sqrt{\gamma^2 - s^2}$ entering the expression (4.2.16) (i.e. under the assumption that (4.2.16) reduces to $\tau = \gamma y$ at $s = 0$), one must take the minus sign in (4.2.17) and to preserve both signs in (4.2.18). Then, we obtain the contour described by (4.2.16) for $\tau > 0, x \geq 0$ (since, taking into account both the evenness of u_y and the oddness of u_x with respect to x, it suffices to consider only the case $x \geq 0$). The bold line in Fig. 4.3 shows the required contour L_2 in the region $\mathrm{Re}\, s > 0$ (for $x \geq 0$), which consists of the segment OM and the curve $N'MN$, and is described by the expressions

$$s = \frac{\tau x - y\sqrt{\gamma^2(x^2+y^2)-\tau^2}}{x^2+y^2} \quad (\gamma y < \tau < \gamma\sqrt{x^2+y^2},\ x \geq 0)\,,$$
$$s = \frac{\tau x \pm \mathrm{i} y\sqrt{\tau^2-\gamma^2(x^2+y^2)}}{x^2+y^2} \quad (\tau > \gamma\sqrt{x^2+y^2},\ x \geq 0)\,. \tag{4.2.19}$$

The plus and minus signs in the latter expression refer to the curves belonging to the upper and the lower half-plane s respectively.

Upon substituting x, y in (4.2.19) by $r\cos\theta, r\sin\theta$ respectively ($0 < \theta < \pi/2$), one gets

$$s(\tau, r, \theta) = \begin{cases} (\tau\cos\theta - \sqrt{\gamma^2 r^2 - \tau^2}\sin\theta)/r & (\gamma y < \tau < \gamma r)\,, \\ (\tau\cos\theta \pm \mathrm{i}\sqrt{\tau^2 - \gamma^2 r^2}\sin\theta)/r & (\tau > \gamma r)\,. \end{cases}$$

Here r and θ are assumed to be fixed as $\tau \to +\infty$. Then, we deduce that s tends to $\tau\exp(\pm\mathrm{i}\theta)/r$, and hence, as shown in Fig. 4.3, the curve L_2 for $\tau \to +\infty$ is bounded by asymptotes issuing from the origin of coordinates at the angles $\pm\theta$ with respect to the real axis. As $s \to \infty$ the expression $p(sx+y\sqrt{\gamma^2-s^2})$ behaves as follows:

$$p(sx + y\sqrt{\gamma^2 - s^2}) = \begin{cases} |p|\,|s| r\,\mathrm{e}^{\mathrm{i}(\omega+\theta+\varphi)} & (\mathrm{Im}\, s < 0)\,, \\ |p|\,|s|\, r\mathrm{e}^{\mathrm{i}(\omega-\theta+\varphi)} & (\mathrm{Im}\, s > 0)\,, \end{cases}$$

where

$$p = |p|\mathrm{e}^{\mathrm{i}\omega} \quad (|\omega| < \pi/2)\,, \qquad s = |s|\mathrm{e}^{\mathrm{i}\varphi}\,.$$

Hence, the power of the exponent, which is equal to $-p(sx + y\sqrt{\gamma^2 - s^2})$, has a negative real part for $|s| \to \infty$ within the region bounded by the curves L, L_2, see Fig. 4.3.

Therefore, the Jordan lemma makes it possible to deform the integration path L in the expressions for $\bar{u}_{x_2}, \bar{u}_{y_2}$ from (4.2.14), into L, being passed in the order $N'MOMN$ where the segment OM of the real axis is passed twice in opposite directions: along the lower and the upper edge. Then, by deforming the path L into L_2 and substituting the variable s by τ with the help of the formula $\tau = sx + y\sqrt{\gamma^2 - s^2}, \bar{u}_{x_2}, \bar{u}_{y_2}$ from (4.2.14) reduce to the form (for $x \geq 0$):

$$\bar{u}_{x2} = \mathrm{i}\int_{L_2} f\,ds = \mathrm{i}\int_{\bar{l}_2} f\,ds + \mathrm{i}\int_{l_2} f\,ds$$
$$= -\frac{2I}{\pi c_1 \mu}\int_{\gamma y}^{+\infty} \mathrm{Im}\left[\frac{s\sqrt{(1-s^2)(\gamma^2-s^2)}}{R(s)}\frac{ds}{d\tau}\right]\mathrm{e}^{-p\tau}d\tau\,, \tag{4.2.20}$$

$$\bar{u}_{y2} = \mathrm{i}\int_{L_2} F\,ds = \mathrm{i}\int_{\bar{l}_2} F\,ds + \mathrm{i}\int_{l_2} F\,ds$$
$$= \frac{2I}{\pi c_1 \mu}\int_{\gamma y}^{+\infty} \mathrm{Im}\left[\frac{s^2\sqrt{1-s^2}}{R(s)}\frac{ds}{d\tau}\right]\mathrm{e}^{-p\tau}d\tau\,, \tag{4.2.21}$$

where

$$f = \frac{I}{\pi c_1 \mu} \frac{s\sqrt{(1-s^2)(\gamma^2-s^2)}}{R(s)} \mathrm{e}^{-p(sx+y\sqrt{\gamma^2-s^2})} ,$$

$$F = \frac{-Is^2\sqrt{1-s^2}\,\mathrm{e}^{-p(sx+y\sqrt{\gamma^2-s^2})}}{\pi c_1 \mu R(s)} .$$

While deriving the expressions (4.2.20, 21), we have taken into account that the branches $\bar{l}_2, l_2$ of the curve L_2 (where $\bar{l}_2, l_2$ denote the curves $N'MO, OMN$, respectively), are symmetric with respect to the real axis, and the integrands

$$\frac{s\sqrt{(1-s^2)(\gamma^2-s^2)}}{R(s)} \frac{ds}{d\tau} , \quad \frac{s^2\sqrt{1-s^2}}{R(s)} \frac{ds}{d\tau}$$

are complex conjugated functions at the complex conjugated points $s, \bar{s}$ at $l_2, \bar{l}_2$ respectively (the function $(sx + y\sqrt{\gamma^2 - s^2})$ has at these points equal real values: $sx + y\sqrt{\gamma^2 - s^2} = \tau$).

Note that the integrands

$$\mathrm{Im}\left[\frac{s^2\sqrt{1-s^2}}{R(s)} \frac{ds}{d\tau}\right] , \quad \mathrm{Im}\left[\frac{s\sqrt{(1-s^2)(\gamma^2-s^2)}}{R(s)} \frac{ds}{d\tau}\right] \tag{4.2.22}$$

entering the real expressions for $\bar{u}_{x2}, \bar{u}_{y2}$ are determined subject to the condition of s residing at the contour l_2, see Fig. 4.4, given by the equation

$$\begin{aligned} s &= \frac{\tau x - y\sqrt{\gamma^2(x^2+y^2)-\tau^2}}{x^2+y^2} \quad (\gamma y < \tau < \gamma\sqrt{x^2+y^2}) , \\ s &= \frac{\tau x + \mathrm{i}y\sqrt{\tau^2-\gamma^2(x^2+y^2)}}{x^2+y^2} \quad (\tau > \gamma\sqrt{x^2+y^2}) . \end{aligned} \tag{4.2.23}$$

The integrals for u_{x1}, u_{y1} from (4.2.14) will be transformed in a similar manner, and the contours L_1, l_1 will be obtained from L_2, l_2 respectively, if γ in (4.2.16, 23) is put equal to unity. As a result, one gets (for $x \geq 0$):

$$\bar{u}_{x1} = \frac{I}{\pi c_1 \mu} \int_y^\infty \mathrm{Im}\left[\frac{s(\gamma^2-2s^2)}{R(s)} \frac{ds}{d\tau}\right] \mathrm{e}^{-p\tau} d\tau , \tag{4.2.24}$$

$$\bar{u}_{y1} = \frac{I}{\pi c_1 \mu} \int_y^\infty \mathrm{Im}\left[\frac{(\gamma^2-2s^2)\sqrt{1-s^2}}{R(s)} \frac{ds}{d\tau}\right] \mathrm{e}^{-p\tau} d\tau , \tag{4.2.25}$$

where s lies at the contour l_1 (see Fig. 4.5) defined by the equations

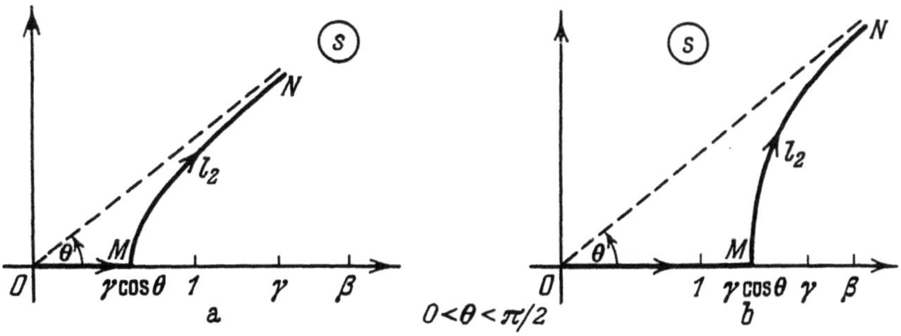

Fig. 4.4. The contour l_2 for $\gamma\cos\theta < 1$ (a) and $\gamma\cos\theta > 1$ (b), $0 < \theta < \pi/2$

$$
\begin{aligned}
s &= \frac{\tau x - y\sqrt{(x^2+y^2)-\tau^2}}{x^2+y^2} \quad (y < \tau < \sqrt{x^2+y^2})\,,\\
s &= \frac{\tau x + \mathrm{i}y\sqrt{\tau^2-(x^2+y^2)}}{x^2+y^2} \quad (\tau > \sqrt{x^2+y^2})\,.
\end{aligned}
\tag{4.2.26}
$$

From (4.2.24–26) it follows that the integrands vanish for real values of s meeting the condition $s < \cos\theta$, or for $\tau < r$. Hence, the expressions (4.2.24, 25) for $x \geq 0, y \geq 0$ may be rewritten in the form:

$$
\begin{aligned}
\bar{u}_{x1} &= \frac{I}{\pi c_1 \mu}\int\limits_0^\infty \mathrm{Im}\left[\frac{s(\gamma^2-2s^2)}{R(s)}\frac{ds}{d\tau}\right] H(\tau-r)\mathrm{e}^{-p\tau}d\tau\,,\\
\bar{u}_{y1} &= \frac{I}{\pi c_1 \mu}\int\limits_0^\infty \mathrm{Im}\left[\frac{(\gamma^2-2s^2)\sqrt{1-s^2}}{R(s)}\frac{ds}{d\tau}\right] H(\tau-r)\mathrm{e}^{-p\tau}d\tau\,,
\end{aligned}
\tag{4.2.27}
$$

where the function s is given by (4.2.26).

The expressions for $\bar{u}_{x_2}, \bar{u}_{y_2}$ at $x \geq 0, y \geq 0$ are obtained in a similar way by considering the cases $\gamma\cos\theta < 1$, $\gamma\cos\theta > 1$, see Fig. 4.4:

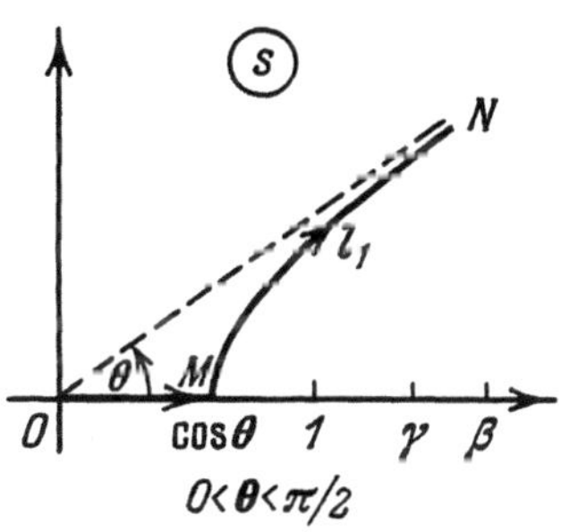

Fig. 4.5. The contour l_1, $0 < \theta < \pi/2$

$$\begin{aligned}
\bar{u}_{x2} &= -\frac{2I}{\pi c_1 \mu} \int_0^\infty \operatorname{Im}\left[\frac{s\sqrt{(1-s^2)(\gamma^2-s^2)}}{R(s)}\frac{ds}{d\tau}\right] \\
&\quad \times \left\{\begin{matrix} H(\tau - r\gamma) \\ H(\tau - x + y\sqrt{\gamma^2-1}) \end{matrix}\right\} e^{-p\tau} d\tau \,, \\
\bar{u}_{y2} &= \frac{2I}{\pi c_1 \mu} \int_0^\infty \operatorname{Im}\left[\frac{s^2\sqrt{1-s^2}}{R(s)}\frac{ds}{d\tau}\right] \\
&\quad \times \left\{\begin{matrix} H(\tau - r\gamma) \\ H(\tau - x + y\sqrt{\gamma^2-1}) \end{matrix}\right\} e^{-p\tau} d\tau \,.
\end{aligned} \tag{4.2.28}$$

Here the upper and the lower lines in each of the expressions for $\bar{u}_{x_2}, \bar{u}_{y_2}$ refer to the cases $\gamma \cos\theta < 1$, $\gamma\cos\theta > 1$, respectively, and the function s is given by formulas (4.2.23).

Referring again to (4.2.27, 28) we see that the expressions for $\bar{u}_{x_1}, \bar{u}_{y_1}, \bar{u}_{x_2}, \bar{u}_{y_2}$ are written in the form of integrals representing the Laplace transformations of the real integrands with respect to τ. Hence, the integrands are the required originals.

Thus, by applying the inverse Laplace transformation with respect to p, to (4.2.27, 28), one gets finally for $x \geq 0$, $y \geq 0$:

$$\begin{aligned}
u_x &= u_{x1} + u_{x2} \,, \\
u_{x1} &= \frac{I}{\pi c_1 \mu} \operatorname{Im}\left[\frac{s(\gamma^2 - 2s^2)}{R(s)}\frac{ds}{d\tau}\right] H(\tau - r) \,, \\
u_{x2} &= -\frac{2I}{\pi c_1 \mu} \operatorname{Im}\left[\frac{s\sqrt{(1-s^2)(\gamma^2-s^2)}}{R(s)}\frac{ds}{d\tau}\right] \left\{\begin{matrix} H(\tau - r\gamma) \\ H(\tau - x - y\sqrt{\gamma^2-1}) \end{matrix}\right\} ,
\end{aligned} \tag{4.2.29}$$

$$\begin{aligned}
u_y &= u_{y1} + u_{y2} \,, \\
u_{y1} &= \frac{I}{\pi c_1 \mu} \operatorname{Im}\left[\frac{(\gamma^2 - 2s^2)\sqrt{1-s^2}}{R(s)}\frac{ds}{d\tau}\right] H(\tau - r) \,, \\
u_{y2} &= \frac{2I}{\pi c_1 \mu} \operatorname{Im}\left[\frac{s^2\sqrt{1-s^2}}{R(s)}\frac{ds}{d\tau}\right] \left\{\begin{matrix} H(\tau - r\gamma) \\ H(\tau - x - y\sqrt{\gamma^2-1}) \end{matrix}\right\} ,
\end{aligned} \tag{4.2.30}$$

where s in the formulas for u_{x_1}, u_{y_1} is given by (4.2.26) and in the formulas for u_{x_2}, u_{y_2} by (4.2.23) with the upper and the lower lines of both the expression for u_{x_2} and the expression for u_{y_2} corresponding to the cases $\gamma\cos\theta < 1$, $\gamma\cos\theta > 1$, respectively.

Note that for $y = 0$, the expressions for u_x, u_y (see (4.2.29, 30)) coincide, to an accuracy of notation, with the expressions for displacements from (3.4.41), obtained by the method of functionally invariant solutions. However, the component u_y from (3.4.41) must be taken with an opposite sign, since the problem was solved for the region $y \leq 0$.

Note. While transforming the integrals (4.2.13) to the form (4.2.12), the parameter p of the Laplace transform was supposed to belong to the right half-plane $\operatorname{Re} p > 0$. However, such transformations may be simplified by assuming that p is real and positive, i.e. $p > 0$. In fact, the assumption does not violate rigorousness of the analysis, since the integrals of the form (4.2.12) obtained from (4.2.13) for $p > 0$, prove to be, due to the properties of the Laplace integrals [4.35], analytic functions within the whole of the right half-plane $\operatorname{Re} p > 0$, and hence, in accordance with the uniqueness theorem, represent the original expressions (4.2.13) within the region $\operatorname{Re} p > 0$. This is the usual way of employing the Cagniard-de Hoop method for solving concrete problems.

Let us now consider briefly the geometric pattern of the wave fronts within the region $x > 0$, see Fig. 4.1. The expressions for u_{x_1}, u_{y_1} describe the contribution of the longitudinal wave P whose front is specified by the equation $r = \tau$, while the expressions for u_{x_2}, u_{y_2} are defined by the contribution of the transverse wave S consisting of the wave with the circular front $r = \tau\gamma^{-1}$ and the head transverse wave with the rectilinear front $\tau - x - y\sqrt{\gamma^2 - 1} = 0$. The head transverse wave S is generated by the traveling longitudinal wave P due to its interaction with the free surface. The front of the head transverse wave touches the circumference $r = \tau\gamma^{-1}$ at the point $\theta = \theta_0$ where $\cos\theta = \cos\theta_0 = \gamma^{-1}$. Hence, the head transverse wave exists at $\theta < \theta_0$. Note that, due to the presence of the function $ds/d\tau$ in (4.2.29, 30), the displacement vector has a singularity of the order of $-1/2$ at the circular fronts of both the longitudinal, $r = \tau$, and the transverse, $r = \tau\gamma^{-1}$, wave. At the front of the transverse wave $r = \tau\gamma^{-1}$ traveling behind the head transverse wave (i. e. for $\theta < \theta_0$), the singularity appears while approaching the front from any direction; however, for $\theta > \theta_0$ it appears only in the case of an approach from the region $r < \tau\gamma^{-1}$.

We must also mention the presence of the displacement vector singularity at the point $x = \tau\beta^{-1}$ of the free surface, traveling at the Rayleigh speed c_R ($\beta = c_1/c_2$). This singularity is of the order of -1 and resides at the free surface only. It appears due to the propagation of the surface Rayleigh wave. However, from the mathematical point of view, the singularity appears due to the function $R(s)$ being zero at the point $s = \beta$ in the denominators of the expressions (4.2.29, 30).

4.3 Diffraction of an Acoustic Wave by a Rigid Sphere

Let us consider the problem of diffraction of a plane acoustic wave by a sphere, which will be solved by the method of integral transforms. For solving the problem, two different approaches will be employed.

It should be mentioned that the problem was originally considered by Kharkevich [4.36] and later by Lefonova and Fedorov [4.37].

Let a rigid stationary sphere of the radius a be immersed into an ideal in compressible fluid at rest, characterized by the initial pressure p_0. The problem will be considered in the spherical system of coordinates r, θ, β ($z = r\cos\theta$,

$x = r\sin\theta\cos\beta, y = r\sin\theta\sin\beta$), see Fig. 4.6 showing the meridional cross-section of the sphere.

Let a plane acoustic pressure wave

$$p_1 = A_0 H(c_0 t + r\cos\theta - a) + p_0 \quad (A_0 = \text{const}, c_0^2 = \lambda/\rho) \tag{4.3.1}$$

be approaching the sphere from infinity, whose front touches the sphere at the instant $t = 0$. According to (2.1.1) (for $\boldsymbol{f} \equiv 0$) the pressure field, $p(r, \theta, t)$, for the axisymmetric problem under consideration, is described by the wave equation

$$\Delta p = \frac{1}{c_0^2}\frac{\partial^2 p}{\partial t^2} \quad (r > a) \tag{4.3.2}$$

subject to the boundary condition

$$\partial p/\partial r = 0 \quad (r = a) \tag{4.3.3}$$

and the initial condition

$$p \equiv p_1 \quad (t \le 0)\,. \tag{4.3.4}$$

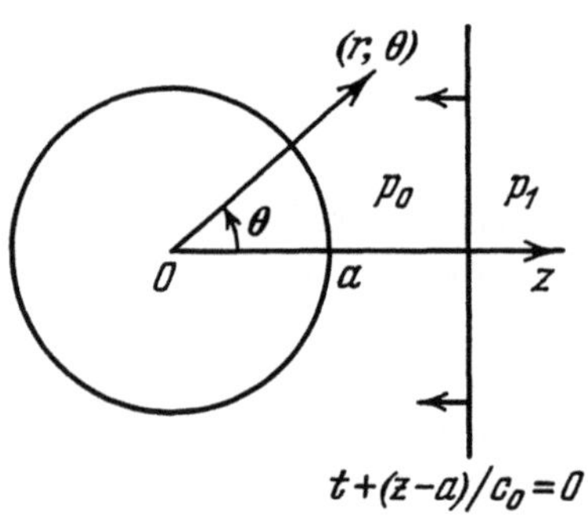

Fig. 4.6. Diffraction of a plane acoustic wave by a sphere

(Condition (4.3.3) follows from the boundary condition $v_r \equiv \dot{u}_r = 0$ $(r = a)$ and the equation (2.1.10) at $\boldsymbol{f} \equiv 0$: $\partial p/\partial r = -\rho\dot{v}_r$.)

By presenting the pressure p in the form

$$p = p_1 + Q\,, \tag{4.3.5}$$

where $Q(r, \theta, t)$ is the disturbance created by the sphere in the pressure field of the incident wave, and introducing the following nondimensional variables

$$\begin{gathered}\bar{r} = \frac{r}{a}\,, \quad \bar{t} = \frac{c_0 t}{a}\,, \quad \bar{Q} = \frac{Q}{A_0}\,, \quad \bar{p}_1 = \frac{p_1 - p_0}{A_0}\,, \\ \bar{p} = \frac{p - p_0}{A_0} = \frac{Q + p_1 - p_0}{A_0}\end{gathered} \tag{4.3.6}$$

we obtain from (4.3.1–6) the following system of equations for determining $\bar{Q}, \bar{p}$ (below the bar is omitted):

$$\Delta Q = \frac{\partial^2 Q}{\partial t^2} \quad (r > 1)\,, \tag{4.3.7}$$

$$\frac{\partial Q}{\partial r} = -\frac{\partial p_1}{\partial r} \quad (r = 1)\,, \tag{4.3.8}$$

$$Q = 0 \quad (t \leq 0) \tag{4.3.9}$$

$$p_1 = H(t + r\cos\theta - 1)\,, \tag{4.3.10}$$

$$p = p_1 + Q\left(\Delta \equiv \frac{\partial^2}{\partial r^2} + \frac{2}{r}\frac{\partial}{\partial r} + \frac{1}{r^2}\frac{\partial^2}{\partial \theta^2} + \frac{\cot\theta}{r^2}\frac{\partial}{\partial\theta}\right)\,. \tag{4.3.11}$$

Then, by applying the Laplace transform with respect to t, to the relationships (4.3.7, 8, 10, 11) and taking into account (4.3.9), one gets

$$\Delta Q^* = s^2 Q^* \quad (r > 1)\,, \tag{4.3.12}$$

$$\frac{\partial Q^*}{\partial r} = -\frac{\partial p_1^*}{\partial r} \quad (r = 1)\,, \tag{4.3.13}$$

$$Q^* \to 0 \quad (r \to \infty)\,, \tag{4.3.14}$$

$$p_1^* = \frac{1}{s}\mathrm{e}^{-s(1-r\cos\theta)}\,, \tag{4.3.15}$$

$$p^* = p_1^* + Q^*\,, \tag{4.3.16}$$

where

$$f^*(r,\theta,s) = \int_{-\infty}^{\infty} f(r,\theta,t)\mathrm{e}^{-st}dt \quad (f = p, Q, p_1)\,. \tag{4.3.17}$$

Since $p_1 = Q = p = 0$ for $t < 1 - r\cos\theta$, the Laplace transform proves to be one-sided, and hence, the complex parameter s appearing in expressions (4.3.12–16) satisfies the inequality $\mathrm{Re}\, s > 0$. Note also that since the disturbance Q is absent for $t < r - 1$, Q^* must decrease as $r \to \infty$; hence, we have the condition (4.3.14).

If the solution to Eq. (4.3.12) is sought in the form of a series expansion in the Legendre polynomials $P_n(\cos\theta)$, then with the help of (4.3.14), one gets

$$Q^* = \sum_{n=0}^{\infty} B_n K_{n+1/2}(sr) P_n(\cos\theta)/\sqrt{r}, \tag{4.3.18}$$

and the incident wave may be presented in the form

$$p_1^* = \frac{1}{s}\mathrm{e}^{-s(1-r\cos\theta)} = \frac{1}{s}\mathrm{e}^{-s}\sqrt{\frac{2\pi}{sr}}\sum_{n=0}^{\infty}\left(n+\frac{1}{2}\right) I_{n+1/2}(sr) P_n(\cos\theta)\,, \tag{4.3.19}$$

where $I_{n+1/2}(sr)$, $K_{n+1/2}(sr)$ are modified Bessel functions of the 1st and the 3rd kind, respectively.

For determining B_n let us use boundary condition (4.3.13). Then, employing (4.3.18, 19) and making use of the Legendre polynomials orthogonality

$$\int_0^\pi P_n(\cos\theta)P_m(\cos\theta)\sin\theta d\theta = \begin{cases} 0\,, & n \neq m\,, \\ \left(n+\frac{1}{2}\right)^{-1}\,, & n = m\,, \end{cases} \tag{4.3.20}$$

one gets from (4.3.13)

$$B_n = -\frac{(n+1/2)}{s^{3/2}} \frac{\left(\sqrt{2\pi/s}I_{n+1/2}(s)\right)'_s}{\left((1/\sqrt{s})K_{n+1/2}(s)\right)'_s} e^{-s}\,.$$

Hence, we have the following expression for Q^*:

$$\begin{aligned} Q^*(r,\theta,s) &= \sum_{n=0}^{\infty} b_n^*(r,s)P_n(\cos\theta)\,, \\ b_n^*(r,s) &= -\frac{(n+1/2)}{s^{3/2}} \frac{\left(\sqrt{2\pi/s}I_{n+1/2}(s)\right)'_s}{\left((1/\sqrt{s})K_{n+1/2}(s)\right)'_s} \frac{K_{n+1/2}(sr)}{\sqrt{r}} e^{-s}\,, \end{aligned} \tag{4.3.21}$$

and its original

$$\begin{aligned} Q(r,\theta,t) &= \sum_{n=0}^{\infty} b_n(r,t)P_n(\cos\theta)\,, \\ b_n(r,t) &= \frac{1}{2\pi i}\int_{\varepsilon-i\infty}^{\varepsilon+i\infty} b_n^*(r,s)e^{st}ds \quad (\varepsilon > 0,\, t > 0)\,. \end{aligned} \tag{4.3.22}$$

In particular, using the relationship [4.38]

$$\left(\frac{1}{\sqrt{s}}I_{n+1/2}(s)\right)'_s \frac{1}{\sqrt{s}}K_{n+1/2}(s) - \left(\frac{1}{\sqrt{s}}K_{n+1/2}(s)\right)'_s \frac{1}{\sqrt{s}}I_{n+1/2}(s) = \frac{1}{s^2} \tag{4.3.23}$$

we get from (4.3.19, 21) the following expression for the transform of the pressure distribution on a sphere ($r = 1$):

$$p^*(1,\theta,s) = p_1^* + Q^* = -\frac{e^{-s}\sqrt{2\pi}}{s^3}\sum_{n=0}^{\infty}\frac{(n+1/2)P_n(\cos\theta)}{\left((1/\sqrt{s})K_{n+1/2}(s)\right)'_s}\,. \tag{4.3.24}$$

For the transform of the nondimensional force F^* (divided by $A_0\pi a^2$) acting onto a sphere, the following expression may be obtained by using the Legendre polynomials orthogonality:

$$\begin{aligned} F^*(s) &= 2\int_0^\pi p^*\cos\theta\sin\theta d\theta \\ &= -\frac{2e^{-s}}{s^3}\sqrt{2\pi}\frac{1}{\left((1/\sqrt{s})K_{3/2}(s)\right)'_s} = \frac{4}{s^2+2s+2}\,. \end{aligned} \tag{4.3.25}$$

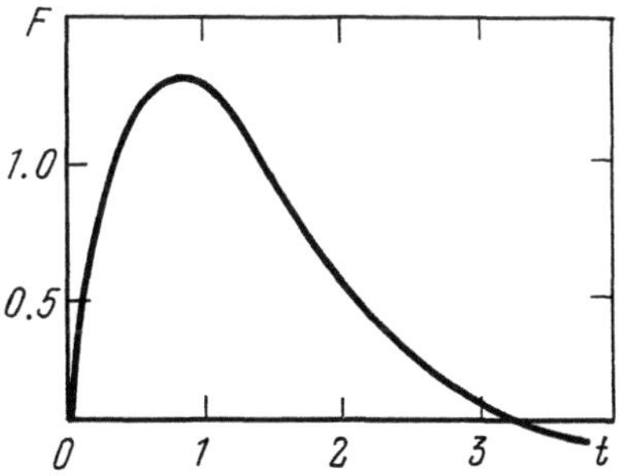

Fig. 4.7. Time dependence of the force applied to the sphere

Hence, by applying the inverse Laplace transform to the latter equality, one arrives at the following formula for the nondimensional force:

$$F(t) = 4e^{-t}\sin t\,. \qquad (4.3.26)$$

The function is plotted in Fig. 4.7. We see that the force attains its maximum value, $F_m = 2\sqrt{2}\exp(-\pi/4)$, at the instant $t = \pi/4$, i.e. before the incident wave embraces the whole of the forward hemisphere.

Thus, we have obtained a closed expression for the force. The simplicity of its form is due to the fact that $F(t)$ depends only on the first harmonic ($n = 1$) of the expansion of the pressure distribution into the series in Legendre polynomials.

As for the pressure distribution on the sphere, $p(1, \theta, t)$ according to (4.3.24) it is given by

$$p(1,\theta,t) = \sum_{n=0}^{\infty} a_n(t) P_n(\cos\theta)\,,$$
$$a_n = -\frac{\sqrt{2\pi}(n+1/2)}{2\pi i}\int_{\varepsilon-i\infty}^{\varepsilon+i\infty}\frac{e^{s(t-1)}ds}{s^3\left((1/\sqrt{s})K_{n+1/2}(s)\right)'_s} \quad (\varepsilon > 0)\,. \qquad (4.3.27)$$

Note that the integrands entering $a_n(t)$ are rational functions, multiplied by $\exp(st)$ and, hence, singular points can be presented by poles only. The terms $a_0(t), a_1(t)$ are readily calculated:

$$a_0(t) = \begin{cases} 2 - \frac{t}{2} - e^{-t} & (0 < t \le 2)\,, \\ 1 - e^{-t} & (t \ge 2)\,, \end{cases}$$
$$a_1(t) = 3e^{-t}\sin t\,. \qquad (4.3.28)$$

Hence, $a_0(t) \to 1, a_1(t) \to 0$ as $t \to \infty$. The denominators of the integrands entering the expressions for $a_n(t)$ at $n > 1$ are polynomials of the $(n+1)^{\text{th}}$ power, whose roots lie, according to the Bessel functions theory [4.38], in the region $\operatorname{Re} s > 0$. Hence, the residues at the poles give functions which decrease as $t \to \infty$. Thus, for $t \to \infty$ one gets

$$p(1,\theta,t) \sim a_0(t) \to 1. \qquad (4.3.29)$$

This conclusion agrees with the physical nature of the phenomenon, since diffraction waves scatter as $t \to \infty$, and the pressure distribution on the surface of a sphere is defined by the incident wave pressure field.

In order to calculate $a_n(t)$ for $n > 1$ one has to know the roots of the $(n+1)^{\text{th}}$ power polynomial, and, hence, comes across the mathematical difficulties of determining roots of such a polynomial (as we know the problem has no general analytic solution for $n \geq 4$). This is the problem one has to deal with while calculating pressure during the initial phase $0 < t < t_0$. In fact, at the front point of a sphere, $\theta = 0$, the series (4.3.27) converges nonhomogeneously with respect to t in the neighborhood of $t = 0$ (for $t < 0$ the series gives $p = 0$, while at $t = +0$ $p \to 2$), and hence, for calculating $p(1, \theta, t)$ with the help of (4.3.27) while approaching $t = 0$, one has to take into account an ever increasing number of terms of the series. However, this is associated with calculating roots of high-power polynomials.

Thus, the solution for the pressure distribution on a sphere, $p(1, \theta, t)$, obtained in the form of (4.3.27) is convenient for calculating pressure at large values of t, and inapplicable at its small values when diffraction is most important. Therefore, for calculating pressure distributions at small values of t, $p(1, \theta, t)$ must be presented in another form. This can be done, for instance, by subjecting (4.3.27) to the Watson transform. Essentially, the transform is a substitution of the series (4.3.27) by an integral over a contour encompassing the integrand's poles with the residues giving the original series. By deforming the integration contour in a certain way one can obtain a solution which is valid for the region under consideration at small values of t. This approach is well known, and its application to the problem of diffraction on a sphere is described, e.g., in [4.39, 40].

Also, one can seek for the $t \to +0$ asymptotic of the solution to the problem of an acoustic wave diffraction on a sphere in a straightforward manner by using asymptotic methods [4.41, 42] which are not treated here.

In order to obtain an expression for $p(1, \theta, t)$ applicable for small values of t, below (see Sect. 4.4) we shall seek for a solution to the problem of a wave diffraction on a sphere in another form prompted by physical considersations.

In conclusion in this section we briefly consider the possibility of obtaining solutions by applying the method of integral transforms to the cases of diffraction of an acoustic wave on a cylinder and of elastic waves on spherical and cylindrical inclusions [4.5, 27, 29, 37, 43, 44, 45].

The case of an acoustic wave diffraction on a cylinder [4.37] differs from the corresponding case for a sphere by the former solution being represented by an expansion in the trigonometric functions of the angular coordinate (instead of the Legendre polynomials), and hence, the solution for a cylinder incorporates modified Bessel functions of the 3rd kind, $K_n(sr)$, with an integer index.

The latter circumstance results in the appearance in the complex plane s not only of the poles, but of branch point $s = 0$, too. This complicates carrying out the inverse Laplace transform and obtaining a simple analytic solution for each of the expansion forms.

As for elastic waves diffraction on a sphere/cylinder, it differs from the corresponding acoustic case in the appearance of two or three unknown functions (the longitudinal and transverse potentials) instead of one unknown function (pressure or the longitudinal potential). However, this does not create principal difficulties in obtaining an analytic solution to the problem, since due to orthogonality of the Legendre polynomials (or trigonometric functions, in the case of a cylinder) on the segment $[0, \pi]$, the boundary conditions for each of the potentials may be separated, and the solution of an elastic problem reduces to solving acoustic problems for each of the potentials.

4.4 Expansion of an Acoustic Wave Solution for a Sphere Over a Time-Dependent Interval

If one returns to the solution (4.3.22) for the disturbance $Q(r, \theta, t)$ obtained above in the form of a series expansion in the Legendre polynomials over the segment $[0, \pi]$, then it becomes obvious that for small values of t, when due to hyperbolicity of the wave equation the pressure distribution is nonzero only at the initial portion, $0 \leq \theta < \cos^{-1}(1-t)$, of the segment $[0, \pi]$ (see region ACA_1 in Fig. 4.8), one must retain a sufficiently large number of the terms of the series in order to approximate the required solution with a satisfactory accuracy. A decrease in t results in the portion of the segment tending to zero ($\cos^{-1}(1-t) \to 0$), and the required solution becomes concentrated near the front point, $\theta = 0$, where $Q \to 0$ as $t \to +0$. As a result, the solution loses its smoothness on the segment $[0, \pi]$, and, hence, convergence of the series deteriorates. The latter circumstance makes it necessary to retain more terms of the series while calculating the pressure distribution on a sphere at small values of t.

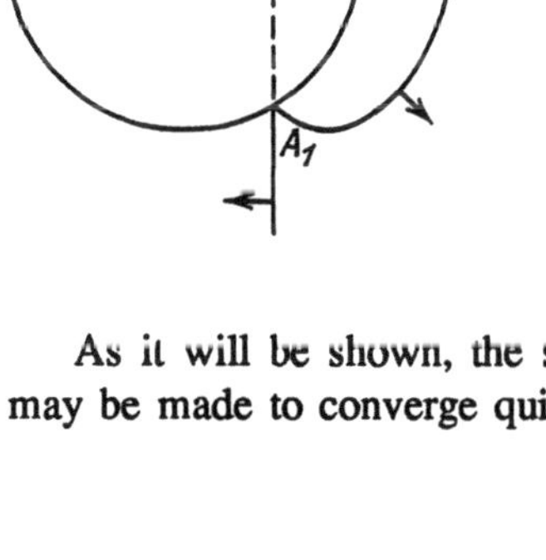

Fig. 4.8. Disturbance domain, ADA_1CA, in the problem on diffraction of a plane acoustic wave by a sphere

As it will be shown, the series representing $Q(r, \theta, t)$ in the expansion in θ may be made to converge quicker by considering the function on the portion of the segment $[0, \pi]$ where it is nonzero, i. e. by expanding the function $Q(r, \theta, t)$ for $0 < t < 2$ within the disturbed region over the time-dependent interval $0 \leq \theta < \cos^{-1}(1-t)$. Below such an expansion will be obtained on the basis of the approach developed by Slepyan [4.25].

Let us select an arbitrary value of α from the interval $0 < \alpha < \pi$. Let also the time t be such that $0 < t < 1 - \cos\alpha < 2$, in other words, let the point $T(1, \alpha)$ belong to the undisturbed region, see Fig. 4.8. According to (4.3.7–10), the disturbance $Q(r, \theta, t)$ satisfies the system of equations

$$\frac{1}{r^2}\frac{\partial}{\partial r}\left(r^2\frac{\partial Q}{\partial r}\right) + \frac{1}{r^2 \sin\theta}\frac{\partial}{\partial \theta}\left(\sin\theta\frac{\partial Q}{\partial \theta}\right) = \frac{\partial^2 Q}{\partial t^2} \quad (t > 0,\ r > 1)\,, \tag{4.4.1}$$

$$\frac{\partial Q}{\partial r} = -\cos\theta\,\delta(t + \cos\theta - 1) \quad (r = 1)\,, \tag{4.4.2}$$

$$Q = \frac{\partial Q}{\partial t} = 0 \quad (t = 0)\,, \tag{4.4.3}$$

whose solution is sought in the form of an expansion in the Legendre functions on the segment $0 \leq \theta \leq \alpha$. Here we shall use the formulas [4.46]:

$$\begin{gathered} Q(r, \theta, t) = \sum_{n=0}^{\infty} a_n P_{\nu_n}(\cos\theta)\,, \\ a_n \equiv a_n(r, t) = \frac{\int_0^\alpha Q(r, \theta, t) P_{\nu_n}(\cos\theta)\sin\theta d\theta}{\int_0^\alpha \left[P_{\nu_n}(\cos\theta)\right]^2 \sin\theta d\theta}\,, \end{gathered} \tag{4.4.4}$$

where ν_n are all the non-negative roots of the equation $P^1_{\nu_n}(\cos\alpha) = 0$, arranged in an increasing order $(0 = \nu_0 < 1 < \nu_1 < \nu_2 < \ldots)$ and $P^\beta_\nu(x)$ is the associated Legendre function of the first kind for the interval $(-1, 1)$ [4.47].

By multiplying both sides of the expressions entering the system (4.4.1–3) by

$$P_{\nu_n}(\cos\theta)\sin\theta \Big/ \int_0^\alpha \left[P_{\nu_n}(\cos\theta)\right]^2 \sin\theta d\theta\,,$$

and integrating with respect to θ between the limits $0, \alpha$ we get

$$\frac{1}{r^2}\frac{\partial}{\partial r}\left(r^2\frac{\partial a_n}{\partial r}\right) - \frac{\nu_n(\nu_n + 1)}{r^2} a_n = \frac{\partial^2 a_n}{\partial t^2} \quad (t < 1 - \cos\alpha,\ r > 1)\,, \tag{4.4.5}$$

$$\frac{\partial a_n}{\partial r} = -\frac{(1-t)P_{\nu_n}(1-t)}{\int_0^\alpha \left[P_{\nu_n}(\cos\theta)\right]^2 \sin\theta d\theta} \quad (r = 1)\,, \tag{4.4.6}$$

$$a_n = \frac{\partial a_n}{\partial t} = 0 \quad (t = 0)\,. \tag{4.4.7}$$

Note that the latter system of equations was obtained by assuming that the required solution $Q(r,\theta,t)$ was a generalized one (thus making it possible to change the order of integration and differentiation with respect to independent variables and time). Besides, (4.4.5) was obtained by using the fact that the medium is undisturbed for $\theta > \cos^{-1}(1-t)$, i.e. $Q(r,\theta,t) \equiv 0$. This is illustrated by Fig. 4.8 where the disturbance Q occupies the region ABA_1CA.

By substituting a_n by $\Phi(r,n,t)/\sqrt{r}$ the system (4.4.5–7) is reduced to

$$\frac{\partial^2\Phi}{\partial r^2}+\frac{1}{r}\frac{\partial\Phi}{\partial r}-\frac{(\nu_n+1/2)^2}{r^2}\Phi=\frac{\partial^2\Phi}{\partial t^2} \quad (t<1-\cos\alpha,\ r>1)\,, \tag{4.4.8}$$

$$\frac{\partial\Phi}{\partial r}-\frac{\Phi}{2}=-\frac{(1-t)P_{\nu_n}(1-t)}{\int\limits_0^\alpha [P_{\nu_n}(\cos\theta)]^2\sin\theta d\theta} \quad (r=1)\,, \tag{4.4.9}$$

$$\Phi=\frac{\partial\Phi}{\partial t}=0 \quad (t=0)\,. \tag{4.4.10}$$

Let us further subject (4.4.8, 10) to the generalized Weber integral transform as described in [4.48] where the coefficients P, Q, ν should be replaced by $\nu_n, 1, \nu_n+1/2$ respectively. Then we get

$$F=\int\limits_1^\infty C_{\nu_n+1/2}(rs,s)\Phi r\,dr\,, \quad \Phi=\int\limits_0^\infty \frac{C_{\nu_n+1/2}(rs,s)Fs\,ds}{|h^{(1)}_{\nu_n+1/2}(s)|^2}\,, \tag{4.4.11}$$

where

$$h_\nu^{(1)}(s)=(\nu-1/2)H_\nu^{(1)}(s)-sH_{\nu+1}^{(1)}(s)\,,$$

$$C_\nu(rs,s)=J_\nu(rs)\left[\left(\nu-\frac{1}{2}\right)Y_\nu(s)-sY_{\nu+1}(s)\right] - Y_\nu(rs)\left[\left(\nu-\frac{1}{2}\right)J_\nu(s)-sJ_{\nu+1}(s)\right]\,,$$

and $J_\nu(s), Y_\nu(s), H_\nu^{(1)}(s)$ are the Bessel functions of the 1st and the 2nd kind and the Hankel function of the 1st kind respectively. Then, (4.4.8, 10) reduce to

$$-s^2F-\frac{2}{\pi}\left(\frac{\partial\Phi}{\partial r}-\frac{\Phi}{2}\right)\bigg|_{r=1}=\frac{d^2F}{dt^2} \quad (t<1-\cos\alpha)\,, \tag{4.4.12}$$

$$F=\frac{dF}{dt}=0 \quad (t=0)\,. \tag{4.4.13}$$

Solving (4.4.12) subject to the initial conditions (4.4.13) we arrive (taking into account the boundary condition (4.4.9) at the following expression for F:

$$F=\frac{2}{\pi s}\int\limits_0^t \frac{\sin[s(t-x)](1-x)P_{\nu_n}(1-x)dx}{\int\limits_0^\alpha [P_{\nu_n}(\cos\theta)]^2\sin\theta d\theta}\,. \tag{4.4.14}$$

Then, subjecting the latter equation to the inverse transform (4.4.11) we obtain Φ, and hence ,

$$a_n = \frac{2}{\pi\sqrt{r}} \times \frac{\int_0^\infty C_{\nu_n+1/2}(rs,s)|h^{(1)}_{\nu_n+1/2}(s)|^{-2}ds \int_0^t (1-x)P_{\nu_n}(1-x)\sin[s(t-x)]dx}{\int_0^\alpha [P_{\nu_n}(\cos\theta))^2 \sin\theta d\theta} . \tag{4.4.15}$$

Therefore, for $0 < t < 1-\cos\alpha$ the disturbance $Q(r,\theta,t)$ is given by the expansion

$$Q(r,\theta,t) = H(\alpha-\theta)\sum_{n=0}^{\infty} a_n P_{\nu_n}(\cos\theta) , \tag{4.4.16}$$

where a_n is defined by (4.4.15) and the Heaviside function (the multiplier before the summation sign) is introduced to preclude an extension of the solution into the region $\alpha < \theta < \pi$ where $Q(r,\theta,t) = 0$. Considering the right-hand side of (4.4.16) as a function of α we deduce that it is independent of α for $\cos^{-1}(1-t) < \alpha < \pi$ (though each term of the series does depend on α), since according to (4.4.16), it gives an exact analytical solution to the problem which is independent of the way the expansion is implemented. Then, upon passing to the limit for $\alpha \to \cos^{-1}(1-t)$ on the right- and left-hand sides of (4.4.16), we get

$$Q(r,\theta,t) = H[\cos^{-1}(1-t)-\theta] \lim_{\alpha\to\cos^{-1}(1-t)} \sum_{n=0}^{\infty} a_n P_{\nu_n}(\cos\theta) . \tag{4.4.17}$$

Passing next to the limit in the expression under the summation sign in (4.4.17) we arrive at the required solution to the problem for $0 < t < 2$ in the form of an expansion over the variable interval $0 \le \theta < \cos^{-1}(1-t)$:

$$\begin{aligned} Q(r,\theta,t) &= H[\cos^{-1}(1-t)-\theta]\sum_{n=0}^{\infty} \bar{a}_n(r,t)P_{\nu_n}(\cos\theta) \\ &= \frac{2}{\pi\sqrt{r}} H[\cos^{-1}(1-t)-\theta]\sum_{n=0}^{\infty}\int_0^\infty C_{\nu_n+1/2}(rs,s)|h^{(1)}_{\nu_n+1/2}(s)|^{-2}ds \\ &\quad \times \frac{P_{\nu_n}(\cos\theta)}{\int_0^{\cos^{-1}(1-t)} [P_{\nu_n}(\cos\theta)]^2\sin\theta d\theta}\int_0^t \sin[s(t-x)]P_{\nu_n}(1-x)(1-x)dx \end{aligned}$$

$$(\bar{a}_n = \lim_{\alpha\to\cos^{-1}(1-t)} a_n) , \tag{4.4.18}$$

where ν_n $(0 = \nu_0 < 1 < \nu_1 < \nu_2 < \ldots)$ are non-negative roots of the equation $P\nu_n^1(1-t) = 0$; hence, the roots depend on t for $n \ge 1$.

For t tending to zero and $n \geq 1$, $\nu = O(t^{-1})$, see, e.g. [4.49]. Then, all the terms of (4.4.18) may be shown to tend to zero (with the exception of the first one ($n = 0$)), and we get :

$$Q(r,\theta,t) \sim \bar{a}_0(r,t)H[\cos^{-1}(1-t)-\theta] \quad (t \to 0).$$

In particular, for $r = 1 (t \to 0)$ one has

$$Q(1,\theta,t) \sim \frac{2}{t}\left(1-\frac{t}{2}-e^{-t}\right)H[\cos^{-1}(1-t)-\theta] \to \begin{cases} 0 & (\theta \neq 0)\,, \\ 1 & (\theta = 0)\,. \end{cases}$$

Hence, the solution (4.4.18) may be conveniently used for calculating at the initial period of time.

For $r = 1$, $C_\nu(rs, s) = 2/\pi$, and we get from (4.4.18)

$$Q(1,\theta,t) = \frac{4}{\pi^2}\sum_{n=0}^{\infty}$$
$$\times \frac{\int\limits_0^\infty |h^{(1)}_{\nu_n+1/2}(s)|^{-2}ds \int\limits_0^t (1-x)P_{\nu_n}(1-x)\sin[s(t-x)]dx}{\int\limits_0^{\cos^{-1}(1-t)} \left[P_{\nu_n}(\cos\theta)\right]^2 \sin\theta d\theta} P_{\nu_n}(\cos\theta)\,. \tag{4.4.19}$$

Let us show that for $t \to 2$ the solution (4.4.18) transforms directly into (4.3.22). In fact, with the help of (4.3.19) we get for $t > 0$, $n \geq 0$:

$$\sqrt{2\pi}\, e^{s(t-1)}\left(\frac{1}{\sqrt{s}}I_{n+1/2}(s)\right)'_s = \int\limits_0^2 P_n(1-x)e^{s(t-x)}(1-x)dx\,. \tag{4.4.20}$$

If $t \geq 2$, then the right-hand side of (4.4.20), and hence, the left-hand one too, may be replaced by

$$2\int_0^2 P_n(1-x)(1-x)\,\mathrm{sh}\,[s(t-x)]dx\,,$$

since the integral of the inverse Laplace transform of the added part vanished for $t \geq 2$.

Then, by putting the path of integration with respect to s onto the imaginary axis (by substituting s by is) and taking into account the relationship between the functions $H^{(1)}_\nu(s)$ and $K_\nu(is)$, (4.3.22) reduces, for $t \geq 2$, to

$$Q(r,\theta,t) = \frac{2}{\pi\sqrt{r}}\sum_{n=0}^{\infty}\left(n+\frac{1}{2}\right)P_n(\cos\theta)$$
$$\times \int\limits_0^\infty \frac{C_{n+1/2}(rs,s)ds}{|h^{(1)}_{n+1/2}(s)|^2}\int\limits_0^2 \sin[s(t-x)](1-x)P_n(1-x)dx\,. \tag{4.4.21}$$

If we now pass to the limit in (4.4.18) as $t \to 2$, then since $\nu \to n$, (4.4.18) reduces, with the help of (4.3.20) to (4.4.21) for $t = 2$.

Since the solution (4.3.22) is convenient for carrying out calculations at large values of t, we shall form a combined solution using the formula (4.4.18) for $0 < t < 2$, and (4.3.22) for $t \geq 2$.

The first term ($n = 0$) of such a combined solution for the pressure distribution on a sphere, $p(1,\theta,t) = H(t-1+\cos\theta)+Q(1,\theta,t)$, has a particularly simple form. Retaining in the first approximation this term only, we get

$$p(1,\theta,t) \approx \begin{cases} \frac{2}{t}(1-\mathrm{e}^{-t})H[\cos^{-1}(1-t)-\theta] & (0 < t < 2)\,, \\ 1-\mathrm{e}^{-t} & (t \geq 2)\,. \end{cases} \tag{4.4.22}$$

The latter expression coincides with the heuristic approximate solution obtained in [4.25].

Figure 4.9 borrowed from [4.25] shows dimensionless pressure $p(1,\theta,t)$ versus time for $\theta = 0, \pi/2, 3\pi/4$. The dependences were obtained by means of finite-difference approximation of the wave equation (the solid curves), by employing (4.4.22) (the dashed curve) and by using the formula $p = a_0(t)$ where $a_0(t)$ was calculated from (4.3.28) (see the dash-point curve). For a constant expansion interval, the zero term $a_0(t)$ from (4.3.28) gives the maximum value of pressure equal to $p_{\max} = (3 - \ln 2)/2$ for $t = \ln 2$. The error in the maximum pressure (which is actually equal to 2 for $t = 0$) amounts to 42%, while (4.4.22) gives the exact maximum value of pressure.

The solution can be further improved by taking into account consequential terms of the expansion in the combined solution constructed above.

The problem of diffraction of a plane acoustic wave on a cylinder was solved in a similar way in [4.50]. Just in the same way one can readily consider the problems of diffraction of elastic waves on a sphere/cylinder, as well as a variety of problems on the interaction of waves with shells and on the propagation of waves in shells subject to the condition that in all the problems one uses hyperbolic systems of equations (the Timoshenko-type theory).

4.5 Diffraction of Acoustic Waves by a Rigid Cone

This section and the next one demonstrate how to apply the integral transformation in terms of spherical coordinates to dynamic problems for cones.

4.5.1 Diffraction of a Plane Acoustic Wave by a Rigid Cone

The problem on diffraction of a plane non-stationary wave by a wedge (see Sect. 3.3) was solved by Sobolev [4.51] already in the 1930s. However, the problem with a cone was an open question for a long time, right up to the 1960s when an exact analytical solution was first offered in [4.52]. In what follows this approach is advanced. The "wave-by-cone" diffraction problem was earlier studied in [4.53].

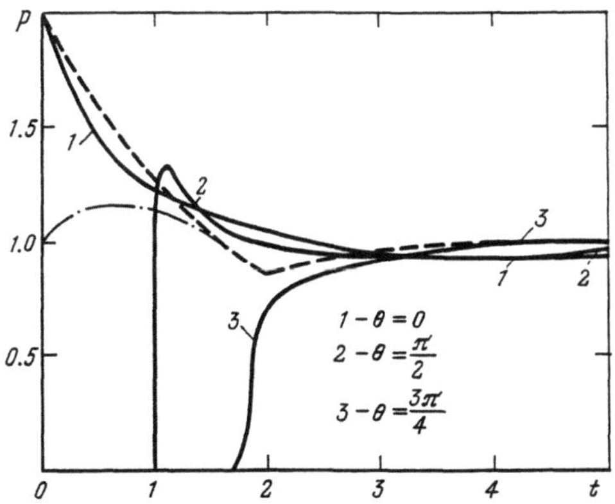

Fig. 4.9. Pressure–time relation for a sphere: numerical solution of the wave equation for three points, $\theta = 0, \pi/2, 3\pi/4$ (*solid lines*); equation (4.4.22) (*dashed line*); formula $p = a_0(t)$ with $a_0(t)$ from equation (4.3.28) (*dash-dot line*)

Assume the plane acoustic wave pressure field to be $p_1 = A_0 f(c_0 t - r\cos\theta) + p_0$ with $A_0 = \text{const}$, $f(t) \equiv 0$ at $t < 0$. Let the infinite circular cone have the vertex angle $2\theta_0 \in (0, \pi)$. The wave is assumed to propagate along the cone axis, zero time ($t = 0$) being taken at the instance of wave front arrival at the cone vertex. The problem is obviously an axisymmetric one. Let us define the disturbance Q as $Q = p_p - p_{1p}$, the non-dimensional pressure p_p as $p_p = (p - p_0)/A_0$, and the auxiliary function p_{1p} as $p_{1p} = (p_1 - p_0)/A_0$. To determine these disturbance and pressure, the following set of equations, initial and boundary conditions must be solved:

$$\begin{aligned} &\Delta Q = \frac{\partial^2 Q}{\partial \tau^2} \quad (\tau = c_0 t)\,, \\ &\frac{\partial Q}{\partial \theta} = -\frac{\partial p_{1p}}{\partial \theta} \quad (\theta = \theta_0)\,, \\ &Q = 0 \quad (\tau < 0)\,, \\ &p_{1p} = f(\tau - r\cos\theta)\,, \quad p_p = Q + p_{1p}\,. \end{aligned} \tag{4.5.1}$$

Here $\{r, \theta\}$ is the spherical coordinate system whose origin is at the cone vertex (Fig. 4.10).

In addition, as equation (2.3.7) establishes, the condition

$$p_p = O(r^\beta)\,, \quad \beta > -1/2\,, \quad r \to 0\,, \tag{4.5.2}$$

must be in the vicinity of the cone vertex in order for the result to be uniquc.

Thus, equations (4.5.1, 2) completely define diffraction of the plane acoustic wave $p_{1p} = f(\tau - r\cos\theta)$ by a cone.

Let us assume $f(\tau)$ to be $f(\tau) = \tau^\alpha H(\tau)$ with $\operatorname{Re}\alpha > -1$; in the case of α being a complex number a physically meaningful solution to the problem would be the real part of the function p_p from the system (4.5.1, 2) with an incident wave pressure field $\operatorname{Re} f(\tau)$. It is obvious that a solution for an arbitrary plane wave may be obtained by means of convolution. With this, apply the Laplace transformation in τ to the system (4.5.1) with $f(\tau) = \tau^\alpha H(\tau)$:

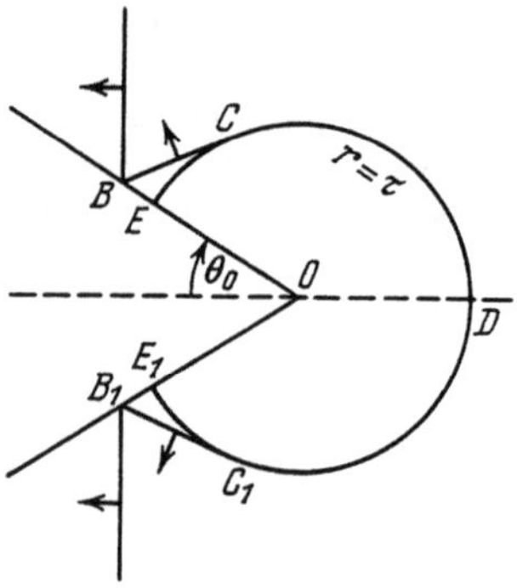

Fig. 4.10. Diffraction of a plane wave by a cone

$$
\begin{aligned}
&\frac{\partial}{\partial r}\left(r^2\frac{\partial Q^*}{\partial r}\right)+\frac{1}{\sin\theta}\frac{\partial}{\partial\theta}\left(\sin\theta\frac{\partial Q^*}{\partial\theta}\right)=r^2s^2Q^*\,,\\
&\frac{\partial Q^*}{\partial\theta}=-\frac{r}{s^\alpha}\sin\theta_0\Gamma(1+\alpha)\,\mathrm{e}^{-rs\sin\theta_0}\quad(\theta=\theta_0)\,,\\
&Q^*\equiv Q^*(s,r,\theta)=\int\limits_0^\infty Q(\tau,r,\theta)\,\mathrm{e}^{-s\tau}d\tau\,,\quad \operatorname{Re}s>0\,.
\end{aligned}
\tag{4.5.3}
$$

Here $\Gamma(.)$ is the gamma-function. Apply the Kontorovich–Lebedev transformation in r (e.g., see [4.35]) to the system (4.5.3) to obtain

$$
\frac{1}{\sin\theta}\frac{d}{d\theta}\left(\sin\theta\frac{dU}{d\theta}\right)+\left(\nu^2-\frac{1}{4}\right)U=0\quad(\operatorname{Im}\nu=0)\,,
\tag{4.5.4}
$$

$$
\frac{dU}{d\theta}=-\frac{\pi\sqrt{\pi}\Gamma(1+\alpha)}{\sqrt{2}\cos(\pi\nu)s^{3/2+\alpha}}P^1_{\nu-1/2}(\cos\theta_0)\quad(\theta=\theta_0)\,,
\tag{4.5.5}
$$

where

$$
U\equiv U(s,\nu,\theta)=\int\limits_0^\infty Q^*K_\nu(sr)\frac{dr}{\sqrt{r}}\,,\quad Q^*=\frac{1}{\pi\mathrm{i}\sqrt{r}}\int\limits_{-\mathrm{i}\infty}^{\mathrm{i}\infty}UI_\nu(sr)\nu d\nu\,.
$$

The solution to equation (4.5.4) has the form

$$
U=A_1P_{\nu-1/2}(\cos\theta)+A_2P_{\nu-1/2}(-\cos\theta)\,.
\tag{4.5.6}
$$

Since the function U cannot be allowed to be singular within the domain $\theta_0\le\theta\le\pi$, conclude that $A_1=0$, in order to get rid of the logarithmic singularity of the function $p_{\nu-1/2}(\cos\theta)$ at $\theta=\pi$. After using the boundary condition (4.5.5) for determining A_2, find

$$
U=\frac{\pi\sqrt{\pi}\Gamma(1+\alpha)P^1_{\nu-1/2}(\cos\theta_0)P_{\nu-1/2}(-\cos\theta)}{\sqrt{2}s^{3/2+\alpha}\cos(\pi\nu)P^1_{\nu-1/2}(-\cos\theta_0)}\,.
\tag{4.5.7}
$$

The inverse Kontorovich–Lebedev transformation in ν applied to (4.5.7) offers

$$Q^* = \frac{\pi\sqrt{\pi}\Gamma(1+\alpha)}{\mathrm{i}\sqrt{2r}s^{3/2+\alpha}} \int_{-\mathrm{i}\infty}^{\mathrm{i}\infty} \frac{P^1_{\nu-1/2}(\cos\theta_0)P_{\nu-1/2}(-\cos\theta)}{\cos(\pi\nu)P^1_{\nu-1/2}(-\cos\theta_0)} I_\nu(sr)\nu d\nu \,. \quad (4.5.8)$$

Here, the right-hand side integral is convergent and analytical in s over the domain specified by the inequality $|\arg s| < \pi/2 - (2\theta_0 - \theta)$. However, the function Q^* is a Laplace transform for the function to be found, so that Q^* must be analytical in the domain $\operatorname{Re} s > 0$. To derive the function Q^* analytical in the whole of the domain $\operatorname{Re} s > 0$ at any θ in the range $\theta_0 \leq \theta \leq \pi$, it is possible to continue analytically the right-hand side of the formula (4.5.8).

When $|\nu| \to \infty$

$$P^m_{\nu-1/2}(\cos\theta) \sim \left(\frac{2}{\pi\sin\theta}\right)^{1/2} \nu^{m-1/2} \sin\left(\nu\theta + \frac{m\pi}{2} + \frac{\pi}{4}\right)$$
$$(\varepsilon \leq \theta \leq \pi - \varepsilon,\ \varepsilon > 0)\,, \quad (4.5.9)$$
$$I_\nu(sr) \sim \frac{1}{\Gamma(1+\nu)}\left(\frac{sr}{2}\right)^\nu \quad \left(\left|\frac{\nu}{sr}\right| \gg 1\right)\,,$$

therefore the integrand in (4.5.8) decreases fast in the right-hand half-plane so we can deform the integration path into L of Fig. 4.11 since the integrand of (4.5.8) is analytical in the half-plane $\operatorname{Re}\nu > 0$ everywhere except for the real axis. Note that the path L in Fig. 4.11 goes to infinity along the real axis properly avoiding the simple poles of the integrand. It is now an easy operation to check that the integral in (4.5.8) taken over the path L is convergent and analytical in the half-plane $\operatorname{Re} s > 0$, thus representing the function Q^* in this half-plane, as the uniqueness theorem [4.35] suggests.

Apply now the inverse Laplace transformation to equation (4.5.8) and change the integration sequence. This results in

$$Q = -\frac{\Gamma(1+\alpha)}{2\sqrt{2\pi r}} \int_L \frac{P^1_{\nu-1/2}(\cos\theta_0)P_{\nu-1/2}(-\cos\theta)}{\cos(\pi\nu)P^1_{\nu-1/2}(-\cos\theta_0)} \nu\, d\nu \int_{c-\mathrm{i}\infty}^{c+\mathrm{i}\infty} I_\nu(sr)\mathrm{e}^{s\tau}\frac{ds}{s^{3/2+\alpha}}$$
$$(c > 0)\,. \quad (4.5.10)$$

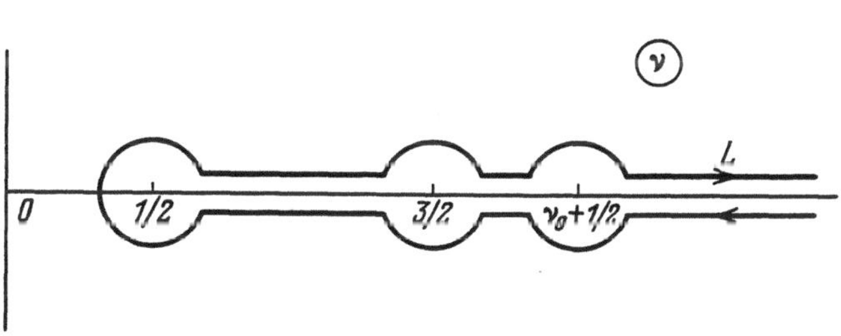

Fig. 4.11. Contour L

Here, the internal integral (in the variable s) may be transformed using the following relationship [4.52] from the operational calculus:

$$\frac{1}{2\pi \mathrm{i}} \int_{c-\mathrm{i}\infty}^{c+\mathrm{i}\infty} \frac{I_\nu(sr)}{s^\beta} \mathrm{e}^{s\tau} ds$$

$$= \begin{cases} \frac{\sqrt{2}\sin[\pi(\nu-\beta)]}{\pi\sqrt{\pi r}} \mathrm{e}^{\pi \mathrm{i}(\beta+1/2)}(\tau^2 - r^2)^{\beta/2-1/4} Q^{1/2-\beta}_{\nu-1/2}(\tau/r) & (r < \tau)\,, \\ \frac{(r^2-\tau^2)^{\beta/2-1/4}}{\sqrt{2\pi r}} P^{1/2-\beta}_{\nu-1/2}(-\tau/r) & (|\tau| < r)\,, \\ 0 & (\tau < -r)\,, \end{cases} \tag{4.5.11}$$

where $Q^\beta_\nu(x)$ is the associated Legendre function of the second kind [4.47], and it is implied that $r > 0$, $\mathrm{Re}\,\beta > -1/2$, $c > 0$.

Making use of (4.5.11) at $\beta = \alpha + 3/2$, re-write (4.5.10) as

$$Q = (r^2 - \tau^2)^{(1+\alpha)/2} \int_L P^{-1-\alpha}_{\nu-1/2}(-\tau/r) F d\nu \quad (|\tau| < r)\,,$$

$$Q = \frac{2}{\pi} \mathrm{e}^{\pi \mathrm{i}\alpha}(\tau^2 - r^2)^{(1+\alpha)/2} \int_L \cos[\pi(\nu - \alpha)] Q^{-1-\alpha}_{\nu-1/2}(\tau/r) F d\nu \quad (\tau > r)\,,$$

$$Q = 0 \quad (\tau < -r)\,,$$

$$F \equiv F(\nu,\theta,r) = \frac{\nu\Gamma(1+\alpha)P^1_{\nu-1/2}(\cos\theta_0)P_{\nu-1/2}(-\cos\theta)}{2\mathrm{i}r\cos\pi\nu P^1_{\nu-1/2}(-\cos\theta_0)}\,. \tag{4.5.12}$$

Employing the asymptotic formula (4.5.9) for the associated Legendre function of the first kind when $|\nu| \to \infty$, we can conclude that, for $2\theta_0 - \theta < \cos^{-1}(\tau/r)$ (and $|\tau| < r$), the integrand in (4.5.12) decreases as an exponential function along any ray (except for the real axis) in the ν-plane, so that the contour L may be deformed into the imaginary axis. Due to the fact that $P^\beta_{\nu-1/2}(x)$ is an even function of ν, the integral obtained has a value of zero. Therefore, if $\tau < 0$, then $Q = 0$ everywhere since $2\theta_0 - \theta < \pi/2 < \cos^{-1}(\tau/r)$. In addition, $Q = 0$ in the domain $\theta > 2\theta_0$ (Fig. 4.10) for $0 < \tau < r$ since $2\theta_0 - \theta < 0 - \cos^{-1} < \tau/r$ there. In the domain $\{\theta_0 \le \theta \le 2\theta_0,\ 0 < \tau < r\}$ the disturbance Q is zero everywhere except for the curvilinear triangle BCE ($B_1C_1E_1$) in which $BC(B_1C_1)$ is a tangent to the circle $r = \tau$, with definition being $2\theta_0 - \theta = \cos^{-1}(\tau/r)$.

As a result, obtain the outline, Fig. 4.10, of the disturbance region in a meridional section of the cone; an exterior of the region $OBCDC_1B_1O$ is not disturbed.

If use is made of the residue theorem, thereafter integrals in (4.5.12) are reduced to series dealing with poles of integrands, then

- a series in zero of $\cos\pi\nu$ for $-r < \tau < r$ and
- the residue at the point $\nu = 1/2$ for $\tau > r$

each supply an incident wave pressure field written with an opposite sign. A proof can be easily performed through application of the inverse Laplace transformation to the Sonin formula – (5) in Sect. 7.10 of [4.54] – and writing the Gegenbauer polynomials in terms of the associated Legendre functions of the first kind – (4) in Sect. 3.15 of [4.47].

Treated in this way the pressure field, $p_p = Q + p_{1p}$, can be represented as

$$
\begin{aligned}
p_p &= (\tau - r\cos\theta)^\alpha H(\tau - r\cos\theta) \quad (\tau < 0)\,,\\
p_p &= \frac{\Gamma(1+\alpha)}{2r\cos^2(\theta_0/2)}(r^2-\tau^2)^{(1+\alpha)/2} P_0^{-1-\alpha}(-\tau/r)\\
&\quad + \frac{\pi}{2r}\Gamma(1+\alpha)(r^2-\tau^2)^{(1+\alpha)/2}\\
&\quad \times \sum_{\nu=\nu_n} \frac{(2\nu+1)P_\nu^1(\cos\theta_0)P_\nu(-\cos\theta)P_\nu^{-1-\alpha}(-\tau/r)}{\sin(\pi\nu)dP_\nu^1(-\cos\theta_0)/d\nu}\\
&(0<\tau<r),\\
p_p &= \frac{\sin(\pi\alpha)\Gamma(1+\alpha)}{\pi r\cos^2(\theta_0/2)}(\tau^2-r^2)^{(1+\alpha)/2} \mathrm{e}^{\pi i\alpha} Q_0^{-1-\alpha}(\tau/r)\\
&\quad + \frac{\Gamma(1+\alpha)}{r}\mathrm{e}^{\pi i(1+\alpha)}(\tau^2-r^2)^{(1+\alpha)/2}\\
&\quad \times \sum_{\nu=\nu_n} \frac{\sin[\pi(\nu-\alpha)]P_\nu^1(\cos\theta_0)P_\nu(-\cos\theta)Q_\nu^{-1-\alpha}(\tau/r)(2\nu+1)}{\sin(\pi\nu)dP_\nu^1(-\cos\theta_0)/d\nu}\\
&(\tau > r)\,.
\end{aligned}
\tag{4.5.13}
$$

Here $\nu_n (n = 0, 1, 2, \ldots)$ are all the positive roots of the following equation,

$$P_\nu^1(-\cos\theta_0) = 0 \quad (\nu_0 < \nu_1 < \nu_2 < \ldots;\ 1 < \nu_0 < 2)\,. \tag{4.5.14}$$

Note that the series written for $0 < \tau < r$ is conditionally convergent for $-1 < \operatorname{Re}\alpha \le 0$ (outside the fronts of both the incident and reflected waves), whereas the series for $\tau > r$ is absolutely convergent.

For the cone surface ($\theta = \theta_0$) the relation (4.5.13) gives

$$
\begin{aligned}
p_p &= 0 \quad (\tau < r\cos\theta_0)\,,\\
p_p &= \frac{\Gamma(1+\alpha)}{2r\cos^2(\theta_0/2)}(r^2-\tau^2)^{(1+\alpha)/2} P_0^{-1-\alpha}(-\tau/r)\\
&\quad - \frac{\Gamma(1+\alpha)}{r\sin\theta_0}(r^2-\tau^2)^{(1+\alpha)/2}\sum_{\nu=\nu_n}\frac{(2\nu+1)P_\nu^{-1-\alpha}(-\tau/r)}{dP_\nu^1(-\cos\theta_0)/d\nu}\\
&(r\cos\theta_0 < \tau < r)\,,\\
p_p &= \frac{\sin(\alpha\pi)\Gamma(1+\alpha)}{\pi r\cos^2(\theta_0/2)}\mathrm{e}^{\pi i\alpha}(\tau^2-r^2)^{(1+\alpha)/2} Q_0^{-1-\alpha}(\tau/r)\\
&\quad + \frac{2\Gamma(1+\alpha)}{\pi r\sin\theta_0}\mathrm{e}^{\pi i\alpha}(\tau^2-r^2)^{(1+\alpha)/2}\sum_{\nu=\nu_n}\frac{(2\nu+1)\sin[\pi(\nu-\alpha)]Q_\nu^{-1-\alpha}(\tau/r)}{dP_\nu^1(-\cos\theta_0)/d\nu}\\
&(\tau > r)\,.
\end{aligned}
\tag{4.5.15}
$$

Also, when $r \to 0$, we derive from equation (4.5.13):

$$p_p(\tau,0,\theta) = \frac{\tau^\alpha}{\cos^2(\theta_0/2)}\ , \qquad \frac{\partial p_p}{\partial r} = \mathrm{O}(r^{\nu_0 - 1})\ . \tag{4.5.16}$$

The expression for $p_p(\tau,0,\theta)$ is confirmed by the results of [4.53] where the pressure at the cone vertex is determined without solving the entire problem.

From (4.5.16) it follows that the solution (4.5.13) is unique.

Let the incident wave be a unit step function: $f(\tau) = H(\tau)$. Appropriate formulae for this situation can be obtained from (4.5.12–16) with α tending to zero. Say, the pressure p_p outside the diffraction region $ECDC_1E_1OE$ (see Fig. 4.10), inside this region, and at $\tau < 0$ is

$$p_p = \frac{1+x}{2\cos^2(\theta_0/2)} + \frac{\pi}{2}\sqrt{1-x^2}\sum_{\nu=\nu_n}\frac{P_\nu^1(\cos\theta_0)P_\nu(-\cos\theta)P_\nu^{-1}(-x)(2\nu+1)}{\sin(\pi\nu)dP_\nu^1(-\cos\theta_0)/d\nu}$$
$$(0 < x < 1)\ ,$$
$$p_p = \frac{1}{\cos^2(\theta_0/2)} - \sqrt{x^2-1}\sum_{\nu=\nu_n}\frac{P_\nu^1(\cos\theta_0)P_\nu(-\cos\theta)Q_\nu^{-1}(x)(2\nu+1)}{dP_\nu^1(-\cos\theta_0)/d\nu}$$
$$(x > 1)\ ,$$
$$p_p = H(x - \cos\theta) \quad (x < 0)\ . \tag{4.5.17}$$

with new variable $x = \tau/r$.

Comparing these formulae, one can see that the function p_p undergoes a jump at the front $BC(B_1C_1)$ of the reflected wave, the jump being

$$[p_p] \equiv p_p^+ - p_p^- = \sqrt{\frac{\sin(2\theta_0-\theta)}{\sin\theta}} \quad (\theta_0 \le \theta \le 2\theta_0)\ , \tag{4.5.18}$$

where signs plus and minus relate to regions behind and ahead of the reflected wave.

On the circle $r = \tau$ being the front of the diffracted wave the function p_p is continuous, while the normal derivative $\partial p_p/\partial r$ is discontinuous. When approaching the arc $EC(E_1C_1)$ from both the diffraction region and the exterior of this region the function $\partial p_p/\partial r$ shows a logarithmic singularity, whereas a finite discontinuity is observed in going across the arc $CD(C_1D_1)$.

In the case when $\tau \to \infty$ at $r =$ const (i.e., when $x \to \infty$), the pressure at each point of the perturbation region tends to $1/\cos^2(\theta_0/2)$, i.e., to the value at the cone vertex.

If $\theta_0 \to \pi/2$, equation (4.5.17) gives the solution for the wave reflection from an unbounded flat wall: $p_p \to 2$; studying the analytical solution (4.5.17) shows that the second term adding a negative pressure is of the order of $\mathrm{O}(\beta)$ where $\beta = \pi/2 - \theta_0$, and takes account for an influence of both the cone vertex and cone surface curvature.

In the case of small θ_0, equation (4.5.17) at $x > 1$ provides an asymptotic expression for the pressure behind the diffracted wave front, far from the arc EC:

$$p_p \sim 1 + \frac{\theta_0^2(x^2-1)}{4(x-\cos\theta)^2}\ .$$

At $\theta = \theta_0$, the formulae (4.5.17) yield

$$p_p = \frac{1+x}{2\cos^2(\theta_0/2)} - \frac{\sqrt{1-x^2}}{\sin\theta_0} \sum_{\nu=\nu_n} \frac{(2\nu+1)P_\nu^{-1}(-x)}{2P_\nu^1(-\cos\theta_0)/d\nu}$$

$$(\cos\theta_0 < x < 1)\,,$$

$$p_p = \frac{1}{\cos^2(\theta_0/2)} + \frac{2\sqrt{x^2-1}}{\pi\sin\theta_0} \sum_{\nu=\nu_n} \frac{(2\nu+1)Q_\nu^{-1}(x)\sin(\pi\nu)}{dP_\nu^1(-\cos\theta_0)/d\nu} \qquad (4.5.19)$$

$$(x > 1)\,,$$

$$p_p = 0 \quad (x < \cos\theta_0)\,.$$

From the formulae (4.5.19), we find the following results:

$$p_p \to \frac{1}{\cos^2(\theta_0/2)} - \frac{2}{\pi\sin\theta_0} \sum_{\nu=\nu_n} \frac{(2\nu+1)\sin(\pi\nu)}{\nu(\nu+1)dP_\nu^1(-\cos\theta_0)/d\nu} \quad (x \to 1)\,,$$

$$p_p \to 2 \quad (x \to \cos\theta_0 + 0)\,.$$

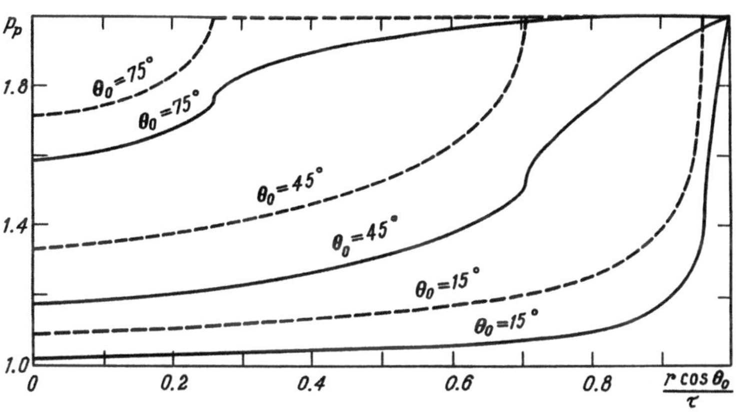

Fig. 4.12. Pressure profiles on cones (*solid lines*) and on wedges (*dashed lines*) with $\theta_0 = 15°, 45°, 75°$

Figure 4.12 taken from [4.52] displays pressures (4.5.19) on the cone surface at angles $\theta_0 = 15, 45, 75$ degrees (solid lines). Roots of the equation $P_\nu^1(-\cos\theta_0) = 0$ for $\theta_0 = 15$ degrees are taken from [4.55], and those for $\theta_0 = 45$ and 75 degrees can be calculated by the McDonald formula [4.56].

For comparison's sake, dashed lines in Fig. 4.12 represent pressures on the wedge with the apex angle of $2\theta_0$ when a step wave impacts symmetrically. These curves rely on the following formulae:

$$p = 2H(x - \cos\theta_0) + \frac{2}{\pi} H(x-1)\tan^{-1}\left[\frac{1-\varepsilon^{2l}}{1+\varepsilon^{2l}} \tan \pi l\right]$$
$$\left(l = \frac{\pi}{2(\pi - \theta_0)},\ \varepsilon = x - \sqrt{x^2-1},\ x = \frac{\tau}{r},\ \left|\tan^{-1} f\right| < \frac{\pi}{2}\right),$$

which is an obvious consequence from known results, see, for example, [4.44]. It is notable that pressures are lower on the cone than on the wedge (at the same value of r/τ and θ_0), which is attributed to three-dimensional effects in the vicinity of the cone. Across the front $r = \tau$, each pressure profile bends suddenly (since $p_p = 2$ in the domain of a reflected wave on the wedge, but the function $\partial p_p/\partial r$ behaves as $(\tau - r)^{-1/2}$ while approaching the front from the diffraction domain) whereas the curves for the cone are smooth and have vertical tangents at the diffracted wave front (once a slope of a tangent varies continuously due to logarithmic behavior of the function $\partial p_p/\partial r$ in going through the diffracted wave front from either side).

4.5.2 Diffraction of a Spherical Acoustic Wave by a Cone

Consider the problem on diffraction of a spherical acoustic pressure wave by a rigid cone [4.57]. Assume that the spherical wave

$$p_1 = A_0 r_0 \frac{f(\tau + r_0 - R)}{R} + p_0 \quad (\tau = c_0 t)$$

($f(\tau) \equiv 0$ for $\tau < 0$) is falling on an infinite circular cone (Fig. 4.13) with an apex angle of $2\theta_0$ ($0 < 2\theta_0 < \pi$). Here, r_0 is the distance from the cone apex to the wave source located at the point E on the cone axis and actuated at the time moment $\tau = -r_0$; $\{R, r, \theta\}$ is the spherical coordinate system originating at the cone apex (so that we can write $R = \sqrt{r^2 + r_0^2 + 2rr_0\cos\theta}$); p_0 and A_0 are constants.

At the time moment $\tau = 0$ when the wave front reaches the cone apex, diffraction begins. The problem under study is axisymmetrical, so the dimensionless pressure $p_s = (p - p_0)/A_0$ meets the wave equation coupled with the following initial and boundary conditions:

$$\frac{\partial}{r^2\partial r}\left(r^2\frac{\partial p_s}{\partial r}\right) + \frac{1}{r^2\sin\theta}\frac{\partial}{\partial\theta}\left(\sin\theta\frac{\partial p_s}{\partial\theta}\right) = \frac{\partial^2 p_s}{\partial\tau^2},$$
$$\frac{\partial p_s}{\partial\theta} = 0 \quad (\theta = \theta_0), \tag{4.5.20}$$
$$p_s = p_{1s} \quad \left(\tau < 0,\ p_{1s} = \frac{p_1 - p_0}{A_0} = \frac{r_0}{R} f(\tau + r_0 - R)\right).$$

Furthermore, in the vicinity of the cone apex the pressure must be

$$p_s = \mathrm{O}(r^\beta), \quad \beta > -1/2, \quad r \to 0, \tag{4.5.21}$$

which ensures that the solution will be unique.

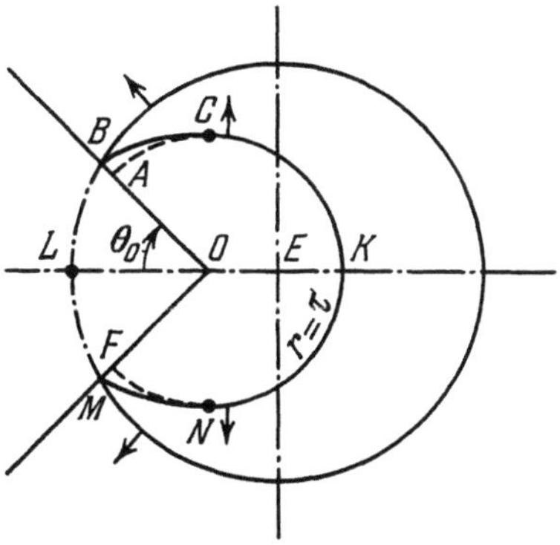

Fig. 4.13. Diffraction of a spherical wave by a cone

To solve this problem, there is no need to solve the set of equations (4.5.1,2) immediately. The much simpler approach is to use the analytical description to the diffraction of a plane acoustic wave by a cone (this has been established above) and employ the formula of [4.58] which relates solutions of the problems on diffraction of spherical and plane waves by a cone. According to [4.58], the solution $v(\tau, r, \theta)$ of the problem on diffraction of the plane wave $v_0 = (4\pi r_0)^{-1}\delta(\tau - r\cos\theta)$ by a cone yields a solution $V(\tau, r, \theta)$ to the problem on diffraction of the spherical wave $V_0(\tau, r, \theta) = (4\pi R)^{-1}\delta(\tau + r_0 - R)$ by the same cone:

$$V(\tau, r, \theta) = v\left(\tau + \frac{\tau^2 - r^2}{2r_0},\ r,\ \theta\right) \tag{4.5.22}$$

(for $\tau < 0$ we have $v \equiv v_0$, $V \equiv V_0$).

The formula (4.5.22) just remains valid for disturbances v' and V' added to the incident waves ($v = v_0 + v'$, $V = V_0 + V'$). This fact and the results obtained in the present section for the plane wave can immediately be applied to diffraction of the spherical acoustic wave $p_{1s} = r_0 f(\tau + r_0 - R)/R$ by a cone to give the pattern in the disturbance domain; note that the boundary of the latter does not depend on the specific function $f(\tau)$. Outside the meridional section $BCKNMOB$ (see Fig. 4.13) bounded by (i) the cone surface (MOB), (ii) the circular arc CKN of radius $r = \tau$, (iii) arcs BC and NM, and (iv) tangents to the circle $r = \tau$ at the points C and N, no disturbances are present: $p_s \equiv p_{1s}$. The arc BC is symmetric to the arc LB about the straight line BO, represents the reflected wave front and can be described by the equation, $2r_0\tau + \tau^2 - r^2 = 2r_0 r\cos(2\theta_0 - \theta)$; from the latter, it follows that the position angle of the point C is $2\theta_0$. The disturbances generated by the cone apex cannot invade the region ABC since they are bounded by the arc AC. We know that the solution $4\pi r_0 v(\tau, r, \theta)$ is derivable from the above-stated solution (4.5.17) for the plane wave, after differentiation with respect to τ. Next, the pressure is a product of convolving the function $4\pi r_0 V$. Finally, using (4.5.22), find

$$\begin{aligned}
p_s =& 4\pi r_0 \int_{-\infty}^{\tau} f(\tau-\zeta) v\left(\zeta+\frac{\zeta^2-r^2}{2r_0},\, r,\, \theta\right) d\zeta \\
=& \int_{-\infty}^{\tau} f(\tau-\zeta)\frac{d}{d\zeta} p_p\left(\zeta+\frac{\zeta^2-r^2}{2r_0},\, r,\, \theta\right)\frac{r_0}{r_0+\zeta} d\zeta \\
=& \frac{f(0) r_0}{r_0+\tau} p_p\left(\tau+\frac{\tau^2-r^2}{2r_0},\, r,\, \theta\right) \\
& - r_0 \int_{-\infty}^{\tau} p_p\left(\zeta+\frac{\zeta^2-r^2}{2r_0},\, r,\, \theta\right)\frac{d}{d\zeta}\left[\frac{f(\tau-\zeta)}{r_0+\zeta}\right] d\zeta\,. \qquad (4.5.23)
\end{aligned}$$

Here, $f(0) = \lim_{\tau\to+0} f(\tau)$; the lower limit of integration in (4.5.23) turns out to be finite since

$$v\left(\tau+\frac{\tau^2-r^2}{2r_0},\, r,\, \theta\right) \equiv 0 \quad \text{as} \quad \tau+\frac{\tau^2-r^2}{2r_0} < r\cos\theta\,,$$

as long as the incident wave is approaching the observation point (r,θ). Then, we should take into account that $p_p(\tau,r,\theta)$ is a homogeneous function of zero order with respect to the variables τ and r; its explicit form is presented by the formulae (4.5.17) where one can mean $p_p(\tau,r,\theta) \equiv p_p(\tau/r,\theta)$. Using the symbol $\eta = \zeta/r + (\zeta^2-r^2)/2r_0 r$, we transform equation (4.5.23) to

$$\begin{aligned}
p_s(\tau,r,\theta) =& \frac{f(0) r_0}{r_0+\tau} p_p(k,\theta) \\
& - r_0 \int_{\cos\theta}^{k} p_p(\eta,\theta)\frac{d}{d\eta}\left[\frac{f\left(\tau+r_0-\sqrt{r^2+r_0^2+2rr_0\eta}\right)}{\sqrt{r^2+r_0^2+2rr_0\eta}}\right] d\eta \\
& \left(k=\frac{\tau}{r}+\frac{\tau^2-r^2}{2r_0 r}\right)\,. \qquad (4.5.24)
\end{aligned}$$

For $\tau < 0$ the latter formula gives $p_s = p_{1s}$ (since, in this case, $k < 0$ and $p_p = H(\tau - r\cos\theta)$). From (4.5.24) and the above results concerning the p_p, obtain for r tending to zero:

$$p_s(\tau,0,\theta) = \frac{f(\tau)}{\cos^2(\theta_0/2)}\,, \qquad \frac{\partial p_s}{\partial r} = O(r^{\nu_0-1})\,, \qquad (4.5.25)$$

where $\nu_0 > 1$; ν_0 is the smallest positive root of the equation, $P_\nu^1(-\cos\theta_0) = 0$. From the asymptotic formula (4.5.25), it is obvious that the solution (4.5.24) satisfies the uniqueness conditions written in (4.5.21).

Consider a particular case: $f(\tau) = H(\tau)$. The formula (4.5.24) and the above investigations into diffraction of a plane wave yield the following results.

The function p_s is discontinuous at fronts of the reflected waves BC and NM (see Fig. 4.13), the jump being

$$[p_s] = p_s^+ - p_s^- = \frac{r_0}{r_0 + \tau} \sqrt{\frac{\sin(2\theta_0 - \theta)}{\sin \theta}} \quad (\theta_0 \leq \theta \leq 2\theta_0) , \tag{4.5.26}$$

where "plus" and "minus" are associated with the domains lying behind and ahead of the reflected wave front, respectively. Prove this. In going across the arc BC where the variable k (in (4.5.24)) takes the value of $\cos(2\theta_0 - \theta)$, the integral varies continuously, once the integrand $p_p(\eta, \theta)$ has a jump only. Therefore,

$$p_s^+ - p_s^- = \frac{f(0) r_0}{r_0 + \tau} (p_p^+ - p_p^-) .$$

With (4.5.18) and $f(0) = 1$, we obtain (4.5.26).

On the circle $r = \tau$, the normal derivative $\partial p_s / \partial r$ is discontinuous: on either sides of the arc AC (NF) this derivative has logarithmic singularity, whereas on the arc CKN this function undergoes a jump.

If $\theta_0 \to \pi/2$, then, relying upon equation (4.5.24), we can take account of the fact that the function $p_p(\eta, \theta)$ tends to 2 in the disturbance domain, thus obtaining the known solution for reflection of a spherical wave from a flat wall:

$$p_s = \frac{r_0}{R} H(\tau + r_0 - R) + \frac{r_0}{\bar{R}} H(\tau + r_0 - \bar{R})$$
$$\left(\bar{R} \equiv \sqrt{r^2 + r_0^2 - 2 r r_0 \cos \theta} \right) . \tag{4.5.27}$$

If the angle θ_0 is small (the case of a slender body), the results can be verified by the retarded potential method.

Figure 4.14 illustrates the formula (4.5.24); pressures p_s on cones with apex angles of $2\theta_0 = 30, 90, 150$ degrees have been calculated for different values of r/r_1 ratio; r_1 is the position radius of the point B in Fig. 4.13. In these calculations, use was made of the numerical results for the function p established above for the case of a step plane wave. Symbols 1 through to 4 in each case mark the curves obtained at $\tau / r_0 = 0.1, 0.2, 0.5, 1.0$, respectively, and the symbol 0 marks the corresponding self-similar solutions, p_p, to the plane wave diffraction problem which appears if $r_0 \to \infty$. Curves from the points $M_1(0, 1.02)$, $M_2(0, 1.17)$, and $M_3(0, 1.59)$ correspond to the apex angles $2\theta_0 = 30, 90, 150$ degrees, respectively.

From the results obtained, it is obvious that the influence of an incident wave-front curvature grows in time, so that the pressure on the front of the spherical wave at $\tau / r_0 = 1$ (symbol 4) is half the pressure on the front of the corresponding plane wave.

These plots clearly indicate that the pressure immediately behind the front of the reflected wave does not depend on the angle θ_0. This is a consequence of the fact that the pressure is equal to $2(1 + \tau / r_0)^{-1}$ at the point B (see Fig. 4.13), behind the front, as equation (4.5.26) with account for the amplitude of the incident wave gives. This, when coupled with plots in Fig. 4.14, suggests that, instead of determining the pressure distributions caused by a spherical wave, use can be made of the pressure profiles produced by a corresponding plane wave; with this, the error for $\tau / r_0 \leq 0.1$ will not exceed 15%.

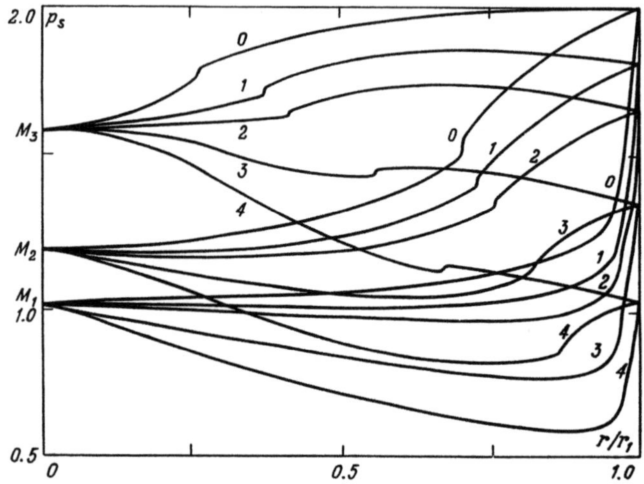

Fig. 4.14. Pressure vs. r/r_1 on cones with apex angles $2\theta_0$ = 30°, 90°, 150° (curves originate at points M_1, M_2, and M_3, respectively). Time moments are $\tau/r_0 = 0.1, 0.2, 0.5, 1.0$ (symbols 1 to 4, respectively). Symbol 0 denotes pressure profiles for the case of plane acoustic wave incidence

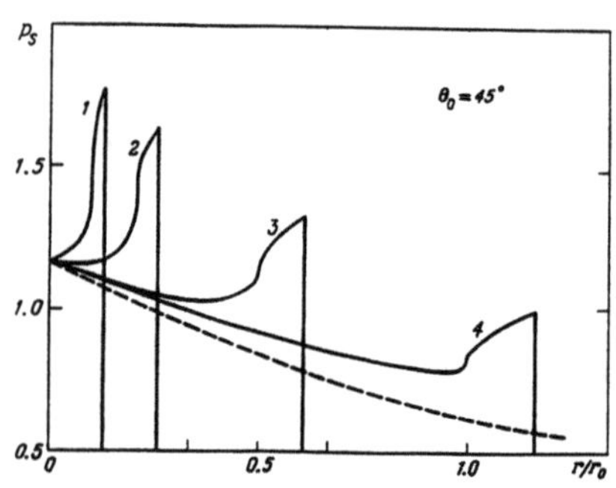

Fig. 4.15. Pressure vs r/r_0 on a cone ($2\theta_0$ = 90°) at the time moments $\tau/r_0 = 0.1, 0.2, 0.5, 1.0$ (symbols 1 to 4, respectively)

Figure 4.15 represents p_s–r/r_0 relations for a cone with the apex angle $2\theta_0$ = 90°. Symbols 1 through to 4 correspond to the above-mentioned values of the parameter τ/r_0. The dashed line indicates the limiting pressure distribution at $\tau = \infty$; when $r/r_0 \to \infty$, this curve tends to the straight line $p_s = 0$. Using these plots, one can get a notion about features of how the pressure distributions form on a cone subjected to a spherical wave $p_{1s} = r_0 H(\tau + r_0 - R)/R$.

4.6 Diffraction of Elastic Waves by a Smooth Rigid Cone

Consider now diffraction of elastic waves by a smooth rigid cone [4.24, 59]

4.6.1 Diffraction of a Plane Longitudinal Wave by a Cone

Let a longitudinal wave with a potential

$$\varphi_0 = -\tfrac{1}{2}H(\tau - r\cos\theta)(\tau - r\cos\theta)^2 \quad (\tau = c_1 t) \tag{4.6.1}$$

be falling on an infinite circular cone freely embedded into an elastic medium (Fig. 4.10). Let $2\theta_0$ $(0 < 2\theta_0 < \pi)$ be the cone apex angle. Assume that the front of the incident wave (being perpendicular to the cone axis) reaches the apex at the time moment $\tau = 0$. Hereinafter, use is made of the spherical coordinates r and θ measured from the cone apex and axis, respectively.

The problem under study is axisymmetric, thus, the motion equations, initial and boundary conditions according to (2.7.41, 53) can be written in terms of the longitudinal (φ) and transverse (ψ) potentials:

$$\begin{aligned} &\Delta\varphi = \frac{\partial^2\varphi}{\partial\tau^2}\,, \quad \Delta\psi = \gamma^2\frac{\partial^2\psi}{\partial\tau^2} \quad \left(\gamma = \frac{c_1}{c_2} > 1\right), \\ &u_\theta = 0\,, \quad \sigma_{\theta r} = 0 \quad (\theta = \theta_0)\,, \\ &\varphi = \varphi_0\,, \quad \psi = 0 \quad (\tau < 0)\,. \end{aligned} \tag{4.6.2}$$

The displacement vector components u_θ and u_r and the stress tensor component $\sigma_{\theta r}$ have the following expressions in terms of the potentials, according to (2.1.7) and (2.7.52, 54):

$$\begin{aligned} &u_\theta = \frac{\partial\varphi}{r\partial\theta} + \frac{\partial}{r\partial r}\left(r\frac{\partial\psi}{\partial\theta}\right), \quad u_r = \frac{\partial\varphi}{\partial r} - \frac{1}{r}\frac{1}{\sin\theta}\frac{\partial}{\partial\theta}\left(\sin\theta\frac{\partial\psi}{\partial\theta}\right), \\ &\sigma_{\theta r} = \mu\left\{2\frac{\partial}{\partial r}\left(\frac{\partial\varphi}{r\partial\theta}\right) + 2\frac{\partial}{\partial r}\left[\frac{1}{r}\frac{\partial}{\partial r}\left(r\frac{\partial\psi}{\partial\theta}\right)\right] - \gamma^2\frac{\partial^3\psi}{\partial\theta\partial\tau^2}\right\}. \end{aligned} \tag{4.6.3}$$

To obtain a unique solution, one should, according to (2.3.6), restrict the displacement vector behavior in the vicinity of the apex:

$$\boldsymbol{u} = \mathrm{O}(r^\varepsilon)\,, \quad \varepsilon > -1/2\,, \quad r \to 0\,. \tag{4.6.4}$$

The latter completes the set of equations (4.6.1–3) in describing a diffraction of elastic wave by a smooth rigid cone.

Using equations (4.6.3) for u_θ and $\sigma_{\theta r}$, one could easily note that the boundary conditions $u_\theta = 0$ and $\sigma_{\theta r} = 0$ $(\theta = \theta_0)$ will be satisfied if

$$\frac{\partial\varphi}{\partial\theta} = 0\,, \quad \frac{\partial\psi}{\partial\theta} = 0 \quad (\theta = \theta_0)\,. \tag{4.6.5}$$

This means that boundary conditions are set for longitudinal and transverse potentials independently. So, we can treat appropriate equations independently as long as the condition (4.6.4) established for the vicinity of the cone apex can be kept apart.

Letting $\varphi = \varphi_0 + \varphi'$, equations (4.6.2, 5) result in two systems for the potentials φ' and ψ coupled by (4.6.4):

$$\Delta\varphi' = \frac{\partial^2\varphi'}{\partial\tau^2}\,, \quad \frac{\partial\varphi'}{\partial\theta} = -\frac{\partial\varphi_0}{\partial\theta} \quad (\theta=\theta_0)\,, \quad \varphi' = 0 \quad (\tau<0)\,; \tag{4.6.6}$$

$$\Delta\psi = \gamma^2\frac{\partial^2\psi}{\partial\tau^2}\,, \quad \frac{\partial\psi}{\partial\theta} = 0 \quad (\theta=\theta_0)\,, \quad \psi = 0 \quad (\tau<0)\,. \tag{4.6.7}$$

Now, the first step is to solve the set of equations (4.6.6) by means of the Laplace transform in τ followed by the Kontorovich–Lebedev transform in r. Subsequent mathematical manipulations are analogous to those performed in Sect. 4.5.1 for the acoustic problem, thus, are allowed to be omitted; it is due to us to multiply the right-hand sides in (4.5.13) by $(-1/2)$ at $\alpha = 2$ to write the final result for the longitudinal potential

$$\begin{aligned}
\varphi &= -\frac{(r+\tau)^3}{12r\cos^2(\theta_0/2)} \\
&\quad -\frac{\pi}{2r}(r^2-\tau^2)^{3/2}\sum_{\nu=\nu_n}\frac{(2\nu+1)P_\nu^1(\cos\theta_0)P_\nu(-\cos\theta)P_\nu^{-3}(-\tau/r)}{\sin(\pi\nu)dP_\nu^1(-\cos\theta_0)/d\nu} \\
&(0<\tau<r)\,, \\
\varphi &= -\frac{r^2+3\tau^2}{6\cos^2(\theta_0/2)} \\
&\quad +\frac{1}{r}(\tau^2-r^2)^{3/2}\sum_{\nu=\nu_n}\frac{(2\nu+1)P_\nu^1(\cos\theta_0)P_\nu(-\cos\theta)Q_\nu^{-3}(\tau/r)}{dP_\nu^1(-\cos\theta_0)/d\nu} \\
&(\tau>r)\,, \\
\varphi &= -\tfrac{1}{2}(\tau-r\cos\theta)^2H(\tau-r\cos\theta) \quad (\tau<0)\,,
\end{aligned} \tag{4.6.8}$$

Here, ν_n $(n = 0,1,\ldots)$ are the entire set of positive roots of the equation, $P_\nu^1(-\cos\theta_0) = 0 (\nu_0 < \nu_1 < \nu_2 < \ldots;\ 1 < \nu_0 < 2)$.

The middle formula in (4.6.8) represents the solution in the domain $ECDC_1E_1OE$ (Fig. 4.10), whereas the upper formula deals with the exterior of this domain. Note that the upper formula gives $\varphi = \varphi_0$ outside the domain $BCDC_1B_10B$; this statement results from the solution of the problem in Sect. 4.5.1.

According to (4.5.16), the function φ from (4.6.8) is such that

$$\frac{\partial\varphi}{\partial r} = \mathrm{O}\left(r^{\nu_0-1}\right), \quad r\to 0\,. \tag{4.6.9}$$

In the same way, we can establish the asymptotic formula

$$\frac{\partial \varphi}{r\partial \theta} = O\left(r^{\nu_0 - 1}\right), \quad r \to 0. \tag{4.6.10}$$

Equations (4.6.9, 10) clearly indicate that, for the condition (4.6.4) to be satisfied, it is sufficient to make use of the longitudinal potential only. We shall therefore take the trivial solution of the system (4.6.7):

$$\psi \equiv 0. \tag{4.6.11}$$

So, no transverse waves appear at axisymmetric incidence of a plane longitudinal wave, and the potential functions obtained are the same as those for the corresponding acoustic problem.

Let us consider the newly derived solution. The disturbance pattern represented by Fig. 4.10 is a result of the mathematical study undertaken in Sect. 4.5.1 for the plane acoustic wave. As the problem is an axisymmetric one, it is sufficient to deal only with the upper half of the meridian section in Fig. 4.10. First of all, note that both the potential φ and all components of the displacement vector $\boldsymbol{u}$ are continuous everywhere around the cone. The incident wave (4.6.1) is such that the stress undergoes a jump at the wave front:

$$[\sigma_{nn}] = \sigma_{nn}^{+} - \sigma_{nn}^{-} = -(\lambda + 2\mu).$$

In the last equation, $\boldsymbol{n}$ is the normal to the front, the signs "plus" and "minus" are associated with the regions behind and ahead of the front, respectively. In going across the front BC of the reflected wave, we observe a jump in stresses:

$$[\sigma_{nn}] = \sigma_{nn}^{+} - \sigma_{nn}^{-} = -(\lambda + 2\mu)\sqrt{\frac{\sin(2\theta_0 - \theta)}{\sin\theta}} \quad (\theta_0 \le \theta \le 2\theta_0) \tag{4.6.12}$$

(since the function $\partial^2\varphi/\partial n^2$ is discontinuous). On the circle $r = \tau$, strains and stresses are continuous, whereas the normal derivative of the strain, $\partial\varepsilon_{rr}/\partial r$, is not (due to discontinuity in $\partial^3\varphi/\partial r^3$). In this case, the function $\partial\varepsilon_{rr}/\partial r$ is logarithmically singular at both sides of the arc EC, whereas across the arc CD this function has a jump.

Figure 4.16 shows diagrams of the stress $(-\sigma_{\theta\theta})$ versus r/r_1 (where r is the position radius of the point B, see Fig. 4.10) along the generating line OB of the cone (at $\lambda = 2\mu$) subjected to the plane compressional wave

$$\varphi_0 = -\frac{(\tau - r\cos\theta)^2 H(\tau - r\cos\theta)}{2(\lambda + 2\mu\sin^2\theta_0)}. \tag{4.6.13}$$

The stress has been calculated using the formula,

$$\sigma_{\theta\theta} = \lambda\frac{\partial^2\varphi}{\partial\tau^2} + 2\mu\left(\frac{\partial^2\varphi}{r^2\partial\theta^2} + \frac{\partial\varphi}{r\partial r}\right), \tag{4.6.14}$$

for which φ is taken from (4.6.8) and normalized by the factor $(\lambda + 2\mu\sin^2\theta_0)^{-1}$. Numbers 1, 2, and 3 correspond to the cone apex angles $2\theta_0$ of 30°, 90°, and 150°, respectively. The results indicate that the stress in the vicinity of the cone

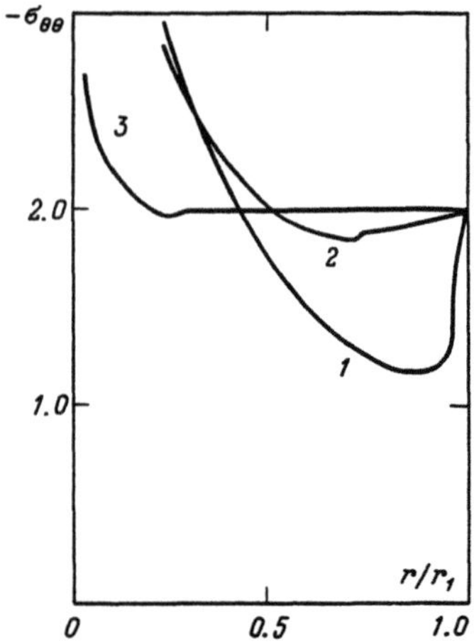

Fig. 4.16. Stress $-\sigma_{\theta\theta}$ vs r/r_1 on cones with apex angles $2\theta_0 = 30°, 90°, 150°$ (curves 1, 2, and 3, respectively)

apex has an integrable singularity of the order of $(\nu_0 - 2)$, and $\nu_0 - 2 = -0.9684$, -0.751, and -0.3331 at apex angles of 30°, 90°, and 150°, respectively. The singularity becomes less strong when the apex angle grows; in the limiting case, if $2\theta_0 \to 180°$, singularity disappears.

4.6.2 Diffraction of a Spherical Elastic Wave by a Cone

It is by analogy with the case of the plane longitudinal incident wave that the solution to the problem with a spherical longitudinal wave

$$\varphi_0 = \frac{r_0}{R} f(\tau + r_0 - R) \quad (f(\tau) = 0 \quad \text{for} \quad \tau < 0) \tag{4.6.15}$$

is the same as the solution to the corresponding acoustic problem (4.5.24), thus, we get the following expressions for the displacement potentials:

$$\begin{aligned} \varphi(\tau, r, \theta) = {} & \frac{f(0) p_p(k, \theta) r_0}{r_0 + \tau} \\ & - r_0 \int\limits_{\cos\theta}^{k} p_p(\eta, \theta) \frac{d}{d\eta} \left[\frac{f\left(\tau + r_0 - \sqrt{r^2 + r_0^2 + 2 r r_0 \eta}\right)}{\sqrt{r^2 + r_0^2 + 2 r r_0 \eta}} \right] d\eta \,, \\ \psi \equiv 0 \quad & \left(k = \frac{\tau}{r} + \frac{\tau^2 - r^2}{2 r r_0}, \ \tau = c_1 t, \ f(0) = \lim_{\tau \to +0} f(\tau) \right) . \end{aligned} \tag{4.6.16}$$

Here $p_p(\eta, \theta)$ is taken from (4.5.17) after substituting η for x. Note that the wave pattern in the disturbance region (Fig. 4.13) has been thoroughly studied in Sect. 4.5.2 concerning the spherical acoustic wave diffraction by a cone.

Let us consider in detail the case of

$$\varphi_0 = -\frac{r_0 (\tau + r_0 - R)^2 H(\tau + r_0 - R)}{R(2\lambda + 4\mu \sin^2\theta_0)} . \tag{4.6.17}$$

One could immediately write the jump in stresses at the incident wave front:

$$[\sigma_{nn}] = \sigma_{nn}^{+} - \sigma_{nn}^{-} = -(\lambda + 2\mu)r_0(\tau + r_0)^{-1}(\lambda + 2\mu\sin^2\theta_0)^{-1} .$$

Everywhere in the disturbance region the potential φ and the displacement vector $\boldsymbol{u}$ are continuous. Across the front of the reflected wave the jump in the stress is

$$[\sigma_{nn}] = \sigma_{nn}^{+} - \sigma_{nn}^{-} = -\frac{r_0(\lambda + 2\mu)}{(\tau + r_0)(\lambda + 2\mu\sin^2\theta_0)}\sqrt{\frac{\sin(2\theta_0 - \theta)}{\sin\theta}} \quad (\theta_0 \le \theta \le 2\theta_0) \tag{4.6.18}$$

(due to discontinuity of the function $\partial^2\varphi/\partial n^2$). On the circle $r = \tau$ both the strains and stresses are continuous, whereas the normal derivative of the strain, $\partial\varepsilon_{rr}/\partial r$, is discontinuous due to discontinuity of the function $\partial^3\varphi/\partial r^3$. Therefore, approaching the arc AC (see Fig. 4.13) from any side reveals a logarithmic discontinuity in $\partial\varepsilon_{rr}/\partial r$, and across the arc CK, a jump.

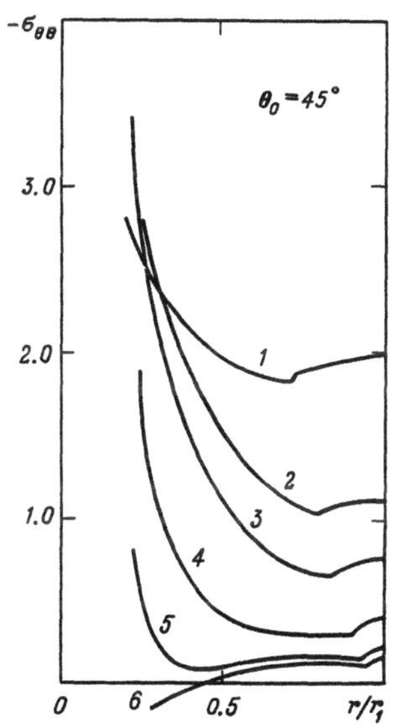

Fig. 4.17. Stress profiles for a cone at time moments $\tau/r_0 = 0.5$, 1.0, 2.5, 5.0, 7.0 (symbols 2 to 6, respectively). Curve 1 is the profile for plane longitudinal wave incidence

Figure 4.17 shows diagrams of the stress $(-\sigma_{\theta\theta})$ versus r/r_1 (where r_1 is the position radius of the point B, see Fig. 4.13) along the generating line OB of the cone. The stress was calculated by equations (4.6.14, 16) for the incident wave (4.6.17) at $2\theta_0 = 90°$ at $\lambda = 2\mu$. The symbol 1 denotes the curve for the plane compressional wave

$$\varphi_0 = -\frac{(\tau - r\cos\theta)^2 H(\tau - r\cos\theta)}{2\lambda + 4\mu\sin^2\theta_0} , \tag{4.6.19}$$

which represents a limiting case of the spherical wave (4.6.17) if $r \to \infty$. The curves 2 through to 6 correspond to stress distributions at $\tau/r_0 = 0.5, 1.0, 2.5, 5.0, 7.0$ on the cone subjected to the wave (4.6.17).

Comparing the results for spherical and plane waves it is seen that the incident wave front curvature has an increasing effect as time goes on. At $\tau/r_0 = 1$ (the curve 3) the stress in the domain $0.6 < (r/r_1 < 1.0)$ on the cone subjected to the plane wave is greater than two times the stress caused by the spherical wave; moreover, tensile stresses are observed in a certain portion of the cone surface at $\tau/r_0 = 7.0$ (the curve 6). Analytical researches show that, for any apex angle $(0 < 2\theta_0 < 180°)$, tensile stresses appear at a certain time moment; as $2\theta_0$ tends to 180° the time moment of tensile stress appearance tends to τ_0 so that in the limiting case of a flat wall and spherical wave the stresses on the wall surface will everywhere be compressive at any $\tau < \tau_0$.

Resembling the case with the plane incident wave, the stress in the vicinity of the cone apex $(r \to 0)$ has an integrable singularity of the same order: $\sigma_{\theta\theta} = O(r^{\nu_0 - 2})$, $1 < \nu_0 < 2$.

As well, note that $\sigma_{\theta\theta} = O(\tau^2)$ at $r = \text{const}$ when $\tau \to \infty$, and $\sigma_{\theta\theta} = O(\tau^{-1})$ at $\tau/r = \text{const}$, when $\tau \to \infty$. These results follow from equations (4.6.14, 16).

Concluding this section, it is worth noting that in an analogous way we can also study the axisymmetric diffraction problems for spherical transverse (torsional) waves and waves due to application of concentrated forces and/or moments.

It is also essential to notice that analytically solving the two above elastodynamics problems turned out to be a possibility due to the fact that boundary conditions for a smooth rigid cone are, in the axisymmetric problem, separable with respect to the potentials, φ and ψ. As a result, every problem has been reduced to two acoustic problems (one for φ and the other for ψ) related to each other by a condition at the cone apex. Moreover, symmetry considerations allowed us to meet the apex condition using the longitudinal potential only; with this, no transverse diffraction waves are present. More detailed studies reveal that boundary conditions for potentials can be separated, and an analytical solution obtained, only if the two following requirements are satisfied simultaneously:

- mixed boundary conditions in problems with cones have the form (2.2.4, 5) and
- complete axial symmetry is specified.

Non-compliance with either of these requirements makes it impossible to separate the boundary conditions for potentials, thus making it difficult to derive an exact analytical solution. The author is not aware of any positive result obtained in such circumstances.

4.7 Impact of a Circular Cylinder on a Stationary Obstacle

It is our aim now to consider waves in a semi-infinite circular cylinder moving at the speed v_0 and striking against an immovable rigid smooth plane (Fig. 4.18).

This problem was solved for the first time in [4.13] using integral transformations. Hereinafter, use will be made of the same technique, but in a somewhat different way.

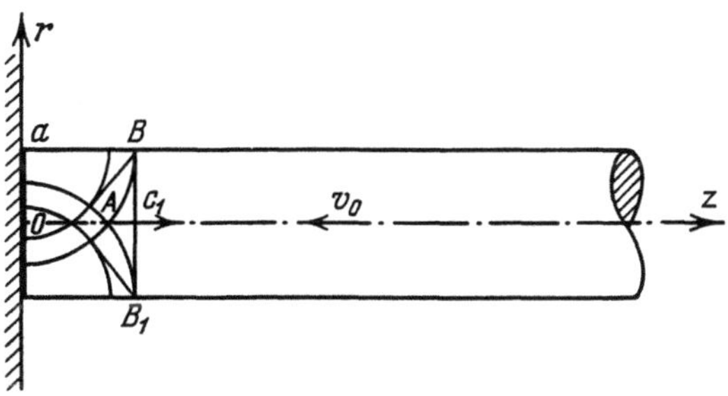

Fig. 4.18. Impact of a circular cylinder against an immovable obstacle

4.7.1 Formulation and Solution of the Problem

Let us introduce the cylindrical coordinate system $\{r, \theta, z\}$ fixed on the immovable obstacle: the z-axis is directed along the cylindrical axis, the origin of the coordinates coincides with the centre of the cylinder end face at the impact time moment, $t = 0$. The axisymmetric dynamic problem under consideration is governed by the Lamé equations (2.1.8) with respect to the displacement vector $\boldsymbol{u}$ (at $\boldsymbol{f} \equiv 0$) subjected to the following boundary conditions:

$$\begin{aligned} &\sigma_{zr} = 0 , \quad \partial u_z/\partial t = 0 \quad (z = 0, 0 \le r < a) , \\ &\sigma_{zr} = 0 , \quad \sigma_{rr} = 0 \quad (r = a, 0 < z < \infty) \end{aligned} \tag{4.7.1}$$

(a is the radius of the cylinder) and initial conditions

$$\boldsymbol{u} = 0 , \quad \partial \boldsymbol{u}/\partial t = -v_0 \boldsymbol{e}_z \quad (t = 0) . \tag{4.7.2}$$

If we try to find $\boldsymbol{u}$ in the form of

$$\boldsymbol{u}(r, z, t) = -v_0 t \boldsymbol{e}_z + \boldsymbol{U}(r, z, t) , \tag{4.7.3}$$

we shall take into account that the vector $\boldsymbol{U} = \{U_r, U_z\}$ (which represents disturbed displacements of the elastic cylinder) satisfies the same Lamé equations but is subjected to somewhat different boundary conditions

$$\begin{aligned} &\sigma_{zr} = 0 , \quad \partial U_z/\partial t = v_0 \quad (z = 0, 0 \le r < a) , \\ &\sigma_{zr} = \sigma_{rr} = 0 \quad (r = a, 0 < z < \infty) \end{aligned} \tag{4.7.4}$$

and zero initial conditions

$$U = \partial U/\partial t = 0 \quad (t = 0)\,. \tag{4.7.5}$$

According to (2.7.48) the disturbance $U(r, z, t)$ to be sought may be written as

$$U = \operatorname{grad}\varphi + \operatorname{rot}\left(-\frac{\partial\psi}{\partial r}\,e_\theta\right)\,. \tag{4.7.6}$$

Then, taking into account the formulas (2.7.36, 46–48, 2.1.7), the solution in terms of potentials φ and ψ must comply with the wave equations

$$\begin{gathered}\Delta\varphi = \frac{\partial^2\varphi}{\partial\tau^2}\,, \quad \Delta\psi = \gamma^2\frac{\partial^2\psi}{\partial\tau^2}\,, \quad \tau > 0\\ \left(\Delta \equiv \frac{\partial^2}{\partial r^2} + \frac{\partial}{r\partial r} + \frac{\partial^2}{\partial z^2}\,, \quad \tau = c_1 t\,, \quad \gamma = \frac{c_1}{c_2}\right)\,,\end{gathered} \tag{4.7.7}$$

and the following boundary conditions for the end face ($z = 0,\ 0 \le r < a$):

$$\frac{\partial^2\varphi}{\partial z\partial\tau} - \frac{\partial}{r\partial r}\left(r\frac{\partial^2\psi}{\partial r\partial\tau}\right) = \frac{v_0}{c_1}\,, \quad 2\frac{\partial^2\varphi}{\partial r\partial z} - \frac{\partial}{\partial r}(\Delta\psi) + 2\frac{\partial^3\psi}{\partial z^2\partial r} = 0\,, \tag{4.7.8}$$

and for the curved surface ($r = a,\ z > 0$):

$$2\frac{\partial^2\varphi}{\partial r\partial z} - \frac{\partial}{\partial r}(\Delta\psi) + 2\frac{\partial^3\psi}{\partial z^2\partial r} = 0\,, \quad (\gamma^2 - 2)\frac{\partial^2\varphi}{\partial\tau^2} + 2\frac{\partial}{\partial r}\left(\frac{\partial\varphi}{\partial r} + \frac{\partial^2\psi}{\partial z\partial r}\right) = 0, \tag{4.7.9}$$

with zero initial conditions:

$$\varphi = \partial\varphi/\partial\tau = 0\,, \quad \psi = \partial\psi/\partial\tau = 0\,. \tag{4.7.10}$$

The set of simultaneous equations (4.7.6–10) completely specifies the solution to the problem for the disturbance U.

Apply the one-sided Laplace transformation (in τ) to the longitudinal potential:

$$\begin{gathered}\bar f = \int_0^\infty f\,e^{-p\tau}\,d\tau\,, \quad f = \frac{1}{2\pi i}\int_{c-i\infty}^{c+i\infty}\bar f\,e^{p\tau}\,dp\\ (f \equiv f(r,z,t)\,, \quad \bar f \equiv \bar f(r,z,p)\,, \quad c > 0\,, \quad \operatorname{Re}p > 0)\,,\end{gathered} \tag{4.7.11}$$

thereafter, the Fourier cosine transformation in z:

$$\begin{gathered}\bar f^{c} = \sqrt{\frac{2}{\pi}}\int_0^\infty \bar f\cos(sz)\,dz\,, \quad \bar f = \sqrt{\frac{2}{\pi}}\int_0^\infty \bar f^{c}\cos(sz)\,ds\\ (\bar f^{c} \equiv \bar f^{c}(r,s,p)\,, \quad \bar f \equiv \bar f(r,z,p))\,.\end{gathered} \tag{4.7.12}$$

Also, the transverse potential, ψ, should be given the Laplace transformation in τ (4.7.11) and the Fourier sine transformation in z:

$$\bar{f}^{\mathrm{s}} = \sqrt{\frac{2}{\pi}} \int\limits_0^\infty \bar{f} \sin(sz) dz \;, \quad \bar{f} = \sqrt{\frac{2}{\pi}} \int\limits_0^\infty \bar{f}^{\mathrm{s}} \sin(sz) ds \tag{4.7.13}$$

$$(\bar{f}^{\mathrm{s}} \equiv \bar{f}^{\mathrm{s}}(r,s,p) \;, \quad \bar{f} \equiv \bar{f}(r,z,p)) \,.$$

Then, equations (4.7.7) with allowance for the boundary conditions (4.7.10) take the following form:

$$\begin{gathered} \tilde{\Delta}\bar{\varphi}^{\mathrm{c}} - p^2\bar{\varphi}^{\mathrm{c}} = \sqrt{\frac{2}{\pi}} \frac{\partial \bar{\varphi}(r,0,p)}{\partial z} \;, \\ \tilde{\Delta}\bar{\psi}^{\mathrm{s}} - \gamma^2 p^2 \bar{\psi}^{\mathrm{s}} = -s\sqrt{\frac{2}{\pi}} \bar{\psi}(r,0,p) \\ \left(\tilde{\Delta} \equiv \frac{d^2}{dr^2} + \frac{1}{r}\frac{d}{dr} - s^2\right) \;, \end{gathered} \tag{4.7.14}$$

and the conditions (4.7.9) for the curved surface and the wave equation for ψ give (after multiplying the first condition in (4.7.9) by $\sin sz$, the second by $\cos sz$, and integrating them with respect to z over the range from 0 to ∞):

$$\begin{gathered} 2s\frac{d\bar{\varphi}^{\mathrm{c}}}{dr} + \gamma^2 p^2 \frac{d\bar{\psi}^{\mathrm{s}}}{dr} + 2s^2 \frac{d\bar{\psi}^{\mathrm{s}}}{dr} = 0 \quad (r = a) \;, \\ (\gamma^2 - 2)p^2 \bar{\varphi}^{\mathrm{c}} + 2\frac{d}{dr}\left(\frac{d\bar{\varphi}^{\mathrm{c}}}{dr} + s\frac{d\bar{\psi}^{\mathrm{s}}}{dr}\right) = 0 \quad (r = a) \,. \end{gathered} \tag{4.7.15}$$

In obtaining equations (4.7.14, 15), use was made of the fact that, when $z \to \infty$, the functions $\{\bar{\varphi}, \bar{\psi}\}$ and their derivatives with respect to z and r are decreasing, because there are no disturbances in φ and ψ for $z > \tau$.

The conditions (4.7.8) yield, after performing the Laplace transformation in τ:

$$\begin{gathered} p\frac{\partial \bar{\varphi}}{\partial z} - p\frac{\partial}{r\partial r}\left(r\frac{\partial \bar{\psi}}{\partial r}\right) = \frac{v_0}{pc_1} \quad (z = 0) \;, \\ 2\frac{\partial^2 \bar{\varphi}}{\partial r \partial z} - \gamma^2 p^2 \frac{\partial \bar{\psi}}{\partial r} + 2\frac{\partial^3 \bar{\psi}}{\partial z^2 \partial r} = 0 \quad (z = 0) \,. \end{gathered} \tag{4.7.16}$$

Solving the first equation with respect to $\partial\bar{\varphi}/\partial z$, substituting the result into the second, and using the equation $\Delta\bar{\psi} = p^2\gamma^2\bar{\psi}$, obtain $\partial\bar{\psi}/\partial r = 0$, which means that

$$\bar{\psi}(r,0,p) = D_0 \;, \quad \frac{\partial \bar{\varphi}(r,0,p)}{\partial z} = \frac{v_0}{c_1 p^2} \quad (D_0 \equiv D_0(p)) \,. \tag{4.7.17}$$

Thereafter, equations (4.7.14) can be written as

$$(\tilde{\Delta} - p^2)\bar{\varphi}^{\mathrm{c}} = \sqrt{\frac{2}{\pi}} \frac{v_0}{c_1 p^2} \;, \quad (\tilde{\Delta} - \gamma^2 p^2)\bar{\psi}^{\mathrm{s}} = -s\sqrt{\frac{2}{\pi}} D_0 \,. \tag{4.7.18}$$

Solutions to these equations are

$$\bar{\varphi}^c = A_0 I_0(\alpha r) + A_1 K_0(\alpha r) - \sqrt{\frac{2}{\pi}} \frac{v_0}{c_1 p^2 \alpha^2} (\alpha = \sqrt{p^2 + s^2}) ,$$
$$\bar{\psi}^s = B_0 I_0(\beta r) + B_1 K_0(\beta r) + s\sqrt{\frac{2}{\pi}} \frac{D_0}{\beta^2} (\beta = \sqrt{\gamma^2 p^2 + s^2}) . \tag{4.7.19}$$

Here, usual measures are taken to provide definiteness in selecting the proper branches of the radicals α and β: cuts are introduced in the s-plane from the points $s = \pm pi$ and $s = \pm\gamma pi$ to infinity; in addition, it is assumed that $\alpha = p$ and $\beta = \gamma p$ if $s = 0$.

As the expression (4.7.6), for the displacement disturbance vector $\boldsymbol{U}$, contains $\partial\psi/\partial r$, we can ignore the constant D_0 in (4.7.19). Furthermore, if we are allowed to assume the displacement vector $\boldsymbol{U}$ to be limited over the axis $r = 0$, then $A_1 = B_1 = 0$. The solutions

$$\bar{\varphi}^c = A_0 I_0(\alpha r) - \sqrt{\frac{2}{\pi}} \frac{v_0}{c_1 p^2 \alpha^2} , \quad \frac{d\bar{\psi}^s}{dr} = B_0 \beta I_1(\beta, r) , \tag{4.7.20}$$

so obtained, may be substituted into the boundary conditions (4.7.15) to give two constants, A_0 and B_0:

$$A_0 = -\sqrt{\frac{2}{\pi}} \frac{(\gamma^2 - 2)v_0}{c_1 \beta F \alpha^2} (\beta^2 + s^2) I_1(\beta\alpha) ,$$
$$B_0 = 2\sqrt{\frac{2}{\pi}} \frac{(\gamma^2 - 2)v_0 s}{c_1 \alpha \beta^2 F} I_1(\alpha a) , \tag{4.7.21}$$
$$F = \frac{1}{\beta}\left[2\gamma^2 p^2 \frac{\alpha}{a} I_1(\beta a) I_1(\alpha a) + 4s^2 \alpha\beta I_1(\alpha a) I_0(\beta a) - (\beta^2 + s^2)^2 I_1(\beta a) I_0(\alpha a)\right]$$

Thus, the solution sought for the vector $\boldsymbol{U}$ introduced in (4.7.6) is described in terms of transforms by the formulae (4.7.20, 21).

It is a consequence of (4.7.6) that

$$\bar{U}_z^s = -s\bar{\varphi}^c - \frac{1}{r}\frac{d}{dr}\left(r\frac{d\bar{\psi}^s}{dr}\right) , \quad \bar{U}_r^c = \frac{d\bar{\varphi}^c}{dr} + s\frac{d\bar{\psi}^s}{dr} , \tag{4.7.22}$$

Then applying the inverse transformations to (4.7.22) and taking into account (4.7.3), we obtain components of the displacement vector $\boldsymbol{u}$:

$$u_z(r, z, t) = -v_0 t + \frac{1}{2\pi \mathrm{i}} \int_{c-\mathrm{i}\infty}^{c+\mathrm{i}\infty} \mathrm{e}^{pc_1 t} dp \sqrt{\frac{2}{\pi}} \int_0^\infty \sin(sz)$$
$$\times \left[s\sqrt{\frac{2}{\pi}} \frac{v_0}{c_1 p^2 \alpha^2} - sA_0 I_0(\alpha r) - B_0 \frac{\beta}{r}\frac{d}{dr}(r I_1(\beta r))\right] ds , \tag{4.7.23}$$

$$u_r(r, z, t) = \frac{1}{2\pi \mathrm{i}} \int_{c-\mathrm{i}\infty}^{c+\mathrm{i}\infty} \mathrm{e}^{pc_1 t} dp \sqrt{\frac{2}{\pi}} \int_0^\infty \cos(sz) \left[A_0 \alpha I_1(\alpha r) + sB_0 \beta I_1(\beta r)\right] ds ,$$

where $0 \le r < a$, $z > 0$, $t > 0$, $c > 0$, the constants A_0 and B_0 are presented by (4.7.21).

Since integrands in integrals with respect to s in (4.7.23) are even functions of the s, we can rewrite (4.7.23) as

$$u_z(r,z,t) = -v_0 t + v_0\left(t - \frac{z}{c_1}\right) H\left(t - \frac{z}{c_1}\right)$$

$$- \frac{1}{2\pi \mathrm{i}} \int\limits_{c-\mathrm{i}\infty}^{c+\mathrm{i}\infty} \mathrm{e}^{pc_1 t} dp \int\limits_{-\infty}^{\infty} \frac{\mathrm{e}^{\mathrm{i}sz}}{\mathrm{i}\sqrt{2\pi}} \left[sA_0 I_0(\alpha r) + B_0 \frac{\beta}{r} \frac{d}{dr}(r I_1(\beta r)) \right] ds \,,$$

$$u_r(r,z,t) = \frac{1}{2\pi \mathrm{i}} \int\limits_{c-\mathrm{i}\infty}^{c+\mathrm{i}\infty} \mathrm{e}^{pc_1 t} dp \int\limits_{-\infty}^{\infty} \frac{\mathrm{e}^{\mathrm{i}sz}}{\sqrt{2\pi}} [\alpha A_0 I_1(\alpha r) + s\beta B_0 I_1(\beta r)]\, ds \,. \quad (4.7.24)$$

Note that the integration sequence in (4.7.24) (with respect to s and p) can be changed.

In the further investigations, one should bear in mind that the expressions in square brackets in (4.7.24) do not have branch points in the p-plane since the functions $\alpha I_1(\alpha r)$, $I_0(\alpha r)$, $\beta I_1(\beta r)$, and $I_0(\beta r)$, when expressed as series in integer powers of the argument, comprise only the even powers of the radicals α and β, thus being the entire functions of the variable p. Hence, it can be only poles that the integrands in (4.7.24) can have as singular points. If $p \to \infty$, the integrands in (4.7.24) (except for the multiplier $\exp(pc_1 t)$ decrease no slower than $O(p^{-5/2})$; therefore, integration with respect to p may be replaced by calculation of residues at the poles of the integrands.

4.7.2 Analysis of the Solution at the Points of a Cylinder's Axis

Let us consider in more detail the strain $\varepsilon_{zz} = \partial u_z / \partial z$ over the axis $r = 0$. Employing (4.7.24), write

$$\varepsilon_{zz}(0,z,t) = \lim_{r \to 0} \frac{\partial u_z(r,z,t)}{\partial z} = -\frac{v_0}{c_1} H\left(t - \frac{z}{c_1}\right)$$

$$+ \frac{v_0(\gamma^2 - 2)}{2\pi^2 c_1 \mathrm{i}} \int\limits_{-\infty}^{\infty} \mathrm{e}^{\mathrm{i}sz} s^2 ds \int\limits_{c-\mathrm{i}\infty}^{c+\mathrm{i}\infty} \frac{(\beta^2 + s^2) I_1(\beta a) - 2\beta\alpha I_1(\alpha a)}{\alpha^2 \beta F} \mathrm{e}^{pc_1 t} dp \,.$$

(4.7.25)

It is a simple operation to demonstrate that the function F defined by (4.7.21) and comprised by the denominator of the integrand in (4.7.25) has no zeroes outside the imaginary axis in the p-plane at real values of s. On the very imaginary axis ($\mathrm{Re}\, p = 0$), we can replace p by $\mathrm{i}k$ to reduce equation (4.7.26),

$$F = 0 \quad (4.7.26)$$

to the well-known Pochhammer equation dealing with frequencies of vibrations in an infinite circular cylinder within which the longitudinal waves propagate. This equation was derived and studied for the first time in [4.60]; further analytical and numerical investigations into this equation have been performed by a number of authors (see, e.g., [4.13, 61]). Equation (4.7.26) is shown to possess an infinite set of roots $p_m = \mathrm{i} f_m(s)$ (where $\operatorname{Im} f_m(s) \equiv 0$, $m = \pm 1, \pm 2, \ldots$) on the axis $\operatorname{Re} p = 0$ at any real value of s; all these roots are simple (except for the roots p_1 and p_{-1} which coincide at $s = 0$), pairwise complex-conjugate ($p_{-m} = -p_m$); when $s \to 0$, they become

$$\begin{aligned} p_m &= \frac{c_3}{c_1} m\mathrm{i}s\left(1 - \frac{\nu^2}{4} a^2 s^2\right) \quad (m = \pm 1)\,, \\ p_m &= \mathrm{i} a_m \quad (a_m \neq 0,\ m = \pm 2, \pm 3, \pm 4, \ldots;\ a_{-m} = -a_m)\,, \end{aligned} \tag{4.7.27}$$

where ν is the Poisson ratio, $c_3 = \sqrt{E/\varrho}$ is the longitudinal wave propagation speed for the rod, according to the elementary theory (E is the Young modulus).

When $p \to \infty$, the fraction in the integrand of (4.7.25) decreases no slower than $\mathrm{O}(p^{-7/2})$. Hence, we can deform the path of integration (with respect to p) into the left-hand half-plane so that equation (4.7.25) becomes as follows:

$$\begin{aligned} &\varepsilon_{zz}(0, z, t) \\ &= \frac{v_0(\gamma^2 - 2)}{\pi c_1} \int\limits_{-\infty}^{\infty} \mathrm{e}^{\mathrm{i}sz} s^2 \left\{ \sum_{p=p_m} \frac{(\beta^2 + s^2) I_1(\beta a) - 2\beta\alpha I_1(\alpha a)}{\beta \alpha^2 \partial F/\partial p} \mathrm{e}^{pc_1 t} \right\} ds\,, \end{aligned} \tag{4.7.28}$$

where $t > 0$, $z > 0$, $m = \pm 1, \pm 2, \ldots$ Note that the function $\alpha^2 = p^2 + s^2$ present in the denominator of (4.7.25) has caused poles at points $p = \pm s\mathrm{i}$, and their residues have cancelled the first term at the right-hand side of (4.7.25). Thus, only residues at zeroes of the function F are contained in the sum in (4.7.28).

If $t \to \infty$, asymptotic behavior of ε_{zz}, equation (4.7.28), can be described using the stationary phase method. With this, rearrange the sum in the integrand as two summands:

$$\sum_{p=p_m} = \sum_{p=\pm p_1} + \sum_{p=\pm p_2, \pm p_3, \ldots} \,. \tag{4.7.29}$$

Here, it is worth noting that the function F has no repeated roots over the p-plane ($p \neq \pm p_1$), so, $\partial F/\partial p$ in the second summand of (4.7.29) cannot be zero at any s ($-\infty < s < +\infty$). The stationary phase method [4.62] establishes that, in such circumstances, the integral in (4.7.28) produced by the second summand of (4.7.29) decreases no slower than $\mathrm{O}(t^{-1/2})$ when $t \to \infty$ ($0 < z \leq c_1 t - \varepsilon$, $\varepsilon > 0$).

Consider the remaining integral of (4.7.28), dealing with residues at the points $p = \pm p_1$. The function $\partial F/\partial p$ behaves as $\mathrm{O}(s^3)$ when $s \to 0$ (since $p_1 = \mathrm{O}(s)$ according to the asymptotic formula (4.7.27)). For this reason, the vicinity of the point $s = 0$ is of maximum value for the integral of the residues at the points $p = \pm p_1$, thus making the main contribution to the solution (4.7.28) when $t \to \infty$. For this contribution to be evaluated, rewrite equation (4.7.28) as

$$\varepsilon_{zz}(0, z, t) = \frac{v_0(\gamma^2 - 2)}{\pi c_1} \int_{-\varepsilon}^{\varepsilon} N(s) \left[e^{i(sz - ip_1 c_1 t)} - e^{i(sz + ip_1 c_1 t)} \right] ds + O(t^{-1/2}) , \tag{4.7.30}$$

$$N(s) = \left[s^2 \frac{(\beta^2 + s^2) I_1(\beta a) - 2\beta\alpha I_1(\alpha a)}{\beta \alpha^2 \partial F / \partial p} \right] \Bigg|_{p = p_1} .$$

Here, $O(t^{-1/2})$ comprises the integrals of the first term of (4.7.30) taken over $(-\infty, -\varepsilon)$ and $(\varepsilon, +\infty)$.

When $s \to 0$ the term $N(s)$ is

$$N(s) = \frac{i}{2s} \frac{c_1}{c_3(\gamma^2 - 1)} + O(s) .$$

With p presented by (4.7.27), we can write (4.7.30) as

$$\varepsilon_{zz}(0, z, t) = \frac{v_0 i}{2\pi c_3} \int_{-\varepsilon}^{\varepsilon} \left[e^{is[z + (c_3 - bs^2)t]} - e^{is[z - (c_3 - bs^2)t]} \right] \frac{ds}{s} + O(t^{-1/2}) , \tag{4.7.31}$$

where $b = \nu^2 a^2 c_3 / 4$, and the $O(t^{-1/2})$ contains the integral of the remainder of the asymptotic formula for $N(s)$.

The integration interval $(-\varepsilon, \varepsilon)$ in (4.7.31) may be extended to the complete axis $(-\infty, \infty)$ which adds $O(t^{-1/2})$ when $t \to \infty$. After simple manipulations, one can obtain

$$\varepsilon_{zz}(0, t, z) = \frac{v_0 i}{2\pi c_3} \int_{-\infty}^{\infty} \left[e^{is[z + (c_3 - bs^2)t]} - e^{is[z - (c_3 - bs^2)t]} \right] \frac{ds}{s} . \tag{4.7.32}$$

which is accurate within $O(t^{-1/2})$ if $t \to \infty$, $r = 0$, $0 < z \le c_1 t - \varepsilon$ $(\varepsilon > 0)$. It is sufficient to retain only the even component (with respect to s) for transforming (4.7.32) into

$$\begin{aligned}
&\varepsilon_{zz}(0,t,z)\\
&= -\frac{v_0'}{2\pi c_3}\int_{-\infty}^{\infty}\left\{\frac{\sin s[z+(c_3-bs^2)t]}{s}-\frac{\sin s[z-(c_3-bs^2)t]}{s}\right\}ds\\
&= \frac{v_0}{2\pi c_3}\int_{-\infty}^{\infty}\left\{\frac{\sin(sz''+s^3bt)}{s}+\frac{\sin(sz'+s^3bt}{s}\right\}ds\\
&= \frac{v_0}{c_3\pi}\left[\int_0^{\infty}\sin\left(\alpha''\eta+\frac{\eta^3}{3}\right)\frac{d\eta}{\eta}+\int_0^{\infty}\sin\left(\alpha'\eta+\frac{\eta^3}{3}\right)\frac{d\eta}{\eta}\right]\\
&= \frac{v_0}{c_3}\left[\frac{1}{3}\int_0^{\alpha''}\mathrm{Ai}\,(\alpha)d\alpha+\int_0^{\alpha'}\mathrm{Ai}\,(\alpha)d\alpha\right]. \qquad (4.7.33)
\end{aligned}$$

In the last formula, $z''=-z-c_3t$, $z'=z-c_3t$, $\alpha''=z''/\sqrt[3]{3bt}$, $\alpha'=z'/\sqrt[3]{3bt}$, $\mathrm{Ai}\,(\alpha)$ is the Airy function (e.g., see [4.49]):

$$\mathrm{Ai}\,(\alpha)=\frac{1}{\pi}\int_0^{\infty}\cos\left(\alpha\eta+\frac{\eta^3}{3}\right)d\eta\,.$$

In addition, use was made of the formula,

$$\frac{1}{\pi}\int_0^{\infty}\sin\frac{\eta^3}{3}\frac{d\eta}{\eta}=\frac{1}{6}\,.$$

We deal with the domain $z>0$, therefore, $\alpha''\to-\infty$ when $t\to\infty$, and

$$\int_0^{\alpha''}\mathrm{Ai}\,(\alpha)d\alpha=-\frac{2}{3}+\mathrm{O}(t^{-1/2})\,.$$

Then, obtain the final expression accurate to $\mathrm{O}(t^{-1/2})$ when $t\to\infty$:

$$\varepsilon_{zz}(0,z,t)=\frac{v_0}{c_3}\left(-\frac{1}{3}+\int_0^{\alpha'}\mathrm{Ai}\,(\alpha)d\alpha\right)\quad(0<z\le c_1t-\varepsilon,\ \varepsilon>0)\,. \qquad (4.7.34)$$

Note that if the second term in the asymptotic formula for the root p_1 is neglected when $s\to0$, i.e., $b=0$ in (4.7.32), the expression (4.7.34) is replaced by

$$\begin{aligned}
\varepsilon_{zz}(0,z,t)&=\frac{v_0}{2\pi c_3}\left[-\int_{-\infty}^{\infty}\sin[s(z+c_3t)]\frac{ds}{s}+\int_{-\infty}^{\infty}\sin[s(z-c_3t)]\frac{ds}{s}\right]\\
&=-\frac{v_0}{c_3}H(c_3t-z)\quad(t\to\infty)\,, \qquad (4.7.35)
\end{aligned}$$

which is the well-known elementary-theory solution to the problem on longitudinal waves in an elastic bar, see [4.6359]. This elementary solution, (4.7.35), can also be written immediately from (4.7.34) for points $z \ll c_3 t$ (i.e. $\alpha' \to -\infty$) and $c_3 t \ll z \le c_1 t - \varepsilon$ (i.e. $\alpha' \to +\infty$) since

$$\lim_{\alpha' \to -\infty} \int_0^{\alpha'} \mathrm{Ai}\,(\alpha) d\alpha = -\frac{2}{3}\,, \qquad \lim_{\alpha' \to +\infty} \int_0^{\alpha'} \mathrm{Ai}\,(\alpha) d\alpha = \frac{1}{3}\,.$$

Consider now points of the z-axis (within the range $0 < z \le c_1 t - \varepsilon$, $\varepsilon > 0$) which move so that the strain ε_{zz} at them remains constant. From (4.7.34), it follows that this will be met if $\alpha' = \alpha_0$ ($\alpha_0 = \text{const}$) and

$$z' = z - c_3 t = \alpha_0 \sqrt[3]{3bt}\,, \tag{4.7.36}$$

In other words, equation (4.7.36) indicates how a point z' must move along the z-axis (from the point $z' = 0$ moving at the speed c_3 along the z-axis) in order to have a constant strain at this point.

Propagation of the wave $\varepsilon_{zz}(0, z, t)$ along the z-axis will be described completely if additional consideration is given to the vicinity $c_1 t - \varepsilon < z < c_1 t$ of the wave front. Physical reasons (as well as asymptotic behavior of integrands in (4.7.24) as $p \to \infty$) show that a domain $\sqrt{c_1^2 t^2 - (a - r)^2} < z < c_1 t$ (i.e., ABB_1 in Fig. 4.18) near the front has no diffraction waves caused by the curved surface of the cylinder; so, a solution for this domain is that for an unbounded elastic medium subjected to a plane longitudinal wave

$$\varepsilon_{zz} = -v_0 c_1^{-1} H(t - z/c_1)\,, \quad \varepsilon_{rr} = \varepsilon_{\theta\theta} = 0\,.$$

When $t \to \infty$, the part of the non-zero-solution domain behind the front tends to the front (the length of this part behaves as $1/t$).

Thus, the impact of a cylinder against a smooth rigid body produces the wave $\varepsilon_{zz}(0, z, t)$ which propagates along the z-axis as illustrated by Fig. 4.19 and can be described as follows:

1) The front of a disturbance has a strain jump on it, $-v_0/c_1$, and propagates along the cylinder at the speed of c_1. At a noticeable distance from the end face $z = 0$ an accompanying wave behind the front becomes a narrow peak carrying only a small portion of energy (which tends to zero when $t \to \infty$). The width of the peak decreases asymptotically in proportion to $1/t$.

2) There is a "quasi-front" which moves at the speed of c_3 along the z-axis. The "quasi-front" diffuses due to the wave dispersion, and the region where strains grow from zero to $-v_0/c_3$ expands proportionally to $t^{1/3}$.

3) Strains between the "quasi-front" and the end face are approximately equal to $-v_0/c_3$, according to the elementary theory.

The above said suggests that, if $t \to \infty$, the main contribution is made by the region behind the "quasi-front"; the solution for this region is described by equation (4.7.34). In Fig. 4.20 (taken from [4.13]) the solid curve represents the

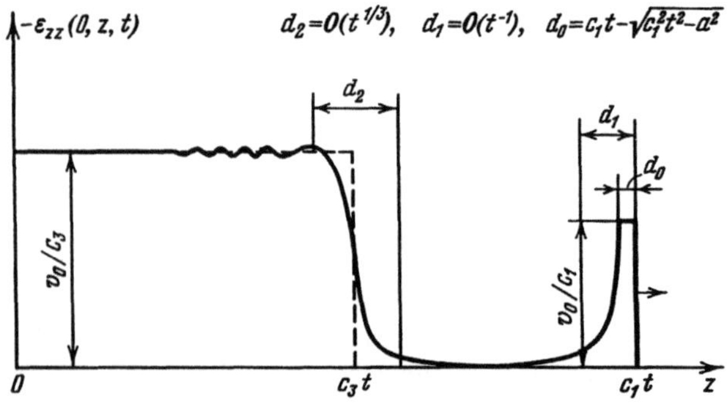

Fig. 4.19. Schematic of a wave $-\varepsilon_{zz}(0,z,t)$ propagating along the z-axis of the cylinder at impact against an immovable obstacle

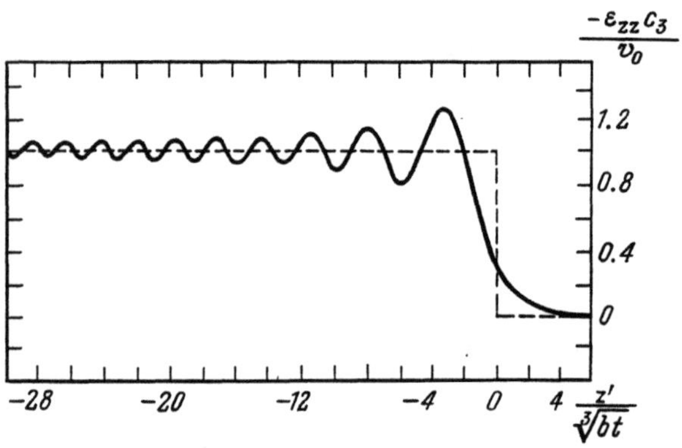

Fig. 4.20. Strain $-\varepsilon_{zz}c_3/v_0$ vs $z'/\sqrt[3]{bt}$, equation (4.7.34)

strain $(-\varepsilon_{zz}c_3/v_0)$ as a function of $z'/\sqrt[3]{bt}$ according to the formula (4.7.34). For comparison, a dashed line in Fig. 4.20 depicts the elementary solution written in (4.7.35). It is worth noting that the solution (4.7.34) oscillates around the elementary one because of the radial inertia.

Consider the strains

$$\varepsilon_{rr}(0,z,t)=\lim_{r\to 0}\frac{\partial u_r}{\partial r}\,,\qquad \varepsilon_{\theta\theta}(0,z,t)=\lim_{r\to 0}\frac{u_r}{r}$$

using the same order of approximation as was used in deriving equation (4.7.34) for $\varepsilon_{zz}(0,z,t)$. One can easily prove that

$$\begin{aligned}\varepsilon_{rr}&=(0,z,t)=\varepsilon_{\theta\theta}(0,z,t)=-\nu\varepsilon_{zz}(0,z,t)\\ &=-\nu\frac{v_0}{c_3}\left(-\frac{1}{3}+\int_0^{\alpha'}\mathrm{Ai}\,(\alpha)d\alpha\right)\end{aligned}\tag{4.7.37}$$

if $t \to \infty$, $0 < z \le c_1 t - \varepsilon$, $\varepsilon > 0$; i.e., when use is made of such approximation, strains are interrelated by the formulas obtained in the elastostatics for axial tension of a rod (e.g., see [4.63]).

It should be noted that numerical solutions to the problem under study is proposed in [4.64, 65]. Another version of the problem is considered in [4.17]: the end face of a cylinder is instantly loaded with normal stresses at zero tangential displacements; a solution is found by the method of integral transforms, as well. Analogous problems are solved in [4.66, 67]. A similar two-dimensional problem concerning collision of semi-infinite strips has been solved in [4.68, 69] using functionally invariant solutions (we should note an unsuccessful name "shear strain function" used in [4.68] for the function $\omega = \partial u_y/\partial x - \partial u_x/\partial y$; this name seems to be the very circumstance which has led the authors of [4.70] to a wrong conclusion that results of [4.68] are incorrect). An interesting two-dimensional problem of colliding of elastic semi-infinite strips was studied in [4.71]. For more information about other problems of elastic waves in semi-infinite strips and cylinders, see [4.72] as well as surveys and monographs [4.25, 27–29, 73].

Concluding this section, note that all the known exact analytical solutions found on the basis of the classical elastodynamics relations applied to the problems with semi-infinite cylinders and strips are obtained for the case of "mixed" boundary conditions where specified at the end face of the cylinder (or the strip) are either (i) the normal stresses and tangential displacements or (ii) shears and normal displacements. In the case when only stresses or only displacements (i.e., "non-mixed" conditions) are defined at the end face, the boundary conditions for the potentials φ and ψ cannot be separated, thus essentially complicating the derivation of the analytical solution. The author is not aware of contributions supplying acurate analytical solutions for the "non-mixed" boundary conditions at the end face.

5. Solution to Three-Dimensional Elastodynamic Problems with Mixed Boundary Conditions for Wedge-Shaped Domains

Dynamic elasticity problems for wedges are of obvious interest in both theory and applications, thus receiving much attention from researchers. Surveys of contributions to this topic can be found in monographs by *Knopoff* [5.1], *Miklowitz* [5.2], and *Dobrushkin* [5.3].

Most difficulties arise if "non-mixed" boundary conditions are imposed, i.e., if either displacements or stresses are specified. No exact theoretical solutions have been obtained to date, even for a rectangular wedge.

In opposition, a set of two-dimensional and three-dimensional dynamic problems have been solved in theory for arbitrary wedges under "mixed" boundary conditions which are comprised of either (I) normal displacements and shear stresses or (II) normal stresses and tangential displacements. Here, one should be aware of the difficulty due to singularity of the longitudinal and transverse displacement potentials at the wedge edge (whereas the displacement field is regular). *Kostrov* [5.4] was the first to solve the plane-strain problem on a transient elastic plane wave diffraction by an arbitrary smooth rigid wedge; use was made of homogeneous solutions. (Note that the earlier result of [5.5] cannot be considered a solution to the problem since the edge boundary condition was not imposed – and satisfied; to be precise, what was solved in [5.5] was a certain diffraction problem with an "edge-mounted" distributed source.) An analogous problem has been treated in [5.6] using the homogeneous solutions under no normal stress and no tangential displacement boundary conditions. The integral transformation method has been applied by *Zemell* [5.7] and the present author [5.8, 9] to diffraction of a cylindrical elastic wave by a smooth rigid wedge. Diffraction of the waves caused by displacements of a buried rectangular foundation is reduced in [5.10, 11] to diffraction of plane waves by a rectangular wedge at the mixed boundary conditions. A plane-strain diffraction problem for an arbitrary elastic wave and a wedge under the mixed boundary conditions has been considered in [5.12]. Solutions to three-dimensional spherical and plane wave diffraction problems have been found by *Poruchikov* [5.9, 13, 14], *Israilov* [5.5], and *Sharipova* [5.6].

The present Chapter deals with the author's method [4.30] of solving the three-dimensional elastodynamic problems for wedges under mixed boundary conditions. Using this method, exact analytical solutions can be obtained to the non-stationary three-dimensional problems of diffraction of spherical and plane waves by an arbitrary smooth rigid wedge.

5.1 Combined Method of Integral Transforms

This Section offers the statement and the solution procedure for three-dimensional dynamic problems at mixed boundary conditions of two kinds: at each face of the wedge under consideration, either normal displacements and shear stresses or normal stresses and tangential displacements are specified.

5.1.1 Problem Formulation

Here, use will be made of a cylindrical coordinate system $\{r, \theta, z\}$. The elastic medium is assumed to occupy a domain specified by relations $\{r > 0,\ 0 < \theta < \pi/l,\ -\infty < z < +\infty\}$. Over boundaries $\theta = 0$, $\theta = \pi/l$ $(1/2 < l < +\infty, l \neq 1)$, the mixed boundary conditions are imposed: either

$$u_\theta = u_\theta^m(t,r,z), \quad \sigma_{\theta r} = \sigma_{\theta r}^m(t,r,z), \quad \sigma_{\theta z} = \sigma_{\theta z}^m(t,r,z) \tag{5.1.1}$$

or an alternative in the following form:

$$u_r = u_r^m(t,r,z), \quad u_z = u_z^m(t,r,z), \quad \sigma_{\theta\theta} = \sigma_{\theta\theta}^m(t,r,z)\,. \tag{5.1.2}$$

Here, the superscript $m = 0$ refers to the boundary $\theta = 0$, and $m = 1$ to $\theta = \pi/l$. Initial conditions can be taken as $\boldsymbol{u} = \partial \boldsymbol{u}/\partial t = 0$ at $t = t_0$.

The problem to be solved under boundary conditions (5.1.1) or (5.1.2) will be hereinafter referred to as the third or the fourth boundary-value problem, respectively; with this, we shall follow the terms of Sect. 2.2. According to equations (2.1.7) and (2.7.36, 37, 45, 46) each of the two problems reduces to the following set of equations, boundary and initial conditions:

$$\begin{aligned}
&\Delta\varphi = \frac{\partial^2\varphi}{\partial\tau^2}\,, && \Delta\psi_j = \gamma^2\frac{\partial^2\psi_j}{\partial\tau^2} \qquad (j = 1,\ 2)\,,\\
&u_\theta = u_\theta^0(\tau,r,z)\,, && \sigma_{\theta r} = \sigma_{\theta r}^0(\tau,r,z)\,,\\
&\sigma_{\theta z} = \sigma_{\theta z}^0(\tau,r,z) && (\theta = 0)\,,\\
&u_\theta = u_\theta^1(\tau,r,z), && \sigma_{\theta r} = \sigma_{\theta r}^1(\tau,r,z)\,,\\
&\sigma_{\theta z} = \sigma_{\theta z}^1(\tau,r,z) && (\theta = \pi/l)\,,\\
&\varphi \equiv \psi_j \equiv 0 && (\tau < \tau_0)\,;
\end{aligned} \tag{5.1.3}$$

$$\begin{aligned}
&\Delta\varphi = \frac{\partial^2\varphi}{\partial\tau^2}, && \Delta\psi_j = \gamma^2\frac{\partial^2\psi_j}{\partial\tau^2} \qquad (j = 1,\ 2)\,,\\
&u_r = u_r^0(\tau,r,z), && u_z = u_z^0(\tau,r,z)\,,\\
&\sigma_{\theta\theta} = \sigma_{\theta\theta}^0(\tau,r,z) && (\theta = 0)\,,\\
&u_r = u_r^1(\tau,r,z)\,, && u_z = u_z^1(\tau,r,z)\,,\\
&\sigma_{\theta\theta} = \sigma_{\theta\theta}^1(\tau,r,z) && (\theta = \pi/l)\,,\\
&\varphi \equiv \psi_j \equiv 0 && (\tau < \tau_0)\,,\\
&(u_z^0(\tau,0,z) = u_z^1(\tau,0,z))\,;
\end{aligned} \tag{5.1.4}$$

the notation being

$$\tau = c_1 t \,, \qquad \tau_0 = c_1 t_0 \,, \qquad \gamma = c_1/c_2 \,,$$

$$u_r = \frac{\partial \varphi}{\partial r} + \frac{\partial \psi_1}{r\partial \theta} + \frac{\partial^2 \psi_2}{\partial r \partial z} \,, \qquad u_z = \frac{\partial \varphi}{\partial z} - \frac{\partial^2 \psi_2}{r^2 \partial \theta^2} - \frac{\partial}{r \partial r}\left(r \frac{\partial \psi_2}{\partial r}\right) \,,$$

$$u_\theta = \frac{\partial \varphi}{r\partial \theta} - \frac{\partial \psi_1}{\partial r} + \frac{\partial^2 \psi_2}{r \partial \theta \partial z} \,,$$

$$\frac{\sigma_{\theta r}}{\mu} = r\frac{\partial}{\partial r}\left[\frac{\partial \varphi}{\partial \theta} - \frac{\partial \psi_1}{r\partial r} + \frac{\partial^2 \psi_2}{r^2 \partial \theta \partial z}\right] + \frac{1}{r}\left[\frac{\partial^2 \varphi}{\partial r \partial \theta} + \frac{\partial^2 \psi_1}{r \partial \theta^2} + \frac{\partial^3 \psi_2}{\partial r \partial z \partial \theta}\right] \,,$$

$$\frac{\sigma_{\theta z}}{\mu} = \frac{1}{r}\left[2\frac{\partial^2 \varphi}{\partial z \partial \theta} - \frac{\partial^3 \psi_2}{r^2 \partial \theta^3} - \frac{\partial}{r \partial r}\left(r \frac{\partial^2 \psi_2}{\partial r \partial \theta}\right)\right] - \frac{\partial^2 \psi_1}{\partial r \partial z} + \frac{\partial^3 \psi_2}{r \partial \theta \partial z} \,,$$

$$\frac{\sigma_{\theta\theta}}{\mu} = (\gamma^2 - 2)\frac{\partial^2 \varphi}{\partial \tau^2} + \frac{2}{r}\left(\frac{\partial^2 \varphi}{r\partial \theta^2} + \frac{\partial \varphi}{\partial r} - \frac{\partial^2 \psi_1}{\partial r \partial \theta} + \frac{\partial \psi_1}{r \partial \theta} + \frac{\partial^3 \psi_2}{r \partial \theta^2 \partial z} + \frac{\partial^2 \psi_2}{\partial r \partial z}\right) .$$

Both sets – (5.1.3) and (5.1.4) – should be solved assuming the edge condition

$$\boldsymbol{u} = \boldsymbol{u}_0(\tau, z) + \mathrm{O}(r^\varepsilon) \,, \quad \varepsilon > 0 \,, \quad r \to 0 \,, \tag{5.1.5}$$

which guarantees both continuity of displacements and integrability of stresses in the vicinity of the edge, as well as uniqueness of solutions to the problems.

Thus the third and the fourth boundary-value problems turned out to be reduced to sets (5.1.3, 5) and (5.1.4, 5), respectively.

It is shown below that these sets may be solved using a combination of the two-sided Laplace transforms (with respect to τ and z), and the techniques of isolating the edge singularities of transforms of the functions to be found.

5.1.2 Problem Solution

Apply the Laplace transformation technique (with respect to τ and z) to equations (5.1.3, 4); account should be taken of the wave equations for the transverse potentials ψ_1 and ψ_3. This gives a new form of boundary conditions:

$$\begin{aligned} &\frac{\partial \bar\varphi^*}{r\partial \theta} - \frac{\partial \bar\psi_1^*}{\partial r} + s\frac{\partial \bar\psi_2^*}{r\partial \theta} = \left(\bar u_\theta^m\right)^* \,, \\ &\bar\psi_1^* = \frac{1}{\gamma^2 p^2 - s^2}\left[\frac{1}{\mu}\left(\bar\sigma_{\theta r}^m\right)^* - 2\frac{d}{dr}\left(\bar u_\theta^m\right)^*\right] \,, \\ &s\frac{\partial \bar\varphi^*}{r\partial \theta} + \left(s^2 - \gamma^2 p^2\right)\frac{\partial \bar\psi_2^*}{r\partial \theta} = \frac{1}{\mu}\left(\bar\sigma_{\theta z}^m\right)^* - s\left(\bar u_\theta^m\right)^* . \end{aligned} \tag{5.1.6}$$

Here, the bar and star over any symbol f (φ, ψ_1, ...) denote, respectively, a Laplace transformation in τ and z:

$$\bar f = \int_{-\infty}^{\infty} \mathrm{e}^{-p\tau} f d\tau \,, \quad \bar f^* = \int_{-\infty}^{\infty} \bar f \mathrm{e}^{-sz} dz \,.$$

Note that $\mathrm{Re}\{p\} > 0$, $\mathrm{Re}\{s\} = 0$, since $f(\tau) \equiv 0$ at $\tau < \tau_0$. It is assumed that $f = \mathrm{O}(\tau^n)$ as $\tau \to \infty$ and that $|\bar{f}|$ is integrable in z.

Eliminate the derivative $\partial\bar{\psi}_1^*/\partial r$ from the first equation of the set (5.1.6) using the second equation and solve the subset of the first and the third equations to determine $\partial\bar{\varphi}^*/\partial\theta$ and $\partial\bar{\psi}_2^*/\partial\theta$. This provides $\partial\bar{\varphi}^*/\partial\theta$, $\bar{\psi}_1^*$, and $\partial\bar{\psi}_2^*/\partial\theta$ at boundaries $\theta = 0$ and $\theta = \pi/l$. Thus, boundary conditions in (5.1.3) separate with respect to longitudinal and transverse potentials, and equations (5.1.3) reduce to the following three sets for $\bar{\varphi}^*$, $\bar{\psi}_1^*$, and $\bar{\psi}_2^*$:

$$\Delta_1\bar{\varphi}^* = (p^2 - s^2)\bar{\varphi}^* \qquad \left(\Delta_1 \equiv \frac{\partial^2}{\partial r^2} + \frac{1}{r}\frac{\partial}{\partial r} + \frac{\partial^2}{r^2\partial\theta^2}\right),$$

$$\frac{\partial\bar{\varphi}^*}{\partial\theta} = U_0(\theta = 0), \qquad \frac{\partial\bar{\varphi}^*}{\partial\theta} = U_1(\theta = \pi/l), \tag{5.1.7}$$

$$\Delta_1\bar{\psi}_1^* = (\gamma^2p^2 - s^2)\bar{\psi}_1^*, \quad \bar{\psi}_1^* = V_0(\theta = 0), \quad \bar{\psi}_1^* = V_1(\theta = \pi/l), \tag{5.1.8}$$

$$\Delta_1\bar{\psi}_2^* = (\gamma^2p^2 - s^2)\bar{\psi}_2^*, \quad \frac{\partial\bar{\psi}_2^*}{\partial\theta} = W_0(\theta = 0), \quad \frac{\partial\bar{\psi}_2^*}{\partial\theta} = W_1(\theta = \pi/l), \tag{5.1.9}$$

where new symbols mean:

$$\begin{aligned}
U_m &= \frac{r}{\gamma^2p^2}\left[(\gamma^2p^2 - s^2)\frac{dV_m}{dr} + \frac{s}{\mu}(\bar{\sigma}_{\theta z}^m)^* + (\gamma^2p^2 - 2s^2)(\bar{u}_\theta^m)^*\right],\\
V_m &= \frac{1}{\gamma^2p^2 - s^2}\left[\frac{1}{\mu}(\bar{\sigma}_{\theta r}^m)^* - 2\frac{d}{dr}(\bar{u}_\theta^m)^*\right],\\
W_m &= \frac{r}{\gamma^2p^2}\left[2s(\bar{u}_\theta^m)^* - \frac{1}{\mu}(\bar{\sigma}_{\theta z}^m)^* + s\frac{dV_m}{dr}\right] \qquad (m = 0,\ 1).
\end{aligned} \tag{5.1.10}$$

Analogously, equations (5.1.4) reduce to three sets for the longitudinal and transverse potentials:

$$\Delta_1\bar{\varphi}^* = (p^2 - s^2)\bar{\varphi}^*, \qquad \bar{\varphi}^* = U_0^0(\theta = 0), \quad \bar{\varphi}^* = U_1^0(\theta = \pi/l), \tag{5.1.11}$$

$$\Delta_1\bar{\psi}_1^* = (\gamma^2p^2 - s^2)\bar{\psi}_1^*, \quad \frac{\partial\bar{\psi}_1^*}{\partial\theta} = V_0^0(\theta = 0), \quad \frac{\partial\bar{\psi}_1^*}{\partial\theta} = V_1^0(\theta = \pi/l), \tag{5.1.12}$$

$$\Delta_1\bar{\psi}_2^* = (\gamma^2p^2 - s^2)\bar{\psi}_2^*, \quad \bar{\psi}_2^* = W_0^0(\theta = 0), \quad \bar{\psi}_2^* = W_1^0(\theta = \pi/l), \tag{5.1.13}$$

with notation

$$\begin{aligned}
U_m^0 &= \frac{2}{\gamma^2p^2}\left[\frac{1}{2\mu}(\bar{\sigma}_{\theta\theta}^m)^* + s(\bar{u}_z^m)^* + \frac{d}{dr}(\bar{u}_r^m)^*\right],\\
V_m^0 &= r\left[(\bar{u}_r^m)^* - \frac{dU_m^0}{dr} - s\frac{dW_m^0}{dr}\right],\\
W_m^0 &= \frac{1}{s^2 - \gamma^2p^2}\left[(\bar{u}_z^m)^* - sU_m^0\right] \qquad (m = 0,\ 1).
\end{aligned} \tag{5.1.14}$$

We can assume validity of the estimate (5.1.5) after Laplace transformation in τ and z to obtain

$$(\bar{\mathbf{u}})^* = \mathrm{const} + \mathrm{O}(r^\varepsilon), \quad \varepsilon > 0, \quad r \to 0 \tag{5.1.15}$$

(under extra assumption of estimate (5.1.15) uniformly valid with respect to θ).

The condition (5.1.15) completes the set (5.1.7–9) for the third boundary-value problem, and the set (5.1.11–13) for the fourth one. These equations show obviously that the longitudinal and the transverse potentials can be sought independently as long as equation (5.1.15) is permitted not to be taken into account.

In our solving equations (5.1.7–9), let us expand functions $\bar{\varphi}^*(p,r,\theta,s)$ and $\bar{\psi}_2^*(p,r,\theta,s)$ in cosine series, and $\bar{\psi}_1^*(p,r,\theta,s)$ in sine series over the interval $0 \le \theta \le \pi/l$. This yields the following equations for factors of these series:

$$\frac{d^2a_n}{dr^2} + \frac{1}{r}\frac{da_n}{dr} - \frac{n^2l^2a_n}{r^2} = \omega^2 a_n + f_n(r)\,, \tag{5.1.16}$$

$$\frac{d^2b_{nj}}{dr^2} + \frac{1}{r}\frac{db_{nj}}{dr} - \frac{n^2l^2b_{nj}}{r^2} = \kappa^2 b_{nj} + f_{nj}(r) \qquad (j = 1,\ 2)\,, \tag{5.1.17}$$

with notation:

$$f_n(r) = \frac{2l}{\pi r^2}\left[U_0 - (-1)^n U_1\right]\,,$$

$$f_{n1}(r) = -\frac{2l^2 n}{\pi r^2}\left[V_0 - (-1)^n V_1\right]\,,$$

$$f_{n2}(r) = \frac{2l}{\pi r^2}\left[W_0 - (-1)^n W_1\right]\,,$$

$$\bar{\varphi}^*(p,r,\theta,s) = \frac{a_0}{2} + \sum_{n=1}^{\infty} a_n \cos(nl\theta)\,, \qquad a_n = \frac{2l}{\pi}\int_0^{\pi/l} \bar{\varphi}^* \cos(nl\theta)d\theta$$

$$(n = 0,1,2,\ldots)\,,$$

$$\bar{\psi}_1^*(p,r,\theta,s) = \sum_{n=1}^{\infty} b_{n1}\sin(nl\theta)\,, \qquad b_{n1} = \frac{2l}{n}\int_0^{\pi/l} \bar{\psi}_1^* \sin(nl\theta)d\theta$$

$$(n = 1,2,3,\ldots)\,,$$

$$\bar{\psi}_2^*(p,r,\theta,s) = \frac{b_{02}}{2} + \sum_{n=1}^{\infty} b_{n2}\cos(nl\theta)\,,$$

$$b_{n2} = \frac{2l}{\pi}\int_0^{\pi/l} \bar{\psi}_2^* \cos(nl\theta)d\theta \qquad (n = 0,1,2,\ldots)\,,$$

$$\omega = \sqrt{p^2 - s^2}\,, \quad \kappa = \sqrt{\gamma^2p^2 - s^2}\,.$$

Branches of the functions ω and κ in the s-plane are separated by cuts from points $s = \pm p$ ($s = \pm\gamma p$ for κ) to the infinity, along radii $\arg s = \arg p$ and $\arg s = \pi + \arg p$, and branches of ω and κ are chosen so that $\omega = p$ and $\kappa = \gamma p$ at $s = 0$. Then, $\mathrm{Re}\{\omega\} > 0$ and $\mathrm{Re}\{\kappa\} > 0$ where $|\mathrm{Re}\{s\}| < \mathrm{Re}\{p\}$. From (5.1.16) and 5.1.17),

$$a_n = A_n K_{nl}(r\omega) + B_n I_{nl}(r\omega) + F_n(r) ,$$

$$F_n(r) = -K_{nl}(r\omega) \int_0^r I_{nl}(x\omega) f_n(x) x dx$$

$$- I_{nl}(r\omega) \int_r^\infty K_{nl}(x\omega) f_n(x) x dx , \qquad (5.1.18)$$

$$b_{nj} = C_{nj} K_{nl}(r\kappa) + D_{nj} I_{nl}(r\kappa) + F_{nj}(r) ,$$

$$F_{nj}(r) = -K_{nl}(r\kappa) \int_0^r I_{nl}(x\kappa) f_{nj}(x) x dx$$

$$- I_{nl}(r\kappa) \int_r^\infty K_{nl}(x\kappa) f_{nj}(x) x dx . \qquad (5.1.19)$$

Treating equations (5.1.11–13) in the same way, we can expand functions $\bar{\varphi}^*$ and $\bar{\psi}_2^*$ in sine series, and $\bar{\psi}_1^*$ in cosine series over the interval $0 \le \theta \le \pi/l$. This yields equations (5.1.16–17) where a_n, b_{nj}, $f_n(r)$, and $f_{nj}(r)$ should be replaced by a_n^0, b_{nj}^0, $f_n^0(r)$, and $f_{nj}^0(r)$, respectively. After solving these, obtain

$$a_n^0 = A_n^0 K_{nl}(r\omega) + B_n^0 I_{nl}(r\omega) + F_n^0(r) , \qquad (5.1.20)$$

$$b_{nj}^0 = C_{nj}^0 K_{nl}(r\kappa) + D_{nj}^0 I_{nl}(r\kappa) + F_{nj}^0(r) , \qquad (5.1.21)$$

with notation

$$\bar{\varphi}^*(p,r,\theta,s) = \sum_{n=1}^\infty a_n^0 \sin(nl\theta) , \qquad a_n^0 = \frac{1}{\pi} \int_0^{\pi/l} \bar{\varphi}^* \sin(nl\theta) d\theta$$

$(n = 1,2,3,\ldots)$,

$$\bar{\psi}_1^*(p,r,\theta,s) = \frac{b_{01}^0}{2} + \sum_{n=1}^\infty b_{n1}^0 \cos(nl\theta) , \qquad b_{n1}^0 = \frac{2l}{\pi} \int_0^{\pi/l} \bar{\psi}_1^* \cos(nl\theta) d\theta$$

$(n = 0,1,2,\ldots)$,

$$\bar{\psi}_2^*(p,r,\theta,s) = \sum_{n=1}^\infty b_{n2}^0 \sin(nl\theta) , \qquad b_{n2}^0 = \frac{2l}{\pi} \int_0^{\pi/l} \bar{\psi}_2^* \sin(nl\theta) d\theta$$

$(n = 1,2,3,\ldots)$,

$$F_n^0(r) = -K_{nl}(r\omega) \int_0^r I_{nl}(x\omega) f_n^0(x) x dx$$

$$- I_{nl}(r\omega) \int_r^\infty K_{nl}(x\omega) f_n^0(x) x dx ,$$

$$F^0_{nj}(r) = -K_{nl}(r\kappa)\int\limits_0^r I_{nl}(x\kappa)f^0_{nj}(x)x\,dx$$
$$- I_{nl}(r\kappa)\int\limits_r^\infty K_{nl}(x\kappa)f^0_{nj}(x)x\,dx \qquad (j=1,2)\,,$$

$$f^0_n(r) = -\frac{2l^2n}{\pi r^2}\left[U^0_0 - (-1)^n U^0_1\right]\,,$$
$$f^0_{nl}(r) = \frac{2l}{\pi r^2}\left[V^0_0 - (-1)^n V^0_1\right]\,,$$
$$f^0_{n2}(r) = -\frac{2l^2n}{\pi r^2}\left[W^0_0 - (-1)^n W^0_1\right]\,.$$

Further, let us assume that the functions f_n and f_{nj} (f^0_n, f^0_{nj}) are limited when $r \to \infty$, whereas functions rf_n and rf_{nj} (rf^0_n and rf^0_{nj}) can be written as const $+\mathrm{O}(r^\varepsilon)$ when $r \to 0$, with $\varepsilon > 0$.

Using the asymptotic behaviour of cylindrical functions

$$K_a(s) \sim \sqrt{\frac{\pi}{2s}}\,\mathrm{e}^{-s}\,, \quad I_a(s) \sim \frac{\mathrm{e}^s}{\sqrt{2\pi s}}\,, \quad \text{as } |s| \to \infty \text{ with } |\arg s| < \pi/2$$

and employing the fact that the functions f_n and f_{nj} (f^0_n, f^0_{nj}) are finite when $r \to \infty$, one can check that functions F_n and F_{nj} (F^0_n, F^0_{nj}) are also finite at the same trend in r. With this, if the factors a_n and b_{nj} (a^0_n, b^0_{nj}) are finite when $r \to \infty$, it is an immediate conclusion from (5.1.18–21) that

$$B_n \equiv D_{nj} \equiv 0 \quad (B^0_n \equiv D^0_{nj} \equiv 0)\,.$$

A_n and C_{nj} of the third boundary-value problem can be determined from the edge condition (5.1.15). For this purpose, expand transforms of $\bar{u}^*_r$ and $\bar{u}^*_z$ in cosine series, and that of $\bar{u}^*_\theta$ in sine series, over the interval $0 \le \theta \le \pi/l$; in so doing, of great value is the expression of the displacement vector in terms of the potentials. As a result the following set of three equations appears from the condition (5.1.15) at $n = 0, 1, 2, \ldots$ taken separately:

$$\begin{aligned}
&\frac{da_n}{dr} + \frac{nl}{r}b_{n1} + s\frac{db_{n2}}{dr} = \text{const} + \mathrm{O}(r^\varepsilon)\,,\\
&sa_n - \kappa^2 b_{n2} = \text{const} + \mathrm{O}(r^\varepsilon)\,, \qquad (5.1.22)\\
&-\frac{nl}{r}a_n - \frac{db_{n1}}{dr} - \frac{snl}{r}b_{n2} = \text{const} + \mathrm{O}(r^\varepsilon) \qquad (r \to 0,\ \varepsilon > 0)\,.
\end{aligned}$$

Here, A_n and C_{nj} are present through (5.1.18–19); at $n = 0$, equations (5.1.22) become the set of two relations with respect to a_0 and b_{02} since $b_{01} = 0$.

An analogous algorithm can be applied to the fourth boundary-value problem: expand $\bar{u}_r$ and $\bar{u}^*_z$ in sine series and $\bar{u}^*_\theta$ in cosine series. If $r \to \infty$, values of A^0_n and C^0_{nj} at $n = 0, 1, 2, \ldots$ will be a result of solving the following set of equations:

$$\frac{da_n^0}{dr} - \frac{nl}{r}b_{n1}^0 + s\frac{db_{n2}}{dr} = \text{ const } + \mathrm{O}(r^\varepsilon) ,$$
$$sa_n^0 - \kappa^2 b_{n2}^0 = \text{ const } + \mathrm{O}(r^\varepsilon) , \qquad (5.1.23)$$
$$\frac{nl}{r}a_n^0 - \frac{db_{n1}^0}{dr} + \frac{snl}{r}b_{n2}^0 = \text{ const } + \mathrm{O}(r^\varepsilon) \qquad (\varepsilon > 0) ,$$

where $a_0^0 = b_{02}^0 = 0$.

As $r \to 0$, asymptotic representations to functions $F_n(r)$ and $F_{nj}(r)$ can be found by using asymptotic formulae for cylindrical functions $I_\alpha(s)$ and $K_\alpha(s)$ as $s \to 0$:

$$I_\alpha(s) = \frac{1}{\Gamma(1+\alpha)}\left(\frac{s}{2}\right)^\alpha + \mathrm{O}\left(s^{2+\alpha}\right) , \quad K_0(s) = -\ln s + \mathrm{O}(1) ,$$
$$K_\alpha(s) = \begin{cases} \dfrac{\Gamma(\alpha)}{2}\left(\dfrac{2}{s}\right)^\alpha + \dfrac{\Gamma(-\alpha)}{2}\left(\dfrac{s}{2}\right)^\alpha + \mathrm{O}\left(s^{2-\alpha}\right) & (0 < \alpha < 1) , \\ \dfrac{\Gamma(\alpha)}{2}\left(\dfrac{2}{s}\right)^\alpha + \mathrm{O}\left(s^{2-\alpha}\right) & (\alpha > 1) \end{cases} \qquad (5.1.24)$$

(usual notation, $\Gamma(\cdot)$, is taken for the gamma-function) in combination with restrictions on behaviour of the functions f_n and f_{nj} when $r \to 0$. Thus,

$$F_0(r) = \text{ const } + \mathrm{O}(r^\varepsilon) , \quad F_1(r) = M_1 r^l + \mathrm{O}(r) , \quad F_n(r) = \mathrm{O}(r) \quad (n \geq 2) ,$$
$$F_{02}(r) = \text{ const } + \mathrm{O}(r) , \quad F_{1j}(r) = M_{1j} r^l + \mathrm{O}(r) , \quad F_{nj}(r) = \mathrm{O}(r) \quad (n \geq 2) ,$$
$$M_1 = -\left(\frac{\omega}{2}\right)^l \frac{H(1-l)}{\Gamma(1+l)} \int_0^\infty K_l(x\omega) f_1(x) x\,dx , \qquad (5.1.25)$$
$$M_{1j} = -\left(\frac{\kappa}{2}\right)^l \frac{H(1-l)}{\Gamma(1+l)} \int_0^\infty K_l(x\kappa) f_1(x) x\,dx .$$

Substituting equations (5.1.18, 19) into equation (5.1.22) and using the asymptotic expressions (5.1.24, 25), one can conclude that equations (5.1.22) are satisfied if $A_n = C_{nj} = 0$ at both $n = 0$ and $n \geq 2$. As well, this is true at $n = 1$ if $l > 1$. But when $n = 1$ and $l < 1$, equations (5.1.26) appear:

$$Sr^{-l-1} + Tr^{l-1} + \mathrm{O}(1) = \text{ const } + \mathrm{O}(r^\varepsilon) ,$$
$$Xr^{-l} + \mathrm{O}(r^l) = \text{ const } + \mathrm{O}(r^\varepsilon) , \qquad (5.1.26)$$
$$Sr^{-l-1} - Tr^{l-1} + \mathrm{O}(1) = \text{ const } + \mathrm{O}(r^\varepsilon) \qquad (r \to 0, \ \varepsilon > 0) ,$$

where

$$S = -2^{l-1}\Gamma(1+l)\left[A_1\omega^{-1} - C_{11}\kappa^{-l} + sC_{12}\kappa^{-l}\right] ,$$
$$T = -2^{-l-1}\Gamma(1-l)\left[A_1\omega^l + C_{11}\kappa^l + sC_{12}\kappa^l\right] + l\left(M_1 + M_{11} + sM_{12}\right) ,$$
$$X = 2^{l-1}\Gamma(l)\left[A_1 s\omega^{-l} - \kappa^{2-l}C_{12}\right] .$$

It is easy to check that $S = T = X = 0$ which yields the following expressions:

$$A_1 = \omega^l \kappa^{2-l} S^{-1} C_{12}\,, \quad C_{11} = \gamma^2 p^2 s^{-1} C_{12}\,,$$
$$C_{12} = \frac{2^{l+1}\sin(\pi l)\Gamma(1+l)s(M_1 + M_{11} + sM_{12})}{\pi[\omega^{2l}\kappa^{2-l} + (s^2+\gamma^2p^2)\kappa^l]}\,. \tag{5.1.27}$$

A_n^0 and C_{nj}^0 of the fourth boundary-value problem are obtainable in the same way:

- asymptotic behaviour of $F_n^0(r)$ and $F_{nj}^0(r)$ for $r \to 0$ is represented by (5.1.25),
- $F_n(r)$, $F_{nj}(r)$ (at $n \geq 1$), M_1, M_{1j}, $f_1(x)$, and $f_{1j}(x)$ should be replaced by $F_n^0(r)$, $F_{nj}^0(r)$ (when $n \geq 1$), M_1^0, M_{1j}^0, $f_1^0(x)$, and $f_{1j}^0(x)$, respectively,
- at $n = 0$, there are no $F_0^0(r)$ and $F_{02}^0(r)$,
- $F_{01}^0(r)$ features an asymptotic behaviour as follows:
  $F_{01}^0(r) =$ const $+ O(r)$ as $r \to 0$, and $F_{01}^0(r) = O(1)$ as $r \to \infty$.

Then, substituting equations (5.1.20, 21) into (5.1.23), obtain identities

$$A_n^0 \equiv C_{nj}^0 \equiv 0 \text{ at both } n = 0 \text{ and } n \geq 2;\,,$$
$$A_1^0 \equiv C_{1j}^0 \equiv 0 \text{ when } l > 1\,,$$

and the following set of equations at $n = 1$, $l < 1$:

$$S^0 r^{-l-1} + T^0 r^{l-1} + O(1) = \text{const} + O(r^\varepsilon)\,,$$
$$X^0 r^{-l} + O(r^l) = \text{const} + O(r^\varepsilon) \qquad (r \to 0,\ \varepsilon > 0)\,, \tag{5.1.28}$$
$$-S^0 r^{-l-1} + T^0 r^{l-1} + O(1) = \text{const} + O(r^\varepsilon)\,,$$

where

$$S^0 = -2^{l-1}\Gamma(1+l)\left[A_1^0\omega^{-l} - C_{11}^0\kappa^{-l} + sC_{12}^0\kappa^{-1}\right]\,,$$
$$T^0 = -2^{-l-1}\Gamma(1-l)\left[A_1^0\omega^l - C_{11}^0\kappa^l + s\kappa^l C_{12}^0\right] + l\left(M_1^0 - M_{11}^0 + sM_{12}^0\right)\,,$$
$$X^0 = 2^{l-1}\Gamma(l)\left[A_l^0 s\omega^{-l} - \kappa^{2-l}C_{12}^0\right]\,.$$

It is the consequence of (5.1.28) that $S^0 = T^0 = X^0 = 0$. This provides formulae for A_1^0, C_{11}^0, and C_{12}^0:

$$A_1^0 = \omega^l\kappa^{2-l}s^{-1}C_{12}^0\,, \quad C_{11}^0 = -\gamma^2p^2s^{-1}C_{12}^0\,,$$
$$C_{12}^0 = \frac{2^{l+1}\sin(\pi l)\Gamma(1+l)s\left(M_1^0 - M_{11}^0 + sM_{12}^0\right)}{\pi\left[\omega^{2l}\kappa^{2-l} + \left(s^2+\gamma^2p^2\right)\kappa^l\right]}\,. \tag{5.1.29}$$

So, transforms of solutions to the two problems are written explicitly.

A solution to the third boundary value problem (at $1/2 < l$, $l \neq 1$) can be written in transforms as follows:

$$\bar{\varphi}^* = \frac{1}{2}F_0(r) + \sum_{n=1}^{\infty} F_n(r)\cos(nl\theta) + \frac{\kappa^{2-l}}{s}\omega^l \cos(l\theta)K_l(r\omega)C_{12}\,,$$

$$\bar{\psi}_1^* = \sum_{n=1}^{\infty} F_{n1}(r)\sin(nl\theta) + \frac{\gamma^2 p^2}{s}\sin(l\theta)K_l(r\kappa)C_{12}\,, \tag{5.1.30}$$

$$\bar{\psi}_2^* = \frac{1}{2}F_{02}(r) + \sum_{n=1}^{\infty} F_{n2}(r)\cos(nl\theta) + \cos(l\theta)K_l(r\kappa)C_{12}\,.$$

Expressions for $f_n(r)$ and $F_{nj}(r)$ are given in (5.1.18, 19) and C_{12} is defined in (5.1.27).

For the fourth bundary problem, a solution (at $1/2 < l,\ l \neq 1$) in transforms can be written as

$$\bar{\varphi}^* = \sum_{n=1}^{\infty} F_n^0(r)\sin(nl\theta) + \frac{\kappa^{2-l}}{s}\omega^l \sin(l\theta)K_l(r\omega)C_{12}^0\,,$$

$$\bar{\psi}_1^* = \frac{1}{2}F_{01}^0(r) + \sum_{n-1}^{\infty} F_{n1}^0(r)\cos(nl\theta) - \frac{\gamma^2 p^2}{s}\cos(l\theta)K_l(r\kappa)C_{12}^0\,, \tag{5.1.31}$$

$$\bar{\psi}_2^* = \sum_{n=1}^{\infty} F_{n2}^0(r)\sin(nl\theta) + \sin(l\theta)K_l(r\kappa)C_{12}^0\,,$$

$F_n^0(r)$ and $F_{nj}^0(r)$ are provided by (5.1.20, 21), C_{12}^0 is presented in (5.1.29).

Final solutions of both the third and fourth boundary problems will be obtained after applying the formulae

$$\varphi = \frac{1}{(2\pi \mathrm{i})^2}\int_{c-\mathrm{i}\infty}^{c+\mathrm{i}\infty} \mathrm{e}^{p\tau}dp \int_{-\mathrm{i}\infty}^{\mathrm{i}\infty} \bar{\varphi}^*\mathrm{e}^{sz}ds\,,$$

$$\psi_j = \frac{1}{(2\pi \mathrm{i})^2}\int_{c-\mathrm{i}\infty}^{c+\mathrm{i}\infty} \mathrm{e}^{p\tau}dp \int_{-\mathrm{i}\infty}^{\mathrm{i}\infty} \bar{\psi}_j^*\mathrm{e}^{sz}ds \qquad (c > 0\,,\ j = 1,\ 2) \tag{5.1.32}$$

to find the original time functions for expressions (5.1.30, 31), respectively.

For a plane-strain state, functions $\bar{\sigma}_{\theta z}^m$ and $\bar{u}_z^m$ involved in (5.1.3, 4) are equal to zero, and other specified functions ($\bar{u}_\theta^m$, $\bar{u}_r^m$, $\bar{\sigma}_{\theta r}^m$, and $\bar{\sigma}_{\theta\theta}^m$) do not depend on z. Sets of equations for the potentials $\varphi(\tau, r, \theta)$ and $\psi(\tau, r, \theta)$ governing the displacement vector:

$$\boldsymbol{u} = \operatorname{grad}\varphi + \operatorname{rot}(\psi \boldsymbol{e}_3)\,,$$

will be obtained from sets (5.1.3, 4) after we nullify functions $\bar{\sigma}_{\theta z}^m$, $\bar{u}_z^m$, ψ_2 and all derivatives with respect to z as well as denote $\psi \equiv \psi_1$. Then, solutions to the third and fourth boundary problems can easily be obtained. As their forms show, these solutions can be formally derived from equations (5.1.30, 31) after,

- first, replacing there the functions $(\bar{u}_\theta^m)^*$, $(\bar{\sigma}_{\theta r}^m)^*$, $(\bar{u}_r^m)^*$, $(\bar{\sigma}_{\theta\theta}^m)^*$, $(\bar{\sigma}_{\theta z}^m)^*$, and $(\bar{u}_z^m)^*$ by $\bar{u}_\theta^m$, $\bar{\sigma}_{\theta r}^m$, $\bar{u}_r^m$, $\bar{\sigma}_{\theta\theta}^m$, 0, and 0, respectively,
- second, passing to the limit as $s \to 0$, and,
- third, adopting $\bar{\varphi} \equiv \lim_{s\to 0} \bar{\varphi}^*$, $\bar{\psi} \equiv \lim_{s\to 0} \psi^*$.

As a result, a solution to the plane-strain dynamic problem with zero initial conditions (at $\tau = \tau_0$) at boundary conditions

$$\begin{aligned} u_\theta &= u_\theta^0(\tau, r)\,, \quad \sigma_{\theta r} = \sigma_{\theta r}^0(\tau, r) \qquad (\theta = 0)\,, \\ u_\theta &= u_\theta^1(\tau, r)\,, \quad \sigma_{\theta r} = \sigma_{\theta r}^1(\tau, r) \qquad (\theta = \pi/l) \end{aligned} \tag{5.1.33}$$

has the following form (in transforms, for $l > 1/2$, $l \neq 1$):

$$\begin{aligned} \bar{\varphi} &= \frac{1}{2} F_0(r) + \sum_{n=1}^{\infty} F_n(r)\cos(nl\theta) + CK_l(pr)\cos(l\theta)\,, \\ \bar{\psi} &= \sum_{n=1}^{\infty} F_{n1}(r)\sin(nl\theta) + C\gamma^l K_l(\gamma pr)\sin(l\theta)\,, \\ C &= \frac{2^{l+1}\sin(\pi l)\Gamma(1+l)(M_1 + M_{11})}{\pi p^l(1+\gamma^{2l})}\,. \end{aligned} \tag{5.1.34}$$

Under the boundary conditions

$$\begin{aligned} u_r &= u_r^0(\tau, r)\,, \quad \sigma_{\theta\theta} = \sigma_{\theta\theta}^0(\tau, r) \qquad (\theta = 0)\,, \\ u_r &= u_r^1(\tau, r)\,, \quad \sigma_{\theta\theta} = \sigma_{\theta\theta}^1(\tau, r) \qquad (\theta = \pi/l)\,, \end{aligned} \tag{5.1.35}$$

and zero initial conditions at $\tau = \tau_0$ the solution to this problem can be written as:

$$\begin{aligned} \bar{\varphi} &= \sum_{n=1}^{\infty} F_n^0(r)\sin(nl\theta) + C^0 K_l(pr)\sin(l\theta)\,, \\ \bar{\psi} &= \frac{1}{2} F_{01}^0(r) + \sum_{n=1}^{\infty} F_{n1}^0(r)\cos(nl\theta) - \gamma^l C^0 K_l(\gamma pr)\cos(l\theta)\,, \\ C^0 &= \frac{2^{l+1}\sin(\pi l)\Gamma(1+l)\left(M_1^0 - M_{11}^0\right)}{\pi p^l\left(1+\gamma^{2l}\right)}\,. \end{aligned} \tag{5.1.36}$$

One should remember that $l > 1/2$ but $l \neq 1$. In equations (5.1.34), we denote

$$F_n = -K_{nl}(pr)\int_0^r I_{nl}(px)f_n(x)x\,dx - I_{nl}(pr)\int_r^\infty K_{nl}(px)f_n(x)x\,dx\,,$$

$$F_{n1} = -K_{nl}(\gamma pr)\int_0^r I_{nl}(\gamma px)f_{n1}(x)x\,dx - I_{nl}(\gamma pr)\int_r^\infty K_{nl}(\gamma px)f_{n1}(x)x\,dx\,,$$

$$M_{11} = -\left(\frac{\gamma p}{2}\right)^{l} \frac{H(1-l)}{\Gamma(1+l)} \int_0^\infty K_l(\gamma p x) f_{11}(x) x\,dx ,$$

$$M_1 = -\left(\frac{p}{2}\right)^{l} \frac{H(1-l)}{\Gamma(1+l)} \int_0^\infty K_l(p x) f_1(x) x\,dx ,$$

$$f_n(r) = \frac{2l}{\pi r^2}\left[U_0 - (-1)^n U_1\right] , \quad f_{n1}(r) = -\frac{2l^2 n}{\pi r^2}\left[V_0 - (-1)^n V_1\right] ,$$

$$U_m = r\bar{u}_\theta^m - \frac{2r}{\gamma^2 p^2}\frac{d^2\bar{u}_\theta^m}{dr^2} + \frac{r}{\gamma^2 \mu p^2}\frac{d\bar{\sigma}_{\theta r}^m}{dr} ,$$

$$V_m = \frac{1}{\gamma^2 \mu p^2}\bar{\sigma}_{\theta r}^m - \frac{2}{\gamma^2 p^2}\frac{d\bar{u}_\theta^m}{dr} \qquad (m = 0,\ 1) . \tag{5.1.37}$$

For the functions F_n^0, F_{n1}^0, M_1^0, and M_{11}^0 involved in (5.1.36) the same formulae remain valid as are for the functions F_n, F_{n1}, M_1, and M_{11}, respectively, if f_n and f_{n1} are replaced by f_n^0 and f_{n1}^0, respectively:

$$f_n^0(r) = -\frac{2l^2 n}{\pi r^2}\left[U_0^0 - (-1)^n U_1^0\right] , \quad f_{n1}^0(r) = \frac{2l}{\pi r^2}\left[V_0^0 - (-1)^n V_1^0\right] ,$$

$$U_m^0 = \frac{\bar{\sigma}_{\theta\theta}^m}{\gamma^2 \mu p^2} + \frac{2}{\gamma^2 p^2}\frac{d\bar{u}_r^m}{dr} , \tag{5.1.38}$$

$$V_m^0 = r\bar{u}_r^m - \frac{r}{\gamma^2 p^2}\left[\frac{d\bar{\sigma}_{\theta\theta}^m}{\mu dr} + 2\frac{d^2\bar{u}_r^m}{dr^2}\right] \qquad (m = 0,\ 1) .$$

Original time functions for the plane-strain problems under consideration can be defined by

$$\varphi = \frac{1}{2\pi i}\int_{c-i\infty}^{c+i\infty} \bar{\varphi} e^{p\tau} dp , \quad \psi = \frac{1}{2\pi i}\int_{c-i\infty}^{c+i\infty} \bar{\psi} e^{p\tau} dp \qquad (c > 0) , \tag{5.1.39}$$

with $\bar{\varphi}$ and $\bar{\psi}$ presented by equations (5.1.34, 36).

One must mention that the above elastodynamics problems require that functions rf_n, rf_{nj}, rf_n^0, and rf_{nj}^0 be limited when $r \to 0$. The functions specified at the boundary will satisfy this requirement if they, when $r \to 0$, behave as follows: for the third boundary problem

$$(\bar{u}_\theta^m)^* = \sum_{j=0}^{2} d_j r^j + O(r^{2+\varepsilon}) , \quad (\bar{\sigma}_{\theta z}^m)^* = d_3 + O(r^\varepsilon) ,$$
$$(\bar{\sigma}_{\theta r}^m)^* = d_4 + d_5 r + O(r^{1+\varepsilon}) \quad (\varepsilon > 0) , \tag{5.1.40}$$

for the fourth boundary problem

$$
\begin{aligned}
(\bar{u}_r^m)^* &= \sum_{j=0}^{2} d_j^0 r^j + \mathrm{O}(r^{2+\varepsilon}) , \\
(\bar{u}_z^m)^* &= d_3^0 + d_4^0 r + \mathrm{O}(r^{1+\varepsilon}) , \\
(\bar{\sigma}_{\theta\theta}^m)^* &= d_5^0 + d_6^0 r + \mathrm{O}(r^{1+\varepsilon}) \qquad (\varepsilon > 0) .
\end{aligned}
\tag{5.1.41}
$$

Here, d_j and d_j^0 ($j = 0, 1, 2, \ldots, 6$) do not depend on r.

According to (5.1.40) it is easily seen in this case (for the third boundary value problem) that only functions rf_n and rf_{n2} are limited when $r \to 0$, whereas the function rf_{n1} behaves as $\mathrm{O}(r^{-1})$, i.e., is unlimited. Nevertheless, this problem is solvable through reduction to the fourth boundary problem where just all of the functions $\{rf_n^0, rf_{nj}^0\}$ are limited when $r \to 0$. To prove this statement, we should differentiate equations for $\bar{\varphi}^*$, $\bar{\psi}_1^*$ and $\bar{\psi}_2^*$ (in sets (5.1.7–9)) with respect to θ, thereafter, by means of the equation for $\bar{\psi}_1^*$, reduce the boundary condition for $\bar{\psi}_1^*$:

$$
\begin{aligned}
\left.\frac{\partial^2 \bar{\psi}_1^*}{\partial \theta^2}\right|_{\theta=0,\pi/l} &= -r^2 \left[\frac{\partial}{r \partial r} \left(r \frac{\partial \bar{\psi}_1^*}{\partial r} \right) - \kappa^2 \bar{\psi}_1^* \right] \Bigg|_{\theta=0,\pi/l} \\
&= -r \frac{d}{dr} \left(r \frac{dV_m}{dr} \right) + r^2 \kappa^2 V_m
\end{aligned}
$$

and, finally, designate

$$
\begin{aligned}
&\bar{\varphi}_1^* \equiv \frac{\partial \bar{\varphi}^*}{\partial \theta} , \quad \bar{\psi}_{11}^* \equiv \frac{\partial \bar{\psi}_1^*}{\partial \theta} , \quad \bar{\psi}_{21}^* \equiv \frac{\partial \bar{\psi}_2^*}{\partial \theta} , \\
&U_m^0 \equiv U_m , \quad V_m^0 \equiv -r \frac{d}{dr} \left(r \frac{dV_m}{dr} \right) + r^2 \kappa^2 V_m , \quad W_m^0 \equiv W_m .
\end{aligned}
$$

Again, these operations give us equations (5.1.11–13) with respect to potentials $\bar{\varphi}_1^*$, $\bar{\psi}_{11}^*$, and $\bar{\psi}_{21}^*$ for the fourth boundary-value problem. Within this problem, functions rf_n^0 and rf_{nj}^0 are obviously limited when $r \to 0$.

Treating analogously the fourth boundary value problem and taking (5.1.41) into account one can conclude that rf_{n1}^0 is limited, but the functions rf_n^0 and rf_{n2}^0 behave as $\mathrm{O}(r^{-1})$ when $r \to 0$. This makes us reduce the solution of the fourth boundary problem to the solution of equations (5.1.7–9) for the third boundary problem (with respect to potentials $\bar{\varphi}_1^*$, $\bar{\psi}_{11}^*$, and $\bar{\psi}_{21}^*$) if the following notation:

$$
\begin{aligned}
&\bar{\varphi}_1^* \equiv \frac{\partial \bar{\varphi}^*}{\partial \theta} , \quad \bar{\psi}_{11}^* \equiv \frac{\partial \bar{\psi}_1^*}{\partial \theta} , \quad \bar{\psi}_{21}^* \equiv \frac{\partial \bar{\psi}_2^*}{\partial \theta} , \\
&U_m \equiv -r \frac{d}{dr} \left(r \frac{dU_m^0}{dr} \right) + r^2 \omega^2 U_m^0 , \quad V_m \equiv V_m^0 , \\
&W_m \equiv -r \frac{d}{dr} \left(r \frac{dW_m^0}{dr} \right) + r^2 \kappa^2 W_m^0 ,
\end{aligned}
$$

is introduced. In these systems, functions rf_n and rf_{nj} are limited when $r \to 0$.

Of course, the estimates (5.1.40, 41) and the utilized technique of reducing one problem to the other remain valid in two-dimensional problems (without a Laplace transformation in z.)

Note that situations with $l = 1/2$ and $l = 1$ which have not been considered can be treated as limiting versions of the above results with $l \to 1/2$ and $l \to 1$, respectively.

If one specifies different mixed conditions at boundaries $\theta = 0$ and $\theta = \pi/l$:

$$\begin{aligned} &u_\theta = u_\theta^0(t,r,z)\,, \quad \sigma_{\theta r} = \sigma_{\theta r}^0(t,r,z)\,, \\ &\sigma_{\theta z} = \sigma_{\theta z}^0(t,r,z) \qquad (\theta = 0)\,, \\ &u_r = u_r^1(t,r,z)\,, \quad u_z = u_z^1(t,r,z)\,, \\ &\sigma_{\theta\theta} = \sigma_{\theta\theta}^1(t,r,z) \qquad (\theta = \pi/l)\,, \end{aligned} \tag{5.1.42}$$

a solution to the corresponding non-stationary dynamic problem can be represented by a superposition of the solutions to the third and fourth problems considered above. It is readily seen that a solution for this problem is a sum of solutions to problems with conditions (5.1.43, 44):

$$\begin{aligned} &u_\theta = 0\,, \quad \sigma_{\theta r} = 0\,, \quad \sigma_{\theta z} = 0 \qquad (\theta = 0)\,, \\ &u_r = u_r^1(t,r,z)\,, \quad u_z = u_z^1(t,r,z)\,, \quad \sigma_{\theta\theta} = \sigma_{\theta\theta}^1(t,r,z) \qquad (\theta = \pi/l)\,, \end{aligned} \tag{5.1.43}$$

$$\begin{aligned} &u_\theta = u_\theta^0(t,r,z)\,, \quad \sigma_{\theta r} = \sigma_{\theta r}^0(t,r,z)\,, \quad \sigma_{\theta z} = \sigma_{\theta z}^0(t,r,z) \qquad (\theta = 0)\,, \\ &u_r = 0\,, \quad u_z = 0\,, \quad \sigma_{\theta\theta} = 0 \qquad (\theta = \pi/l)\,. \end{aligned} \tag{5.1.44}$$

But zero boundary conditions in equations (5.1.43) at $\theta = 0$ give $\partial\varphi/\partial\theta = \psi_1 = \partial\psi_2/\partial\theta = 0$ $(\theta = 0)$ according to the formulae (5.1.10) for U_m, V_m, and W_m; on the other hand, $\varphi = \partial\psi_1/\partial\theta = \psi_2 = 0$ as a consequence of equations (5.1.44) and the formulae (5.1.14) for U_m^0, V_m^0, and W_m^0 $(\theta = \pi/l)$. Then the problem with conditions (5.1.43) may be transformed into the fourth problem by means of continuing (through the boundary $\theta = 0$) the potentials φ and ψ_2 as even functions and the potential ψ_1 as an odd function; the boundary values for the domain $|\theta| < \pi/l$ are:

$$u_r = u_r^1(t,r,z)\,, \quad u_z = u_z^1(t,r,z)\,, \quad \sigma_{\theta\theta} = \sigma_{\theta\theta}^1(t,r,z) \qquad (\theta = \pm\pi/l)\,.$$

By analogy, potentials φ and ψ_2 should be continued through the boundary $\theta = \pi/l$ as odd functions, and ψ_1 as an even function, to reduce the problem with the boundary condition (5.1.44) to the third boundary problem for the domain $0 < \theta < 2\pi/l$:

$$u_\theta = u_\theta^0(t,r,z)\,, \quad \sigma_{\theta r} = \sigma_{\theta r}^0(t,r,z)\,, \quad \sigma_{\theta z} = \sigma_{\theta z}^0(t,r,z) \qquad (\theta = 0,\ 2\pi/l)\,.$$

This technique of solving the three-dimensional non-stationary dynamic problem remains valid for harmonic problems if, in the above formulae, we replace p by $ik + \varepsilon$ $(\mathrm{Im}\{k\} = 0,\ \varepsilon > 0)$ and pass to the limit $(\varepsilon \to 0)$.

Concluding this section, we should note that, using the present method, exact theoretical solutions to the two- and three-dimensional problems on diffraction of elastic plane, cylindrical, and spherical waves by an arbitrary smooth rigid wedge have been obtained and examined, see [5.8, 9, 14, 16]. Below, some of these problems are discussed.

5.2 Diffraction of a Spherical Elastic Wave by a Smooth Rigid Wedge

Let us consider the problem [5.13] on diffraction of a spherical longitudinal wave by a smooth rigid wedge (Fig. 5.1a, b). An elastic medium is assumed to fill a domain $\{r > 0,\ 0 < \theta < \pi/l, -\infty < z < +\infty\}(1/2 \leq l)$ and makes contact with a wedge $(\pi/l < \theta < 2\pi)$. At the wedge faces defined by the equations, $\theta = 0$ and $\theta = \pi/l$, conditions $u_\theta = 0$, $\sigma_{\theta r} = \sigma_{\theta z} = 0$ are specified. Here, r, θ, z are the cylindrical coordinates (with z-axis running along the edge of the wedge). At the instant $\tau = -r_0 (\tau = c_1 t, r_0 > 0)$ a source of a spherical elastic longitudinal wave appears at the point $(r_0, \theta_0, 0)$. Within this wave, the displacement $\boldsymbol{u}_0$ is defined by the potential

$$\varphi_0 = \frac{r_0}{R} f(\tau + r_0 - R) \qquad (\boldsymbol{u}_0 = \operatorname{grad} \varphi_0)\ . \tag{5.2.1}$$

A wave front arrives at the surface of the wedge at an instant τ_0. Here, we assume $-r_0 < \tau_0 \leq 0$; $f(\cdot)$ is an arbitrary function satisfying the applicability conditions for the Laplace transformation ($f(\tau) \equiv 0$ at $\tau < 0$);

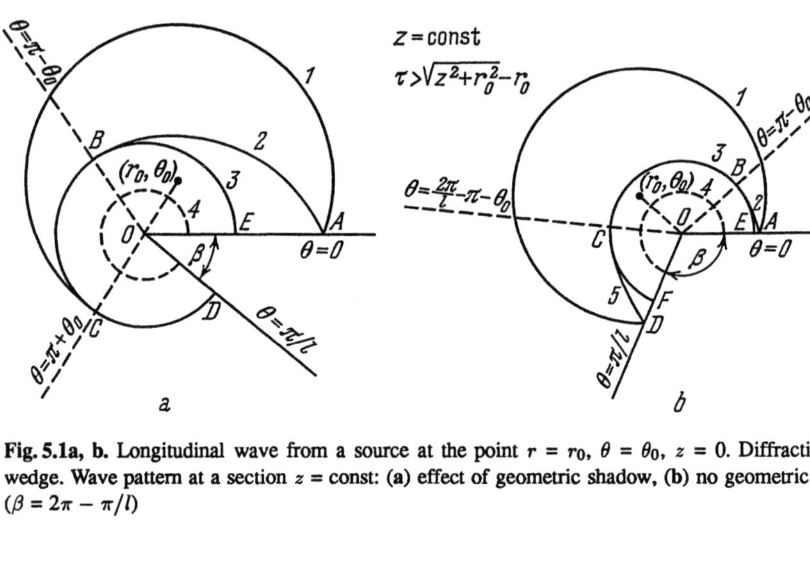

Fig. 5.1a, b. Longitudinal wave from a source at the point $r = r_0$, $\theta = \theta_0$, $z = 0$. Diffraction by a wedge. Wave pattern at a section z = const: (a) effect of geometric shadow, (b) no geometric shadow ($\beta = 2\pi - \pi/l$)

$$R = \left[z^2 + r^2 + r_0^2 - 2rr_0 \cos(\theta - \theta_0)\right]^{1/2}, \quad 0 < \theta_0 < \pi/l .$$

The wedge introduces disturbances $\boldsymbol{w} = \boldsymbol{u} - \boldsymbol{u}_0$ (Fig. 5.1a, b) into an incident wave displacement field; they can be described in terms of the longitudinal (φ_1) and transverse (ψ_1, ψ_2) potentials as follows:

$$\boldsymbol{w} = \operatorname{grad} \varphi_1 + \operatorname{curl}(\psi_1 e_3) + \operatorname{curl}\operatorname{curl}(\psi_2 e_3) . \tag{5.2.2}$$

5.2.1 Solution to the Problem

For determining the disturbed motion, there are the following equations for Laplace transforms in τ and z, according to equations (5.1.7–9) and (5.1.15):

$$\Delta_1 \bar{\varphi}_1^* = (p^2 - s^2)\bar{\varphi}_1^*, \quad \frac{\partial \bar{\varphi}_1^*}{\partial \theta} = -\frac{\partial \bar{\varphi}_0^*}{\partial \theta}(\theta = 0), \quad \frac{\partial \bar{\varphi}_1^*}{\partial \theta} = -\frac{\partial \bar{\varphi}_0^*}{\partial \theta}(\theta = \pi/l), \tag{5.2.3}$$

$$\Delta_1 \bar{\psi}_1^* = (\gamma^2 p^2 - s^2)\bar{\psi}_1^*, \quad \bar{\psi}_1^* = 0(\theta = 0), \quad \bar{\psi}_1^* = 0(\theta = \pi/l), \tag{5.2.4}$$

$$\Delta_1 \bar{\psi}_2^* = (\gamma^2 p^2 - s^2)\bar{\psi}_2^*, \quad \frac{\partial \bar{\psi}_2^*}{\partial \theta} = 0(\theta = 0), \quad \frac{\partial \bar{\psi}_2^*}{\partial \theta} = 0(\theta - \pi/l), \tag{5.2.5}$$

$$\bar{w}^* = \text{const} + \mathrm{O}(r^\varepsilon), \quad \varepsilon > 0, \; r \to 0 . \tag{5.2.6}$$

It is shown in Sect. 5.1 that a solution to the system (5.2.1–5) is given (at $l > 1/2$ and $l \neq 1$) by the general formulae (5.1.27, 30, 32) where we must specify $F_{n1} \equiv F_{n2} \equiv 0$, $M_{11} = M_{12} = 0$ and derive M_1 with due account of equation (5.2.7):

$$f_n(r) = \frac{2l}{\pi r^2}\left[(-1)^n \left.\frac{\partial \bar{\varphi}_0^*}{\partial \theta}\right|_{\theta = \pi/l} - \left.\frac{\partial \bar{\varphi}_0^*}{\partial \theta}\right|_{\theta = 0}\right], \tag{5.2.7}$$

where

$$\begin{aligned} \bar{\varphi}_0^*(p, r, \theta, s) &= \bar{f}(p) \mathrm{e}^{pr_0} r_0 \int_{-\infty}^{\infty} \mathrm{e}^{sz - pR} \frac{dz}{R} \\ &= 2\bar{f}(p) r_0 \mathrm{e}^{pr_0} K_0(\rho\omega), \quad \omega = \sqrt{p^2 - s^2}, \\ \rho = \sqrt{r^2 + r_0^2 - 2rr_0 \cos(\theta - \theta_0)}, &\quad f(\tau) \rightleftharpoons \bar{f}(p) . \end{aligned} \tag{5.2.8}$$

As a result, the function M_1 takes the following form:

$$M_1 = \frac{8r_0}{\Gamma(l)} \bar{f}(p) \mathrm{e}^{pr_0} \cos(l\theta_0) K_l(r_0\omega) \left(\frac{\omega}{2}\right)^l H(1 - l) . \tag{5.2.9}$$

In this case, it is easy to prove that the following asymptotic formulas are observed for the function $f_n(r)$: $f_n(r) \to 0$ when $r \to \infty$, $r f_n(r) = \mathrm{O}(1)$ when $r \to 0$, i.e., the function $f_n(r)$ complies with the constraints.

So, obtain

$$\begin{aligned}
\bar{\varphi}_1^* &= \frac{1}{2}F_0(r) + \sum_{n=1}^{\infty} F_n(r)\cos(nl\theta) + \frac{\kappa^{2-l}}{s}\omega^l \cos(l\theta) K_l(r\omega) C_{12}\,, \\
\bar{\psi}_1^* &= \frac{\gamma^2}{s} p^2 \sin(l\theta) K_l(r\kappa) C_{12}\,, \\
\bar{\psi}_2^* &= \cos(l\theta) K_l(r\kappa) C_{12}\,, \quad \kappa = \sqrt{\gamma^2 p^2 - s^2}\,,
\end{aligned} \tag{5.2.10}$$

where

$$C_{12} = \frac{16 l r_0 \sin(l\pi) f(p) \cos(l\theta_0) K_l(r_0\omega) s\omega^l}{\pi\left[\omega^{2l}\kappa^{2-l} + (s^2+\gamma^2p^2)\kappa^l\right]} e^{pr_0} H(1-l)\,.$$

The function $F_n(r)$ is given by equation (5.1.18) with $f_n(x)$ established in equation (5.2.7).

It is necessary to note that a sum of the terms which comprise $F_n(n = 0,1,2,\ldots)$ in the equation for $\bar{\varphi}_1^*$ in the set (5.2.10) (let us denote it as $\bar{\varphi}_{1a}^*$) is the Laplace transform for a disturbed solution to the corresponding acoustic problem. The simplest proof of this is as follows. It is an obvious consequence of the solution procedure employed that the inverse transform of this sum, φ_{1a}, satisfies the system

$$\Delta\varphi = \frac{\partial^2\varphi}{\theta\tau^2}\,, \quad \frac{\partial\varphi}{\partial\theta} = -\frac{\partial\varphi_0}{\partial\theta}(\theta = 0,\ \pi/l)\,, \quad \varphi \equiv 0(\tau < \tau_0)$$

as well as satisfying the conditions:

$$r\frac{\partial\varphi}{\partial r} = O(r^\varepsilon)\,, \quad \frac{\partial\varphi}{\partial\tau} = O(1)\,, \quad \varepsilon > 0,\ r \to 0$$

(see [5.18]) which provide uniqueness of the solution to the acoustic problem. Of course, this can be shown immediately, by reducing the inverse transform φ_{1a} to the known "acoustic" solution [5.19].

Finally, after adding the transform of the incident spherical wave and using inverse Laplace transforms with respect to s and p, the solution appears:

$$\begin{aligned}
\varphi &= \varphi_a + \frac{\cos(l\theta)}{2\pi i}\int_{c-i\infty}^{c+i\infty} e^{p\tau}dp \int_{-i\infty}^{i\infty} \kappa^{2-l}\omega^l C_{12} K_l(r\omega) e^{sz}\frac{ds}{s}\,, \\
\psi_1 &= \frac{\sin(l\theta)}{(2\pi i)^2}\int_{c-i\infty}^{c+i\infty} e^{p\tau}dp \int_{-i\infty}^{i\infty} \gamma^2 p^2 C_{12} K_l(r\kappa) e^{sz}\frac{ds}{s}\,, \\
\psi_2 &= \frac{\cos(l\theta)}{(2\pi i)^2}\int_{c-i\infty}^{c+i\infty} e^{p\tau}dp \int_{-i\infty}^{i\infty} C_{12} K_l(r\kappa) e^{sz} ds\,,
\end{aligned} \tag{5.2.11}$$

where $\varphi = \varphi_0 + \varphi_1$, $\varphi_a = \varphi_0 + \varphi_{1a}$, $c > 0$. This solution turns out to be valid for any $l \geq 1/2$.

Thus, for a wedge angle $\beta \geq \pi$ (i.e., $l \geq 1$) potentials of both the elastic and acoustic problems are the same (since $C_{12} = 0$). Therefore, in other words, a response to the longitudinal wave does not show transverse reflection/diffraction waves: $\psi_1 \equiv \psi_2 \equiv 0$.

In opposition, presence of an edge in a wedge with an angle $0 \leq \beta < \pi$ (i.e., $1/2 \leq l < 1$) causes additional diffraction waves: the longitudinal one $(\varphi - \varphi_a)$ and the transverse one (ψ_1 and ψ_2). This means that, in such a case, transverse waves appear in the diffraction domain.

Note that, outside the diffraction domain, the transverse waves do not appear, in either case: a longitudinal incident wave generates only a longitudinal reflection wave. This result could be predicted well in advance if we take into account the known results [5.20, 21].

Now, let us consider in detail the most interesting situation which occurs when the wedge angle, β, is $0 \leq \beta < \pi (1/2 \leq l < 1)$. In this case, the acoustic solution, φ_a, may be written as

$$\begin{aligned} \varphi_a = {} & \frac{f(+0)}{1+\tau/r_0} U\left(\tau + \frac{\tau^2 - r^2 - z^2}{2r_0}, r, \theta\right) \\ & - \int_{R-r_0}^{\tau} \frac{d}{dx}\left[\frac{f(\tau - x)}{1 + x/r_0}\right] U\left(x + \frac{x^2 - r^2 - z^2}{2r_0}, r, \theta\right) dx , \end{aligned} \tag{5.2.12}$$

employing the formula by Filippov for a wedge [5.19]. Here $U(\tau, r, \theta)$ is the solution to the acoustic problem on diffraction of a plane wave $U_0 = H[\tau + r\cos(\theta - \theta_0)]$ by a wedge; this solution, according to results of [5.22], is

$$\begin{aligned} U(\tau, r, \theta) = {} & H(r - \tau)\{\sigma(\theta - \theta_0) H[\tau + r\cos(\theta - \theta_0)] \\ & + \sigma(\theta + \theta_0) H[\tau + r\cos(\theta + \theta_0)^*]\} \\ & + \frac{H(\tau - r)}{\pi}\left[\tan^{-1} \frac{(1 - y^{2l})\sin(l\pi)}{(1 + y^{2l})\cos(l\pi) - 2y^l \cos[l(\theta - \theta_0)]}\right. \\ & \left. + \tan^{-1} \frac{(1 - y^{2l})\sin(l\pi)}{(1 + y^{2l})\cos(l\pi) - 2y^l \cos[l(\theta + \theta_0)]}\right] . \end{aligned} \tag{5.2.13}$$

In the latter, $y = \tau/r - [(\tau/r)^2 - 1]^{1/2}$, values of $\tan^{-1}$ are assumed to be within the interval $(0, \pi)$, and the symbol $(\theta + \theta_0)^*$ is a substitute for $\theta + \theta_0 + 2\pi m/l$ where the integer $m (m = 0, -1)$ should be so chosen that the inequality, $-\pi/l < (\theta + \theta_0)^* \leq \pi/l$, be valid at any specific point of a physical space. The function $\sigma(\theta)$ present in equation (5.2.13) equals unity for $|\theta| < \pi$, equals zero for $\pi < |\theta| < \pi/l$, and is considered periodic everywhere: $\sigma(\theta + 2\pi/l) = \sigma(\theta)$.

From (5.2.11) it follows that, as long as $1/2 \leq l < 1$, the elastic terms added to the acoustic solution disappear only at $\theta_0 = \pi/2l$ (symmetry about the bisector plane of the wedge under consideration). In the limit, when $l \to 1 - 0$, the elastic terms of an asymmetrical case ($\theta_0 \neq \pi/2l$) disappear as well ($\varphi \to \varphi_a$, $\psi_1 \to 0$, $\psi_2 \to 0$) to give a solution to the problem of the reflection of a spherical wave from a flat plate.

By using equation (5.2.11) it can be shown that, in the vicinity of the edge, the displacements are continuous and the stresses are integrable.

After changing the variable — $q = s/p$ — equations (5.2.11) can be reduced to

$$\begin{aligned}
\varphi &= \varphi_a + \frac{\cos(l\theta)}{(2\pi i)^2} \int_{c-i\infty}^{c+i\infty} \bar{f}(p) p e^{p(r_0+\tau)} dp \int_L K_l\left(pr\sqrt{1-q^2}\right) \\
&\quad \times (1-q^2)^{l/2} W_0 e^{pqz} dq\,, \\
\psi_1 &= \frac{\gamma^2 \sin(l\theta)}{(2\pi i)^2} \int_{c-i\infty}^{c+i\infty} \bar{f}(p) p e^{p(r_0+\tau)} dp \int_L K_l\left(pr\sqrt{\gamma^2-q^2}\right) \\
&\quad \times (\gamma^2-q^2)^{l/2-1} W_0 e^{pqz} dq\,, \qquad (5.2.14) \\
\psi_2 &= \frac{\cos(l\theta)}{(2\pi i)^2} \int_{c-i\infty}^{c+i\infty} \bar{f}(p) e^{p(r_0+\tau)} dp \int_L K_l\left(pr\sqrt{\gamma^2-q^2}\right) \\
&\quad \times (\gamma^2-q^2)^{l/2-1} e^{pqz} W_0 q\, dq\,, \\
W_0 &\equiv W_0(p,q) = \frac{16 l r_0 \cos(l\theta_0) \sin(l\pi) K_l\left(pr_0\sqrt{1-q^2}\right)(1-q^2)^{l/2}}{\pi\left[(1-q^2)^l + (q^2+\gamma^2)(\gamma^2-q^2)^{l-1}\right]}
\end{aligned}$$

The contour L is shown in Fig. 5.2; there, $\omega_0 = \pi/2 = \arg p$, and cuts made in the s-plane became cuts in the q-plane (along the real axis, from points $\pm\gamma$ and ± 1 to infinity; note that $\sqrt{\gamma^2-q^2} = \gamma$, $\sqrt{1-q^2} = 1$ at $q = 0$).

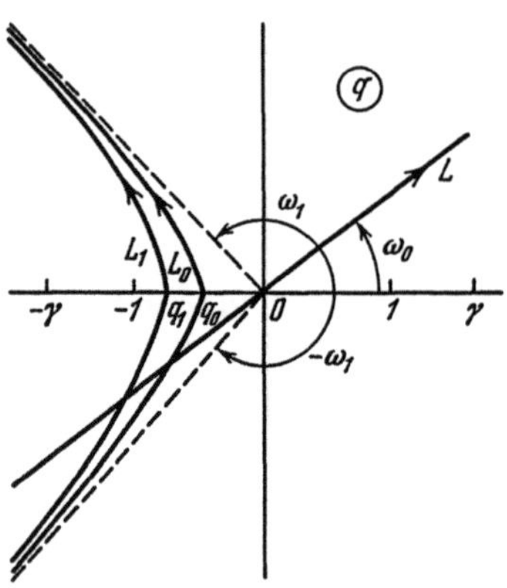

Fig. 5.2. Paths L, L_0, and L_1

It is sufficient to consider the formulae (5.2.14) at $z \geq 0$ only, since φ and ψ_1 are even functions whereas ψ_2 is known to be an odd function (one can establish this fact easily by using explicit expressions (5.2.14)). Then, at $x > 0$, simple operations can demonstrate that the contour L employed in the equation for φ can be deformed into the curve L_0, and the contour employed in expressions for ψ_1 and ψ_2, into the curve L_1. Points of these curves satisfy the following equations:

$$\begin{aligned} &L_0 : \ \mathrm{Im}\ f_0(q) = 0 \\ &\left(f_0(q) = qz - (r + r_0)\sqrt{1-q^2}\right) , \\ &L_1 : \ \mathrm{Im}\ f_1(q) = 0 \\ &\left(f_1(q) = qz - r\sqrt{\gamma^2 - q^2} - r_0\sqrt{1-q^2}\right) . \end{aligned} \tag{5.2.15}$$

The two curves shown in Fig. 5.2 (for the q-plane) are symmetric about the real axis and, on moving away from an origin of the coordinate system, have the same asymptotes at angles $\pm\omega_1$ to the real axis: $\tan\omega_1 = -(r + r_0)/z$. The points q_0 and q_1 where these curves intersect the real axis ($-1 < q_m < 0$), $m = 0, 1$) are the saddle points of the corresponding curves and should be determined from equations

$$f_0'(q) = 0 , \quad f_1'(q) = 0 , \tag{5.2.16}$$

respectively. With this, the function $f_0(q)$ being real-valued over the curve L_0 takes the maximum value at the point $q_0(f_0(q_0) = -[z^2 + (r + r_0)^2]^{1/2})$; $f_0(q)$ decreases monotonical from $f_0(q_0)$ down to $-\infty$ in moving away from q_0 along L_0 in the upper (or lower) half-plane. Analogously, the function $f_1(q)$ being real over the curve L_1 takes the maximum value at the point $q_1(f_1(q_1) = -R_1,\ R_1 \equiv R_1(z, r))$; $f_1(q)$ decreases monotonical from $f_1(q_1)$ down to $-\infty$ in moving away from q_1 along L_1 within the upper (or lower) half-plane. As $z \to 0$, values of q_0 and q_1 tend to zero, whereas the curves L_0 and L_1 go to the imaginary axis.

As a result, since the function W has no poles in the q-plane with the cuts $(-\infty, -1]$, $[1, \infty)$, we may reduce equations (5.2.14) to the following form at $z \geq 0$:

$$\varphi = \varphi_a + \cos(l\theta) \int\limits_{L_0} (1 - q^2)^{l/2-1/4} dq \int\limits_{-0}^{x_0} f'(x) S_0 dx ,$$

$$\psi_1 = \gamma^2 \sin(l\theta) \int\limits_{L_1} (\gamma^2 - q^2)^{l/2-5/4} dq \int\limits_{-0}^{x_1} f'(x) S_1 dx ,$$

$$\psi_2 = \cos(l\theta) \int\limits_{L_1} (\gamma^2 - q^2)^{l/2-5/4} q dq \int\limits_{0}^{x_1} f(x) S_1 dx ,$$

$$S_j = \frac{4l\sqrt{r_0}\cos(l\theta)\sin(l\pi)(1-q^2)^{l/2-1/4} F_j}{\pi i\sqrt{r}\left[(1-q^2)^l + (q^2+\gamma^2)(\gamma^2-q^2)^{l-1}\right]} \qquad (j = 0, 1) ,$$

$$F_0 = P_{l-1/2}\left[1+\frac{(x_0-x)^2+2(x_0-x)(r+r_0)\sqrt{1-q^2}}{2rr_0(1-q^2)}\right],$$

$$F_1 = P_{l-1/2}\left[1+\frac{(x_1-x)^2+2(x_1-x)\left(r\sqrt{\gamma^2-q^2}+r_0\sqrt{1-q^2}\right)}{2rr_0\sqrt{(1-q^2)(\gamma^2-q^2)}}\right],$$

$$x_0 = \tau+r_0+qz-(r+r_0)\sqrt{1-q^2}\,,$$

$$x_1 = \tau+r_0+qz-r\sqrt{\gamma^2-q^2}-r_0\sqrt{1-q^2}\,, \qquad (5.2.17)$$

where $P_{l-1/2}(x)$ is the Legendre function of the first kind; the derivative $f'(x)$ in the expressions for φ and ψ_1 is assumed to be a generalized derivative (the lower limit -0 is introduced on purpose to take into account the jump in the function $f(x)$ at the point $x=0$). In formulating equations (5.2.17), use has been made of the operational calculus formula for the transform $K_l(ps)K_l(p\alpha)\exp\{p(s+\alpha)\}$:

$$K_l(ps)K_l(p\alpha)e^{p(s+\alpha)} \doteqdot \frac{\pi}{2\sqrt{s\alpha}}P_{l-1/2}\left[\frac{(\tau+2s)(\tau+2\alpha)}{2s\alpha}-1\right]H(\tau) \qquad (5.2.18)$$

(according to the results of [5.13]) where $|\arg s|<\pi$, $|\arg\alpha|<\pi$; l represents any complex number.

5.2.2 Analysis of the Obtained Solution

From equations (5.2.17) at $\max x_0(q) < 0$ ($\max x_0(q) = x_0(q_0) = \tau + r_0 - \sqrt{z^2+(r+r_0)^2}$) we find $\varphi=\varphi_a$, whereas at $\max x_1(q)<0$ ($\max x_1(q)=x_1(q_1)=\tau+r_0-R_1$) we have $\psi_1=\psi_2=0$. Hence, equations $\tau+r_0=\sqrt{z^2+(r+r_0)^2}$ and $\tau+r_0=R_1$ represent frontal surfaces of longitudinal and transverse diffraction waves, respectively. Patterns of disturbance domains at the section $z=\text{const}$ ($\tau>\sqrt{z^2+r_0^2}-r_0$) with and without a shadow are shown in Fig. 5.1a and b, respectively; wave fronts 1 to 5 are given by the following equations:

1) $\tau+r_0=R(\theta)\,,$

2) $\tau+r_0=R(-\theta)\,,$

3) $\tau+r_0=\sqrt{z^2+(r_0+r)^2}\,,$

4) $\tau+r_0=R_1\,,$

5) $\tau+r_0=R(2\pi/l-\theta)\,,$

$$R(\theta)\equiv R=\left[z^2+r^2+r_0^2-2rr_0\cos(\theta-\theta_0)\right]^{1/2}.$$

Note once again that presence of the above edge condition causes, first of all, a principal difference between the solution to the elastic problem and that to the corresponding acoustic problem: not only a supplementary longitudinal wave $\varphi-\varphi_a$ appears but also diffraction waves (of the two types ψ_1 and ψ_2) differing with respect to the polarization directions of the displacement vector. The additional disturbances $\varphi - \varphi_a$, ψ_1, and ψ_2 describe elasticity effects.

Let us consider the incident wave (5.2.1) in detail for $f(\tau) = -\tau^2 H(\tau)/2$. The stress at the front of such a compression wave undergoes a jump:

$$[\sigma_{nn}] = \sigma_{nn}^+ - \sigma_{nn}^- = -(\lambda + 2\mu) r_0 (\tau + r_0)^{-1} \ .$$

As $r \to \infty$, this incident wave becomes a plane wave whose potential is

$$\varphi_0 = -[\tau + r\cos(\theta - \theta_0)]^2 H[\tau + r\cos(\theta - \theta_0)]/2 \ .$$

To investigate the solution, it is sufficient to consider conditions of $0 < \theta_0 < \pi/l - \pi$ (Fig. 5.1,a); within this situation, all presumable disturbed motion areas are present:

- the area of a reflected wave ($0 < \theta < \pi - \theta_0$, $\sqrt{(\tau + r_0)^2 - z^2} - r_0 < r$, $\tau + r_0 - R(-\theta) > 0$),
- the diffraction area ($r < \sqrt{(\tau + r_0)^2 - z^2} - r_0$), and
- the shadow area ($\pi + \theta_0 < \theta < \pi/l$).

In going through the frontal surface of the reflection wave 2 (Fig. 5.1a) the potential of the reflected wave

$$\varphi = -r_0 [2R(-0)]^{-1} [\tau + r_0 - R(-0)]^2 H[\tau + r_0 - R(-0)]$$

is added to the potential of the incident wave. There appears a jump in the normal stress at the front of the reflected wave: $[\sigma_{nn}] = -(\lambda + 2\mu) r_0 (\tau + r_0)^{-1}$ (this is due to discontinuity in the function $\partial^2 \varphi_a / \partial n^2$).

In going through the front of the longitudinal diffraction wave 3, the strain field (and stresses) is continuous, whereas the derivative of the strains with respect to the normal to the front, $\partial \varepsilon_{nn}/\partial n$, undergoes an infinite discontinuity (since the derivative $\partial^3 \varphi / \partial n^3$ undergoes such a discontinuity). Note that this derivative is finite if we arrive at the front out of the region ahead of the front, whereas it has a singularity of the order of $\varepsilon^{-1/2}$ if we approach to the front from behind. In this case, both the acoustic solution φ_a and the supplementary elastic term have singularities of the order of $\varepsilon^{-1/2}$.

At the front of the transverse wave 4, strains are continuous, whereas normal derivatives of strains, $\partial \varepsilon_{n\theta}/\partial n$ and $\partial \varepsilon_{n\nu}/\partial n$, undergo an infinite discontinuity (ν is measured along the line of intersection of the transverse wave front and the plane where θ = const). This is because of discontinuity of the derivatives $\partial^3 \psi_1 / \partial n^3$ and $\partial^4 \psi_2 / \partial n^4$, respectively, at this front. Note that these derivatives are finite on approach from outside ($\tau + r_0 < R_1$), whereas they have a singularity of the order of $\varepsilon^{-1/2}$ on approach from inside ($\tau + r_0 > R_1$).

It can be seen from the investigations undertaken that the supplementary e-lastic part of the solution is comparable in value with the diffraction part of the acoustic solution not only in the vicinity of the edge but also near the front of the diffraction wave 3, so, the elastic problem differs essentially from the acoustic one not only near the edge but, in general, over the whole of the diffraction domain

$r < \sqrt{(\tau + r_0)^2 - z^2} - r_0$. Some numerical results obtained with equations (5.2.17) are presented in [5.23].

Similarly, diffraction problems for spherical transverse waves can be considered [5.9, 16].

5.3 Diffraction of an Arbitrary Incident Plane Elastic Wave by a Rigid Smooth Wedge

In this Section, we shall consider problems of diffraction of plane elastic waves approaching a wedge at an arbitrary angle to its edge [5.14, 16].

For solving these three-dimensional problems, solutions to the corresponding problems with spherical waves can be used. First, let us demonstrate how to derive solutions to the three-dimensional problems on a plane wave diffraction from solutions to spherical wave diffraction problems.

Assume a source of the spherical wave

$$\varphi_n = r_0 f(\tau/\gamma_n + r_0 - R)/R \qquad (n = 0, 1, 2) \tag{5.3.1}$$

to be applied at a point $(r_0, \theta_0, 0)$ at a time moment $\tau = r_0\gamma_n$. Here, we take $f(\tau) \equiv 0$ if $\tau < 0$,

$$R = \sqrt{z^2 + r^2 + r_0^2 - 2rr_0 \cos(\theta - \theta_0)}\ ,$$

$$\varphi_n = \begin{cases} \varphi_0, & n = 0, \\ \psi_{10}, & n = 1, \\ \psi_{20}, & n = 2, \end{cases} \qquad \gamma_n = \begin{cases} 1, & n = 0, \\ \gamma, & n = 1, 2 \end{cases}$$

(φ_0, ψ_{10}, and ψ_{20} represent the incident waves to be described by potentials φ, ψ_1, and ψ_2, respectively).

Use will be made of new independent variables z', τ'

$$\begin{aligned} &z' = z + \rho_0 \cos\alpha_0, \qquad \tau' = \tau + \gamma_n(r_0 - \rho_0) \\ &(\rho_0 = r_0/\sin\alpha_0, \quad 0 < \alpha_0 < \pi)\ . \end{aligned} \tag{5.3.2}$$

In the new, shifted cylindrical coordinate system $\{r, \theta, z'\}$ with an origin at the point O' (Fig. 5.3) the source has coordinates $(\rho_0 \sin\alpha_0, \theta_0, \rho_0 \cos\alpha_0)$. This means that ρ_0, α_0, and θ_0 are the coordinates of the source in the spherical system $\{\rho, \alpha, \theta\}$ related to the cylindrical system $\{r, \theta, z'\}$ by the formulae $z' = \rho\cos\alpha,\ r = \rho\sin\alpha$, and $\tau' = 0$ corresponds to the time moment when the incident wave (5.3.1) arrives at the point $O'(z' = 0,\ r = 0)$. If we pass now to the limit as $\rho_0 \to \infty$ (considering that $\tau', r, \theta, z', \alpha_0$, and θ_0 are fixed) then

$$\tau/\gamma_n + r_0 - R \to \tau'/\gamma_n + r\sin\alpha_0 \cos(\theta - \theta_0) + z' \cos\alpha_0\ ,$$

since

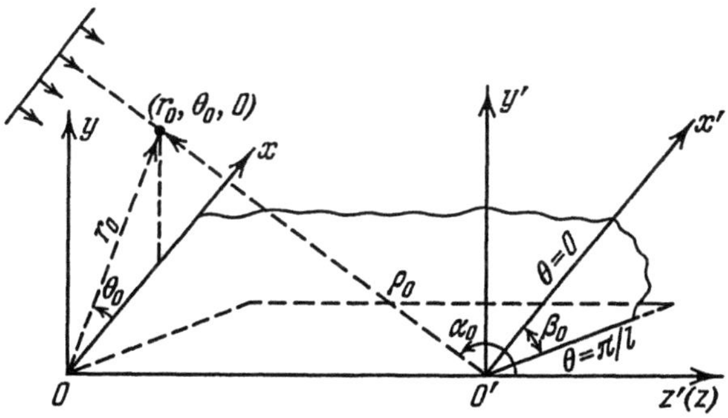

Fig. 5.3. Wave–wedge interaction: a spherical elastic wave from the point $\{r = r_0,\ \theta = \theta_0,\ z = 0\}$ and a plane elastic wave

$$R = \rho_0 - r \sin\alpha_0 \cos(\theta - \theta_0) - z' \cos\alpha_0 + o(1) .$$

With this,

$$\varphi_n / \sin\alpha_0 \to f(\tau'/\gamma_n + r \sin\alpha_0 \cos(\theta - \theta_0) + z' \cos\alpha_0) .$$

In so doing, we find ourselves dealing with a plane wave which went out of an infinite point whose spherical coordinates are $\{\alpha_0, \theta_0\}$.

Thus, a passage to the limit as $\rho_0 \to \infty$ in diffraction problems for spherical waves on a wedge offers solutions (if these exist) to three-dimensional problems for the corresponding plane waves.

As an example, let us consider the case of a longitudinal wave represented by a Heaviside function

$$\varphi_0 = H(\tau' + r \sin\alpha_0 \cos(\theta - \theta_0) + z' \cos\alpha_0) . \tag{5.3.3}$$

A solution of this problem will be a result of properly applying the solution of the diffraction problem for the spherical wave (5.2.1) at $f(x) = H(x)$ as stated in the previous Section. Indeed, by means of equation (5.2.17) (at $f(x) = H(x)$, $z \geq 0$) the solution for the spherical wave $\varphi_0 = r_0 H(\tau + r_0 - R)/R$ can, in the variables τ, r, z, θ at $1/2 \leq 1$, be written in the following form:

$$\varphi = \varphi_a + \cos(l\theta) \int\limits_{L_0} A(1 - q^2)^{l/2-1/4} T_0(x_0, q)\, dq ,$$

$$\psi_1 = \gamma^2 \sin(l\theta) \int\limits_{L_1} A(\gamma^2 - q^2)^{l/2-5/4} T_1(x_1, q)\, dq ,$$

$$\psi_2 = \cos(l\theta) \int\limits_{L_1} A(\gamma^2 - q^2)^{l/2-5/4} q\, dq \int\limits_0^{x_1} T_1(x, q)\, dx ,$$

$$T_0(x,q) = P_{l-1/2}\left[1 + \frac{x^2 + 2x(r+r_0)\sqrt{1-q^2}}{2rr_0(1-q^2)}\right] H(x)\,,$$

$$T_1(x,q) = P_{l-1/2}\left[1 + \frac{x^2 + 2x\left(r\sqrt{\gamma^2-q^2} + r_0\sqrt{1-q^2}\right)}{2rr_0\sqrt{(1-q^2)(\gamma^2-q^2)}}\right] H(x)\,,$$

$$A = \frac{4l\sqrt{r_0}\sin(\pi l)\cos(l\theta_0)(1-q^2)^{l/2-1/4}H(1-l)}{\pi i\sqrt{r}\left[(1-q^2)^l + (q^2+\gamma^2)(\gamma^2-q^2)^{l-1}\right]}\,,$$

$$x_0 = \tau + r_0 + qz - (r+r_0)\sqrt{1-q^2}\,,$$

$$x_1 = \tau + r_0 + qz - r\sqrt{\gamma^2-q^2} - r_0\sqrt{1-q^2}\,. \tag{5.3.4}$$

The integration paths L_0 and L_1 are defined by equations (5.2.15, 16).

Consider the contour L_1 in more detail because analogous results for L_0 can be derived from results for L_1 through substituting formally a unity for the parameter γ.

It was mentioned in the previous Section that the maximum value of the function $x_1(\alpha)$ over L_1 is at the point q_1 which has been shown to be a saddle point and be determined from the following equation,

$$f_1'(q) = 0 \qquad \left(f_1(q) = qz - r\sqrt{\gamma^2-q^2} - r_0\sqrt{1-q^2}\right)\,. \tag{5.3.5}$$

From (5.3.5) it follows that this point is on the real axis within the interval $(-1, 0)$, and the function $f_1(q)$ decreases monotonical from $f_1(q)$ down to $-\infty$ at moving away from q_1 along L_1 in the upper (and/or lower) half-plane. For this reason and due to features of the Heaviside function presented in integrands for ψ_1 and ψ_2, it is permissible to consider the curve L_1 finished at conjugate points $q_1^{(1,2)}$ which satisfy the equation $x_1 = 0$, i.e., equation

$$f_1(q) = -\tau - r_0\,. \tag{5.3.6}$$

Now, let us use new variables τ', z' and parameters ρ_0, α_0 specified by equations (5.3.2). Without loss of generality in the latter, one can adopt $\pi/2 \le \alpha_0 < \pi$ in order to satisfy the condition $z \ge 0$ essential for equations (5.3.4) when $\rho_0 \to \infty$. It is permissible to seek a solution to equations (5.3.5, 6), as $\rho_0 \to \infty$, by means of the expressions,

$$q_1 = a_0 + a_1\rho_0^{-\nu} + o(\rho_0^{-\nu})\,, \tag{5.3.7}$$

$$q_1^{(1,2)} = b_0 + b_1\rho_0^{-\varepsilon} + o(\rho_0^{-\varepsilon})\,. \tag{5.3.8}$$

Substituting (5.3.7) and (5.3.8) into (5.3.5) and (5.3.6), respectively, we obtain $\nu = 1$, $\varepsilon = 1/2$ and thereafter determine the coefficients a_n and b_n.

Similar results for analogous points q_0 and $q_0^{(1,2)}$ on the contour L_0 can be found through substituting unity for γ in relations for q_1 and $q_1^{(1,2)}$.

Thus, we may bring results for L_0 and L_1 together to obtain (when $\rho_0 \to \infty$)

$$q_0 = \cos\alpha_0 + \mathrm{O}\left(\rho_0^{-1}\right) , \quad q_1 = \cos\alpha_0 + \mathrm{O}\left(\rho_0^{-1}\right) ,$$

$$q_0^{(1,2)} = \cos\alpha_0 \pm \mathrm{i}\sqrt{\frac{2}{\rho_0}} \sin\alpha_0 \sqrt{\tau' + z'\cos\alpha_0 - r\sin\alpha_0} + \mathrm{O}\left(\rho_0^{-1}\right) ,$$

$$q_1^{(1,2)} = \cos\alpha_0 \pm \mathrm{i}\sqrt{\frac{2}{\rho_0}} \sin\alpha_0 \sqrt{\tau' + z'\cos\alpha_0 - r\sqrt{\gamma^2 - \cos^2\alpha_0}} + \mathrm{O}\left(\rho_0^{-1}\right) .$$

In (5.3.4), we can now use new variables introduced by (5.3.2) (where $n = 0$ and $\pi/2 \le \alpha_0 < \pi$). Change of the variable: $s = \sqrt{\rho_0}(q - \cos\alpha_0)/(\sqrt{2}\sin\alpha_0)$ within the integrals in q and expansion of the integrands in series, as $\rho_0 \to \infty$, reduce equations (5.3.4) to the form

$$\varphi = \varphi_a + \cos(l\theta)(\sin\alpha_0)^{l-1/2}\Phi \int\limits_{\bar{\eta}_0}^{\eta_0} P_{l-1/2}(y_0)ds + \mathrm{O}\left(\rho_0^{-3/2}\right) ,$$

$$\psi_1 = \gamma^2 \sin(l\theta)\left(\gamma^2 - \cos^2\alpha_0\right)^{l/2-5/4} \Phi \int\limits_{\bar{\eta}_1}^{\eta_1} P_{l-1/2}(y_1)ds + \mathrm{O}\left(\rho_0^{-3/2}\right) ,$$

$$\psi_2 = \cos(l\theta)\left(\gamma^2 - \cos^2\alpha_0\right)^{l/2-5/4} \Phi \int\limits_0^{\tau'} d\tau' \int\limits_{\bar{\eta}_1}^{\eta_1} P_{l-1/2}(y_1)ds + \mathrm{O}\left(\rho_0^{-3/2}\right) ,$$

$$\Phi = \frac{4\sqrt{2}l\sin(\pi l)\cos(l\theta_0)(\sin\alpha_0)^{l+1}H(1-l)}{\pi\mathrm{i}\sqrt{r}\left[\sin^{2l}\alpha_0 + \left(\gamma^2 + \cos^2\alpha_0\right)\left(\gamma^2 - \cos^2\alpha_0\right)^{l-1}\right]} ,$$

$$y_n = \frac{s^2}{r\sqrt{\gamma^2 - \cos^2\alpha_0}} + \frac{z'\cos\alpha_0 + \tau'}{r\sqrt{\gamma_n^2 - \cos^2\alpha_0}} + \mathrm{O}\left(\rho_0^{-1}\right) ,$$

$$\eta_n = \mathrm{i}\left[\tau' + z'\cos\alpha_0 - r\sqrt{\gamma_n^2 - \cos^2\alpha_0}\right]^{1/2} + \mathrm{O}\left(\rho_0^{-1/2}\right) ,$$

$$\gamma_n = \begin{cases} 1, & n = 0, \\ \gamma, & n = 1, \end{cases} \tag{5.3.9}$$

where $\bar{\eta}_n$ is a complex conjugate to η_n.

According to equation (29) in Section 3.7 of [5.24], we find when $\rho_0 \to \infty$:

$$\int\limits_{\bar{\eta}_n}^{\eta_n} P_{l-1/2}(y_n)ds = \left(r\sqrt{\gamma_n^2 - \cos^2\alpha_0}\right)^{1/2} \int\limits_{-\nu_n \mathrm{i}}^{\nu_n \mathrm{i}} P_{l-1/2}\left(s^2 + \nu_n^2 + 1\right) ds$$

$$= \mathrm{i}\left(r\sqrt{\gamma_n^2 - \cos^2\alpha_0}\right)^{1/2} \int\limits_1^{d_n} P_{l-1/2}(x)\frac{dx}{\sqrt{d_{n-x}}}$$

$$= \frac{\mathrm{i}}{l\sqrt{2}}\left(r\sqrt{\gamma_n^2 - \cos^2\alpha_0}\right)^{1/2} F(l, d_n) .$$

Here, new symbols mean

$$\nu_n = \sqrt{d_n - 1}\,, \quad d_n = \left(\tau' + z'\cos\alpha_0\right)/r\sqrt{\gamma_n^2 - \cos^2\alpha_0}\,,$$
$$F(l, d_n) = \left[\left(d_n + \sqrt{d_n^2 - 1}\right)^l - \left(d_n + \sqrt{d_n^2 - 1}\right)^{-l}\right] H(d_n - 1)\,.$$

After multiplication by $1/\sin\alpha_0$ on both sides of expressions for φ, ψ_1, and ψ_2 in (5.3.9) and after a subsequent passage to the limit ($\rho_0 \to \infty$), a closed-form solution for the incident wave (5.3.3) appears:

$$\varphi(r, \theta, z', \tau') = \varphi_a^p + B\cos(l\theta)F(l, d_0)\,,$$
$$\psi_1(r, \theta, z', \tau') = B\beta_0^{l-2}\left(\beta_0^2 + \cot^2\alpha_0\right)\sin(l\theta)F\left(l, d_0/\beta_0\right)\,,$$
$$\psi_2(r, \theta, z', \tau') = \frac{r}{2}\beta_0^{l-1}B\cos(l\theta)\cot\alpha_0\left[\frac{F(l+1, d_0/\beta_0)}{l+1} - \frac{F(l-1, d_0/\beta_0)}{l-1}\right]\,,$$
$$\beta_0 = \frac{\sqrt{\gamma^2 - \cos^2\alpha_0}}{\sin\alpha_0}\,, \quad B = \frac{4\cos(l\theta_0)\sin(\pi l)H(1-l)}{\pi\left[1 + \beta_0^{2l} + 2\beta_0^{2l-2}\cot^2\alpha_0\right]}\,. \tag{5.3.10}$$

Here, φ_a^p is the corresponding solution to the acoustic problem and, according to [5.22], has the following form:

$$\varphi_a^p = U\left(\tau' + z'\cos\alpha_0,\ r\sin\alpha_0,\ \theta\right)\,.$$

The right-hand-side function $U(\tau, r, \theta)$ represents a solution to the acoustic problem of diffraction of a plane wave $U_0 = H[\tau + r\cos(\theta - \theta_0)]$ and is to be determined using equation (5.2.13) at $1/2 \le l < 1$.

The solution (5.3.10) has been obtained for $\pi/2 \le \alpha_0 < \pi$ but remains valid at any $0 < \alpha_0 < \pi$) which is readily verified by means of change of variables: $\alpha_1 = \pi - \alpha_0$, $z_1 = -z'$. At $\alpha_0 = \pi/2$, equations (5.3.10) give the known two-dimensional solution found by *Kostrov* [5.4]. Notice that some inaccuracies need to be corrected in the latter: in formulae (2.9) of [5.4] a lost multiplier 1/2 must be present in the terms being supplementary to an acoustic solution; in an expression for φ, one of the multipliers $\cos k\theta_0$ should be replaced by $\cos k\theta$, and in the expression for ψ the multiplier $(b/a)^k$ must be replaced by the term $(b/a)^{-k}$. Finally,

$$\varphi = \varphi_a^p + \frac{4\sin(l\pi)\cos(l\theta)\cos(l\theta_0)}{\pi(1+\gamma^{2l})}\left[N\left(\frac{\tau'}{r}\right) - \frac{1}{N(\tau'/r)}\right]H(\tau' - r)\,,$$
$$\psi_1 = \frac{4\gamma^l\sin(l\pi)\cos(l\theta_0)\sin(l\theta)}{\pi(1+\gamma^{2l})}\left[N\left(\frac{\tau'}{r\gamma}\right) - \frac{1}{N(\tau'/r\gamma)}\right]H(\tau' - r\gamma)\,, \tag{5.3.11}$$
$$N(x) = \left[x + \sqrt{x^2 - 1}\right]^l\,.$$

The three-dimensional problem of diffraction of the plane longitudinal wave considered herein has also been studied in [5.15] using another method in which a coordinate system is moving along the edge at the speed of the front of an incident wave (unfortunately, there are some inaccuracies in the final formulae of

[5.15]). Problems on diffraction of plane transverse waves approaching a wedge at an angle to its edge were considered by Sharipova [5.16] using the technique employed above.

To conclude this Section, let us consider briefly how a plane longitudinal wave $\varphi = \bar{\varphi}_0$ carrying a uniform compressional stress field behind the front interacts with a smooth wedge with the apex angle of $0 \leq \beta < \pi$:

$$\bar{\varphi}_0 = -\frac{x^2 H(x)}{2(\lambda+2\mu)} \quad (x = \tau + z\cos\alpha_0 + r\sin\alpha_0\cos(\theta-\theta_0),\ \tau = c_1 t)\ . \tag{5.3.12}$$

In this case, displacement potentials $\bar{\varphi}(r,\theta,z,\tau)$, $\bar{\psi}_1(r,\theta,z,\tau)$, and $\bar{\psi}_2(r,\theta,z,\tau)$ result from the corresponding expressions in (5.3.10) (where $\tau' = \tau$ and $z' = z$) as the convolution product:

$$\begin{aligned}
\bar{\varphi}(r,\theta,z,\tau) &= -\frac{r^2\sin^2\alpha_0}{\lambda+2\mu}\int_{-\infty}^{y}(y-\eta)\varphi(\theta,\eta)d\eta\ ,\\
\bar{\psi}_1(r,\theta,z,\tau) &= -\frac{r^2\left(\gamma^2-\cos^2\alpha_0\right)}{(\lambda+2\mu)\beta_0^2}\int_{-\infty}^{y}(y-\eta)\psi_1\left(\theta,\eta/\beta_0\right)d\eta\ ,\\
\bar{\psi}_2(r,\theta,z,\tau) &= -\frac{r^3\left(\gamma^2-\cos^2\alpha_0\right)}{(\lambda+2\mu)\beta_0^2}\int_{-\infty}^{y}(y-\eta)\psi_2\left(\theta,\eta/\beta_0\right)d\eta\ .
\end{aligned} \tag{5.3.13}$$

In (5.3.13) the following notation is introduced:

$$\begin{aligned}
&\varphi(\theta,y) \equiv \varphi(r,\theta,z,\tau)\ , \quad \psi_1(\theta,y/\beta_0) \equiv \psi_1(r,\theta,z,\tau)\ ,\\
&\psi_2(\theta,y/\beta_0) \equiv \psi_2(r,\theta,z,\tau)/r\ , \quad y = (\tau + z\cos\alpha_0)/(r\sin\alpha_0)\ .
\end{aligned}$$

Lower limits of integration with respect to η in (5.3.13) will indeed be finite; this is a consequence of equations (5.3.10). Then, using (5.3.10) and (5.3.13), obtain the following expression for the component $\sigma_{\theta\theta}$ of the stress tensor at $1/2 \leq l < 1$:

$$\begin{aligned}
\sigma_{\theta\theta} &= \lambda\frac{\partial^2\bar{\varphi}}{\partial\tau^2}\\
&+\frac{2\mu}{r}\left[\frac{1}{r}\frac{\partial^2\bar{\varphi}}{\partial\theta^2}+\frac{\partial\bar{\varphi}}{\partial r}-\frac{\partial^2\bar{\psi}_1}{\partial r\partial\theta}+\frac{1}{r}\frac{\partial\bar{\psi}_1}{\partial\theta}+\frac{\partial^2\bar{\psi}_2}{\partial r\partial z}+\frac{1}{r}\frac{\partial^3\bar{\psi}_2}{\partial\theta^2\partial z}\right]=\sum_{m=0}^{3}\sigma_{\theta\theta}^{(m)}\ ,
\end{aligned}$$

$$\sigma_{\theta\theta}^{(0)} = -\frac{\lambda+2\mu(1-y^2)\sin^2\alpha_0}{\lambda+2\mu}\varphi_{\mathrm{a}}^{p}(\theta,y) - \frac{4\mu\sin^2\alpha_0}{\lambda+2\mu}\int_{-\infty}^{y}\varphi_{\mathrm{a}}^{p}(\theta,\eta)\eta d\eta\ ,$$

$$\begin{aligned}
\sigma_{\theta\theta}^{(1)} &= \frac{B\cos(l\theta)}{\lambda+2\mu}\left\{-F(l,y)\left[\lambda+2\mu(1-y^2)\sin^2\alpha_0\right]\right.\\
&\left.+2\mu\sin\alpha_0\left[\Phi_1(l,y)-\Phi_2(l,y)\right]\right\}\ ,
\end{aligned}$$

$$\sigma_{\theta\theta}^{(2)} = \frac{\mu}{\lambda+2\mu} B\cos(l\theta)\beta_0^l l\left(\beta_0^2+\cot^2\alpha_0\right)\sin^2\alpha_0 \times\left[\Phi_1(l,y/\beta_0)-\Phi_2(l,y/\beta_0)\right] ,$$

$$\sigma_{\theta\theta}^{(3)} = -\frac{\mu}{\lambda+2\mu} B\cos(l\theta)\beta_0^l\cos^2\alpha_0\left[(2-l^2)\Phi_1(l,y/\beta_0)-\Phi_2(l,y/\beta_0)\right] ,$$

$$\Phi_1(l,y) = \frac{F(l+2,y)}{2(l+1)(l+2)}+\frac{F(l-2,y)}{2(l-1)(l-2)}-\frac{F(l,y)}{l^2-1} ,$$

$$\Phi_2(l,y) = y\left[\frac{F(l+1,y)}{l+1}-\frac{F(l-1,y)}{l-1} ,\right]$$

$$F(l,y) = \left[\left(y+\sqrt{y^2-1}\right)^l-\left(y+\sqrt{y^2-1}\right)^{-l}\right]H(y-1) , \tag{5.3.14}$$

$\varphi_{\mathrm{a}}^0(\theta,y)$ is taken from (5.3.13) (at $\tau'=\tau$ and $z'=z$).

If the wedge occupies a quarter of the space ($l=2/3$) and $\theta_0=\pi/2$, then equations (5.3.14) for stresses $\sigma_{\theta\theta}$ on faces $\theta=0$, $\theta=3\pi/2$ can be written in elementary functions. For example, in a particular case of $\lambda=2\mu$ the component $\sigma_{\theta\theta}^{(0)}$ takes the form

$$\sigma_{\theta\theta}^{(0)} = -\left(1+\sin^2\alpha_0\right)\left[1-\frac{H(y-1)}{\pi}\tan^{-1}\frac{\sqrt{3}(1-\omega^{2/3})}{1+\omega^{2/3}}\right] + \frac{\sqrt{3}}{8\pi}\sin^2\alpha_0\left[\frac{1}{2}(\omega^{4/3}-\omega^{-4/3})+\omega^{2/3}-\omega^{-2/3}\right]H(y-1) \quad (\theta=0) ,$$

$$\sigma_{\theta\theta}^{(0)} = \left\{-\left(1+\sin^2\alpha_0\right)\left[1-\frac{1}{\pi}\tan^{-1}\frac{\sqrt{3}(1+\omega^{2/3})}{1-\omega^{2/3}}\right] + \frac{\sin^2\alpha_0}{2}\left[1-\frac{\sqrt{3}}{4\pi}\left(\frac{\omega^{4/3}}{2}-\frac{\omega^{-4/3}}{2}-\omega^{2/3}+\omega^{-2/3} + \frac{8}{\sqrt{3}}\tan^{-1}\frac{1+2\omega^{2/3}}{\sqrt{3}}-\frac{8\pi}{3\sqrt{3}}\right)\right]\right\}H(y-1) \quad (\theta=3\pi/2) ,$$

$$\omega = y-\sqrt{y^2-1} .$$

For this case ($l=2/3$, $\theta_0=\pi/2$, $\lambda=2\mu$), Fig. 5.4a, b gives distributions of the stress $-\sigma_{\theta\theta}$ at the faces $\theta=0$, $\theta=3\pi/2$; the variable $1/y=r\sin\alpha_0/(\tau+z\cos\alpha_0)$ is considered as argument of these functions. These curves are obtained on the basis of equations (5.3.14). The numbers 1 to 6 denote curves for $\alpha_0=15°$, $30°$, $45°$, $60°, 75°, 90°$. As studies show, the resulting stress $\sigma_{\theta\theta}$ has an integrable singularity $r^{-2/3}$, although the component $\sigma_{\theta\theta}^{(0)}$ from the acoustic part of the solution behaves as $r^{-4/3}$ if $r\to 0$, and each of the components $\sigma_{\theta\theta}^{(1)}$, $\sigma_{\theta\theta}^{(2)}$, and $\sigma_{\theta\theta}^{(3)}$ (resulting from the supplementary potentials $\varphi-\varphi_{\mathrm{a}}$, Ψ_1, and Ψ_2, respectively) behaves as $r^{-8/3}$.

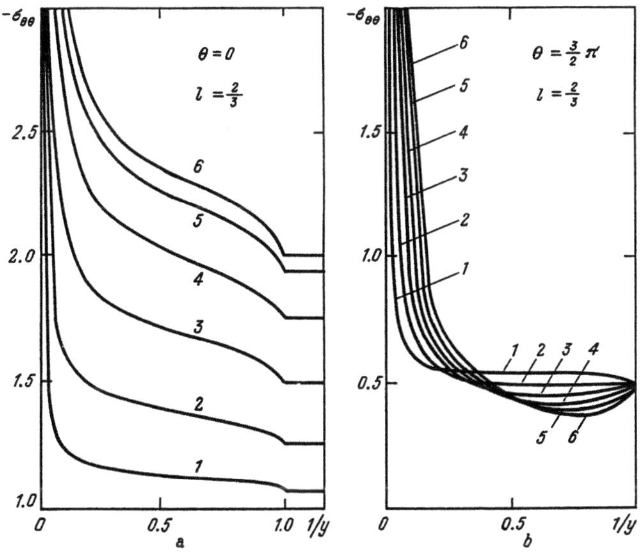

Fig. 5.4a,b. Stress ($-\sigma_{\theta\theta}$) vs. $1/y$ on a quarter of a space: plane waves at angles α_0=15, 30, 45, 60, 75, 90 degrees (curves 1 to 6, respectively): **(a)** $\theta = 0$, **(b)** $\theta = 3\pi/2$

6. Wiener–Hopf Method in Elastodynamics

The Wiener–Hopf method (the factorization method) is widely used for solving certain integral equations and boundary-value problems by means of the Laplace, Fourier, Mellin integral transformations and a number of others. This method was first proposed by *Wiener* and *Hopf* in [6.1] to solve homogeneous integral equations with the kernel depending on a difference of arguments

$$u(x) = \int_0^\infty u(\xi)k(x-\xi)d\xi \qquad (x > 0) .$$

Subsequently, a significant contribution in developing this method has been made by *Fok* [6.2, 3] who studied a non-homogeneous equation

$$u(x) = \int_0^\infty u(\xi)k(x-\xi)d\xi + f(x) \qquad (x > 0) .$$

A detailed discussion of the Wiener–Hopf method is presented in [6.4, 5].

6.1 Problems with a Stationary Boundary

To solve the elastodynamics problems, the Wiener–Hopf method was, for the first time, applied by *Maue* [6.6, 7] to (i) the harmonic problem of diffraction at a semi-infinite cut with free edges and (ii) the study of waves occurring at an instantaneous appearance of a semi-infinite crack. In these problems, we shall consider mixed boundary conditions specified over two semi-infinite intervals, and one boundary condition imposed over the whole of the infinite interval. The boundary point separating the mixed boundary conditions is assumed to be fixed. Flitman was the first to apply the Wiener–Hopf method (in [6.8, 9]) to the problems where mixed boundary conditions are specified over three fixed parts of the infinite interval and one condition is universal for all infinite interval. Below, we illustrate the use of this technique in a two-dimensional problem [6.8] of indentation of a smooth punch into an elastic medium.

6.1.1 A Semi-Infinite Punch

To get a better understanding of the principal ideas of the Wiener–Hopf method as applied to solving dynamic problems, we shall first consider the situation with a semi-infinite punch. Assume that, at instance $t = 0$, a smooth punch begins its indentation into an elastic half-plane $y \geq 0$, at a semi-infinite interval $x \geq 0$ of its boundary. Let the law of motion of this punch be specified by equation $u_y = f_0(x,t)$, $t > 0$, where $f_0(x,t)$ is assumed to be a limited function with a finite number of discontinuity lines at $x \geq 0$, $t \geq 0$. The medium had no displacements before the motion of the punch. Thus, to determine the displacement vector $\boldsymbol{u}$, we shall work with the following boundary and initial conditions:

$$\sigma_{xy} = 0 \qquad (-\infty < x < \infty,\ y = 0)\ , \tag{6.1.1}$$

$$\sigma_{yy} = 0 \qquad (x < 0,\ y = 0)\ , \tag{6.1.2}$$

$$u_y = f_0(x,t) \qquad (x \geq 0,\ y = 0)\ , \tag{6.1.3}$$

$$\boldsymbol{u} = \partial \boldsymbol{u}/\partial t = 0 \qquad (t = 0)\ . \tag{6.1.4}$$

In addition, to ensure the uniqueness of the solution it is necessary to restrict the behavior of the displacement vector in the vicinity of the edge of the punch ($x = 0$):

$$\boldsymbol{u}(x,y,t) = \boldsymbol{u}_0(t) + \mathrm{O}(r^{\varepsilon})\ , \quad \varepsilon > 0\ , \quad r \to 0 \qquad \left(r = \sqrt{x^2+y^2}\right)\ . \tag{6.1.5}$$

The system of conditions (6.1.1–5) and the Lamé equations (2.1.8) (at $\boldsymbol{f} \equiv 0$) define completely a solution to our problem.

Prior to solving this system, let us change the homogeneous boundary condition (6.1.2) to a more general non-homogeneous one:

$$\sigma_{yy} = F_0(x,t) \qquad (x < 0,\ y = 0)\ , \tag{6.1.6}$$

where $F_0(x,t)$ is assumed to be equal to zero at any $x < -c_1 t - k (k \geq 0)$ and limited everywhere except for the point $x = 0$; as an argument approaches this point, the function F_0 must, according to (6.1.5), satisfy the constraint

$$F_0(x,t) = \mathrm{O}\left(x^{-1+\varepsilon}\right)\ , \qquad \varepsilon > 0, \quad x \to -0\ . \tag{6.1.7}$$

The boundary condition (6.1.6) will serve us in future, in solving dynamic problems involving punches of finite width ($0 \leq x < 2l$). A solution to the problem with the semi-infinite punch ($0 \leq x < \infty$) appears if we assign $F_0\ (x,t) \equiv 0$ in the solution to the problem with conditions (6.1.1, 3–7). If use is made of the representation

$$\boldsymbol{u} = \operatorname{grad} \varphi + \operatorname{rot}\,(\psi \boldsymbol{e}_3)\ , \tag{6.1.8}$$

the Lamé dynamic equations with conditions (6.1.1, 3, 4, 6, 7) reduce to the following set:

$$\Delta\varphi = \frac{\partial^2 \varphi}{\partial \tau^2}, \quad \Delta\psi = \gamma^2 \frac{\partial^2 \psi}{\partial \tau^2} \quad \left(\tau = c_1 t,\ \gamma = \frac{c_1}{c_2},\ \Delta \equiv \frac{\partial^2}{\partial x^2} + \frac{\partial^2}{\partial y^2}\right), \tag{6.1.9}$$

$$2\frac{\partial^2 \varphi}{\partial x \partial y} + \frac{\partial^2 \psi}{\partial y^2} - \frac{\partial^2 \psi}{\partial x^2} = 0 \qquad (y = 0,\ -\infty < x < \infty)\,, \tag{6.1.10}$$

$$(\gamma^2 - 2)\Delta\varphi + 2\left(\frac{\partial^2 \varphi}{\partial y^2} - \frac{\partial^2 \psi}{\partial x \partial y}\right) = \frac{1}{\mu} F(x,\tau) \qquad (y = 0,\ x < 0)\,, \tag{6.1.11}$$

$$\frac{\partial \varphi}{\partial y} - \frac{\partial \psi}{\partial x} = f(x,\tau) \qquad (y = 0,\ x \geq 0)\,, \tag{6.1.12}$$

$$\varphi = \frac{\partial \varphi}{\partial \tau} = \psi = \frac{\partial \psi}{\partial \tau} = 0 \qquad (\tau = 0)\,, \tag{6.1.13}$$

where $f(x,\tau) \equiv f_0(x,t)$, $F(x,\tau) \equiv F_0(x,t)$. The set of equations (6.1.9–13) becomes complete utilizing the condition (6.1.5) with the help of the representation (6.1.8).

The Laplace transformation with respect to τ and the two-sided Laplace transformation with respect to x applied in sequence to (i) equations (6.1.9), (ii) the boundary condition (6.1.10), and (iii) relations

$$\frac{1}{\mu}\sigma_{yy} = (\gamma^2 - 2)\Delta\varphi + 2\left(\frac{\partial^2 \varphi}{\partial y^2} - \frac{\partial^2 \psi}{\partial x \partial y}\right), \qquad u_y = \frac{\partial \varphi}{\partial y} - \frac{\partial \psi}{\partial x}\,,$$

reduce (6.1.9–13) to the following set of equations and boundary conditions for $\bar{\varphi}^*$ and $\bar{\psi}^*$:

$$\frac{d^2 \bar{\varphi}^*}{dy^2} = (p^2 - s^2)\bar{\varphi}^*, \qquad \frac{d^2 \bar{\psi}^*}{dy^2} = (\gamma^2 p^2 - s^2)\bar{\psi}^*\,, \tag{6.1.14}$$

$$2s\frac{d\bar{\varphi}^*}{dy} + \frac{d^2 \bar{\psi}^*}{dy^2} - s^2 \bar{\psi}^* = 0 \qquad (y = 0)\,, \tag{6.1.15}$$

$$\bar{\sigma}^+_{yy} + \bar{\sigma}^-_{yy} = (\gamma^2 - 2)p^2 \bar{\varphi}^* + 2\left(\frac{d^2 \bar{\varphi}^*}{dy^2} - s\frac{d\bar{\psi}^*}{dy}\right) \qquad (y = 0)\,, \tag{6.1.16}$$

$$\bar{u}^+_y + \bar{u}^-_y = \frac{d\bar{\varphi}^*}{dy} - s\bar{\psi}^* \qquad (y = 0)\,, \tag{6.1.17}$$

where the new symbols mean:

$$\bar{\sigma}^+_{yy} = \int\limits_0^\infty e^{-sx} dx \int\limits_0^\infty \frac{\sigma_{yy}}{\mu} e^{-p\tau} d\tau\,, \quad \bar{\sigma}^-_{yy} = \int\limits_{-\infty}^0 e^{-sx} dx \int\limits_0^\infty \frac{F(x,\tau)}{\mu} e^{-p\tau} d\tau\,,$$

$$\bar{u}^+_y = \int\limits_0^\infty e^{-sx} dx \int\limits_0^\infty f(x,\tau) e^{-p\tau} d\tau\,, \quad \bar{u}^-_y = \int\limits_{-\infty}^0 e^{-sx} dx \int\limits_0^\infty u_y(x,\tau) e^{-p\tau} d\tau\,,$$

$$\bar{\Phi} = \int\limits_0^\infty e^{-p\tau} \Phi d\tau\,, \quad \bar{\Phi}^* = \int\limits_{-\infty}^\infty e^{-sx} \bar{\Phi} dx \qquad (\Phi = \varphi,\ \psi,\ \mathrm{Re}\, p > 0)\,. \tag{6.1.18}$$

Functions $\bar{\sigma}^+_{yy}$ and $\bar{u}^-_y$ are regarded as unknown.

There is need to determine a convergence domain in the s-plane for the integrals in (6.1.18). Since all disturbances are, at least, limited when $x \to +\infty$ and are absent if $x < -\tau - k$, then

$$\bar{\bar{\Phi}} = \begin{cases} \mathrm{O}\,(1) & (x \to +\infty), \\ \mathrm{O}\,(\mathrm{e}^{px}) & (x \to -\infty)\,. \end{cases}$$

Therefore, the integral $\bar{\Phi}^*$ is convergent and analytic in the domain $0 < \mathrm{Re}\{s\} < \mathrm{Re}\{p\}$. For the same reason, functions $\bar{\sigma}^+_{yy}$ and $\bar{u}^+_y$ are analytic in s on the domain $\mathrm{Re}\{s\} > 0$, whereas $\bar{\sigma}^-_{yy}$ and $\bar{u}^-_y$ are analytic in the domain $\mathrm{Re}\{s\} < \mathrm{Re}\{p\}$. Thus, $0 < \mathrm{Re}\{s\} < \mathrm{Re}\{p\}$ for equations (6.1.14–18).

By solving equations (6.1.14) and dealing only with solutions decaying as $y \to +\infty$, we find

$$\bar{\varphi}^* = A\mathrm{e}^{-y\sqrt{p^2-s^2}}\,, \qquad \bar{\psi}^* = B\mathrm{e}^{-y\sqrt{\gamma^2p^2-s^2}}\,. \tag{6.1.19}$$

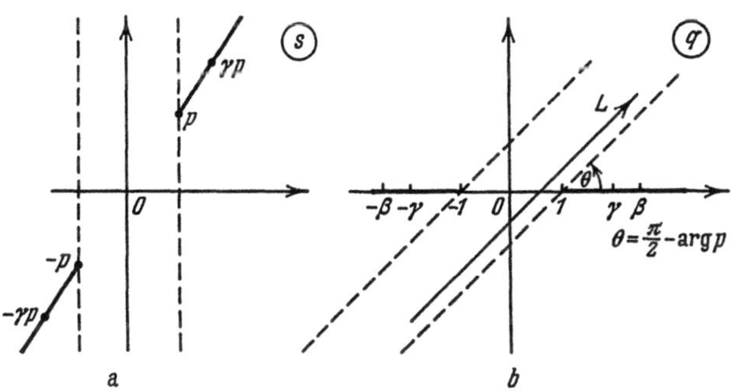

Fig. 6.1a,b. (a) Cuts in s-plane; (b) Cuts and the contour L in q-plane

Here, for specifying the branches of radicals $\sqrt{p^2-s^2}$ and $\sqrt{\gamma^2p^2-s^2}$ in the s-plane, cuts from points $s = \pm p$ and $s = \pm\gamma p$ to infinity go along two radii: $\arg s = \arg p$ and $\arg s = \pi + \arg p$ (Fig. 6.1a). It is assumed that, at $s = 0$, equalities $\sqrt{p^2-s^2} = p$ and $\sqrt{\gamma^2p^2-s^2} = \gamma p$ are observed. Then, the following inequalities take place:

$$\mathrm{Re}\sqrt{p^2-s^2} > 0, \ \mathrm{Re}\sqrt{\gamma^2p^2-s^2} > 0 \qquad (|\mathrm{Re}\, s| < \mathrm{Re}\, p)\ .$$

Substituting (6.1.19) into boundary conditions (6.1.15–17), obtain three algebraic equations for determining four unknonws A, B, $\bar{\sigma}^+_{yy}$, and $\bar{u}^-_y$:

$$\begin{aligned}
&-2s\sqrt{p^2-s^2}A + \left(\gamma^2p^2 - 2s^2\right)B = 0\,, \\
&\left(\gamma^2p^2 - 2s^2\right)A + 2s\sqrt{\gamma^2p^2-s^2}B = \bar{\sigma}^+_{yy} + \bar{\sigma}^-_{yy}\,, \\
&-\sqrt{p^2-s^2}A - sB = \bar{u}^-_y + \bar{u}^+_y\,.
\end{aligned}$$

After eliminating A and B, the following equation relates $\bar{\sigma}_{yy}^{+}$ and $\bar{u}_y^{-}$:

$$\frac{\gamma^2 p^2 \sqrt{p^2 - s^2}}{R(p,s)} \left(\bar{\sigma}_{yy}^{+} + \bar{\sigma}_{yy}^{-}\right) + \bar{u}_y^{+} + \bar{u}_y^{-} = 0 \qquad (0 < \operatorname{Re} s < \operatorname{Re} p)\,,$$
$$R(p,s) = \left(\gamma^2 p^2 - 2s^2\right)^2 + 4s^2 \sqrt{\left(\gamma^2 p^2 - s^2\right)\left(p^2 - s^2\right)}\,. \tag{6.1.20}$$

Consider the properties of the functions present in equation (6.1.20). By means of the relation (6.1.5), we find that the Laplace transforms (in τ) for functions σ_{yy} and u_y (at $y = 0$) must behave as

$$\bar{\sigma}_{yy} = \mathrm{O}(x^{-1+\varepsilon}) \quad (\varepsilon > 0)\,, \qquad \bar{u}_y = \mathrm{O}(1), \quad x \to 0 \tag{6.1.21}$$

when $x \to 0$. On the basis of Laplace transformation properties [6.5], conclude that the function $\bar{\sigma}_{yy}^{+}$ being analytic in the domain $\operatorname{Re}\{s\} > 0$ tends to zero in this domain when $s \to \infty$. Also, the function $\bar{\sigma}_{yy}^{-}$ tends to zero (when $s \to \infty$) in the domain $\operatorname{Re}\{s\} < \operatorname{Re}\{p\}$ where this function is analytic. In the same way, it may be proven that the function $\bar{u}_y^{+}$ being analytic in the domain $\operatorname{Re}\{s\} > 0$ decreases at least as s^{-1} in this domain if $s \to \infty$, and $\bar{u}_y^{-}$ is analytic in the half-plane $\operatorname{Re}\{s\} < \operatorname{Re}\{p\}$ and decreases as rapidly as s^{-1} in this half-plane when $s \to \infty$.

To clarify the properties of the function $K(p,s) \equiv \gamma^2 p^2 \sqrt{p^2 - s^2}/R(p,s)$, note that

$$R(p,s) = p^4 R(s/p) = p^4 R(q) \qquad (q = s/p)\,,$$

where $R(q) \equiv (\gamma^2 - 2q^2)^2 + 4q^2\sqrt{(\gamma^2 - q^2)(1 - q^2)}$ is the right-hand side of the Rayleigh equation ($R(q) = 0$) being a single-valued analytic function in the q-plane with cuts $[-\gamma, -1]$ and $[1, \gamma]$ and with two real roots $q_{1,2} = \pm\beta (\beta > \gamma)$. In addition, we notice the asymptotic behaviour

$$R(q) = -2q^2\left(\gamma^2 - 1\right) + \mathrm{O}(1) \qquad (q \to \infty)\,. \tag{6.1.22}$$

This leads to the conclusion that the function $K(p,s)$ is analytic in the strip $|\operatorname{Re}\{s\}| < \operatorname{Re}\{p\}$, does not have zeros here, and decreases as s^{-1} in this strip when $s \to \infty$.

By means of the Cauchy formula, we can write $K(p,s)$ (in the strip $|\operatorname{Re}\{s\}| < \operatorname{Re}\{p\}$) as

$$K(p,s) = -\frac{\gamma^2\sqrt{1-q^2}}{2p\left(\gamma^2 - 1\right)\left(q^2 - \beta^2\right)} \exp\left\{-\ln\left[\frac{R(q)}{2\left(1-\gamma^2\right)\left(q^2-\beta^2\right)}\right]\right\}$$
$$= K_+(p,s)K_-(p,s)\frac{\gamma^2}{2p\left(1-\gamma^2\right)}\,,$$
$$K_+(p,s) \equiv K_+(q) \tag{6.1.23}$$
$$= \frac{\sqrt{1+q}}{q+\beta} \exp\left\{\frac{1}{2\pi\mathrm{i}} \int_{L_1} \ln\left[\frac{R(z)}{2\left(1-\gamma^2\right)\left(z^2-\beta^2\right)}\right] \frac{dz}{z-q}\right\}\,,$$

$$K_-(p, s) \equiv K_-(q)$$
$$= \frac{\sqrt{1-q}}{q-\beta} \exp\left\{ -\frac{1}{2\pi i} \int_{L_2} \ln\left[\frac{R(z)}{2(1-\gamma^2)(z^2-\beta^2)} \right] \frac{dz}{z-q} \right\} .$$

Note that the use of the change $q = s/p$ "contracts" the s-plane by the factor $|p|$ and turns the plane by an angle $-\arg p$. So, the strip $|\mathrm{Re}\{s\}| < \mathrm{Re}\{p\}$ (Fig. 6.1a) becomes an inclined strip in the q-plane (Fig. 6.1b). Contours L_1 and L_2 are presented in Fig. 6.2a, branches of radicals $\sqrt{1 \pm q}$ have been selected such that they are positive at $q = 0$, and a branch of the logarithmic function has been selected such that its argument equals zero at $z = 0$.

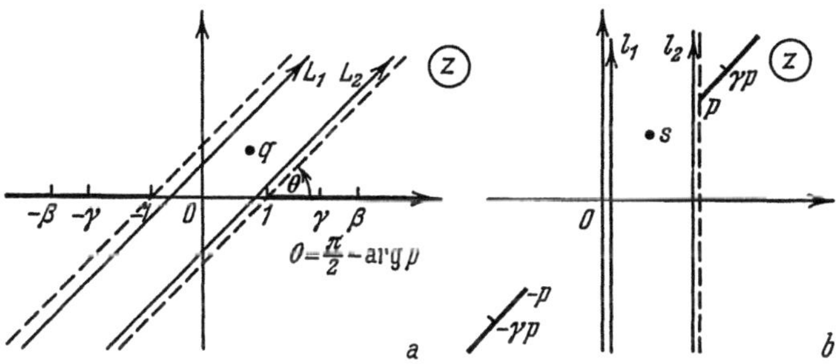

Fig. 6.2a,b. (a) Contours L_1 and L_2; (b) Contours l_1 and l_2

The logarithmic integrands in (6.1.23) are analytic in the z-plane everywhere outside the cuts $[-\gamma, -1]$, $[1, \gamma]$ and decay as $O(z^{-2})$ when $z \to \infty$. Therefore, after deforming the contours L_1 and L_2 so that they go along the negative and positive real semi-axes respectively, and taking into account that the logarithmic function turns out to be conjugate at the upper and the lower sides of the cuts $[-\gamma, -1]$ and $[1, \gamma]$, and considering that $\ln(x + iy) = \ln\sqrt{x^2 + y^2} + i\tan^{-1}(y/x)$, one obtains, as a result, the following expressions for $K_+(p, s)$ and $K_-(p, s)$:

$$K_+(p, s) \equiv K_+(q) = \frac{\sqrt{1+q}}{q+\beta} \exp\left[-\frac{1}{\pi} \int_{-\gamma}^{-1} g(z) \frac{dz}{z-q} \right] ,$$

$$K_-(p, s) \equiv K_-(q) = \frac{\sqrt{1-q}}{q-\beta} \exp\left[\frac{1}{\pi} \int_{1}^{\gamma} g(z) \frac{dz}{z-q} \right] , \tag{6.1.24}$$

$$g(z) = \tan^{-1} \frac{4z^2\sqrt{(\gamma^2 - z^2)(z^2 - 1)}}{(\gamma^2 - 2z^2)^2} ;$$

the radical in $g(z)$ is assumed to be arithmetic.

The function $K_+(q)$ is analytic everywhere except the cut drawn along the real axis $(-\infty, -1]$ and does not have zeros there, $K_-(q)$ is analytic everywhere except the cut drawn along the real axis $[1, +\infty)$ and does not have zeros there, either. If $q \to \infty$, the functions $K \pm (q)$ decay as $q^{-1/2}$. So, (I) $K_+(p, s)$ is analytic in the s-plane outside the cut which goes from the point $s = -p$ along the half-line $\arg s = \pi + \arg p$ to the infinity, and (II) the function $K_-(p, s)$ is analytic in the s-plane outside the cut which goes from the point $s = p$ along the half-line $\arg s = \arg p$ to infinity. The functions $K \pm (p, s)$ have no zeros at finite points of the s-plane around these cuts, whereas behave as $s^{-1/2}$ at infinity. (Note that decomposing the function $K(p, s)$ into the functions $K_+(p, s)$ and $K_-(p, s)$ is conventionally called "factorization").

By using (6.1.23), one can represent (6.1.20) in the form which follows:

$$\begin{aligned} &\frac{\sigma^+_{yy} K_+(p,s)\gamma^2}{2p\left(\gamma^2-1\right)} - N(p,s) = \frac{\bar{u}^-_y}{K_-(p,s)} \qquad (0 < \operatorname{Re} s < \operatorname{Re} p)\,, \\ &N(p,s) = -\frac{\bar{\sigma}^-_{yy} K_+(p,s)\gamma^2}{2p\left(\gamma^2-1\right)} + \frac{\bar{u}^+_y}{K_-(p,s)}\,. \end{aligned} \tag{6.1.25}$$

The function $N(p, s)$ being analytic in the strip $0 < \operatorname{Re}\{s\} < \operatorname{Re}\{p\}$ decreases as $s^{-1/2}$ if $s \to \infty$. By means of the Cauchy-type integrals this function may be written as $N(p, s) = N_+(p, s) + N_-(p, s)$:

$$\begin{aligned} &N_+(p,s) = -\frac{1}{2\pi i}\int_{l_1} \frac{N(p,z)dz}{z-s}\,, \\ &N_-(p,s) = \frac{1}{2\pi i}\int_{l_2} \frac{N(p,z)dz}{z-s}\,, \qquad 0 < \operatorname{Re} s < \operatorname{Re} p\,. \end{aligned} \tag{6.1.26}$$

The contours l_1 and l_2 (Fig. 6.2b) are within the strip $0 < \operatorname{Re}\{z\} < \operatorname{Re}\{p\}$ at its left and right edges, respectively.

It is not a complicated operation to demonstrate that (I) $N_+(p, s)$ is analytic in the half-plane $\operatorname{Re}\{s\} > 0$ and decreases within this half-plane as $s^{-1/2-\varepsilon} (\varepsilon > 0)$ when $s \to \infty$, and (II) $N_-(p, s)$ is analytic in the domain $\operatorname{Re}\{s\} < \operatorname{Re}\{p\}$ and decreases as $s^{-1/2}$ in this domain when $s \to \infty$.

Now, let us re-write (6.1.26):

$$N_-(p,s) + \frac{\bar{u}^-_y}{K_-(p,s)} = \frac{\bar{\sigma}^+_{yy} K_+(p,s)\gamma^2}{2p\left(\gamma^2-1\right)} - N_+(p,s) \quad (0 < \operatorname{Re} s < \operatorname{Re} p)\,. \tag{6.1.27}$$

Taking into account the whole of the abovesaid consideration, conclude that the rigth-hand side of (6.1.27) is regular at $\operatorname{Re}\{s\} > 0$ and decreases in this half-plane when $s \to \infty$, whereas the left-hand side is regular in the domain $\operatorname{Re}\{s\} < \operatorname{Re}\{p\}$ and decreases in this domain when $s \to \infty$. From the equality of these expressions in the strip $0 < \operatorname{Re}\{s\} < \operatorname{Re}\{p\}$ it follows that in the complex s-plane the analytic function $F(s)$ exists; this function is the same as the right-hand side of equation (6.1.27) in the domain $\operatorname{Re}\{s\} > 0$ and is the same as the left-hand side of equation (6.1.27) in the domain $\operatorname{Re}\{s\} < \operatorname{Re}\{p\}$. Since this function is limited, conclude that

$F(s) \equiv$ const according to the Liouville theorem [6.4]. Moreover, since $F(s) \to 0$ as $s \to \infty$, this constant is equal to zero, i.e., $F(s) \equiv 0$. Therefore, taking into account that, in the half-planes $\mathrm{Re}\{s\} > 0$ and $\mathrm{Re}\{s\} < \mathrm{Re}\{p\}$, the function $F(s)$ can be represented by the right- and the left-hand sides of (6.1.27), respectively, we find

$$\frac{\bar{\sigma}_{yy}^{+} K_{+}(p,s)\gamma^2}{2p\left(\gamma^2-1\right)} - N_{+}(p,s) = 0 \qquad (\mathrm{Re}\, s > 0)\,, \tag{6.1.28}$$

$$N_{-}(p,s) + \frac{\bar{u}_y^{-}}{K_{-}(p,s)} = 0 \qquad (\mathrm{Re}\, s < \mathrm{Re}\, p)\,. \tag{6.1.29}$$

Consequently,

$$\bar{\sigma}_{yy}^{+} = \frac{2p\left(\gamma^2-1\right)}{\gamma^2 K_{+}(p,s)} N_{+}(p,s) \qquad (\mathrm{Re}\, s > 0)\,, \tag{6.1.30}$$

$$\bar{u}_y^{-} = -K_{-}(p,s) N_{-}(p,s) \qquad (\mathrm{Re}\, s < \mathrm{Re}\, p)\,. \tag{6.1.31}$$

From properties of the functions $N_{\pm}(p,s)$ and $K_{\pm}(p,s)$ it follows that the functions $\bar{\sigma}_{yy}^{+}$ and $\bar{u}_y^{-}$ satisfy the requirements of analyticity and behavior at infinity as formulated for them above.

This technique for solving the functional equation (6.1.20) is referred to as the Wiener–Hopf method, or the factorization method.

Inverse Laplace transforms in s and p applied to (6.1.30–31) yield the original time functions

$$\sigma_{yy}(x,0,\tau) = \frac{\mu\left(\gamma^2-1\right))}{\gamma^2 \pi \mathrm{i}} \int\limits_{c-\mathrm{i}\infty}^{c+\mathrm{i}\infty} \mathrm{e}^{p\tau} p\, dp \frac{1}{2\pi \mathrm{i}} \int\limits_{d-\mathrm{i}\infty}^{d+\mathrm{i}\infty} \frac{N_{+}(p,s)}{K_{+}(p,s)} \mathrm{e}^{sx} ds \quad (x > 0)\,, \tag{6.1.32}$$

$$u_y(x,0,\tau) = -\frac{1}{2\pi \mathrm{i}} \int\limits_{c-\mathrm{i}\infty}^{c+\mathrm{i}\infty} \mathrm{e}^{p\tau} dp \frac{1}{2\pi \mathrm{i}} \int\limits_{d-\mathrm{i}\infty}^{d+\mathrm{i}\infty} N_{-}(p,s) K_{-}(p,s) \mathrm{e}^{sx} ds \quad (x < 0)$$

$$(c > d > 0)\,. \tag{6.1.33}$$

To solve the problem for a semi-infinite punch (with boundary and initial conditions (6.1.1–4)) we must nullify the function $F(x,\tau)$ in (6.1.18); this gives $\bar{\sigma}_{yy}^{-} = 0$ for $N(p,s)$ in (6.1.25).

Below, let us consider in detail the problem for a semi-infinite smooth, rigid punch at $f_0(x,t) = f_0(t) \equiv f(\tau)$; the corresponding wave fronts are presented in Fig. 6.3. Under these specific conditions, equation (6.1.25) for $N(p,s)$ comprises $\bar{\sigma}_{yy}^{-} = 0$ and $\bar{u}_y^{+} = \bar{f}(p)/s$. Taking into account that $K_{\pm}(p,s) \equiv K_{\pm}(q)$, we find from (6.1.26):

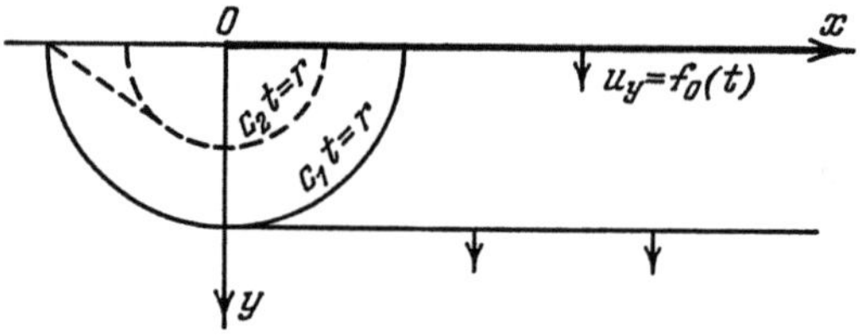

Fig. 6.3. Wave pattern in the problem of forcing the smooth rigid punch into an elastic medium ($\sigma_{xy}=0$, $u_y=f_0(t)$ for $x>0$, $y=0$)

$$N(p,s)=\frac{\bar f(p)}{sK_-(q)}\,,\qquad N_+(p,s)=\frac{\bar f(p)}{sK_-(0)}\,,$$
$$N_-(p,s)=\frac{\bar f(p)}{s}\left[\frac{1}{K_-(q)}-\frac{1}{K_-(0)}\right]\qquad\left(q=\frac{s}{p}\right)\,. \tag{6.1.34}$$

Thereafter, the formulae (6.1.32, 33) give

$$\sigma_{yy}(x,0,\tau)=\frac{1}{2\pi\mathrm{i}}\int\limits_{c-\mathrm{i}\infty}^{c+\mathrm{i}\infty}p\bar f(p)M_+(px)\mathrm{e}^{p\tau}\,dp\qquad(x>0)\,,$$
$$u_y(x,0,\tau)=\frac{1}{2\pi\mathrm{i}}\int\limits_{c-\mathrm{i}\infty}^{c+\mathrm{i}\infty}\bar f(p)M_-(px)\mathrm{e}^{p\tau}\,dp\qquad(x<0)\,, \tag{6.1.35}$$

$$M_+(px)=\frac{\mu\left(\gamma^2-1\right)}{\gamma^2K_-(0)\pi\mathrm{i}}\int\limits_L\frac{\mathrm{e}^{pxq}\,dq}{qK_+(q)}\,,$$
$$M_-(px)=\frac{1}{2\pi\mathrm{i}}\int\limits_L\left[\frac{K_-(q)}{k_-(0)}-1\right]\frac{\mathrm{e}^{pxq}}{q}\,dq\,. \tag{6.1.36}$$

The contour L is shown in Fig. 6.1b. We shall deform the contour L along the cut $[1,+\infty]$ for $M_-(px)$ and along the cut $(-\infty,-1]$ in the integral for $M_+(px)$, thereafter replace q by $(-q)$ in the function $M_+(px)$. This reduces (6.1.36) to the following form:

$$\frac{\gamma^2K_-(0)}{2\mu\left(\gamma^2-1\right)}M_+(px)=\frac{1}{K_+(0)}$$
$$-\frac{1}{\pi}\int\limits_1^\infty\frac{\beta-q}{\sqrt{q-1}}\mathrm{e}^{-\alpha-pxq}\left\{1+\left[\cos g(q)-1\right]H(\gamma-q\right\}\frac{dq}{q}\qquad(x>0)\,,$$

$$M_-(px)$$

$$= -\frac{1}{\pi}\int\limits_1^\infty \frac{\sqrt{q-1}}{q-\beta} e^{\alpha+pxq}\left\{1+[\cos g(q)-1]\,H(\gamma-q)\right\}\frac{dq}{q} \quad (x<0)\,,$$

$$\alpha \equiv \alpha(px,q) = \frac{1}{\pi}\int\limits_1^\gamma \frac{g(z)dz}{z-q}\,. \tag{6.1.37}$$

Herein, for integrals from 1 to $+\infty$ and the integral from 1 to γ (at $1<q<\gamma$) in the expression for α, Cauchy principal values will be taken; radicals in (6.1.37) are arithmetic ones. In addition, note that

$$K_-(0) = -K_+(0) = -\gamma^{-2}\sqrt{2\left(\gamma^2-1\right)}\,.$$

Using (6.1.37), equations (6.1.35) yield:

$$\sigma_{yy}(x,0,\tau) = -\mu\gamma^2 f'(\tau) + \frac{\mu\sqrt{2\left(\gamma^2-1\right)}}{\pi} H(\tau/x-1) \tag{6.1.38}$$

$$\times \frac{\partial}{\partial\tau}\int\limits_1^{\tau/x} \frac{\beta-q}{\sqrt{q-1}} e^{-\alpha}\left\{1+[\cos g(q)-1]\,H(\gamma-q)\right\} f(\tau-xq)\frac{dq}{q}$$

$$(x>0)\,,$$

$$u_y(x,0,\tau) = \frac{\gamma^2}{\pi\sqrt{2\left(\gamma^2-1\right)}} H(-1-\tau/x) \tag{6.1.39}$$

$$\times \int\limits_1^{-\tau/x} \frac{\sqrt{q-1}}{q-\beta} e^{\alpha}\left\{1+[\cos g(q)-1]\,H(\gamma-q)\right\} f(\tau+xq)\frac{dq}{q}$$

$$(x<0)\,.$$

In (6.1.38), the derivative $f'(\tau)$ is meant as a generalized one. In a particular case, when $f(\tau)=H(\tau)$, we have $f'(\tau)=\delta(\tau)$.

6.1.2 Analysis of the Punch Solution

By means of expressions (6.1.38, 39), we may clear up some properties of the problem with the semi-infinite smooth rigid punch. First, consider the expression for the vertical displacement over the boundary $\{y=0,\ x<0\}$. Here, $u_y=0$ for $\tau+x<0$ which is in compliance with the zero initial conditions; $\tau+x=0$ is the equation of the front of the longitudinal wave at a free part of the boundary $y=0$. At this front, functions u_y, $\partial u_y/\partial\tau$ and $\partial u_y/\partial x$ are equal to zero. At the front of the transverse wave ($\tau+x\gamma=0$) the function $u_y(x,0,\tau)$ and its first derivatives with respect to x and τ are continuous as well as at the front of the longitudinal wave. Using (6.1.39) it can be shown that the function $u_y(x,0,\tau)$ has a logarithmic singularity at $\tau=\beta x$, i.e., there is a logarithmic discontinuity in vertical displacements at the free part of the boundary, this discontinuity propagating at a speed of the Rayleigh waves (once $\beta = c_1/c_R$). It is easy to check that $u_y(x,0,\tau)$ is continuous at the point $x = 0$, whereas $\partial u_y/\partial x$ turns out to be unlimited when $x \to -0$, its asymptotic behaviour at small x being

$$\partial u_y/\partial x \sim M(\tau)(-x)^{1/2} .$$

The formula (6.1.38) for the stress $\sigma_{yy}(x,0,\tau)$ immediately under the punch ($x > 0$) gives

$$\sigma_{yy}(x,0,\tau) = -\mu\gamma^2 f'(\tau) , \qquad (6.1.40)$$

i.e., if a disturbance from the edge $x = 0$ did not arrive at a point the stress at the point is as if $\sigma_{xy} = 0$ and $u_y = f(\tau)$ over the whole of the boundary; this is well predictable for the hyperbolic equations of the elastodynamics.

At the instance $\tau = x$, the front of the longitudinal wave arrives at the point x under the punch. There is no jump in stress at this front ($[\sigma_{yy}] = 0$) if $f(\tau) = O(\tau^\alpha)$ as $\tau \to 0(\alpha > 1/2)$, and the supplementary stress $\sigma_{yy} + \mu\gamma^2 f'(\tau)$ behind the front behaves as $f_1(x)(\tau - x)^{\alpha-1/2}$ if $f(\tau) \sim c\tau^\alpha$ as $\tau \to 0$, $0 \le \alpha \le 1/2$; $f_1(x)$ tends to zero as $x^{-1/2}$ when $x \to +\infty$. Also, in equation (6.1.38) a front of a transverse wave is present (see a dashed line in Fig. 6.3). It is at these boundary conditions under the punch ($x > 0$) that the longitudinal (transverse) wave generates only a longitudinal (transverse, respectively) wave at its interaction with the boundary [6.10, 11]. Thus, the stress σ_{yy} is the stress in the longitudinal wave if $\tau/\gamma < x < \tau$, and is the total stress due to the longitudinal and transverse waves if $0 < x < \tau/\gamma$. The point $x = 0$ features an integrable singularity of order of $x^{-1/2}$:

$$\sigma_{yy} = -\frac{\mu\sqrt{2(\gamma^2-1)}}{\pi\sqrt{x}} \frac{\partial}{\partial\tau} \int_0^\tau \frac{f(q)dq}{\sqrt{\tau - q}} \qquad (x \to +0) ,$$

which is analogous to that of the corresponding static problem.

It should be mentioned that, if the common point of the mixed boundary conditions moves along the x-axis (at $t > 0$) at a constant velocity v, there is a need to employ a new, moving coordinate system $\{x', y'\}$ introduced with the formulae, $x' = x - vt$, $y' = y$. Then the non-stationary problem (in variables $\{t, x', y'\}$) appears again where the common point of the mixed boundary conditions is fixed and wave equations for φ and ψ are transformed. A solution of this problem has no principal differences from the one considered above, therefore, we shall not develop this.

6.2 A Finite-Width Punch

For a finite-width punch ($0 < x < l$) a solution can be obtained by employing a superposition of solutions for semi-infinite punches. This result is based on the fact that the elastodynamics equations are hyperbolic, so, any disturbance propagates at a finite speed. Therefore, as long as diffraction waves from an opposite edge have not arrived at the region under consideration, the solution for the semi-infinite punch is valid.

6.2.1 Solution to the Problem

Relying upon the abovesaid, let us construct a solution of a dynamic problem for a smooth rigid punch of finite width ($0 < x < 2l$) which indents into a half-plane $y \geq 0$ at $u_y = f_0(t) \equiv f(\tau)$ (the boundary $y = 0$ outside the punch is free of stresses, as was considered earlier). The punch is symmetrical, therefore, the solution to be found is symmetrical with respect to the straight line $x = l$:

$$u_y(2l - x, y, \tau) = u_y(x, y, \tau), \quad \sigma_{yy}(2l - x, y, \tau) = \sigma_{yy}(x, y, \tau) .$$

Hence, suffices it to find the solution in the region $\{x \leq l, y \geq 0\}$.

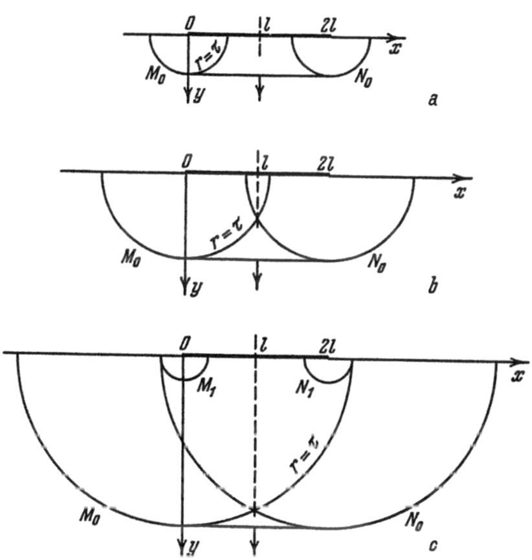

Fig. 6.4a–c. Disturbances for (a) $0 < \tau < l$, (b) $l < \tau < 2l$, (c) $2l < \tau < 3l$

First, consider the problem for $0 < \tau < l$. Throughout this interval the diffraction disturbances M_0 and N_0 from opposite edges of the punch ($x = o, x = 2l$) do not interact (Fig. 6.4a), so, the solution at $x \leq l$ will be the same as in the problem on penetration by the semi-infinite punch ($0 < x < \infty$) (the fronts of diffraction transverse waves are not shown on the Fig. 6.4. Then, the stress under the punch $(0 < x(2l)$ for $0 < \tau < l$ may be written as

$$\sigma_{yy}(x,\tau) = -\mu\gamma^2 f'(\tau) + \sigma^0_{yy}(x,\tau) + \sigma^0_{yy}(2l - x,\tau) , \tag{6.2.1}$$

where $\sigma^0_{yy}(x,\tau) = \tilde{\sigma}^0_{yy}(x,\tau) + \mu\gamma^2 f'(\tau)$ is the disturbance supplied by the diffraction wave from the edge $x = 0$, with $\sigma^0_{yy}(x,\tau)$ being the solution for a semi-infinite punch $(0 < x < \infty)$ and $(-\mu\gamma^2 f'(\tau))$ being the solution to the one-dimensional problem on indentation by an infinite punch $-\infty < x < \infty)$.

For $l \le \tau < 2l$ (Fig. 6.4b) it follows from the linearity of the problem that the stress under the punch is a result of superimposing the disturbances $\sigma^0_{yy}(x,\tau)$ and $\sigma^0_{yy}(2l - x,\tau)$ propagating from the left and right edges, respectively, so, equation (6.2.1) remains valid throughout this time interval.

For $2l \le \tau < 3l$, a new pair of diffraction waves, M_1 and N_1, occurs (Fig. 6.4c). These new disturbances must behave outside the punch so as to nullify the disturbances M_0 and N_0 which come from opposite edges and violate the boundary conditions on the axis $y = 0$ outside the punch.

For $2l \le \tau < 3l$, when these disturbances do not interact, supplementary disturbances σ^1_{yy} and u^1_y within the domain $x \le l$ (which are caused by the appearance of the diffraction wave M_1 near the left edge at $\tau_1 > 0, \tau_1 = \tau - 2l)$ are governed by the following boundary conditions on the axis $y = 0$:

$$\begin{aligned} &\sigma^1_{yy} = -\sigma^0_{yy}(2l - x,\tau_1) \qquad &&(x < 0,\ y = 0) ,\\ &u^1_y = 0 &&(x > 0,\ y = 0) . \end{aligned}$$

At $3l \le \tau < 4l$, supplementary disturbances in the stress under the punch are a result of superposition of the disturbances $\sigma^1_{yy}(x,\tau_1)$ and $\sigma^1_{yy}(2l - x,\tau_1)$ caused by the diffraction waves M_1 and N_1 symmetrical with respect to the line $x = l$. The same is valid over a wider range $2l \le \tau < 4l$.

Thus, once $\tau(\tau = c_1 t)$ exceeds $2nl(n = 1,2,3,\ldots)$, there appears a new (nth)pair of diffraction waves, M_n and N_n. They carry supplementary disturbances symmetrical with respect to the line $x = l$, with the disturbance $\boldsymbol{u}^n$ (from the wave M_n) satisfying the Lamé equations (in x, y, and τ_n) at the following boundary and initial conditions:

$$\begin{aligned} &\sigma^n_{yy} = -\sigma^{n-1}_{yy}(2l - x,\tau_n) \qquad &&(x < 0,\ y = 0) ,\\ &u^n_y = 0 &&(x > 0,\ y = 0) ,\\ &\boldsymbol{u}^n = \partial\boldsymbol{u}^n/\partial\tau_n = 0 &&(\tau_n = 0),\ \tau_n = \tau - 2nl \ge 0 . \end{aligned} \tag{6.2.2}$$

At $y = 0$, a solution to the Lamé equations at the boundary and initial conditions (6.2.2) and the condition (6.1.5) for $\boldsymbol{u}^n$ may be written as (6.1.32, 33) after changing $\tau, F(x,\tau), f(x,\tau), \sigma_{yy}(x,0,\tau)$, and $u_y(x,0,\tau)$ by $\tau_n, -\sigma^{n-1}_{yy}(2l - x,\tau_n), 0,$ $\sigma^n_{yy}(x,\tau_n)$, and $u^n_y(x,\tau_n)$ respectively.

As a result, the stress under the punch $0 < x < 2l$ at $2kl \le \tau < 2l(k + 1)$ $(k = 0,1,2,\ldots)$ is

$$\sigma_{yy}(x,\tau) = -\mu\gamma^2 f'(\tau) + \sum_{n=0}^{k} \left[\sigma_{yy}^n(x,\tau_n) + \sigma_{yy}^n(2l-x,\tau_n)\right] , \tag{6.2.3}$$

where $\tau_n = \tau - 2nl$, $\sigma_{yy}^0(x,\tau) = \bar{\sigma}_{yy}^0(x,\tau) + \mu\gamma^2 f'(\tau)$ and $\bar{\sigma}_{yy}^0(x,\tau)$ is the right-hand side of equation (6.1.32); the function $\sigma_{yy}^n(x,\tau_n)$ is represented by the right-hand side of (6.1.32) in which the variable τ and the functions $F(x,\tau)$ and $f(x,\tau)$ (in $N(p,s)$) should be replaced by $\tau_n, -\sigma_{yy}^{n-1}(2l-x,\tau_n), 0$, respectively.

6.2.2 The Acoustic Case

Consider now a particular case: indentation of an acoustic half-plane by the rigid punch displaced at a constant speed v_0. The solution to this problem — a distribution of stresses (or a pressure) under the punch — can be obtained from the formulae (6.1.32, 38, 42) after passage to the limit as $\mu \to 0$ with a specific function $f(\tau) = v_0\tau H(\tau)/c_1$. One should take into account that the following asymptotic formulas (as $\mu \to 0$, i.e., as $\gamma \to \infty$) take place for components of equations (6.1.24, 26, 36):

$$K_\pm(q) \sim \pm\sqrt{2(1\pm q)}/\gamma ,$$

$$M_+(px) \sim -\lambda\gamma \left[1 - \frac{1}{\pi}\int_1^\infty e^{-px\tau}\frac{d\tau}{\sqrt{\tau-1}}\right] \qquad \left(\sqrt{\tau-1} > 0,\ x > 0\right) ,$$

$$N_+(p,s) \sim \frac{\gamma\sqrt{2}}{4\pi\lambda p^{3/2}\mathrm{i}} \int_{\varepsilon-\mathrm{i}\infty}^{\varepsilon+\mathrm{i}\infty} \frac{\sqrt{p+z}}{z-s}dz \int_{-\infty}^{0} e^{-z\xi}\bar{\sigma}_{yy}^{n-1}(2l-\xi,p)d\xi \quad (\varepsilon > 0) .$$

In addition, $c_1 \to c_0 = \sqrt{\lambda/\rho}$. The disturbance σ_{yy}^n to be found can be written as

$$\sigma_{yy}^n(x,\tau_n) = \int_{c-\mathrm{i}\infty}^{c+\mathrm{i}\infty} \bar{\sigma}_{yy}^n(x,p)e^{p\tau_n}dp \qquad (c > 0) ,$$

$$\tau_n = \tau - 2nl \qquad (\tau = c_0 t) , \tag{6.2.4}$$

$$\bar{\sigma}_{yy}^n(x,p) = -\frac{1}{4\pi^2}\int_{d-\mathrm{i}\infty}^{d+\mathrm{i}\infty} \frac{e^{sx}dx}{\sqrt{p+s}} \int_{\varepsilon-\mathrm{i}\infty}^{\varepsilon+\mathrm{i}\infty} \frac{\sqrt{p+z}dz}{z-s}\int_{-\infty}^{0} e^{-z\xi}\bar{\sigma}_{yy}(2l-\xi,p)d\xi$$

$$(x > 0, \quad c > d > \varepsilon > 0) . \tag{6.2.5}$$

Let us deform the contour $(\varepsilon - \mathrm{i}\infty), \varepsilon + \mathrm{i}\infty)$ along the cut drawn from the point $z = -p$, and replace $p + z = z'$ and $p + s = s'$. After removing the primes, obtain

$$\sigma_{yy}^n(x,p) \tag{6.2.6}$$

$$= -\frac{e^{-px}}{2\pi^2\mathrm{i}}\int_{d-\mathrm{i}\infty}^{d+\mathrm{i}\infty} e^{sx}\frac{ds}{\sqrt{s}}\int_0^\infty \frac{\sqrt{z}dz}{z+s}\int_{-\infty}^{0} e^{-(z+p)\xi}\bar{\sigma}_{yy}^{n-1}(2l-\xi,p)d\xi$$

$$(d > 0) .$$

Using the relation

$$\frac{1}{2\pi i}\int\limits_{d-i\infty}^{d+i\infty}\frac{e^{sx}\,ds}{\sqrt{s}(s+z)}=\frac{2e^{-zx}}{i\sqrt{\pi z}}\int\limits_{0}^{i\sqrt{zx}}e^{-u^2}du\,,$$

equation (6.2.6) can be simplified to give

$$\bar{\sigma}^n_{yy}(x,p)=\frac{\sqrt{x}}{\pi}\int\limits_{-\infty}^{0}\bar{\sigma}^{n-1}_{yy}\frac{e^{-p(x-\xi)}}{(\xi-x)\sqrt{-\xi}}d\xi\,. \tag{6.2.7}$$

The use of (6.2.7) in (6.2.4), the subsequent replacement $\xi=-\tau$, and the fact that

$$\bar{\sigma}^{n-1}_{yy}(2l-\xi,p)=\int\limits_{-\xi}^{\infty}\sigma^{n-1}_{yy}(2l-\xi,\tau)e^{-p\tau}d\tau\,,$$

make it possible to obtain

$$\sigma^n_{yy}(x,\tau_n)=-\frac{1}{\pi}\int\limits_{0}^{(\tau_n-x)/2}\sigma^{n-1}_{yy}(2l+\tau,\tau_n-x-\tau)\sqrt{\frac{x}{\tau}}\frac{d\tau}{x+\tau}\,. \tag{6.2.8}$$

After all, the notations $P_n(x,\tau_n)=-\sigma^n_{yy}(x,\tau_n)/(\rho c_0 v_0)$, $c_0=\sqrt{\lambda/\rho}$ yield the following expression for the pressure $p(x,\tau)=-\sigma_{yy}(x,\tau)/(\rho c_0 v_0)$ under the punch at $2kl\le\tau<2(k+1)l$, $0<x<2l$:

$$\begin{aligned}
&P(x,\tau)=1+\sum_{n=0}^{k}[P_n(x,\tau_n)+P_n(2l-x,\tau_n)]\,,\\
&P_0(x,\tau_0)=-\frac{2}{\pi}\tan^{-1}\sqrt{\frac{\tau_0-x}{x}}H(\tau_0-x)\,,\\
&P_n(x,\tau_n)=-\frac{1}{\pi}\int\limits_{0}^{(\tau_n-x)/2}P_{n-1}(2l+\tau,\tau_n-x-\tau)\sqrt{\frac{x}{\tau}}\frac{d\tau}{x+\tau}\\
&(\tau_n=\tau-2ln,\quad n=0,1,2,\dots,k)\,.
\end{aligned} \tag{6.2.9}$$

This result is the same as the one of [6.12] and differs in notation only. From the recurrent formula (6.2.9) it can be seen that P_n is negative at even values of n, whereas positive at odd ones — signs of the diffraction waves alternate with each other. Waves with even numbers represent rarefaction waves, waves with odd numbers represent compression waves.

The resulting unit load

$$F=\int\limits_{0}^{2l}P\,dx/(2l)$$

takes the following form for $0 < \tau \leq 4l$:

$$F = 1 - \frac{\eta}{2} + \frac{1}{\pi}\left[(\eta - 2)\cos^{-1}\frac{2}{\eta} + \sqrt{\eta^2 - 4} - 2\,\mathrm{arch}\,\frac{\eta}{2}\right] H(\eta - 2) \tag{6.2.10}$$

$$(\eta = \tau/l = c_0 t/l)\ .$$

In Fig. 6.5, analytical results correspond to equation (6.2.10) with due account for next two pairs of diffraction waves at $\tau > 4l$.

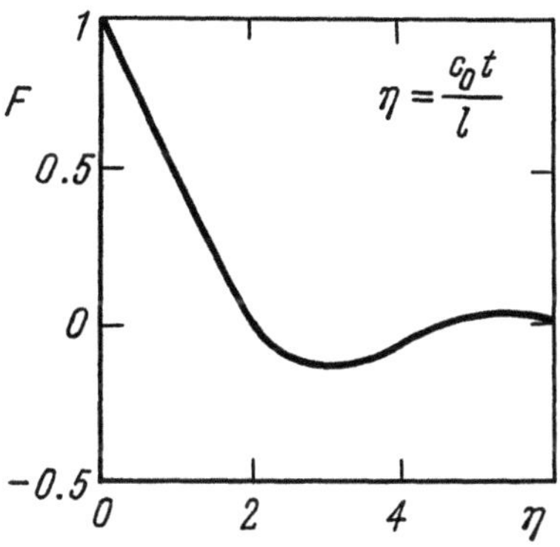

Fig. 6.5. Unit load vs. time in acoustic case

Note that the considered two-dimensional problem for an acoustic solid and numerous analogous two- and three-dimensional acoustic problems on diffraction and penetration were considered in a number of ways in [6.12, 13].

Analogous two- and three-dimensional non-stationary problems for an elastic medium were studied in [6.14–17].

It should be said here that no closed form solution has been found to analogous three-dimensional problems; this is a notable difference from the two-dimensional case considered above. Only asymptotic solutions (when $t \to 0$) have been obtained for three-dimensional problems in [6.12] by means of the Wiener–Hopf method. Moreover, analytical solutions are not yet obtained even for the two-dimensional case with more complicated mixed boundary conditions amongst which no one is imposed throughout an infinite interval. In [6.18] a solution of such a problem has been reduced to a solution of integral equations whose kernels are of first-order singularity at one end of an integration interval.

6.3 Problems with Moving Boundary Edges

It is due to investigations into the two-dimensional problem on variable-speed dynamic crack propagation that methods for solving dynamic problems were advanced last years. From the mathematical point of view, these problems reduce to a solution of the set of hyperbolic equations under mixed boundary conditions specified over a straight line; one of the conditions is a mixed condition, and the other is effective over the entire line. The point separating the mixed boundary conditions on a straight line travels along this line at a speed which, in general, is time-dependent.

For the first time the problem with variable-speed propagation of a crack at an antiplane deformation has been solved by *Kostrov* ([6.19]) using reduction to the Abel integral equation. Later, *Kostrov* ([6.20, 21]) used the same technique to find an exact analytical solution to a more complicated problem concerning the variable-speed propagation of a crack under conditions of plane strain at arbitrary time-dependent loads. The technique used by Kostrov has been further improved by Slepyan and his colleagues [6.16, 22–24]; this has, in particular, reduced the number of quadratures in a general solution, and the approach itself took a perfect form. The method developed in these contributions differs from the usual Wiener–Hopf method: after factorization, consideration is carried out for a plane of physical variables, i.e., original time functions are considered rather than their transforms. Hereinafter, we shall discuss this improved method on the basis of the above contributions.

6.3.1 Problem Formulation and Application of Integral Transforms

Assume that an elastic isotropic medium fills an unbounded space outside a crack occupying the domain $\{x_1 = 0, x_2 \le l(t)\}$ (x_1, x_2, and x_3 are the Cartesian coordinates). Let all the forces applied to the medium be independent of the x_3 coordinate (a two-dimensional non-stationary problem). Then, the motion equations (2.1.1) (at $f_i \equiv 0$) and Hooke's law (2.1.7) may be written as

$$\sigma_{\alpha\beta,\beta} = \rho \ddot{u}_\alpha \,, \qquad \sigma_{3\alpha,\alpha} = \rho \ddot{u}_3 \,, \tag{6.3.1}$$

$$\sigma_{\alpha\beta} = \mu \left[\delta_{\alpha\beta} \left(\gamma^2 - 2\right) u_{\varepsilon,\varepsilon} + u_{\alpha,\beta} + u_{\beta,\alpha}\right] \,, \qquad \sigma_{3\alpha} = \mu u_{3,\alpha} \,, \tag{6.3.2}$$

the notation being

$$\alpha, \beta, \varepsilon = 1, 2, \quad \gamma = c_1/c_2 \,, \quad \delta_{\alpha\beta} = \begin{cases} 1, & \alpha = \beta, \\ 0, & \alpha \neq \beta. \end{cases}$$

In the motion equations (6.3.1), effects of body forces are not included. If the body forces are present and initial conditions are non-zero, then due to linearity of the equations one can do as follows. Let us use u_i^0 and σ_{ik}^0 to designate the solution corresponding to the same body forces and initial conditions but for an unbounded elastic medium without cracks. Formulating such a solution offers no difficulties. If we assume the solution for the solid with a crack to be a sum $\boldsymbol{u}+\boldsymbol{u}^0$, then the solution u satisfies (I) the motion equations (without body forces), (II) the zero initial conditions

$$u \equiv 0 \qquad (t < 0) \tag{6.3.3}$$

and (III) new boundary conditions at the crack $x_1 = 0$ (which have been changed after addition of the solution $(-u^0)$). Let the crack be subjected to the stress $p_j^0(t, x_2)$ (of identical values at the upper and the lower sides), $j = 1, 2, 3$. For σ_{ik}, boundary conditions at the crack take then the form

$$\sigma_j(t, x_2) \equiv \sigma_{j1}(t, 0, x_2) = p_j(t, x_2) \qquad (x_2 < l(t)) , \tag{6.3.4}$$

where $j = 1, 2, 3$,

$$p_j(t, x_2) = p_j^0(t, x_2) - \sigma_{j1}^0(t, 0, x_2) .$$

Equations (6.3.1, 2) will be solved by means of the Laplace transformation in time t and the Fourier transformation in x_2-coordinate (the so-called LF-transformation):

$$f^{\mathrm{LF}}(s, q) = \int_{-\infty}^{\infty}\int_{0}^{\infty} f(t, x_2) \mathrm{e}^{-st+\mathrm{i}qx_2} dt\, dx_2 , \tag{6.3.5}$$

$$f(t, x_2) = \frac{1}{4\pi^2 \mathrm{i}} \int_{-\infty}^{\infty}\int_{a-\mathrm{i}\infty}^{a+\mathrm{i}\infty} f^{\mathrm{LF}}(s, q) \mathrm{e}^{st-\mathrm{i}qx_2} ds\, dq$$

$$(f = u_j , \sigma_{jk} ; \quad a > 0) . \tag{6.3.6}$$

In so doing, it is convenient to express the unknown displacements through the stresses at the plane $x_1 = 0$: in the problem under consideration the stresses are continuous across this plane. For these purposes, we shall solve equations (6.3.1, 2) as follows. According to (2.7.36, 37), a solution of these equations for the displacement vector $u = (u_1, u_2, u_3)$ which is independent of x_3 can be reduced to solving the wave equations

$$\Delta\varphi = c_1^{-2}\ddot{\varphi} , \quad \Delta\psi_\alpha = c_2^{-2}\ddot{\psi}_\alpha \qquad \left(\Delta \equiv \frac{\partial^2}{\partial x_1^2} + \frac{\partial^2}{\partial x_2^2} , \quad \alpha = 1, 2\right) \tag{6.3.7}$$

with respect to potentials φ, ψ_1 and ψ_2 related to the vector u by the formula,

$$u = \operatorname{grad} \varphi + \operatorname{curl} (\psi_1 e_3) + \operatorname{curl} \operatorname{curl} (\psi_2 e_3) . \tag{6.3.8}$$

After applying the LF-transformation (in t and x_2) to (6.3.7) and solving the equations for the transforms φ^{LF}, ψ_1^{LF}, and ψ_2^{LF} in the half-space $x_1 > 0$, we find

$$\varphi^{\mathrm{LF}} = A\mathrm{e}^{-x_1\sqrt{c_1^{-2}s^2+q^2}} , \quad \psi_\alpha^{\mathrm{LF}} = B_\alpha \mathrm{e}^{-x_1\sqrt{c_2^{-2}s^2+q^2}} \qquad (\alpha = 1, 2) . \tag{6.3.9}$$

Thereafter, components of the stress, $\sigma_{j1}(t,0,x_2)$, can be considered as specified values over the plane $x_1=0$. Hooke's law, equation (6.3.2), provides the following boundary conditions at $x_1=0$ for determining the vector $\boldsymbol{u}$:

$$\sigma_\alpha(t,x_2)=\mu\left[\delta_{\alpha 1}\left(\gamma^2-2\right)u_{\varepsilon,\varepsilon}+u_{\alpha,1}+u_{1,\alpha}\right], \quad \sigma_3(t,x_2)=\mu u_{3,1} \quad (\alpha=1,2) \tag{6.3.10}$$

By means of the three equations of (6.3.10) the constants A, B_1, B_2 present in (6.3.9) are determined unambiguosly. Thus, a solution in the half-space $x_1>0$ for transforms of the displacements u_i^{LF} is expressed in terms of transforms of the stresses $\sigma_j^{\mathrm{LF}}(s,q)$ at the plane $x_1=0$.

Treating analogously equations (6.3.7) in the domain $x_1<0$, find

$$\varphi^{\mathrm{LF}}=A^0 \mathrm{e}^{x_1\sqrt{c_1^{-2}s^2+q^2}}, \quad \psi_\alpha^{\mathrm{LF}}=B_\alpha^0 \mathrm{e}^{x_1\sqrt{c_1^{-2}s^2+q^2}}. \tag{6.3.11}$$

The constants A^0, B_1^0, B_2^0 are also deriveable out of the boundary conditions (6.3.10) at $x_1=0$, so, $u_j^{\mathrm{LF}}(s,x_1,q)$ for the half-space $x_1<0$ may be written in terms of the functions $\sigma_j^{\mathrm{LF}}(s,q)$.

As a result, the solution, $u_j(s,x_1,q)$, can be represented in the following formulae valid at both $x_1<0$ and $x_1>0$:

$$\begin{aligned} u_1^{\mathrm{LF}}&(s,x_1,q)\\ &=-\frac{1}{\mu R(s,q)}\Big[\mathrm{e}^{-|x_1|n_1}\left(2iqn_1n_2\sigma_2(s,q)+\left(n_2^2+q^2\right)n_1\sigma_1^{\mathrm{LF}}(s,q)\operatorname{sgn}x_1\right)\\ &\quad -iq\mathrm{e}^{-|x_1|n_2}\left(\sigma_2^{\mathrm{LF}}(s,q)\left(n_2^2+q^2\right)-2qin_1\sigma_1^{\mathrm{LF}}(s,q)\operatorname{sgn}x_1\right)\Big], \end{aligned} \tag{6.3.12}$$

$$\begin{aligned} u_2^{\mathrm{LF}}&(s,x_1,q)\\ &=-\frac{1}{\mu R(s,q)}\Big[iq\mathrm{e}^{-|x_1|n_1}\left(\sigma_1^{\mathrm{LF}}(s,q)\left(n_2^2+q^2\right)+2qin_2\sigma_2^{\mathrm{LF}}(s,q)\operatorname{sgn}x_1\right)\\ &\quad -\mathrm{e}^{-|x_1|n_2}\left(2qin_1n_2\sigma_1^{\mathrm{LF}}(s,q)-\left(n_2^2+q^2\right)n_2\sigma_2^{\mathrm{LF}}(s,q)\operatorname{sgn}x_1\right)\Big], \end{aligned} \tag{6.3.13}$$

$$u_3^{\mathrm{LF}}(s,x_1,q)=-\frac{1}{\mu n_2}\mathrm{e}^{-|x_1|n_2}\sigma_3^{\mathrm{LF}}(s,q)\operatorname{sgn}x_1. \tag{6.3.14}$$

Within expressions (6.3.12–14) we denote

$$R(s,q)=\left(n_2^2+q^2\right)^2-4q^2n_1n_2, \quad n_\alpha=\sqrt{c_\alpha^{-2}s^2+q^2} \quad (\alpha=1,2)$$
$$\operatorname{sgn}x_1=\begin{cases}1, & x_1>0,\\ -1, & x_1<0.\end{cases} \tag{6.3.15}$$

The use of the notation

$$u_j^{\mathrm{LF}}(s,q)=\tfrac{1}{2}\left[u_j^{\mathrm{LF}}(s,+0,q)-u_j^{\mathrm{LF}}(s,-0,q)\right] \tag{6.3.16}$$

produces, from (6.3.12–15),

$$u_j^{\mathrm{LF}}(s,q)=S_j^{\mathrm{LF}}(s,q)\sigma_j^{\mathrm{LF}}(s,q) \qquad (j=1,2,3), \tag{6.3.17}$$

where

$$S_\alpha^{\mathrm{LF}}(s,q) = -\frac{c_2^{-2}s^2\sqrt{c_\alpha^{-2}s^2+q^2}}{\mu R(s,q)} \qquad (\alpha = 1,2)\,,$$
$$S_3^{\mathrm{LF}}(s,q) = -\frac{1}{\mu\sqrt{c_\alpha^{-2}s^2+q^2}}\,. \tag{6.3.18}$$

Equations (6.3.12–15) could provide a solution to the problem if from the boundary conditions the functions $\sigma_j^{\mathrm{LF}}(s,q)$ could be derived, i.e., if the stresses were known over the entire x_2-axis. However, the stresses are only specified over the crack, $x_2 < l(t)$. But the displacements are continuous on a continuation of this crack, so,

$$u_j(t,x_2) = \tfrac{1}{2}\left[u_j(t,+0,x_2) - u_j(t,-0,x_2)\right] = 0 \tag{6.3.19}$$

at $x_2 > l(t)$. Thus, equations (6.3.17) relate LF-transforms $\sigma_j^{\mathrm{LF}}(s,q)$ and $u_j^{\mathrm{LF}}(s,q)$ whose original time functions $\sigma_j(t,x_2)$ and $u_j(t,x_2)$ satisfy the conditions (6.3.4, 19).

Note that equations (6.3.17) represent the solution to a two-dimensional problem in crack dynamics. This solution can be understood as superposition of the following three items.

I) A solution to the plane-strain problem of the propagation of a crack subjected to a normal loading which is described by equation (6.3.17) at $j = 1$ with conditions for σ_1 and u_1:

$$\sigma_1(t,x_2) = p_1(t,x_2) \quad (x_2 < l(t))\,, \qquad u_1(t,x_2) = 0 \quad (x_2 > l(t)) \tag{6.3.20}$$

(in this case, $\sigma_2(t,x_2) \equiv \sigma_3(t,x_2) \equiv 0$ in equations (6.3.12–14)).

II) A solution to the plane-strain problem on propagation of a crack subjected to transverse shear which is described by equation (6.3.17) at $j = 2$ with conditions for σ_2 and u_2:

$$\sigma_2(t,x_2) = p_2(t,x_2) \quad (x_2 < l(t))\,, \qquad u_2(t,x_2) = 0 \quad (x_2 > l(t)) \tag{6.3.21}$$

(for this problem, $\sigma_1(t,x_2) \equiv \sigma_3(t,x_2) \equiv 0$ in equations (6.3.12–14)).

III) A solution to the problem of the dynamics of a crack subjected to an antiplane deformation which is described by equation (6.3.17) at $j = 3$ with conditions

$$\sigma_3(t,x_2) = p_3(t,x_2) \quad (x_2 < l(t))\,, \qquad u_3(t,x_2) = 0 \quad (x_2 > l(t)) \tag{6.3.22}$$

(in this case, $\sigma_1(t,x_2) \equiv \sigma_2(t,x_2) \equiv 0$ in equations (6.3.12, 13)).

It should be mentioned that the function $S_j^{\mathrm{LF}}(s,q)$ is the LF-transform of a fundamental solution for the jth component of the displacement vector $u_j(t,+0,x_2)$ ($u^{\mathrm{LF}}(s,+0,q) = S_j^{\mathrm{LF}}(s,q)$) in the problem of the behaviour of a continuum in the domain $x_1 > 0$ after a sudden application of the jth component of the stress vector — $\sigma_{j1}(t,0,x_2) = \delta(t)\delta(x_2)$ — to a boundary. This is a consequence from equations (6.3.12–14) if $x_1 \to +0$. In particular, $S_1^{\mathrm{LF}}(s,q)$ is the LF-transform of a fundamental solution to the Lamb problem.

A solution of every (jth) equation in (6.3.17) should be based on factorization, i.e., on the following decomposition of a fundamental solution transform S_j^{LF} (index j is removed hereinafter):

$$S^{\mathrm{LF}}(s,q) = S_+^{\mathrm{LF}}(s,q) S_-^{\mathrm{LF}}(s,q) .$$

In terms of original time functions, this is equivalent to convolution:

$$S(t,x_2) = \int_{-\infty}^{\infty} \int_{-0}^{t+0} S_+(\tau,\xi) S_-(t-\tau, x_2-\xi) d\tau \, d\xi , \tag{6.3.23}$$

or, in common brief notation,

$$S(t,x_2) = S_+(t,x_2) * S_-(t,x_2) .$$

The objective of the factorization procedure is to decompose the fundamental solution $S(t,x_2)$ into functions $S_\pm(t,x_2)$ describing the directed wave disturbances.

Note. In the integral with respect to τ in equation (6.2.23), integration limits -0 and $t+0$ are stated since the generalized functions present in these convolution products can be singular at the points $\tau = 0$ and $\tau = t$, so account will be taken of their contribution. In other words, integration is carried out over the whole of the carrier of the integrand (see, for example, [6.25]). These limits can be replaced by $-\infty$ and $+\infty$, respectively, without affecting the value of the integral. Hereinafter, we shall write 0 and t in our taking account of the contribution of the points $\tau = 0$ and $\tau = t$, however, we shall also bear in mind the abovesaid.

6.3.2 Splitting of Fundamental Solutions

Consider the transform $S^{\mathrm{LF}}(s,q)$ being an analytical function in variables s and q and having singular points:

$$\begin{gathered} s_m = iqv_m, \quad q_m = is/v_m, \quad v_m = \mathrm{const} \neq 0 , \\ m = 1,2,\ldots,n; \quad v_1 < v_2 < \ldots < v_n . \end{gathered} \tag{6.3.24}$$

Let us assume that this function can be written as a product

$$S^{\mathrm{LF}}(s,q) = \prod_{m=1}^{n} S_m^{\mathrm{LF}}(s,q) , \tag{6.3.25}$$

where each factor S_m^{LF} has only one singular point $s = s_m (q = q_m)$. Assume that functions S^{LF} and S_m^{LF} increase slowly over the q-plane (i.e., there exists the number α such that $|S^{\mathrm{LF}}| < |q|^\alpha$, $|S_m^{\mathrm{LF}}| < |q|^\alpha$ as $|q| \to \infty$). Returning to original functions (at $x_2 \equiv x$); prove that $S_m(t,x)$ is a generalized function whose carrier is on a ray $x = v_m t$. From conditions (6.3.24) it follows that there are no singular points on the real axis in the q-plane. In the formula for inversion,

$$S_m(t,x) = \frac{1}{4\pi^2 \mathrm{i}} \int_{-\infty}^{\infty} \int_{a-\mathrm{i}\infty}^{a+\mathrm{i}\infty} S_m^{\mathrm{LF}}(s,q) \mathrm{e}^{st-\mathrm{i}qx}\, ds\, dq \qquad (a>0) ,$$

use $s = \mathrm{i}qc + p$. Then

$$S_m(t,x) = \frac{1}{4\pi^2 \mathrm{i}} \int_{-\infty}^{\infty} \int_{a-\mathrm{i}\infty}^{a+\mathrm{i}\infty} S_m^{\mathrm{LF}}(p+\mathrm{i}qc,q) \Phi\, dp\, dq , \qquad (6.3.26)$$

$$\Phi = \exp\left[pt - \mathrm{i}q(x-ct)\right] .$$

A singular point on the q-plane, see (6.3.24), is defined by equation $q_m = -\mathrm{i}(p + \mathrm{i}q_m c)/v_m$ whence

$$q_m = -\frac{\mathrm{i}p}{v_m - c} \qquad (\mathrm{Re}\, p > 0) . \qquad (6.3.27)$$

If $c < v_m$, then we may deform the path of integration in the inversion formula (6.3.26) (for the Fourier transformation) so that integration will be performed in the upper half-plane where no singular points exist. A proper treatment shows that $s_m = 0$ for $x < ct$ (within the bounds of the latter inequality the integrand decreases as an exponential function when $q \to \mathrm{i}\infty$). In the opposite case of $c > v_m$, the path of integration can be deformed into the lower half-plane (the singular point is located in the upper half-plane) to show $S_m = 0$ for $x > ct$. The value of c is an arbitrary number, thus, conclude

$$S_m = 0 \qquad (x < v_m t, \quad x > v_m t) . \qquad (6.3.28)$$

Hence, the carrier of the S_m is localized on the ray $x = v_m t$, and following [6.26], we obtain

$$S_m = \sum_{k=0}^{N} g_{m_k}(t) \delta^{(k)}(x - v_m t) ,$$
$$S_m^{\mathrm{LF}} = \sum_{k=0}^{N} (-\mathrm{i}q)^k g_{m_k}^L (s - \mathrm{i}qv_m) \qquad \left(g_{m_k}^L \risingdotseq g_{m_k}(t)\right) . \qquad (6.3.29)$$

In particular, if $S_m^{\mathrm{LF}} = (s - \mathrm{i}qv_m)^\alpha$, then

$$S_m = \begin{cases} \dfrac{t_+^{-\alpha-1}}{\Gamma(-\alpha)} \delta(v_m t - x) & (\alpha \neq 0, 1, \ldots), \\ D_{t+x/v_m}^k & (\alpha = k = 0, 1, \ldots). \end{cases} \qquad (6.3.30)$$

Here, we denote

$$D^k_{t+x/v_m} = \sum_{n=0}^{k} \delta^{(n)}(t)\delta^{(k-n)}(x)v_m^{k-n}\binom{k}{n},$$

$$D^k_{t+x/v_m} * f(t,x) = \left(\frac{\partial}{\partial t} + v_m\frac{\partial}{\partial x}\right)^k f(t,x), \tag{6.3.31}$$

$$\binom{k}{n} = \frac{k(k-1)\dots(k-n+1)}{1\cdot 2\dots n}, \quad t_+^\alpha = \begin{cases} t^\alpha & (t>0), \\ 0 & (t<0). \end{cases}$$

The carrier of the generalized function D^k_{t+x/v_m} is limited to the point $t = x = 0$ being on the ray $x = v_m t$.

Consider now the convolution product of several functions:

$$S_\Omega = S_{m_1} * S_{m_2} * \dots * S_{m_r}, \quad m_1 < m_2 < \dots < m_r,$$
$$S_\Omega^{\mathrm{LF}} = S_{m_1}^{\mathrm{LF}} S_{m_2}^{\mathrm{LF}} \dots S_{m_r}^{\mathrm{LF}},$$

The singular points of the function S_Ω^{LF} are on the segment connecting the points $q = -\mathrm{i}s/v_{m_r}$ and $q = -\mathrm{i}s/v_{m_1}$, i.e., they comply with the inequality

$$\frac{1}{v_{m_r}} \le \frac{\mathrm{i}q}{s} \le \frac{1}{v_{m_1}} \qquad (\mathrm{Re}\, s > 0) \tag{6.3.32}$$

($\mathrm{i}q/s$ is a real number).

Re-write (6.3.26) with respect to S_Ω, use the latter in couple with equation (6.3.27), and repeate reasoning made for (6.3.28) at both $c < v_{m_1}$ and $c > v_{m_r}$, to obtain

$$S_\Omega = 0 \qquad (x < v_{m_1}t, \quad x > v_{m_r}t). \tag{6.3.33}$$

Thus, the carrier of the convolution product for multiple functions is a union of respective carriers and areas between them:

$$v_{m_1}t \le x \le v_{m_r}t. \tag{6.3.34}$$

Each of the functions subject to convolution describes some concentrated disturbance (differing from zero at any time moment only at one point) which moves at the speed v_m along the x-axis. Convolving this function with another function of the same type ($v = v_k$) corresponds to the fact that every point of the ray $x = v_m t (x = v_k t)$ generates a disturbance which moves at the speed $v_k(v_m)$; as a result, disturbances occupy a sector

$$\min(v_m, v_k)t \le x \le \max(v_m, v_k)t$$

on the (x,t)-plane (Fig. 6.6). A convolution with the third function ($v = v_p$) corresponds to radiation which propagates at a speed v_p from every point of a sector, etc. Note that inequalities of (6.3.34) take place regardless of specific multipliers in decomposition for the function S_Ω^{LF}. It is sufficient for this function not to have other singular points in addition to those stated by (6.3.32).

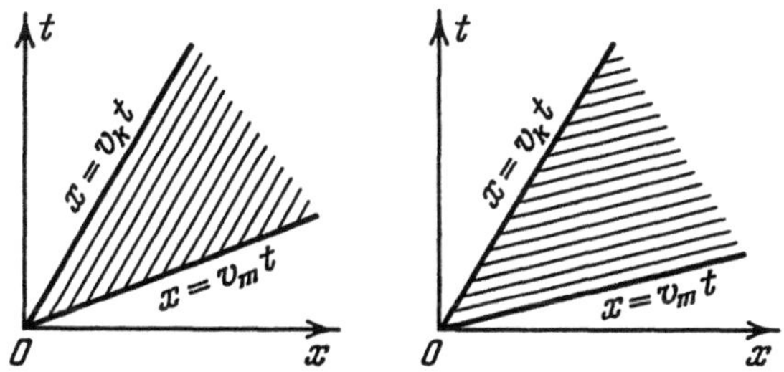

Fig. 6.6. Disturbance region on the (x, t)-plane

6.3.3 Solution of the Two-Dimensional Dynamic Problem

Consider now the equation

$$u^{\mathrm{LF}}(s,q) = S^{\mathrm{LF}}(s,q)\sigma^{\mathrm{LF}}(s,q)\ , \tag{6.3.35}$$

in which the functions $\{u^{\mathrm{LF}}, \sigma^{\mathrm{LF}}, S^{\mathrm{LF}}\}$ are of the same sense as used in equation (6.3.17); they should be defined specifically, depending on a choice of a problem — I, II, or III, see the formulae (6.3.18, 20, 22). There exist other mixed dynamic problems which can be reduced to solving equation (6.3.35). Below, a general solution of this equation is presented.

Let us introduce functions $u_{\pm}(t,x)$ and $\sigma_{\pm}(t,x)$

$$f_+ = \begin{cases} f & (x > l(t)), \\ 0 & (x < l(t)), \end{cases} \qquad f_- = \begin{cases} 0 & (x > l(t)), \\ f & (x < l(t)) \end{cases} \qquad (f = u, \sigma)\ . \tag{6.3.36}$$

Hereinafter, we assume that $x_2 = x$. With this, a displacement u and a stress σ may be written as

$$u = u_+ + u_-, \qquad \sigma = \sigma_+ + \sigma_-\ . \tag{6.3.37}$$

According to the formulae (6.3.20–22), it follows from (6.3.36) that $u_+ \equiv 0$, however, in order for generality of a solution of equation (6.3.35) to be maintained, we shall not take this equality into account. Let us assume that functions σ_- and u_+ are specified, and the functions σ_+ and u_- are required to be found. In this case, equation (6.3.35) can be formulated as

$$u_+^{\mathrm{LF}} + u_-^{\mathrm{LF}} = S^{\mathrm{LF}}\left(\sigma_+^{\mathrm{LF}} + \sigma_-^{\mathrm{LF}}\right)\ . \tag{6.3.38}$$

Assume now the function S^{LF} to be such that the factorization considered in this Section results in the representation $S^{\mathrm{LF}} = S_+^{\mathrm{LF}} S_-^{\mathrm{LF}}$ such that inverse transforms for functions $S_{\pm}^{\mathrm{LF}}$ and $P_{\pm}^{\mathrm{LF}} (P_{\pm}^{\mathrm{LF}} = 1/S_{\pm}^{\mathrm{LF}})$ satisfy the following conditions:

$$\begin{aligned} &S_+(t,x) = P_+(t,x) = 0 \qquad (x < v_+ t), \\ &S_-(t,x) = P_-(t,x) = 0 \qquad (x > v_- t), \quad v_- < v_+\ . \end{aligned} \tag{6.3.39}$$

Let us assume that the speed of motion of the point $x = l(t)$ where the boundary conditions are separated satisfies the inequality

$$v_- < \dot{l}(t) < v_+ \, . \tag{6.3.40}$$

Then, dividing both sides of equation (6.3.38) by S_-^{LF}, we re-write it in the form

$$P_-^{\mathrm{LF}}\left(u_+^{\mathrm{LF}} + u_-^{\mathrm{LF}}\right) = S_+^{\mathrm{LF}}\left(\sigma_+^{\mathrm{LF}} + \sigma_-^{\mathrm{LF}}\right) \, . \tag{6.3.41}$$

Applying the inverse LF-transformation to (6.3.41) one can obtain

$$\begin{aligned} &P_- * u_- + \left[P_- * u_+ - S_+ * \sigma_-\right] H\left[l(t) - x\right] \\ &\quad = S_+ * \sigma_+ - \left[P_- * u_+ - S_+ * \sigma_-\right] H\left[x - l(t)\right] \, . \end{aligned} \tag{6.3.42}$$

In a subsequent treatment, account will be taken of the fact that, from conditions (6.3.39, 40), the following equations yield:

$$P_- * u_- = 0 \quad (x > l(t)) \, , \qquad S_+ * \sigma_+ = 0 \quad (x < l(t)) \, . \tag{6.3.43}$$

Indeed, an original function for $P_-^{\mathrm{LF}} u_-^{\mathrm{LF}}$ should be written as

$$\int_{-\infty}^{\infty}\int_0^t P_-(\tau,\xi) u_-(t-\tau, x-\xi) H(v_-\tau - \xi) H\left[l(t-\tau) - x + \xi\right] d\tau \, d\xi \, , \tag{6.3.44}$$

where multipliers (Heaviside functions) emphasize that carriers of $P_-(t,x)$ and $u_-(t,x)$ are limited. The convolution product (6.3.44) differs from zero only within those domains on the plane (x,t) where arguments of the Heaviside functions satisfy inequalities $v_-\tau \geq \xi$, $\xi \geq x - l(t-\tau)$. Hence, if $x > l(t)$, one finds

$$v_-\tau > l(t) - l(t-\tau) \geq \left[\dot{l}(t)\right]_{\min} \tau \, ,$$

i.e., $v_- > [\dot{l}(t)]_{\min}$. However, this is in contradiction with the condition (6.3.40). Therefore, $P_- * u_- = 0$ for $x > l(t)$. It can be analogously proven that $S_+ * \sigma_+ = 0$ for $x < l(t)$. From (6.3.42, 43) it follows that the left-hand side of (6.3.42) is equal to zero if $x > l(t)$, whereas the right-hand side becomes zero when $x < l(t)$. But they are equal to each other over the entire x-axis, therefore, each of them can differ from zero only at a point of the axis $x = l(t)$. Thus, each of them is equal to some generalized function C whose carrier is the point $x = l(t)$. Finally, obtain

$$\begin{aligned} P_- * u_- &= (S_+ * \sigma_- - P_- * u_+) H\left[l(t) - x\right] + C \, , \\ S_+ * \sigma_+ &= -(S_+ * \sigma_- - P_- * u_+) H\left[x - l(t)\right] + C \, . \end{aligned} \tag{6.3.45}$$

As is known [6.26], such a generalized function C can be written in the form,

$$C \equiv C\left(x - l(t), t\right) = \sum_{m=0}^{n} f_m(t) \delta^{(m)}\left[x - l(t)\right] \, , \tag{6.3.46}$$

where $\delta_{(m)}$ is the mth derivative of the delta-function.

Now, apply again the LF-transformation to (6.3.45), divide the first equation by P_-^{LF} and the second by S_+^{LF}, thereafter perform an inverse transformation, to obtain the solution

$$\begin{aligned} u_-(t,x) &= S_- * \left\{ \left[\sigma_- * S_+ - u_+ * P_- \right] H\left[l(t) - x \right] + C \right\} , \\ \sigma_+(t,x) &= -P_+ * \left\{ \left[\sigma_- * S_+ - u_+ * P_- \right] H\left[x - l(t) \right] - C \right\} . \end{aligned} \tag{6.3.47}$$

Expressions within figure brackets equal zero if $x > l(t)$ (for u_-) and if $x < l(t)$ (for σ_+), and functions S_- and P_+ satisfy conditions of (6.3.39), therefore, the latter solutions comply with the necessary conditions

$$u_- = 0 \quad (x > l(t)) , \qquad \sigma_+ = 0 \quad (x < l(t)) .$$

The generalized function C must be defined by additional conditions (for example, by the condition of a limited energy flux to an edge $x = l(t)$ of a crack in the corresponding problem). Note that the function C produces homogeneous solutions. Indeed, if $\sigma_- = u_+ \equiv 0$ while $C \not\equiv 0$, then from (6.3.46, 47) it follows that

$$\begin{aligned} u = u_- &= S_- * C \\ &= \int_{-\infty}^{\infty} \int_0^t S_-(t-\tau, x-\xi) * \sum_{m=0}^{n} f_m(\tau)\delta^{(m)}\left[\xi - l(\tau)\right] d\tau\, d\xi \\ &= \sum_{m=0}^{n} \frac{\partial^m}{\partial x^m} \int_0^t S_-\left[t-\tau, x - l(\tau)\right] f_m(\tau) d\tau \not\equiv 0 , \\ \sigma = \sigma_+ &= \sum_{m=0}^{n} \frac{\partial^m}{\partial x^m} \int_0^t P_+\left[t-\tau, x - l(\tau)\right] f_m(\tau) d\tau \not\equiv 0 . \end{aligned} \tag{6.3.48}$$

The fact that the carriers of the functions u and σ stated in (6.3.48) indeed satisfy the conditions

$$u = 0 \quad \text{for} \quad x > l(t), \qquad \sigma = 0 \quad \text{for} \quad x < l(t)$$

is a consequence of the circumstances used in deriving the formulae (6.3.47, 48). However, we may perform a proof immediately. By means of the inequalites (6.3.40), obtain

$$x - l(t) + v_-(t-\tau) \le x - l(\tau) = x - l(t) + l(t) \quad l(\tau) \le x - l(t) + v_+(t-\tau) .$$

Hence,

$$\begin{aligned} &x - l(\tau) - v_-(t-\tau) > 0 \quad \text{for} \quad x > l(t) , \\ &v_+(t-\tau) - \left[x - l(\tau)\right] > 0 \quad \text{for} \quad x < l(t) . \end{aligned}$$

This, when coupled with equations (6.3.39), gives

$$S_-\left[t-\tau,\ x - l(\tau)\right] = 0 \quad (x > l(t)), \quad P_+\left[t-\tau,\ x - l(\tau)\right] = 0 \quad (x < l(t))$$

Thus, both functions, u and σ, satisfy the same equations.

Since the functions $S_\pm$ and $P_\pm$ are known (they can be determined independently of the functions $\sigma_-(t,x)$, $u_+(t,x)$, $l(t)$), the solution (6.3.47) is, in general, expressed through a quadrupled integration. An analysis simplifies essentially when carriers of the functions $S_\pm$ and $P_\pm$ are localized on rays in the (x,t)-plane. A convolution with the delta-function or its kth derivative (a convolution in terms of one variable) is known to correspond to an identity transformation or a k-fold differentiation, respectively, therefore, double integrals are the maximum difficulties which are likely to occur in the functions u_- and σ_+.

6.4 Some Crack and Punch Problems

In order to exemplify application of the method expounded in Sect. 6.3, we shall now solve tow-dimensional problems where semi-infinite cracks subjected to either plane or antiplane load conditions propagate at variable speeds; in addition, a solution for the two-dimensional acoustic problem on a punch penetration will be found.

6.4.1 The Plane-Strain Problem of a Semi-Infinite Crack Propagation

Consider propagation of a semi-infinite crack $x_2 < l(t)$ subjected to normal loads from a quiescent state. This Problem I is shown to lead to equations (6.3.17) (at $j = 1$) with conditions (6.3.20). In this particular case, the analytical function $S^{\mathrm{LF}}(s,q)$:

$$S^{\mathrm{LF}}(s,q) \equiv S_1^{\mathrm{LF}}(s,q) = -\frac{c_2^{-2}s^2\sqrt{c_1^{-2}s^2+q^2}}{\mu R(s,q)} \tag{6.4.1}$$

has the following singular points in the q-plane: $q = \pm\mathrm{i}s/c_1$, $q = \pm\mathrm{i}s/c_2$ (branch points), $q = \pm\mathrm{i}s/c_R$ (simple poles), according to (6.3.18). Singular points in the s-plane correspond to those in the q-plane. Thus, all singular points of the function $S^{\mathrm{LF}}(s,q)$ are of the same kind as (6.3.24) indicates, so, the carrier of $S(t,x)$ is the sector $-c_1 t \le x \le c_1 t$ (hereinafter, we change x_2 by x), whereas the solution to equation (6.3.17) with conditions (6.3.20) is given by (6.3.47) where one must assume $\sigma_- = p_1(t,x)$, $u_+ \equiv 0$.

Factorization, $S^{\mathrm{LF}} = S_+^{\mathrm{LF}}S_-^{\mathrm{LF}}$, can be carried out in a conventional way, with factors $S_\pm^{\mathrm{LF}}$ depending on the range in which the crack propagation speed $\dot{l}(t)$ is. If $0 \le \dot{l}(t) < c_R$, all singular points of the function S_+^{LF} must be in the lower half-plane of the q-plane, whereas those of the S_-^{LF}, in the upper one. In case of $c_R < \dot{l}(t) < c_2$, one must carry out the factorization differently. Namely, S_+^{LF} is required to have singular points $q = -\mathrm{i}s/c_1$ and $q = -\mathrm{i}s/c_2$, and the function S_-^{LF}, all the remaining points. Lastly, for a crack speed satisfying the inequality $c_2 < \dot{l}(t) < c_1$ the S_+^{LF} after factorization must have one singular point $q = -\mathrm{i}s/c_1$ (and S_-^{LF} must have all the remaining points).

These conclusions follow from the conditions (6.3.39, 40). Consider, for example, the case $0 \le \dot{l}(t) < c_R$. If S_+^{LF} has all its singular points in the lower half-plane, whereas S_-^{LF} is singular only in the upper one, the following sectors in the (x, t)-plane will be carriers for the functions S_+, P_+, S_-, and P_-:

$$
\begin{aligned}
&c_R t \le x \le c_1 t(S_+), \qquad c_2 t \le x \le c_1 t(P_+)\,,\\
&-c_1 t \le x \le -c_R t(S_-), \quad -c_1 t \le x \le -c_2 t(P_-)\,,
\end{aligned}
$$

respectively (note that carriers for P_+ and P_- are narrower than those for S_+ and S_-, respectively, since the points $q = \pm \mathrm{i}s/c_R$ are not singular points of $P_\pm^{LF}$). Hence, $S_+(t,x) = P_+(t,x) = 0$ for $x < c_R t$ and $S_-(t,x) = P_-(t,x) = 0$ for $x > -c_R t$. So, if $|\dot{l}(t)| < c_R$, a solution to this problem is written by relations (6.3.47); this complies with the conditions (6.3.39, 40) if one assumes that $v_+ = c_R$ and $v_- = -c_R$.

Thus, as a result of factorization for the case $0 \le \dot{l}(t) < c_R$, we obtain

$$
\begin{aligned}
&S_+^{LF} = \frac{c_R}{\sqrt{c_1}} \frac{\sqrt{s - \mathrm{i}c_1 q}}{s - \mathrm{i}q c_R} D_+^{LF}\,, \quad S_-^{LF} = -\frac{(1-\nu)c_R}{\mu\sqrt{c_1}} \frac{\sqrt{s + \mathrm{i}c_1 q}}{s + \mathrm{i}q c_R} D_-^{LF}\,,\\
&P_\pm^{LF} = \frac{1}{S_\pm^{LF}}\,,
\end{aligned}
\tag{6.4.2}
$$

where

$$
\begin{aligned}
&D_\pm^{LF} = \exp\left[\frac{1}{\pi}\int_{c_1^{-1}}^{c_2^{-1}} \frac{\varphi(\alpha)d\alpha}{\alpha \mp \mathrm{i}q/s}\right]\,,\\
&\varphi(\alpha) = \tan^{-1}\frac{4\alpha^2\sqrt{c_2^{-2} - \alpha^2}\sqrt{\alpha^2 - c_1^{-2}}}{\left(2\alpha^2 - c_2^{-2}\right)^2}\,,
\end{aligned}
\tag{6.4.3}
$$

ν is Poisson's ratio, $\varphi(\alpha)$ comprises arithmetic radicals. These expressions for $D_\pm^{LF}$ are obtained in [6.23] after decomposing the function

$$
D^{LF} = \frac{c_2^{-2}s^2}{1-\nu}\frac{\left(s^2/c_R^2 + q^2\right)}{R(s,q)}
$$

by means of the Cauchy-type integral:

$$
\begin{aligned}
&D^{LF} = D_+^{LF} D_-^{LF}\,,\\
&D_\pm^{LF}(s,q) = \exp\left\{\pm\frac{1}{2\pi\mathrm{i}}\int_{-\infty}^{\infty} \ln\left[D^{LF}(s, s\xi)\right] \frac{d\xi}{\xi - q/s}\right\}
\end{aligned}
\tag{6.4.4}
$$

and modifying the expressions in (6.4.4): integration for $D_+^{LF}(D_-^{LF})$ requires $\mathrm{Im}\{q\}$ to be positive (negative, respectively).

The concluding operations in obtaining a solution to the Problem I for $0 \le \dot{l}(t) < c_R$ are to use expressions (6.4.2) in equations (6.3.47), write $p_1(t,x)$ and 0 in place of σ_- and u_+, respectively, and determine the generalized function C.

It should be mentioned that this solution obtained by means of (6.3.47) is too cumbersome and, for $l(t) \not\equiv$ const, is expressed by a quadrupled integral. For this reason, we do not consider it in detail here; the reader is referred to *Kostrov* [6.20, 21] and *Slepyan* [6.23]. It follows from these contributions that, as long as $0 \le \dot{l}(t) < c_R$, the stress in the vicinity of a crack edge has an integrable singularity of the order of $-1/2$ (therefore, displacements are limited) if the stress is found after neglecting the homogeneous solution generated by the function C. Hence, in order for the energy flux on crack edge $x = l(t)$ to be limited, the generalized function C must be set to zero (if $C \not\equiv 0$, unlimited displacements and nonintegrable stresses will be at the crack edge, resulting in an unlimited energy flux).

Note that a solution of the two-dimensional problem under consideration simplifies if $\dot{l}(t) \equiv v =$ const. Study this situation. We can write $s = \mathrm{i}qv + s'$ to reduce the relation (6.3.35) to

$$u_0^{\mathrm{LF}}(s',q) = S_0^{\mathrm{LF}}(s',q)\sigma_0^{\mathrm{LF}}(s',q)\,,$$

where all expressions

$$f_0^{\mathrm{LF}}(s',q) \equiv f^{\mathrm{LF}}(\mathrm{i}qv + s',q) \qquad (f \equiv u, S, \sigma)$$

are indeed transforms of the corresponding original time functions obtained by means of LF-transformation in the moving coordinate system $\{\xi, y, t\}(\xi = x - vt)$, for example,

$$u_0^{\mathrm{LF}}(s',q) = \int_{-\infty}^{\infty}\int_0^{\infty} u(t, vt + \xi)\mathrm{e}^{-s't + \mathrm{i}q\xi} dt\, d\xi\,.$$

The point separating the boundary conditions is motionless in the new coordinate system ($\xi = 0$), therefore the function $S_0^{\mathrm{LF}}(s',q)$ may be ordinarily factorized in the q-plane, within the bounds of a traditional usage of the Wiener–Hopf method (see Sect. 6.1.1 where the correspondence of the double Laplace transformation with the LF-transformation is given by the formulae, $p \to s'$, $s \to -\mathrm{i}q$).

6.4.2 The Antiplane Problem of a Semi-Infinite Crack Propagation

Consider the non-stationary Problem III in antiplane crack dynamics. According to equations (6.3.17) (at $j = 3$) and (6.3.2) a solution to this can be described by equation

$$u^{\mathrm{LF}}(s,q) = S^{\mathrm{LF}}(s,q)\sigma^{\mathrm{LF}}(s,q) \tag{6.4.5}$$

where

$$S^{\mathrm{LF}}(s,q) = -\frac{1}{\mu\sqrt{c_2^{-2}s^2+q^2}}\,, \tag{6.4.6}$$

and in expressions for $u^{\mathrm{LF}} = u_+^{\mathrm{LF}} + u_-^{\mathrm{LF}}$ and $\sigma^{\mathrm{LF}} = \sigma_+^{\mathrm{LF}} + \sigma_-^{\mathrm{LF}}$ one must assume

$$u_+(t,x) = 0, \quad \sigma_-(t,x) = p_3(t,x) \tag{6.4.7}$$

(hereinafter, x_2 is replaced by x).

From (6.4.6) it follows that the function S^{LF} has two singular points $q = \pm \mathrm{i}s/c_2$ (branch points) which pertain to the type (6.3.24). For the sake of definiteness, assume $0 \le \dot{l}(t) < c_2$. In this case, a factorization for the function S^{LF} may be performed so that the function S_+^{LF} will possess a singular point in the lower half-plane ($q = -\mathrm{i}s/c_2$) whereas the function S_-^{LF} eill be singular in the upper half-plane ($q = \mathrm{i}s/c_2$). The factorization can be carried out without difficulty:

$$S^{\mathrm{LF}} = S_+^{\mathrm{LF}} S_-^{\mathrm{LF}}, \quad S_+^{\mathrm{LF}} = \frac{\sqrt{c_2}}{\sqrt{s-\mathrm{i}qc_2}}, \quad S_-^{\mathrm{LF}} = -\frac{\sqrt{c_2}}{\mu\sqrt{s+\mathrm{i}qc_2}}\,. \tag{6.4.8}$$

In future, there will also be need for functions $P_\pm^{\mathrm{LF}} = 1/S_\pm^{\mathrm{LF}}$.

Original time functions arc defined by equation (6.3.30) at $\alpha = \pm 1/2$:

$$\begin{aligned}
S_- &= -\frac{\sqrt{c_2}}{\mu}\frac{t_+^{-1/2}}{\sqrt{\pi}}\delta(c_2t+x), \qquad P_- = \frac{\mu}{\sqrt{c_2}}\frac{t_+^{-3/2}}{2\sqrt{\pi}}\delta(c_2t+x)\,,\\
S_+ &= \frac{\sqrt{c_2}}{\sqrt{\pi}}t_+^{-1/2}\delta(c_2t-x) \qquad (\delta(-c_2t-x) = \delta(c_2t+x))\,,\\
P_+ &= -\frac{t_+^{-3/2}}{2\sqrt{\pi c_2}}\delta(c_2t-x)\,.
\end{aligned} \tag{6.4.9}$$

After factorizing the function $S^{\mathrm{LF}}(s,q)$, its inverse transform

$$S(t,x) = -\frac{c_2}{\pi\mu}\frac{H(c_2t-|x|)}{\sqrt{c_2^2t^2-x^2}} = S_+ * S_-$$

(with a carrier occupying a sector $|x| \le c_2t$) turns out to be decomposed into two functions, S_+ and S_-, each of which corresponding to a directed wave disturbance. The carrier of the function $S_+(t,x)$ is the ray $x = c_2t$, and the carrier of $S_-(t,x)$ is the ray $x = -c_2t$. A solution to this problem involving the functions from (6.4.8) at conditions (6.4.7) can be described by the formulae (6.3.47) with the function C being still unknown (note that the solution of this mixed problem is valid for $|\dot{l}(t)| < c_2$ as conditions (6.3.39,40) give at $v_- = -c_2$, $v_+ = c_2$). In order to simplify integrals in (6.3.47), first assume that $\sigma_- = \delta(x-\xi)\delta(t-\tau)$, $\xi < l < \tau$, $\tau \ge 0$. Then

$$\begin{aligned}
\sigma_- * S_+ &= \delta(x-\xi)\delta(t-\tau) * \frac{\delta(c_2t-x)\sqrt{c_2}}{\sqrt{\pi}}t_+^{-1/2}\\
&= \sqrt{\frac{c_2}{\pi}}\delta\left[c_2(t-\tau)-(x-\xi)\right](t-\tau)_+^{-1/2}\,.
\end{aligned}$$

Without taking account of the function C, obtain

$$\begin{aligned}
\sigma_+(t,x) &= -P_+ * \left[\left(\sigma_- * S_+\right) H\left(x - l(t)\right)\right] \\
&= \frac{1}{2\pi}\left[t_+^{-3/2}\delta\left(c_2 t - x\right)\right] * \Big[\delta\left(c_2(t-\tau)\right. \\
&\qquad \left.-(x-\xi)\right)(t-\tau)_+^{-1/2} H\left(x - l(t)\right)\Big] \\
&= \frac{1}{2\pi}\int_0^t \int_{-\infty}^{\infty} (t-t')_+^{-3/2}\delta\left[c_2(t-t') - x + x'\right](t'-\tau)_+^{-1/2} \\
&\quad \times \delta\left[c_2(t'-\tau) - x' + \xi\right] H\left(x' - l(t')\right) dx'dt' \qquad (6.4.10)\\
&= \frac{1}{2\pi}\delta\left[c_2(t-\tau) - (x-\xi)\right]\int_0^t (t-t')_+^{-3/2}(t'-\tau)_+^{-1/2} \\
&\quad \times H\left[x - c_2(t-t') - l(t')\right] dt' .
\end{aligned}$$

In specifying the integration limits in (6.4.10), the Note to the formula (6.3.23) is taken into account.

The Heaviside function undergoes the unit jump at

$$x - c_2(t-t') - l(t') = 0 . \qquad (6.4.11)$$

Since $\dot{l}(t) < c_2$, the left-hand side of equations (6.4.11) increases as t' grows. Let us denote the root of equation (6.4.11) as $t_* = t_*(t,x)$. It is obvious that the left-hand side in (6.4.11) is positive if $t' > t_*$, therefore, integration is carried out in effect from $t' = t_*$ instead of zero.

Taking into account the properties of generalized functions, the integral in (6.4.10) may be reduced to

$$\begin{aligned}
\int_{t_*}^{t} (t-t')_+^{-3/2}(t'-\tau)_+^{-1/2} dt' &= -2\frac{\partial}{\partial t}\int_{t_*}^{t} (t-t')_+^{-1/2}(t'-\tau)_+^{-1/2} dt' \\
&= -\frac{2}{t-\tau}(t_*-\tau)_+^{1/2}(t-t_*)_+^{-1/2} ,
\end{aligned} \qquad (6.4.12)$$

so, expression (6.4.10) can be written as

$$\begin{aligned}
&\sigma_+(t,x) \\
&= -\frac{1}{\pi(t-\tau)}\sqrt{\frac{t_*-\tau}{t-t_*}} H(t_*-\tau)H(t-t_*)\delta\left[c_2(t-\tau) - (x-\xi)\right] .
\end{aligned} \qquad (6.4.13)$$

In the coefficient at $\delta[c_2(t-\tau)-(x-\xi)]$, we can specify $c_2 t - x = c_2\tau - \xi$, and after taking account of equation (6.4.11) (which is valid at $t' = t_*$), can re-write equation (6.4.13) finally as

$$\sigma_+(t,x) \tag{6.4.14}$$
$$= -\frac{1}{\pi(x-\xi)}\sqrt{\frac{l(t_*)-\xi}{x-l(t_*)}}H\left[l(t_*)-\xi\right]H\left[x-l(t_*)\right]\delta\left[c_2(t-\tau)-(x-\xi)\right] .$$

At arbitrary stress $\sigma_- = p_3(t,x)$ (with $p_3(0,x)=0$), it suffices to multiply the right-hand side of (6.4.14) by $p_3(\tau,\xi)$ and to integrate the product in both $\xi \epsilon (-\infty, l(\tau))$ and $\tau \epsilon (0,\infty)$. As a result, find

$$\sigma_+(t,x) = -\frac{H[x-l(t)]}{\pi\sqrt{x-l(t_*)}}\int_{x-c_2t}^{l(t_*)} p_3\left(t-\frac{x-\xi}{c_2},\xi\right)\frac{\sqrt{l(t_*)-\xi}\,d\xi}{x-\xi} . \tag{6.4.15}$$

Here, the expression $x - l(t_*)$ within the argument of the Heaviside function H in (6.4.14) is replaced by $x - l(t)$ since $t_* \to t$ when $x \to l(t_*)$, see (6.4.11).

The solution (6.4.15) has been derived by *Kostrov* [6.19].

Using analogously equation (6.3.47) for u_-, obtain

$$\begin{aligned} u_-(t,x) = -\frac{c_2}{\pi\mu}\Bigg[&\int_0^{t_0} d\tau \int_{x-c_2(t-\tau)}^{l(t_0)-c_2(t_0-\tau)} p_3(\tau,\xi)\frac{d\xi}{F} \\ &+ \int_{t_0}^{t} d\tau \int_{x-c_2(t-\tau)}^{x+c_2(t-\tau)} p_3(\tau,\xi)\frac{d\xi}{F}\Bigg] H\left[l(t)-x\right] , \end{aligned} \tag{6.4.16}$$

where $t_0 = t_0(t,x)$ is the root of the equation, $l(t_0) = x + c_2(t-t_0)$ (note that $t > t_0$ for any $t > 0$),

$$F = \{[c_2(t-\tau)-(x-\xi)]\,[c_2(t-\tau)+(x-\xi)]\}^{1/2} .$$

Consider briefly one interesting property of the solution (6.4.14) mentioned by *Slepyan* in [6.23]. It can be seen from (6.4.14) that the stress on the continuation of the crack represents the concentrated load which moves at a speed c_2 of shear waves and depends on only two dimensions: (i) the distance $x-\xi$ from this point to the point where the external pulse has been applied and (ii) the distance $l(t_*)-\xi$ between the point where the pulse has been applied and the crack edge position at $t = t_*$ (note that the expression $x - l(t_*)$ can be written as a difference of the distances $x-\xi$ and $l(t_*)-\xi$). In this case, from equation (6.4.11) and from the equation $c_2t - x = c_2\tau - \xi$ (used to derive (6.4.14)) it follows that t_* is the time moment when the shear wave caused by the pulse arrives at the crack edge. Now, using equation (6.4.14) it is not difficult to determine displacements and stresses in the half-plane $x_1 > 0$ since the stress is known over the whole of the axis $x_2 = x$; due to symmetry in the solution — $\sigma_{31}(t,-x_1,x_2) = \sigma_{31}(t,x_1,x_2)$, $u_3(t,-x_1,x_2) = -u_3(t,x_1,x_2)$ — the solution for the whole (x_1,x_2)-plane can be written.

In the domain $x_1 > 0$ the solution in terms of transforms is a result of using equation (6.3.15):

$$u_3^{\mathrm{LF}}(s,x_1,q) = -\mathrm{e}^{-x_1 n_2}\left(\sigma_+^{\mathrm{LF}} + \sigma_-^{\mathrm{LF}}\right)\frac{1}{\mu n_2}, \quad n_2 = \sqrt{c_2^{-2}s^2 + q^2}\,, \tag{6.4.17}$$

where σ_-^{LF} is the LF-transform of the pulse $\sigma_- = \delta(x_2 - \xi)\delta(t-\tau)$, σ_+^{LF} is the transform of the solution (6.4.14) (at $x = x_2$). After performing the inverse transformation for (6.4.17) with due account for the fact that the function

$$\pi^{-1}H(t - r/c_2)\Big/\sqrt{t^2 - r^2/c_2^2}\left(r = \sqrt{x_1^2 + x_2^2}\right)$$

is the original time function for $\exp(-x_1 n_2)/n_2$, obtain

$$\begin{aligned} u_3(t,x_1,x_2) &= -\left[f(t,x_2,x_2)\right]^{-1/2} \\ &\quad\times\left\{H\left[f(t,x_1,x_2)\right] - H\left[f(t-t_*,x_1,x_2-c_2t_*)\right]\right\}\,, \\ f(t,x_1,x_2) &= (t-\tau)^2 - c_2^{-2}\left[(x_2-\xi)^2 + x_1^2\right]\,. \end{aligned} \tag{6.4.18}$$

Thus, when sides of the crack are subjected to the pulse $\sigma = \delta(x_2 - \xi)\delta(t-\tau)$, crack propagation has an effect only through the parameter t_* which represents the time moment when the shear wave from the pulse arrives at the crack edge. Subsequent propagation of the crack ($0 \le \dot{l} < c_2$) does not affect the solution.

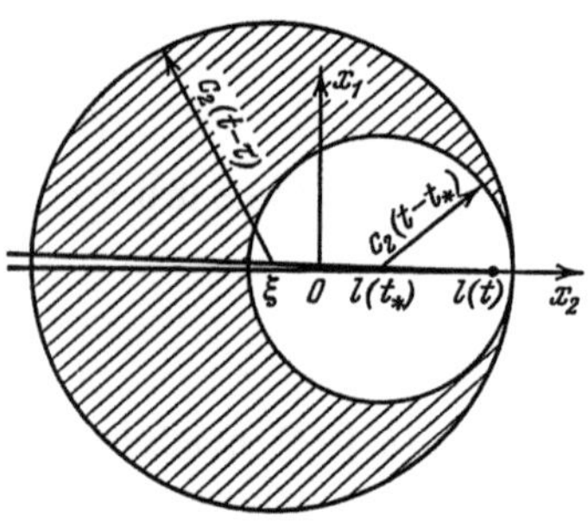

Fig. 6.7. Wave pattern near the crack edge at instance t

From (6.4.18), it follows that displacements in the (x_1, x_2)-plane are defined as a sum of the shear wave due to pulse application (see a shaded field in Fig. 6.7) and the wave reflected from the edge of the crack (see an unshaded circle in Fig. 6.7). The incident wave is described by the solution including the first term within the brackets in (6.4.18), whereas the reflected wave is introduced by the second term within these brackets. The reflected wave arises at the instance $t = t_*$ when the incident wave arrives at the crack edge. Equations (6.4.18) show that u_3 — consequently, the stress — both approach zero if the inequality,

$$(x_2 - \xi - c_2 t_*)^2 + x_1^2 < c_2^2(t - \tau - t_*)^2\,,$$

takes place. Since we have $c_2 t - x_2 = c_2\tau - \xi$ and equation (6.4.11) is effective (at $x = x_2$, $t' = t_*$, this inequality is an equivalent to the other one:

$$[x_2 - l(t_*)]^2 + x_1^2 < c_2^2(t-t_*)^2 \,, \tag{6.4.19}$$

which defines the interior of a circle of radius $c_2(t-t_*)$ with the center point $x_2 = l(t_*)$ on the crack. This means that there are no disturbances within the circle (6.4.19). Thus, the shear wave emitted at the instance $t = \tau$ by the pulse $\sigma_- = \delta(x_2 - \xi)\delta(t-\tau)$ at the point $x_2 = \xi(\tau \geq 0, \xi < l(\tau))$

- diffracts in the vicinity of an edge of the propagating crack,
- suppresses disturbances within the crack edge influence domain (6.4.19), and
- appears as a concentrated load (6.4.14) (at $x = x_2$) which moves along a crack continuation together with the boundary of this influence domain (i.e., with the point $x_2 = c_2(t-t_*) + l(t_*)$).

Consider the asymptotic behaviour of the stress at a crack edge ($\eta = x - l(t) \to +0$). At $x = l(t) + 0$, the equality $t' = t_* = t$ satisfies equation (6.4.11). Therefore, it is permissible to adopt that $l(t') \sim l(t) - v(t)(t-t')$, $v(t) = \dot{l}(t)$, $\eta \to +0$. Substituting this asymptotic equality into equation (6.4.11), obtain

$$t' = t_* \sim t - \frac{\eta}{c_2 - v(t)}, \quad l(t_*) = x - \frac{c_2\eta}{c_2 - v(t)} \,.$$

Turning to (6.4.15), one can find

$$\sigma_+ \sim -\frac{\sqrt{1 - v(t)/c_2}}{\pi\sqrt{\eta}} \int\limits_{l(t)-c_2 t}^{l(t)} p_3\left(t - \frac{l(t)-\xi}{c_2}, \xi\right) \frac{d\xi}{\sqrt{l(t)-\xi}} \,. \tag{6.4.20}$$

Hence it appears that the generalized function C in (6.3.47) whose carrier is localized on the line $x = l(t)$ would indeed be put equal to zero since the stress σ_+ turned out to be an integrable function; this circumstance provides boundedness of the energy flux on the crack edge.

The expression (6.4.20) may be transformed to give

$$\begin{aligned} \sigma_+(t,x) &= \frac{k_3}{\sqrt{2\pi(x - l(t))}} + \mathrm{O}(1) \qquad (x \to l(t) + 0) \,, \\ k_3 &= -\sqrt{\frac{2}{\pi}\left(1 - \frac{v(t)}{c_2}\right)} \int\limits_0^{c_2 t} p_3\left(t - \frac{\xi}{c_2},\, l(t) - \xi\right) \frac{d\xi}{\sqrt{\xi}} \end{aligned} \tag{6.4.21}$$

where $k_3 = k_3(t)$ represents the stress intensity factor which plays an important role in fracture mechanics investigations. As is known from [6.27, 28] the Griffith energy criterion for a perfectly brittle linear elastic medium in the case of dynamic propagation of an antiplane crack may be written in terms of the stress intensity factor k_3:

$$\Gamma(v) = \frac{k_3^2}{4\mu\sqrt{1 - v^2/c_2^2}} \,, \tag{6.4.22}$$

here, k_3 comes from (6.4.21), $P(v)$ is the effective surface energy which is assumed to be a function of the crack propagation speed, is characteristic of the material under consideration, and must be determined either experimentally or from some physical assumptions concerning the fracture mechanism. With $P(v)$ known, one can use equation (6.4.22) to derive a crack propagation law. Such investigations have been performed in [6.19].

6.4.3 The Acoustic Problem of a Punch

Consider the problem where an acoustic half-plane $x_1 > 0$ is penetrated by the semi-infinite punch $x_2 < l(t)$ (Fig. 6.8) at a speed $v_0(t, x_2)$. Assuming $\mu = 0$, $f \equiv 0$ in expressions (2.7.1–4) and taking account of (2.1.9), obtain

$$\Psi \equiv 0, \quad \boldsymbol{v} \equiv \dot{\boldsymbol{u}} = \operatorname{grad} \dot{\varphi}, \quad p = -\lambda \Delta \varphi = -\rho \ddot{\varphi} . \tag{6.4.23}$$

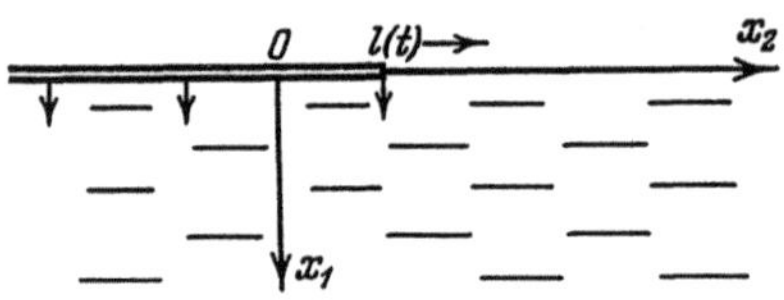

Fig. 6.8. Forcing the punch into an acoustic half-space

Thus, it is an immediate consequence of (6.4.23) that a solution (in variables t, x_1, x_2) to this two-dimensional problem can be described by the wave equation for the longitudinal potential φ, with zero initial conditions and the following conditions at the boundary:

$$\varphi = 0 \quad (x_2 > l(t)), \qquad \frac{\partial \dot{\varphi}}{\partial x_1} = v_0(t, x_2) \quad (x_2 < l(t)) .$$

Using the LF-transformation the following equation appears which relates the transform of the potential φ^{LF} to the normal speed $v^{\mathrm{LF}} = (\partial \dot{\varphi}/\partial x_1)^{\mathrm{LF}}$ at the boundary $x_1 = 0$:

$$s\varphi^{\mathrm{LF}}(s, q) = -\frac{1}{\sqrt{s^2/c_0^2 + q^2}} v^{\mathrm{LF}}(s, q) . \tag{6.4.24}$$

Comparing equations (6.4.5) and (6.4.24) it follows that the solution of equation (6.4.24) may be obtained from that for (6.4.5) by using the formulae,

$$\dot{\varphi}(t, x) = \mu u(t, x), \qquad v(t, x) = \sigma(t, x) ,$$

with replacement in (6.4.15, 16) for u_- and σ_+: c_2 and $p_3(t, x)$ must be changed by c_0 (the sonic speed in the acoustic medium) and $v_0(t, x)$, respectively. Thus, by means of equations (6.4.15, 16), obtain the following expressions for the solution at the boundary $x_1 = 0$ ($x_2 = x$, $\dot{l}(t) < c_0$):

$$v(t,x) = -\frac{1}{\pi\sqrt{x-l(t_*)}} \int\limits_{x-c_0 t}^{l(t_*)} v_0\left[t - \frac{x-\xi}{c_0}, \xi\right] \frac{\sqrt{l(t_*)-\xi}}{x-\xi} d\xi \quad (x > l(t)) ,$$

$$\dot{\varphi}(t,x) = -\frac{c_0}{\pi}\left[\int\limits_0^{t_0} d\tau \int\limits_{x-c_0(t-\tau)}^{l(t_0)-c_0(t-\tau)} v_0(\tau,\xi)\frac{d\xi}{\Phi} + \int\limits_{t_0}^{t} d\tau \int\limits_{x-c_0(t-\tau)}^{x+c_0(t-\tau)} v_0(\tau,\xi)\frac{d\xi}{\Phi}\right]$$

$$(x < l(t)) ,$$

$$\Phi \equiv \{[c_0(t-\tau)-(x-\xi)]\,[c_0(t-\tau)+(x-\xi)]\}^{1/2} , \tag{6.4.25}$$

the notation being: t_* = root of the equation $x - c_0(t-t_*) - l(t_*) = 0$, t_0 = root of the equation $x + c_0(t-t_0) - l(t_0) = 0$. In the particular case when $v_0(t,x) \equiv v_0 = \text{const}$, we can find from the formulae (6.4.25) for the speed v at the free surface and the pressure p under the punch:

$$v(t,x) = -\frac{2v_0}{\pi}\left[\sqrt{\frac{l(t_*)-x+c_0 t}{x-l(t_*)}} - \tan^{-1}\sqrt{\frac{l(t_*)-x+c_0 t}{x-l(t_*)}}\right] (x > l(t)) ,$$

$$p(t,x) = -\rho\ddot{\varphi} = \rho c_0 v_0 \tag{6.4.26}$$

$$\times\left[1 - \frac{2}{\pi}\tan^{-1}\sqrt{\frac{x+c_0 t - l(t_0)}{l(t_0)-x}} + \frac{2\dot{l}(t_0)}{\pi\left[c_0+\dot{l}(t_0)\right]}\sqrt{\frac{x+c_0 t-l(t_0)}{l(t_0)-x}}\right]$$

$$\times (x < l(t)) .$$

Note that $\dot{l}(t) < c_0$.

In the case of punch with a motionless boundary $l(t) \equiv a$, one can use (6.4.26) to obtain known results (see Sect. 6.2.2):

$$v(t,x) = -\frac{2v_0}{\pi}\left[\sqrt{\frac{a-x+c_0 t}{x-a}} - \tan^{-1}\sqrt{\frac{a-x+c_0 t}{x-a}}\right] \quad (x > a) ,$$

$$p(t,x) = \rho c_0 v_0\left[1 - \frac{2}{\pi}\tan^{-1}\sqrt{\frac{x+c_0 t-a}{a-x}}\right] \quad (x < a) . \tag{6.4.27}$$

In the case of a finite width of either the crack or the punch, a solution of the problem may be obtained for any time moment by using the fact that the Lamé equations are hyperbolic. It is obvious that, as long as waves from any particular edge of the crack (or the punch) have not reached the other edge, we face the problem with mixed boundary conditions specified for semi-infinite intervals, so, use may be made of the method [6.20, 23] considered above.

7. Homogeneous Solutions to Dynamic Problems for Anisotropic Elastic Media (Willis' Method)

In recent years, non-stationary dynamic problems involving anisotropic elastic bodies have attracted much attention from researchers because a number of important applied problems need be solved.

7.1 Studies in Elastodynamics for Anisotropic Media

We have to note that fundamental results on wave propagation in anisotropic media can be found in monographs by *Fedorov* [7.1], *Musgrave* [7.2], *Petrashen'* [7.3] and numerous investigations into this area are outlined in surveys [7.4, 5].

The non-stationary dynamic problems of the anisotropic elasticity theory were being systematically considered by Sveklo from the 1940s, see [7.6–8]. In his first works, [7.6, 8], an attempt was made to generalize the Smirnov–Sobolev method so as to consider the case of anisotropic media subjected to plane-strain conditions (6 independent constants c_{ijkl}). It has been shown that the general solution to the sets of simultaneous equations of motion in the case when the solution is a homogeneous function of zero degree is governed by two branches of an arbitrary analytic function of the complex variable specified over the two-sheeted Riemann surface. Sveklo used these results to solve some specific problems on wave propagation/diffraction. The ideas suggested by him for generalizing functionally invariant solutions have also been used in [7.9] where consideration is given to point wave-sources in some specific conditions.

Extensive investigations into the dynamics of anisotropic media including the development of theoretical foundations, formulation of solution procedures, and solutions of specific problems are reported in [7.10–16], from the 1950s. From the results of [7.1–3, 9, 11, 12] it follows, that the disturbances emitted by a point source in an anisotropic space have more complex geometry of wave fronts than those in the isotropic medium. Note that the classical Lamb problem (which becomes more complex for anisotropic media than for isotropic bodies) is investigated in [7.8, 13–16].

The results of [7.17] indicate that solving dynamic problems with mixed boundary conditions in the case of a pronounced anisotropy meets some more difficulties. In this connection, we must remember the essential contribution of *Burridge* and *Willis* [7.18, 19] (see also [7.20]) who developed methods of treating non-stationary, dynamic problems for anisotropic media, in particular, problems with mixed boundary conditions.

According to [7.19] and [7.20], the method presented by Burridge in his unpublished paper (see [7.19]) is suggested as a solution to the two-dimensional non-stationary self-similar problems of propagation of a plane crack (or a cut) from a point in an arbitrary anisotropic medium (21 independent constants c_{kjkl}), the load applied to sides of the crack being constant. The principle of the method is to represent the accelerations by means of superposition of plane waves. This approach makes it possible to generalize the Smirnov–Sobolev method to deal with the case of arbitrary anisotropic media. *Odintsev* [7.20] used this approach for solving two-dimensional mixed problems for a half-plane with boundary conditions described by arbitrary homogeneous functions of the nth degree (n is integer).

Contribution [7.19] was the first step in overcoming the difficulties of obtaining homogeneous solutions depending on the three space variables. It was the first time that an accurate analytical solution was derived for the self-similar problem on propagation of the plane elliptic crack (cut) from a point in an arbitrarily anisotropic medium; no solutions to this problem were reported earlier even for an isotropic medium.

In the fundamental research by *Willis* [7.18], another method of obtaining non-stationary homogeneous solutions (of integer degree with respect to x_1, x_2, x_3, t) to two- and three-dimensional problems is proposed for the general anisotropic medium. This regular method is based on the use of Fourier transforms in coordinates x_1, x_2 and the time t and makes it possible to solve non-stationary dynamic problems for anisotropic media in a simpler way than all those proposed by the techniques and approaches known previously.

Below, the Willis method is set forth with minor changes. In addition, the new result obtained by means of the Willis method is given: in Sect. 7.6, formulae represent the elastic solution in terms of displacement/stress jumps at the plane separating two different homogeneous anisotropic half-spaces. These expressions generalize those reported earlier in works [7.18, 21].

7.2 Solution to the First Boundary Value Problem

According to (2.1.6), the equations of motion for a linear anisotropic medium free of body forces can be written in the form

$$c_{ijkl}u_{k,lj} = \rho \ddot{u}_i \, , \qquad (i,j,k,l = 1,2,3) \, , \tag{7.2.1}$$

where the constants c_{ijkl} satisfy equations (2.1.5). Hooke's law is

$$\sigma_{ij} = c_{ijkl}u_{k,l} \, , \tag{7.2.2}$$

see (2.1.2, 3). In the matrix notation, equations (7.2.1) take the form

$$\mathbf{K}\left(\frac{\partial}{\partial t}, \nabla\right) \boldsymbol{u} = 0 \qquad \left(\nabla = \sum_{i=1}^{3} \boldsymbol{e}_i \frac{\partial}{\partial x_i}\right), \tag{7.2.3}$$

where $\mathbf{K}(\omega, \xi)$ is the matrix with components

$$K_{ik}(\omega, \boldsymbol{\xi}) = c_{ijkl}\xi_j\xi_l - \rho\omega^2\delta_{ik} \qquad (\boldsymbol{\xi} = \{\xi_1, \xi_2, \xi_3\}) \tag{7.2.4}$$

(note that, unless otherwise stated, the repeating indices in a monomial expression imply summation to be carried out from 1 to 2 for Greek indices and from 1 to 3 for Roman ones).

7.2.1 Three-Dimensional Case

Let the elastic anisotropic half-space $x_3 > 0$ be at rest at $t = 0$. Assume that at the boundary $x_3 = 0$ the stress vector $\boldsymbol{\tau}(t, x_\alpha) = \tau_1, \tau_2, \tau_3$ is specified as a homogeneous function of the nth (integer) degree with respect to the variables t, x_1, x_2, i.e.,

$$\boldsymbol{\tau}(t, x_\alpha) = t^n \boldsymbol{\tau}(\boldsymbol{y}) \qquad (t > 0), \tag{7.2.5}$$

where $\boldsymbol{y} = \{y_1, y_2\}$, $y_\alpha = x_\alpha/t$ ($\alpha = 1, 2$). Then, the displacement field, $\boldsymbol{u}(t, \boldsymbol{x})$ generated by the load $\boldsymbol{\tau}(t, x_\alpha)$ is a homogeneous function of the $(n+1)$th degree with respect to t, x_1, x_2 and satisfies the differential equations

$$\mathbf{K}\left(\frac{\partial}{\partial t}, \nabla\right) \boldsymbol{u} = 0 \qquad (x_3 > 0), \tag{7.2.6}$$

as well as the initial and boundary conditions

$$\boldsymbol{u}(0, \boldsymbol{x}) = \dot{\boldsymbol{u}}(0, \boldsymbol{x}) = 0 \quad (x_3 \geq 0), \tag{7.2.7}$$

$$\mathbf{C}_3(\nabla)\boldsymbol{u}(t, \boldsymbol{x}) = c_{i3kl}u_{k,l} = \tau_i(t, x_\alpha) \quad (x_3 \to 0), \tag{7.2.8}$$

$$\boldsymbol{u}(t, \boldsymbol{x}) \to 0 \quad (|\boldsymbol{x}| \to \infty), \tag{7.2.9}$$

here, the matrix $\mathbf{C}_j(\boldsymbol{\xi})$ is comprised of the components $C^j_{ik}(\boldsymbol{\xi})$ defined by the formula,

$$C^j_{ik}(\boldsymbol{\xi}) = c_{ijkl}\xi_l \,. \tag{7.2.10}$$

The condition (7.2.9) imposes restrictions on the behaviour of $\boldsymbol{\tau}(t, x_\alpha)$ as $(x_1^2 + x_2^2) \to \infty$; below, we shall assume that the function $\boldsymbol{\tau}(y)$ introduced in (7.2.5) is integrable along the line $\eta_\alpha y_\alpha = p$.

A formal solution of the problem (7.2.6–9) can be performed by means of the Fourier transformation in spatial coordinates and time. The conditions stated for this problem allow us to search the transform to the function $\boldsymbol{u}(t, \boldsymbol{x})$ assuming the transform to be integrable over any bounded area and, as arguments tend to infinity, be increasing as slowly as a polynomial, i.e., the transform can be sought in the class of generalized functions of slow growth (see, for example, [7.22, 23]).

Consequently, the original time function $\boldsymbol{u}(t, \boldsymbol{x})$ will be a generalized solution to the problem (7.2.6–9) which is unique in the class of the generalized functions of slow growth. Let us define the Fourier transform, $f^*(\omega, \xi_\alpha, x_3)$, for the function $f(t, \boldsymbol{x})(f(t, \boldsymbol{x}) \equiv 0$ if $t < 0)$ as

$$f^*(\omega, \xi_\alpha, x_3) = (2\pi)^{-3/2} \int_0^\infty dt \int_{-\infty}^{\infty} \int_{-\infty}^{\infty} f(t, \boldsymbol{x}) \exp\left[\mathrm{i}(\omega t + \xi_\alpha x_\alpha)\right] dx_1\, dx_2 \,. \tag{7.2.11}$$

Here, we assume that $\mathrm{Im}\{\omega\} > 0$, $\mathrm{Im}\{\xi_1\} = \mathrm{Im}\{\xi_2\} = 0$, so, equation (7.2.11) at constant values of ξ_1, ξ_2, x_3 defines $f^*(\omega, \xi_\alpha, x_3)$ which is a function of ω and is analytic in the upper half-plane of ω. The inversion formula for (7.2.11) has the form of

$$f(t, \boldsymbol{x}) = (2\pi)^{-3/2} \int_{-\infty}^{\infty} \int_{-\infty}^{\infty} d\xi_1 d\xi_2 \int_{-\infty+0\mathrm{i}}^{+\infty+0\mathrm{i}} f^*(\omega, \xi_\alpha, x_3) \exp\left[-\mathrm{i}(\omega t + \xi_\alpha x_\alpha)\right] d\omega \,. \tag{7.2.12}$$

The latter gives $f(t, \boldsymbol{x}) = 0$ for any $t < 0$.

In the problem under consideration, analyticity of the functions $\boldsymbol{u}(\omega, \xi_\alpha, x_3)$ in the region $\mathrm{Im}\{\omega\} > 0$ may be derived form the assumption of existence of the solution $\boldsymbol{u}(t, \boldsymbol{x})$ which is zero for $t < 0$ and pertains to the class of the generalized functions of slow growth. The transform $\boldsymbol{u}^*(\omega, \xi_\alpha, x_3)$ can be obtained explicitly from the solution of the problem (7.2.6–9). In this case, one can verify immediately both analyticity of the function $\boldsymbol{u}^*(\omega, \xi_\alpha, x_3)$ for $\mathrm{Im}\{\omega\} > 0$ and the statement that $\boldsymbol{u}^*(\omega_1 + 0\mathrm{i}, \xi_\alpha, x_3)$ pertains to the class of the generalized functions of slow growth with respect to real variables ω_1, ξ_1, ξ_2 at any constant value of x_3. Therefore, we shall not further consider substantiation of these properties.

Applying the transformation (7.2.11) to equations (7.2.1), find:

$$\begin{aligned} &\mathbf{K}\left(\omega, \xi_1, \xi_2, \mathrm{i}\partial/\partial x_3\right) \boldsymbol{u}^*(\omega, \xi_\alpha, x_3) = 0 \qquad (x_3 > 0)\,, \\ &\left(\boldsymbol{u}^* = \{u_1^*, u_2^*, u_3^*\}\right) \,. \end{aligned} \tag{7.2.13}$$

The general solution of ordinary differential equations (7.2.13) can be found in the form:

$$\boldsymbol{u}^* = \boldsymbol{v} \exp\left\{-\mathrm{i}\xi_3 x_3\right\} \,, \tag{7.3.14}$$

where the vector $\boldsymbol{v}$ does not depend on x_3. The expression (7.2.14) satisfies (7.2.13) if

$$\mathbf{K}(\omega, \boldsymbol{\xi})\boldsymbol{v} = 0 \,. \tag{7.2.15}$$

Equation (7.2.15) has non-trivial solutions if

$$\det K_{ik}(\omega, \boldsymbol{\xi}) = 0 \,. \tag{7.2.16}$$

However, equation (7.2.16) has six roots $\xi_3 = \xi_3^N(\omega,\xi_\alpha)$ $(N = 1,\dots,6)$ for every triplet (ω,ξ_1,ξ_2) comprised of fixed numbers; amongst these roots, there can only be real values and pairs of complex-conjugate values if $\mathrm{Im}\{\omega\} = \mathrm{Im}\{\xi_1\} = \mathrm{Im}\{\xi_2\} = 0$. According to this, the general solution of equation (7.2.13) can be written as

$$\boldsymbol{u}^*(\omega,\xi_\alpha,x_3) = \sum_{N=1}^{6} \operatorname{adj}\mathbf{K}\left(\omega,\boldsymbol{\xi}^N\right)\boldsymbol{b}^N(\omega,\xi_\alpha)\mathrm{e}^{-\mathrm{i}\xi_3^N x_3}\,, \tag{7.2.17}$$

where $\boldsymbol{\xi}^N = \{\xi_1,\xi_2,\xi_3^N(\omega,\xi_\alpha)\}$, $\operatorname{adj}\mathbf{A}$ is the adjoint matrix for the matrix $\mathbf{A}$; the vector $\boldsymbol{b}^N$ must be determined from the boundary conditions (7.2.8).

To meet the conditions (7.2.9), we shall remove from (7.2.17) the summands with those values of N for which the terms $\xi_3^N(\omega,\xi_\alpha)$ have positive imaginary parts (i.e., we remove the terms increasing as exponential function when $x_3 \to +\infty$). In order to find out the appropriate values of N, examine briefly the Riemann surface of the algebraic function $\xi_3^N(\omega,\xi_\alpha)$ defined by equation (7.2.16) which is of the sixth degree in ξ_3. The function $\xi_3(\omega,\xi_\alpha)$ is single-valued if ω goes through all six sheets $\sum^N$ $(N = 1,\dots,6)$ of the Riemann surface; let us assume that the function $\xi_3(\omega,\xi_\alpha)$ takes the "value" $\xi_3^N(\omega,\xi_\alpha)$ on the sheet $\sum^N$. All these sheets are pasted to each other along the lines connecting the branch point of the functions $\xi_3^N(\omega,\xi_\alpha)$; equation (7.2.16) indicates that the function ξ_3^N remains finite at branch points for finite ξ_1,ξ_2,ω.

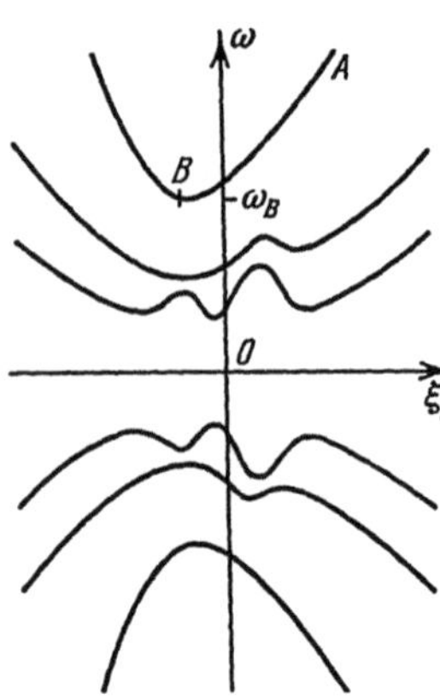

Fig. 7.1. Likely curves $\omega = \omega(\xi_3)$ representing the solutions to equation (7.2.16) for real values of ξ_3

The most important branch points are on the real axis in the ω-plane. These points could be examined thoroughly by means of Fig. 7.1 representing the likely curves defined by equation (7.2.16) with ξ_1 and ξ_2 fixed whereas ω and ξ_3 are variable and real. It is worth noting that these curves are symmetric about the axis $\omega = 0$, since equation (7.2.16) includes ω^2 only. Note that the existence of six real curves $\omega = \omega(\xi_3)$ is an immediate consequence of the structure of the hyperbolic equations (7.2.3). Prove this statement. Let us seek for a solution to equations (7.2.16) by using the formula:

$$\omega(\boldsymbol{\xi}) = \pm\left(\xi_1^2+\xi_2^2+\xi_3^2\right)^{1/2} c(\boldsymbol{\zeta}), \tag{7.2.18}$$

where $\boldsymbol{\zeta} = \boldsymbol{\xi}/(\xi_1^2+\xi_2^2+\xi_3^2)^{1/2}$ is the unit vector of the direction $\boldsymbol{\xi}$, $c(\boldsymbol{\zeta})$ is one of the likely plane wave propagation speeds for this direction. Substituting (7.2.18) into (7.2.16) provides:

$$\det K_{ik}(\omega,\boldsymbol{\xi}) = \left(\xi_1^2+\xi_2^2+\xi_3^2\right)^3 \det\left[c_{ijkl}\zeta_j\zeta_l - \rho c^2(\boldsymbol{\zeta})\delta_{ik}\right] = 0 \tag{7.2.19}$$

This means that $c^2(\boldsymbol{\zeta})$ is a characteristic root of the symmetric matrix $a_{ik} = c_{ijkl}\zeta_j\zeta_l$. This matrix is known to be positive definite. The latter statement follows from the fact that the elastic potential W of (2.1.4) is a positive definite quadratic function. With this, we can assume $\varepsilon_{kl} = \eta_k\zeta_l + \zeta_k\eta_l$ in (2.1.4) and use the relation (2.1.15) to obtain from (2.1.14) that

$$W = 2c_{ijkl}\eta_i\eta_k\zeta_j\zeta_l = 2a_{ik}\eta_i\eta_k > 0\,.$$

This completes our proof. So, equation (7.2.19) has three real, positive roots $c_i^2(\boldsymbol{\zeta})(i = 1,2,3)$, and we can write: $c_1(\boldsymbol{\zeta}) \geq c_2(\boldsymbol{\zeta}) \geq c_3(\boldsymbol{\zeta}) > 0$. As a result, obtain six real curves $\omega = \omega(\xi_3)$ described by (7.2.18).

If roots $\xi_3^N = \xi_3^N(\omega_0,\xi_\alpha)$ are real, then they are defined by the points where the straight line $\omega = \omega_0$ intersects the curves (7.2.16) depicted in Fig. 7.1. If $|\omega_0|$ is great enough, then there are six real roots $\xi_3^N(\omega_0,\xi_\alpha)$; three of them correspond to those interception points where $\partial\omega/\partial\xi_3 > 0$, and the three remaining points, to those with $\partial\omega/\partial\xi_3 < 0$ (the existence of six real roots ξ_3^N in the case of great modulus of a real ω_0 can be derived from asymptotic behaviour of the six-order equation, (7.2.16), by letting $\omega_0 \to \pm\infty$). We shall allocate three roots of the first kind ($\partial\omega/\partial\xi_3 > 0$) to the Riemann sheets $\sum^N$ numbered with $N = 1,2,3$, and the three roots of the second kind to those with $N = 4,5,6$. Branch points correspond to the points where the straight line $\omega = \omega_0$ is tangential to curves (7.2.16). If we go through any such point (ω_0,ξ_3), a certain pair of real roots (say, $\xi_3^N(\omega,\xi_\alpha)$ and $\xi_3^M(\omega,\xi_\alpha)$ becomes a pair of complex-conjugate roots. (In order to prove this, note that if ω becomes less than ω_B marked in Fig. 7.1, then the horizontal line $\omega = \omega_B$ can intersect the curves at four points only, which means that equation (7.2.16) being of the sixth degree in ξ_3 has only four real roots, whereas the remaining two roots are necessarily complex-conjugate since coefficients of the equation are real when ω, ξ_1, and ξ_2 are real). If there exists a branch point (ω_0,ξ_3) on the curves symmetric about the axis $\omega = 0$, then there is another branch point, $(-\omega_0,\xi_3)$.

With this, the cut must connect two points, (ω_0,ξ_3) and $(-\omega_0,\xi_3)$. In a similar way, we should introduce a cut from any point of tangency, where $\omega_0 > 0$, to the corresponding point, where $\omega_0 < 0$. Some of these cuts can turn out to be unnecessary if a pair of roots becomes again real (at maxima of the curves $\omega = \omega(\xi_3)$ in Fig. 7.1 for $\omega_0 > 0$); nevertheless, we outlined the most important cuts in the Riemann surface.

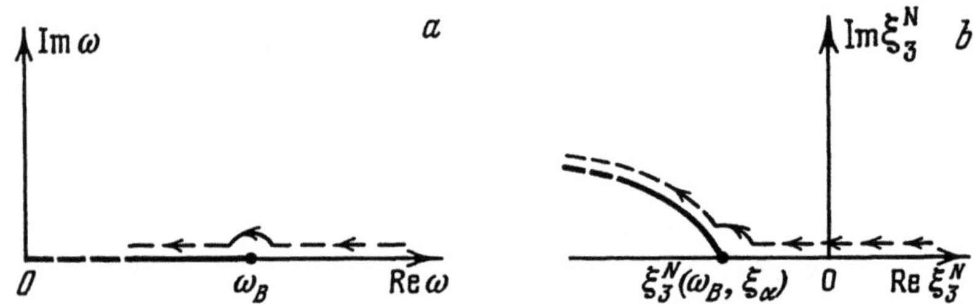

Fig. 7.2a,b. (a) The path around the point $\omega = \omega_B$ in the complex ω-plane. (b) The path around the point $\xi_3^N = \xi_3^N(\omega_B, \xi_\alpha)$ in the ξ_3^N-plane

Consider now the behaviour of the function $\xi_3^N(\omega, \xi_\alpha)$ in the case when ω varies in the upper half-plane from the right to the left along the real axis, Fig. 7.2a. Such variation of ω corresponds to movement up–down along the ω-axis in Fig. 7.1. The root $\xi_3^N(\omega, \xi_\alpha)$ which corresponds to this sheet of the Riemann surface $\sum^N$ (with N being in the range $1, 2, 3$), remains positive and decreases (together with ω) if ω is positive and great enough. When we reach a branch point (assumed to be simple, such as the point ω_B on the curve AB) we shall go around this point in the upper half-plane (Fig. 7.2a) — counter-clockwise once we go above the real axis. But following the conformal mapping theory, the same orientation must turn out to be kept for this branch of the analytical function $\xi_3^N = \xi_3^N(\omega, \xi_\alpha)$ when we go around the corresponding branch point $\xi_3^N(\omega_B, \xi_\alpha)$ which is on the real axis of the complex ξ_3^N-plane. Thus, we are to go above this point so that our path turns right from the real axis of the ξ_3^N-plane around the point $\xi_3^N(\omega_B, \xi_\alpha)$, that is, our path must come into the region $\mathrm{Im}\{x_3^N\} > 0$ (Fig. 7.2b). In this connection, note that the inequality, $\mathrm{Im}\{\xi_3^N\} > 0$, thakes place for the whole of the upper half-plane $\mathrm{Im}\{\omega\} > 0$. Prove this. the imaginary part of the continuous function $\xi_3^N(\omega, \xi_\alpha)$ can change its sign only after going through the very ξ_3^N for which $\mathrm{Im}\{\xi_3^N\} = 0$. However, no one such ξ_3^N is present in the region $\mathrm{Im}\{\omega\} > 0$; otherwise a complex root ω_N would appear at real ξ_3^N which must be judged to be impossible since equation (7.2.16) with real ξ_1, ξ_2, ξ_3 has only real roots $\pm\omega_1$, $\pm\omega_2$, and $\pm\omega_3$, see Fig. 7.1.

Proceeding in the same way, we find that the imaginary part of each branch ξ_3^N with $N = 4, 5, 6$ is negative if the complex variable ω is in its upper half-plane. Therefore, the branch points of the function $\xi_3^N(\omega, \xi_\alpha)$ off the real axis $\mathrm{Im}\{\omega\} = 0$ (within the region $\mathrm{Im}\{\omega\} > 0$) connect such sheets $\sum^N$ and $\sum^M$ for which both N and M pertain to the same range of integers: either $(1, 2, 3)$ or $(4, 5, 6)$.

From the above consideration, it follows that the sum in (7.2.17) must only comprise the terms with $N = 4, 5, 6$. Substituting (7.2.17) into the Fourier transform for the boundary condition (7.2.8), obtain

$$\sum_{N=4}^{6} \mathbf{C}_3\left(\boldsymbol{\xi}^N\right) \operatorname{adj}\mathbf{K}\left(\omega, \boldsymbol{\xi}^N\right) \boldsymbol{b}^N(\omega, \xi_\alpha) = \mathrm{i}\boldsymbol{\tau}^*(\omega, \xi_\alpha)\,, \tag{7.2.20}$$

where

$$\boldsymbol{\tau}^*(\omega, \xi_\alpha) = (2\pi)^{-3/2} \int\limits_0^\infty dt \int\limits_{-\infty}^{\infty} \int\limits_{-\infty}^{\infty} \boldsymbol{\tau}(t, x_\alpha) \exp[\mathrm{i}(\omega t + \xi_\alpha x_\alpha)]\, dx_1\, dx_2 \,. \tag{7.2.21}$$

Equation (7.2.20) can be solved by finding the vector $\boldsymbol{b}^N$ independent of N:

$$\boldsymbol{b}^N(\omega, \xi_\alpha) = \mathrm{i} \left[\sum_{M=4}^{6} \mathbf{C}_3 \left(\boldsymbol{\xi}^M\right) \operatorname{adj} \mathbf{K}\left(\omega, \boldsymbol{\xi}^M\right) \right]^{-1} \boldsymbol{\tau}^*(\omega, \xi_\alpha) \,, \tag{7.2.22}$$

where superscript "-1" denotes the matrix inversion operation. So,

$$\boldsymbol{u}_*(\omega, \xi_\alpha, x_3) = \sum_{N=4}^{6} \mathbf{B}^N(\omega, \xi_\alpha) \boldsymbol{\tau}^*(\omega, \xi_\alpha) \exp\left[-\mathrm{i}\xi_3^N x_3\right] \,, \tag{7.2.23}$$

where the matrix $\mathbf{B}^N$ is

$$\mathbf{B}^N(\omega, \xi_\alpha) = \mathrm{i} \operatorname{adj} \mathbf{K}\left(\omega, \boldsymbol{\xi}^N\right) \left[\sum_{M=4}^{6} \mathbf{C}_3 \left(\boldsymbol{\xi}^M\right) \operatorname{adj} \mathbf{K}\left(\omega, \boldsymbol{\xi}^M\right) \right]^{-1} . \tag{7.2.24}$$

Note that the formulae (7.2.23, 24) rely upon the assumption that all roots $\xi_3^N (N = 4, 5, 6)$ differ from each other. The situation of multiple roots (e.g., in the case of an isotropic medium where two roots coincide) can be treated by passing to the limit in (7.2.23) with ξ_3^M tending to ξ_3^N. Also this could be considered immediately, after replacing (7.2.20) with another expression taking into account the multiplicity of roots. Since $\boldsymbol{\tau}(t, x_\alpha)$ is a homogeneous function of the nth order with respect to $\{t, x_\alpha\}$, then $\boldsymbol{\tau}^*(\omega, \xi_\alpha)$ is homogeneous with respect to ω and ξ_α, of order of $-(n+3)$. From (7.2.24), it follows, as well, that $\mathbf{B}^N(\omega, \xi_\alpha)$ is the homogeneous function of order of -1 in $\{\omega, \xi_\alpha\}$ because the $\xi_3^N(\omega, \xi\alpha)$ is homogeneous in these variables, of the first order. In the light of these properties, it is more convenient to find, at first, the function $\partial_t^{(n+2)} \boldsymbol{u}(t, \boldsymbol{x}) (\partial_t^{(n)} f \equiv \partial^n f / \partial t^n)$ whose Fourier transform can be expressed from (7.2.23) after multiplication by $(-\mathrm{i}\omega)^{n+2}$:

$$\begin{aligned} \partial_t^{(n+2)} \boldsymbol{u}(t, \boldsymbol{x}) &= (-\mathrm{i})^{n+2} (2\pi)^{-3/2} \sum_{N=4}^{6} \int\limits_{-\infty}^{\infty} \int\limits_{-\infty}^{\infty} d\xi_1\, d\xi_2 \int\limits_{-\infty+0\mathrm{i}}^{+\infty+0\mathrm{i}} \omega^{n+2} \mathbf{B}^N(\omega, \xi_\alpha) \\ &\quad \times \boldsymbol{\tau}^*(\omega, \xi_\alpha) \exp\left\{-\mathrm{i}\left(\omega t + \boldsymbol{\xi}^N \cdot \boldsymbol{x}\right)\right\} d\omega \,. \end{aligned} \tag{7.2.25}$$

This equation should be considered in terms of generalized functions; this means that (7.2.25) may be written as

$$\partial_t^{(n+2)} \boldsymbol{u}(t, \boldsymbol{x})$$
$$= (-\mathrm{i})^{n+2}(2\pi)^{-3/2} \sum_{N=4}^{6} \lim_{\varepsilon \to +0} \int_{-\infty}^{\infty} \int_{-\infty}^{\infty} d\xi_1\, d\xi_2 \int_{-\infty+0\mathrm{i}}^{\infty+0\mathrm{i}} \omega^{n+2} \tag{7.2.26}$$
$$\times \mathbf{B}^N(\omega, \xi_\alpha)\boldsymbol{\tau}^*(\omega, \xi_\alpha) \exp\left\{-\mathrm{i}\left(\omega t + \boldsymbol{\xi}^N \cdot \boldsymbol{x} - \mathrm{i}\varepsilon|\boldsymbol{\xi}|\right)\right\} d\omega\,.$$

where $|\boldsymbol{\xi}| = (\xi_1^2 + \xi_2^2)^{1/2}$. In order to simplify these integrals, let us replace ω by the new variable $\Omega = \omega/|\boldsymbol{\xi}|$, introduce $\eta_\alpha = \xi_\alpha/|\boldsymbol{\xi}|$, and make use of a polar coordinate system on the plane (ξ_1, ξ_2). Due account for the homogeneity stated above allows us to bring (7.2.26) to the following form:

$$\partial_t^{(n+2} \boldsymbol{u}(t, \boldsymbol{x})$$
$$= (-\mathrm{i})^{n+2}(2\pi)^{-3/2} \sum_{N=4}^{6} \lim_{\varepsilon \to +0} \oint_{|\eta|=1} ds \int_{-\infty+0\mathrm{i}}^{\infty+0\mathrm{i}} \Omega^{n+2}\mathbf{B}^N(\Omega, \boldsymbol{\eta})\boldsymbol{\tau}^*(\Omega, \boldsymbol{\eta}) d\Omega$$
$$\times \int_0^\infty \exp\left\{-\mathrm{i}|\boldsymbol{\xi}|\left(\Omega t + \boldsymbol{\xi}^N(\Omega, \boldsymbol{\eta}) \cdot \boldsymbol{x} - \mathrm{i}\varepsilon\right)\right\} d|\boldsymbol{\xi}| \tag{7.2.27}$$
$$\left(\boldsymbol{\eta} = \{\eta_1, \eta_2\}, \quad |\boldsymbol{\eta}| = \sqrt{\eta_1^2 + \eta_2^2}\right)\,.$$

This, after integration in $|\boldsymbol{\xi}|$, produces

$$\partial_t^{(n+2)} \boldsymbol{u}(t, \boldsymbol{x}) = -(-\mathrm{i})^{n+1}(2\pi)^{-3/2} \tag{7.2.28}$$
$$\times \sum_{N=4}^{6} \lim_{\varepsilon \to +0} \oint_{|\eta|=1} ds \int_{-\infty+0\mathrm{i}}^{+\infty+0\mathrm{i}} \frac{\Omega^{N+2}\mathbf{B}^N(\Omega, \boldsymbol{\eta})\boldsymbol{\tau}^*(\Omega, \boldsymbol{\eta})}{\Omega t + \eta_\alpha x_\alpha + \xi_3^N(\Omega, \boldsymbol{\eta})x_3 - \mathrm{i}\varepsilon} d\Omega\,.$$

Integrals with respect to Ω in equation (7.2.28) can be calculated by means of (I) closing the path in the upper half-plane of the complex variable Ω and (II) applying the Cauchy residue theorem. In so doing, we shall take into account that the integrals over the cuts that go out of likely branch points within the region $\mathrm{Im}\{\Omega\} > 0$ (and separate single-valued branches of the analytic function $\xi_3^N(\Omega, \boldsymbol{\eta})$) make no contribution to the sum in (7.2.28) — these integrals cancel. This results from the fact established above: the likely branch points in the region $\mathrm{Im}\{\Omega\} > 0$ (and, therefore, the sides of the cuts connecting these points) must belong to the sheets $\sum^N$ and $\sum^M$ with superscripts N and M from the set $\{4, 5, 6\}$. Prove this statement. Let branch points C and D in the region $\mathrm{Im}\{\Omega\} > 0$ pertain to sheets $\sum^N$ and $\sum^M$. Let these points be connected by the cuts CD, the upper side of the cut in $\sum^N$ being pasted to the lower side of the cut in $\sum^M$ whereas the lower side of the cut in $\sum^N$ being pasted to the upper side of the cut in $\sum^M$. In this case, integrals over the upper side of the cut CD in $\sum^N$ and over the lower one in $\sum^M$ cancel, as well as integrals over the lower and upper sides of

CD in $\sum^N$ and $\sum^M$, respectively; this is due to the fact that these integrals differ from each other only in path orientations. In other words, the functions $B^N(\Omega, \boldsymbol{\eta})$ present in (7.2.28) are analytic within the region $\mathrm{Im}\{\Omega\} > 0$.

It is obvious now, that, after deforming the integration path within the region $\mathrm{Im}\{\Omega\} > 0$, integrals of (7.2.28) take contributions only of zeros of functions

$$\Phi^N(\Omega) = \Omega t + \eta_\alpha x_\alpha + \xi_3^N(\Omega, \boldsymbol{\eta})x_3 - \mathrm{i}\varepsilon \,, \tag{7.2.29}$$

provided that zeros exist within the region $\mathrm{Im}\{\Omega\} > 0$. If $t \leq 0$, the function $\Phi^N(\Omega)$ is not equal to zero within the region $\{\mathrm{Im}\{\Omega\} > 0, x_3 \geq 0\}$, since $\mathrm{Im}\{\xi_3^N(\Omega), \boldsymbol{\eta})\} < 0$ here, thus, $\mathrm{Im}\{\Phi^N(\Omega)\} < 0$. With this, there exists such non-negative t_N that $\Phi^N(\Omega)$ has no zeros within the region $\mathrm{Im}\{\Omega\} > 0$ when $t \leq t_N$. The t_N can be evaluated as follows. Assume that Ω^N is a root of the equation

$$\Omega^N t + \eta_\alpha x_\alpha + \xi_3^N\left(\Omega^N, \boldsymbol{\eta}\right) x_3 = 0 \,. \tag{7.2.30}$$

Then, once ε is of arbitrary small magnitude, a root Ω of the equation,

$$\Omega t + \eta_\alpha x_\alpha + \xi_3^N(\Omega, \boldsymbol{\eta})x_3 - \mathrm{i}\varepsilon = 0 \tag{7.2.31}$$

can be represented by

$$\Omega = \Omega^N + \varepsilon\Omega_1 \,; \tag{7.2.32}$$

Ω_1 introduced here satisfies the equation

$$\Omega_1\left[t + \xi^N_{3,\Omega}\left(\Omega^N, \boldsymbol{\eta}\right) x_3\right] = \mathrm{i} \tag{7.2.33}$$

related to (7.2.30); here, $\xi^N_{3,\Omega} \equiv \partial\xi_3^N/\partial\Omega$. If Ω^N is in the upper half-plane, Ω must belong to the same half-plane, as well, as ε is arbitrarily small. However, equation (7.2.30) possesses a real root if t is small enough, then Ω is in the upper half-plane only if the condition

$$t + \xi^N_{3,\Omega}\left(\Omega^N, \boldsymbol{\eta}\right) x_3 > 0 \tag{7.2.34}$$

is met. This inequality can be immediately studied for any medium under investigation. However, $\Phi^N(\Omega) = 0$ at $x_3 = 0$ if $\Omega = (-\eta_\alpha x_\alpha + \mathrm{i}\varepsilon)/t$; this value of Ω is in the upper half plane for all positive t and becomes singular when t tends to zero. One can check that the case of positive x_3 can be studied with $|\Omega|$ tending to infinity. If $\mathrm{Im}\{\Omega^N\} = 0$ and $|\Omega^N|$ is great, then from the asymptotic behavior of equation (7.2.16) we find:

$$\xi_3^N\left(\Omega^N, \eta\right) \sim -\Omega^N/c(0,0,1), \quad \xi^N_{3,\Omega}\left(\Omega^N, \eta\right) \sim -1/c(0,0,1) \,, \tag{7.2.35}$$

where $c(0,0,1)$ is one of the speeds at which the plane waves can propagate along the x_3-axis. From (7.2.35) it follows that inequality (7.2.34) becomes valid if $t > t_N$ where

$$t_N = x_3/c\,(0,0,1) \,. \tag{7.2.36}$$

The statement that this t_N is the least value with which the function $\Phi^N(\Omega)$ has a root in the upper half-plane can be validated through the use of continuity with respect to the variable x_3.

Evaluating the integral with respect to Ω in (7.2.28) by means of the residue theorem, thereafter passing to the limit as $\varepsilon \to 0$, we find

$$\begin{aligned} \partial_t^{(n+2)} \boldsymbol{u}(t,\boldsymbol{x}) &= (2\pi)^{-1/2}(-\mathrm{i})^{n+2} \sum_{N=4}^{6} H(t-t_N) \\ &\times \oint_{|\eta|=1} \left(\Omega^N\right)^{n+2} \frac{\mathbf{B}^N(\Omega^N,\boldsymbol{\eta})\boldsymbol{\tau}^*(\Omega^N,\boldsymbol{\eta})}{t+\xi^N_{3,\Omega}(\Omega^N,\boldsymbol{\eta})x_3} ds\,. \end{aligned} \tag{7.2.37}$$

The variable Ω^N satisfies equation (7.2.30) with $t > t_N$. Equation (7.2.37) is the main result; the function $\boldsymbol{u}(t,\boldsymbol{x})$ can be obtained by means of $(n+2)$-fold integration in t.

In a similar way, we may apply the Fourier transformation to equation (7.2.2) to obtain

$$\boldsymbol{\sigma}_j^*(\omega,\xi_\alpha,x_3) = C_j\left(-\mathrm{i}\xi_1,-\mathrm{i}\xi_2,\partial/\partial x_3\right)\boldsymbol{u}(\omega,\xi_\alpha,x_3)\,. \tag{7.2.38}$$

where $\boldsymbol{\sigma}^*(\omega,\xi_\alpha,x_3)$ is the transform for the stress vector $\boldsymbol{\sigma}_j(t,\boldsymbol{x}) = \{\sigma_{1j},\sigma_{2j},\sigma_{3j}\}$. Substitution of (7.2.23) into (7.2.38) gives

$$\begin{aligned} &\boldsymbol{\sigma}_j^*(\omega,\boldsymbol{\xi}_\alpha,x_3) \\ &= \sum_{N=4}^{6} \mathbf{C}_j\left(-\mathrm{i}\xi^N\right)\mathbf{B}^N(\omega,\xi_\alpha)\boldsymbol{\tau}^*(\omega,\xi_\alpha)\exp\left\{-\mathrm{i}\xi_3^N x_3\right\} \\ &= -\mathrm{i}\sum_{N=4}^{6} \mathbf{C}_j\left(\boldsymbol{\xi}^N\right)\mathbf{B}^N(\omega,\xi_\alpha)\boldsymbol{\tau}^*(\omega,\xi_\alpha)\exp\left\{-\mathrm{i}\xi_3^N x_3\right\}\,. \end{aligned} \tag{7.2.39}$$

In the particular case of $j = 3$, equations (7.2.22, 39) indicate that $\boldsymbol{\sigma}_3^*(\omega,\xi_\alpha,+0) = \boldsymbol{\tau}^*(\omega,\xi_\alpha)$ at $x_3 = 0 (x_3 \to +0)$.

By analogy with (7.2.37), find from (7.2.39):

$$\begin{aligned} &\partial_t^{(n+1)}\boldsymbol{\sigma}_j(t,\boldsymbol{x}) \\ &= \frac{(-\mathrm{i})^{n+2}}{\sqrt{2\pi}} \sum_{N=4}^{6} H(t-t_N) \\ &\times \oint_{|\eta|=1} (\Omega^N)^{n+1} \frac{\mathbf{C}_j(\Omega^N,\boldsymbol{\eta})\mathbf{B}^N(\Omega^N,\boldsymbol{\eta})}{t+\xi^N_{3,\Omega}(\Omega^N,\boldsymbol{\eta})x_3}\boldsymbol{\sigma}^*\left(\Omega^N,\boldsymbol{\eta}\right) ds\,. \end{aligned} \tag{7.2.40}$$

The stress will be a result of $(n+1)$-fold integration in t (the Ω^N here is that of (7.2.37)).

To demonstrate the close relationship of solution procedures for the three-dimensional case and its two-dimensional analog, represent the $\boldsymbol{\tau}(\omega,\xi_\alpha)$ as a

Cauchy-type integral. Using the variables $y_\alpha = x_\alpha/t$ and equation (7.2.5), rewrite (7.2.21) as

$$\tau^*(\omega,\xi_\alpha) = (2\pi)^{-3/2}\int_0^\infty t^{n+2}dt \int_{-\infty}^{\infty}\int_{-\infty}^{\infty} \tau(y)\exp\{it(\omega+\xi_\alpha y_\alpha)\}\,dy_1\,dy_2\,. \tag{7.2.41}$$

After integrating equation (7.2.41) in t (for $\mathrm{Im}\{\omega\} > 0$), the formula (7.2.42) appears:

$$\tau_*(\omega,\xi_\alpha) = \frac{(-\mathrm{i})^{n+1}}{(2\pi)^{3/2}}\partial_\omega^{(n+2)} \int_{-\infty}^{\infty}\int_{-\infty}^{\infty} \frac{\tau(y)dy_1dy_2}{\omega+\xi_\alpha y_\alpha}\,, \tag{7.2.42}$$

which can be integrated along the line $\xi_\alpha y_\alpha = p$ to give finally

$$\tau_*(\omega,\xi_\alpha) = \frac{(-\mathrm{i})^{n+1}}{(2\pi)^{3/2}}\partial_\omega^{(n+2)} \int_{-\infty}^{\infty} \frac{\tau_0(p,\xi_\alpha)dp}{\omega+p}\,, \tag{7.2.43}$$

where the vector

$$\tau_0(p,\xi_\alpha) = \int_{-\infty}^{\infty}\int_{-\infty}^{\infty} \tau(y)\delta(p-\xi_\alpha y_\alpha)\,dy_1\,dy_2 \tag{7.2.44}$$

is the Radon transform for the vector $\tau(y)$ (see Sect. 7.2.3 below). Substituting (7.2.43) into (7.2.37) and (7.2.40), we obtain the following representation:

$$\partial_t^{(n+2)}u(t,x) = -\frac{1}{2\pi}\sum_{N=4}^{6} H(t-t_N) \oint_{|\eta|=1} (\Omega^N)^{n+2}\frac{\mathbf{B}^N(\Omega^N,\eta)F^{(n+2)}(-\Omega^N,\eta)}{t+\xi_{3,\Omega}^N(\Omega^N,\eta)x_3}ds\,, \tag{7.2.45}$$

$$\begin{aligned}\partial_t^{(n+1)}\sigma_j(t,x) &= -\frac{1}{2\pi}\sum_{N=4}^{6} H(t-t_N)\\ &\times \oint_{|\eta|=1} (\Omega^N)^{n+1}\mathbf{C}_j(\Omega^N,\eta)\frac{\mathbf{B}^N(\Omega^N,\eta)F^{(n+2)}(-\Omega^N,\eta)}{t+\xi_{3,\Omega}^N(\Omega^N,\eta)x_3}ds\,,\end{aligned} \tag{7.2.46}$$

where Ω^N is the root of equation (7.2.30) ($\mathrm{Im}\{\Omega^N\} \geq 0$ for $t > t_N$) and the function F is defined as

$$F^{(n+2)}(z,\eta) = \frac{1}{2\pi\mathrm{i}}\partial_z^{n+2} \int_{-\infty}^{\infty} \frac{\tau_0(p,\eta)dp}{p-z}\,. \tag{7.2.47}$$

(It must be concluded that, in the expression for $\boldsymbol{F}^{(n+2)}(-\Omega^N, \boldsymbol{\eta})$, the contour of integration in p goes above the point $p = -\Omega^N$).

Replacing the $\boldsymbol{\eta}$ by $-\boldsymbol{\eta}$, equations (7.2.45,46) can be shown to define real functions. Prove this. If $\xi_3^N(\omega, \boldsymbol{\eta})$ $(N = 4,5,6)$ is one of roots of equation $\det\{K_{lm}(\omega, \boldsymbol{\eta}, \xi_3)\} = 0$ then $-\overline{\xi_3^N(\omega, \boldsymbol{\eta})}$ is one of roots of equation $\det\{K_{lm}(-\bar{\omega}, -\boldsymbol{\eta}, \xi_3)\} = 0$; here, the bar sign over the function symbol identifies a complex conjugate value. Consequently, if ξ_3^N is a complex variable, then for some M, unnecessarily equal to N but being in the same range (4,5,6), we have:

$$\xi_3^M(-\bar{\omega}, -\boldsymbol{\eta}) = -\overline{\xi_3^N(\omega, \boldsymbol{\eta})} \,. \tag{7.2.48}$$

The analytic continuation operation confirms that this result remains valid when ξ_3^N is real. Consequently, we can use complex conjugate values of both sides of (7.2.30) to obtain, after multiplying them by -1,

$$-\overline{\Omega^N(\boldsymbol{\eta})}t - \eta_\alpha x_\alpha + \xi_3^M\left(-\overline{\Omega^N(\boldsymbol{\eta})}, -\boldsymbol{\eta}\right) x_3 = 0 \,.$$

This relationship and equation (7.2.30) (with $N = M$ and $\boldsymbol{\eta} = -\boldsymbol{\eta}$):

$$\Omega^M(-\boldsymbol{\eta})t + (-\eta_\alpha)x_\alpha + \xi_3^M\left(\Omega^M(-\boldsymbol{\eta}), -\boldsymbol{\eta}\right) x_3 = 0$$

make it possible to write sequentially:

$$\Omega^M(-\boldsymbol{\eta}) = -\overline{\Omega^N(\boldsymbol{\eta})}, \quad \xi_3^M\left(\Omega^M(-\boldsymbol{\eta}), -\boldsymbol{\eta}\right) = -\overline{\xi_3^N\left(\Omega^N(\boldsymbol{\eta}), \boldsymbol{\eta}\right)} \,. \tag{7.2.49}$$

On the basis of equations (7.2.49), we can immediately demonstrate that the right-hand side in each of the expressions (7.2.45,46) is the same as its own conjugate value; in this case, account will be taken of the evenness of the Radon transformation.

Note that although it is impossible to completely get out of integrals when deriving $\boldsymbol{u}(t, \boldsymbol{x})$ and $\boldsymbol{\sigma}_j(t, \boldsymbol{x})$ from (7.2.45,46), one may nevertheless simplify these expressions by integrating them in t. For this purpose, differentiate equation (7.2.30) in t to find

$$\frac{d\Omega^N}{\Omega^N} = -\frac{dt}{t + \xi^N_{3,\Omega}(\Omega^N, \boldsymbol{\eta})x_3} \,. \tag{7.2.50}$$

After integrating equations (7.2.45,46) in t from t_N to t and introducing the new variable of integration Ω^N with account for both equation (7.2.50) and the fact that $\Omega^N(t, \boldsymbol{x}, \boldsymbol{\eta}) = \infty$, obtain:

$$\begin{aligned} &\partial_t^{(n+1)}\boldsymbol{u}(t, \boldsymbol{x}) \\ &= -\frac{1}{2\pi}\sum_{N=4}^{6} \oint_{|\eta|=1} ds\, H(t - t_N) \int\limits_{\Omega^N(t,\boldsymbol{x},\boldsymbol{\eta})} \Omega^{n+1}\mathbf{B}^N(\Omega, \boldsymbol{\eta})\boldsymbol{F}^{(n+2)}(-\Omega, \boldsymbol{\eta})d\Omega \,, \end{aligned} \tag{7.2.51}$$

$$\begin{aligned} &\partial_t^{(n)}\boldsymbol{\sigma}_j(t, \boldsymbol{x}) \\ &= -\frac{1}{2\pi}\sum_{N=4}^{6} \oint_{|\eta|=1} ds\, H(t - t_N) \int\limits_{\Omega^N(t,\boldsymbol{x},\boldsymbol{\eta})} \Omega^{n}\mathbf{C}_j(\Omega, \boldsymbol{\eta})\mathbf{B}^N(\Omega, \boldsymbol{\eta})\boldsymbol{F}^{(n+2)}(-\Omega, \boldsymbol{\eta})d\Omega \,. \end{aligned} \tag{7.2.52}$$

Here, integration Ω is performed along a path defined by equation (7.2.30) and lying within the region $\operatorname{Im}\{\Omega\} > 0$. It is known that

$$\begin{gathered} \mathbf{C}_j(\Omega, \boldsymbol{\eta}) = \mathrm{O}(\Omega), \quad \mathbf{B}^N(\Omega, \boldsymbol{\eta}) = \Omega^{-1}\mathbf{B}^N(1, \boldsymbol{\eta}/\Omega) = \mathrm{O}(\Omega^{-1}), \\ \boldsymbol{F}^{(n+2)}(-\Omega, \boldsymbol{\eta}) = \partial_\Omega^{(n+2)}\mathrm{O}(\Omega^{-1}) = \mathrm{O}\left(\Omega^{-(n+3)}\right), \end{gathered} \tag{7.2.53}$$

when $\Omega \to \infty$. Thus, as $|\Omega| \to \infty$, integrands in (7.2.51), 52) behave as $\mathrm{O}(\Omega^{-3})$. As these integrands are analytical functions in the region $\operatorname{Im}\{\Omega\} > 0$, it is permissible, for integration in Ω, to make use of any path (in the region $\operatorname{Im}\{\Omega\} > 0$) which is drawn from $\Omega^N(t, \boldsymbol{x}, \boldsymbol{\eta})$ to infinity.

7.2.2 Two-Dimensional Case

Let us consider a particular case for (7.2.5): the specified stress vector $\boldsymbol{\tau}(t, x_\alpha)$ is constant along every line

$$\nu_\alpha x_\alpha = pt . \tag{7.2.54}$$

where the vector $\boldsymbol{\nu} = \{\nu_1, \nu_2\}$ of the (x_1, x_2)-plane is constant. Consequently,

$$\boldsymbol{\tau}(t, x_\alpha) = t^n \boldsymbol{\tau}(p) \qquad (t > 0) . \tag{7.2.55}$$

So, the vector $\boldsymbol{u}(t, \boldsymbol{x})$ depends only on $\{t, pt, x_3\}$, thus being comprised of homogeneous functions of $(n+1)$th degree in these variables. Not to solve this problem directly, let us consider the above formulae (7.2.37, 40). There is need to find only the form to be taken by the Fourier transform, $\boldsymbol{\tau}^*(\Omega, \boldsymbol{\eta})$, for the load (7.2.55). With this, obtain

$$\begin{aligned} &\boldsymbol{\tau}^*(\Omega, \boldsymbol{\eta}) \\ &\quad = (2\pi)^{-3/2} \int_{-\infty}^{\infty} t^n dt \int_{-\infty}^{\infty}\int_{-\infty}^{\infty} \boldsymbol{\tau}(p) \exp\{\mathrm{i}\Omega t + \eta_\alpha x_\alpha)\}\, dx_1\, dx_2 . \end{aligned} \tag{7.2.56}$$

It is convenient for us to replace the coordinate q by the projection of $\boldsymbol{x}/t$ onto the unit vector $\boldsymbol{\lambda} = \{\lambda_1, \lambda_2\}$ being in the (x_1, x_2)-plane and perpendicular to the $\boldsymbol{\nu}$. So, (7.2.56) can be re-written as

$$\begin{aligned} \boldsymbol{\tau}^*(\Omega, \boldsymbol{\eta}) = (2\pi)^{-3/2} &\int_{-\infty}^{\infty} t^{n+2} dt \int_{-\infty}^{\infty}\int_{-\infty}^{\infty} \boldsymbol{\tau}(p) \\ &\times \exp\{\mathrm{i}t\,(\Omega + \eta_\alpha \nu_\alpha p + \eta_\alpha \lambda_\alpha q)\}\, dp\, dq . \end{aligned} \tag{7.2.57}$$

Integrating equation (7.2.56) in q, taking into account the fact that

$$\int_{-\infty}^{\infty} \mathrm{e}^{\mathrm{i}qt\eta_\alpha\lambda_\alpha} dq = 2\pi\delta(\lambda_\alpha\eta_\alpha)/t , \tag{7.2.58}$$

thereafter integrating in t, find

$$\tau^*(\Omega,\eta) = \frac{(-\mathrm{i})^n}{\sqrt{2\pi}}\delta(\lambda_\alpha\eta_\alpha)\partial_\Omega^{(n+2)}\int_{-\infty}^{\infty}\frac{\tau(p)dp}{\Omega+\eta_\alpha\nu_\alpha p} = \frac{(-\mathrm{i})^n}{\sqrt{2\pi}}\delta(\lambda_\alpha\eta_\alpha)$$
$$\times\,\partial_\Omega^{(n+1)}\left[H(\nu_\alpha\eta_\alpha)\int_{-\infty}^{\infty}\frac{\tau(p)dp}{p+\Omega} - H(-\nu_\alpha\eta_\alpha)\int_{-\infty}^{\infty}\frac{\tau(p)dp}{p-\Omega}\right]. \tag{7.2.59}$$

The latter should be substituted into (7.2.37); integration along the path $|\eta| = 1$ should be performed with due account for the fact that the presence of the function $\delta(\lambda_\alpha\eta_\alpha)$ permits only the points $\eta = \pm\nu$ to make a contribution to the integral (note that they contribute complex conjugate values). This results in the formula,

$$\partial_t^{(n+2)}\boldsymbol{u}(t,\boldsymbol{x}) \tag{7.2.60}$$
$$= \frac{(-1)^{n+1}}{\pi}\sum_{N=4}^{6}\mathrm{Re}\left\{\frac{(\Omega^N)^{n+2}\mathbf{B}^N(\Omega^N,\boldsymbol{\nu})}{t+\xi_{3,\Omega}^N(\Omega^N,\boldsymbol{\nu})x_3}\partial_{\Omega^N}^{(n+1)}\int_{-\infty}^{\infty}\frac{\tau(p)dp}{p+\Omega^N}\right\}H(t-t_N)$$
$$\left(\Omega^N \equiv \Omega^N(\boldsymbol{\nu}), \mathrm{Im}\Omega^N \geq 0\right)\ ,$$

where the integration path goes above the singular point $p = -\Omega^N$.

Proceeding in the same way, equation (7.2.40) becomes

$$\partial_t^{(n+1)}\boldsymbol{\sigma}_j(t,\boldsymbol{x}) = \frac{(-1)^{n+1}}{\pi}\sum_{N=4}^{6}\mathrm{Re}\left\{\frac{(\Omega^N)^{n+1}\mathbf{C}_j(\Omega^N,\boldsymbol{\nu})}{t+\xi_{3,\Omega}(\Omega^N),\boldsymbol{\nu})x_3}\right.$$
$$\left.\times\,\mathbf{B}^N(\Omega^n,\boldsymbol{\nu})\partial_{\Omega^N}^{(n+1)}\int_{-\infty}^{\infty}\frac{\tau(p)dp}{p+\Omega^N}\right\}H(t-t_N)\,. \tag{7.2.61}$$

Equations (7.2.60, 61) can be written as (7.2.45, 46) if the function

$$G^{(n+1)}(z,\boldsymbol{\nu}) = \frac{1}{2\pi\mathrm{i}}\partial_z^{(n+1)}\int_{-\infty}^{\infty}\frac{\tau(p)dp}{p-z} \qquad (\mathrm{Im}\, z < 0) \tag{7.2.62}$$

is introduced. Substitution into (7.2.60, 61) results in

$$\partial_t^{(n+2)}\boldsymbol{u}(t,\boldsymbol{x}) = -2\sum_{N=4}^{6}\mathrm{Im}\left\{\frac{(\Omega^N)^{n+2}\mathbf{B}^N(\Omega^N,\boldsymbol{\nu})}{t+\xi_{3,\Omega}^N(\Omega^N,\boldsymbol{\nu})x_3}\right.$$
$$\left.\times\,G^{(n+1)}\left(-\Omega^N,\boldsymbol{\nu}\right)\right\}H(t-t_N)\,, \tag{7.2.63}$$

$$\partial_t^{(n+1)}\boldsymbol{\sigma}_j(t,\boldsymbol{x}) = -2\sum_{N=4}^{6}\mathrm{Im}\left\{\frac{(\Omega^N)^{n+1}\mathbf{C}_j(\Omega^N,\boldsymbol{\nu})}{t+\xi_{3,\Omega}^N(\Omega^N,\boldsymbol{\nu})x_3}\right.$$
$$\left.\times\,\mathbf{B}^N\left(\Omega^N,\boldsymbol{\nu}\right)G^{(n+1)}\left(-\Omega^N,\boldsymbol{\nu})x_3\right)\right\}H(t-t_N) \tag{7.2.64}$$

$$\left(\Omega^N \equiv \Omega^n(\boldsymbol{\nu}),\ \operatorname{Im}\Omega^N \geq 0\right) .$$

Integrating these equations once in t, make them analogous to equations (7.2.51,52):

$$\partial_t^{(n+1)} \boldsymbol{u}(t, \boldsymbol{x}) \tag{7.2.65}$$
$$= -2 \sum_{N=4}^{6} \operatorname{Im} \left\{ \int_{\Omega^N(t,\boldsymbol{x},\boldsymbol{\nu})}^{\infty} \Omega^{n+1} \mathbf{B}^N(\Omega, \boldsymbol{\nu})\, G^{(n+1)}(-\Omega, \boldsymbol{\nu}) d\Omega \right\} H(t - t_N) ,$$

$$\partial_t^{(n)} \boldsymbol{\sigma}_j(t, \boldsymbol{x})$$
$$= -2 \sum_{N=4}^{6} \operatorname{Im} \left\{ \int_{\Omega^N(t,\boldsymbol{x},\boldsymbol{\nu})}^{\infty} \Omega^N C_j(\Omega, \boldsymbol{\nu}) \mathbf{B}^N(\Omega, \boldsymbol{\nu}) G^{(n+1)}(-\Omega, \boldsymbol{\nu}) d\Omega \right\} \tag{7.2.66}$$
$$\times H(t - t_N) .$$

Here, $\operatorname{Im}\{\Omega\} > 0$, and the functions to be integrated behave as $O(\Omega^{-2})$ when $|\Omega| \to \infty$, since in this case

$$G^{(n+1)}(-\Omega, \boldsymbol{\nu}) = O(\Omega^{-n-2}) .$$

One must mention that, using these representations of the solutions for the two- and three-dimensional cases, behaviour of solutions near the wave fronts could be studied. Since similar research for the anisotropic medium is notably cumbersome, this will not be considered herein, and the reader is asked to refer to the primary source [7.18].

7.2.3 Radon Transform

Some important properties of the Radon transformation have been obtained in [7.24] using the close relationship between the Fourier and Radon transformations. The latter decomposes an arbitrary function into plane waves of an arbitrary type (while the Fourier transformation is focused on exponential plane waves). Let us consider briefly those properties of the Radon transformation which are a necessity hereinafter.

The Radon transform $f_0(p, \boldsymbol{\xi})$ for the function $f(\boldsymbol{x})$ is, by definition, the integral along the line $\boldsymbol{\xi} \cdot \boldsymbol{x} = p(\boldsymbol{\xi} = \{\xi_1, \xi_2\},\ \boldsymbol{x} = \{x_1, x_2\})$:

$$f_0(p, \boldsymbol{\xi}) = \int_{-\infty}^{\infty} \int_{-\infty}^{\infty} f(\boldsymbol{x}) \delta(\boldsymbol{\xi} \cdot \boldsymbol{x} - p) dx_1\, dx_2 . \tag{7.2.67}$$

It is obvious that the function $f_0(p, \boldsymbol{\xi})$ is an even and homogeneous, of order of -1, i.e.,

$$f_0(\lambda p, \lambda \boldsymbol{\xi}) = |\lambda|^{-1} f_0(p, \boldsymbol{\xi}) . \tag{7.2.68}$$

Also, it is apparent from (7.2.67) that

$$\int_{-\infty}^{\infty} p^m f_0(p,\boldsymbol{\xi})dp = \int_{-\infty}^{\infty}\int_{-\infty}^{\infty} f(\boldsymbol{x})(\boldsymbol{\xi}\cdot\boldsymbol{x})^m dx_1\, dx_2 \,, \tag{7.2.69}$$

so the left-hand side in (7.2.69) is a homogeneous polynomial of the mth order in components of the vector $\boldsymbol{\xi}$. If use is made of the unit vector $\boldsymbol{\eta} = \boldsymbol{\xi}/|\boldsymbol{\xi}|$, equation (7.2.69) can be re-written in the following equivalent form:

$$\int_{-\infty}^{\infty} p^m f_0(p,\boldsymbol{\eta})dp = P_m(\boldsymbol{\eta}) \,, \tag{7.2.70}$$

where $P_m(\boldsymbol{\eta})$ is a homogeneous polynomial whose order in coordinates of the vector $\boldsymbol{\eta}$ does not exceed m. It is worth reporting one property related to the carrier of the function $f_0(p,\boldsymbol{\eta})$. From (7.2.67) it follows that if $f(\boldsymbol{x}) = 0$ for $r > a (r \equiv \sqrt{x_1^2+x_2^2})$ then

$$f_0(p,\boldsymbol{\eta}) = 0 \quad \text{for} \quad |p| > a \,. \tag{7.2.71}$$

As the work [7.24] establishes, the properties (7.2.68,70,71) formulate the sufficient condition of whether the function $f(p,\boldsymbol{\eta})$ defines the Radon transform of the function $f(\boldsymbol{x})$ whose carrier is the region $r \le a$. To conclude, write the Radon transform inversion formula:

$$f(\boldsymbol{x}) = -\frac{1}{4\pi^2}\oint_{|\eta|=1} ds \int_{-\infty}^{\infty} \frac{f_0(p,\boldsymbol{\eta})dp}{(p-\boldsymbol{\eta}\cdot\boldsymbol{x}+0\mathrm{i})^2} \,, \tag{7.2.72}$$

and mention that the Radon transformation for a vector implies the Radon transformation for each one of components of this vector.

7.3 Solution of the Second Boundary-Value Problem

Herein, consideration will be given to the second boundary problem, with a particular anticipation of the problems of Sect. 7.6.

7.3.1 Three-Dimensional Case

In the case that the displacement vector

$$\boldsymbol{u}(t,\boldsymbol{x}) = \boldsymbol{w}(t,x_\alpha) = t^{n+1}\boldsymbol{w}(\boldsymbol{y}) \qquad (x_3 \to 0,\ y_\alpha = x_\alpha/t,\ t>0) \tag{7.3.1}$$

is specified at the boundary $x_3 = 0$ of the half-space $x_3 > 0$, equations (7.2.6) with conditions (7.2.7,9, 7.3.1) can be solved by using the method expounded in the previous Section. In terms of Fourier transforms, according to (7.2.17), we can write for this boundary:

$$\sum_{N=4}^{6} \operatorname{adj} \mathbf{K}(\omega, \boldsymbol{\xi}^N) \boldsymbol{b}^N(\omega, \xi_\alpha) = \boldsymbol{w}^*(\omega, \xi_\alpha) . \tag{7.3.2}$$

The vector $\boldsymbol{b}^N(\omega, \xi_\alpha)$ may be determined independently of N if we assume

$$\boldsymbol{b}^N(\omega, \xi_\alpha) = \left[\sum_{M=4}^{6} \operatorname{adj} \mathbf{K}\left(\omega, \xi^M\right)\right]^{-1} \boldsymbol{w}^*(\omega, \xi_\alpha) . \tag{7.3.3}$$

Then, from (7.2.17, 38),

$$\boldsymbol{u}^*(\omega, \xi_\alpha, x_3) = (-\mathrm{i}) \sum_{N=4}^{6} \mathbf{D}^N(\omega, \xi_\alpha) \boldsymbol{w}^*(\omega, \xi_\alpha) \exp\left\{-\mathrm{i}\xi_3^N x_3\right\} , \tag{7.3.4}$$

$$\boldsymbol{\sigma}_j^*(\omega, \xi_\alpha, x_3) = -\sum_{N=4}^{6} \mathbf{C}_j\left(\xi^N\right) \mathbf{D}^N(\omega, \xi_\alpha) \cdot \boldsymbol{w}^*(\omega, \xi_\alpha) \exp\left\{-\mathrm{i}\xi_3^N x_3\right\} \tag{7.3.5}$$

where the matrix D^N is introduced as

$$\mathbf{D}^N(\omega, \xi_\alpha) = \mathrm{i} \operatorname{adj} \mathbf{K}(\omega, \boldsymbol{\xi}^N) \left[\sum_{M=4}^{6} \operatorname{adj} \mathbf{K}(\omega, \boldsymbol{\xi}^M)\right]^{-1} . \tag{7.3.6}$$

Expressions (7.3.4, 5) can be inverted in the same way as (7.2.23, 39): equations (7.3.4, 5) are obtained from (7.2.23, 39), respectively, by replacing the combination $\mathbf{B}^N(\omega, \xi_\alpha)\boldsymbol{\tau}^*(\omega, \xi_\alpha)$ by $(-\mathrm{i})\mathbf{D}^N(\omega, \xi_\alpha)\boldsymbol{w}^*(\omega, \xi_\alpha)$ which is of the same order of homogeneity, $-(n+4)$, in variables ω, ξ_1, ξ_2. Thus,

$$\begin{aligned} &\partial_t^{(n+2)} \boldsymbol{u}(t, \boldsymbol{x}) \\ &= \frac{(-\mathrm{i})^{n+3}}{\sqrt{2\pi}} \sum_{N=4}^{6} H(t - t_N) \oint_{|\boldsymbol{\eta}|=1} \frac{(\Omega^N)^{n+2} \mathbf{D}^N(\Omega^N, \boldsymbol{\eta})}{t + \xi_{3,\Omega}^N(\Omega^N, \boldsymbol{\eta}) x_3} \boldsymbol{w}^*(\Omega^N, \boldsymbol{\eta}) ds , \end{aligned} \tag{7.3.7}$$

$$\begin{aligned} \partial_t^{(n+1)} \boldsymbol{\sigma}_j(t, \boldsymbol{x}) &= \frac{(-\mathrm{i})^{n+3}}{\sqrt{2\pi}} \sum_{N=4}^{6} H(t - t_N) \\ &\times \oint_{|\boldsymbol{\eta}|=1} \frac{(\Omega^N)^{n+1} \mathbf{C}_j(\Omega^N, \boldsymbol{\eta})}{t + \xi_{3,\Omega}^N(\Omega^N, \boldsymbol{\eta}) x_3} \mathbf{D}^N(\Omega^N, \boldsymbol{\eta}) \boldsymbol{w}^*(\Omega^N, \boldsymbol{\eta}) ds \end{aligned} \tag{7.3.8}$$

where Ω^N is the root of equation (7.2.30) with $\operatorname{Im}\{\Omega^N\} \geq 0$ for $t > t_N$.

Treated analogously, the latter expressions, (7.3.7, 8), can be represented as (7.2.45, 46):

$$\partial_t^{(n+2)} \boldsymbol{u}(t, \boldsymbol{x}) \tag{7.3.9}$$
$$= -\frac{1}{2\pi} \sum_{N=4}^{6} H(t - t_N) \oint_{|\eta|=1} \frac{(\Omega^N)^{n+2} \mathbf{D}^N(\Omega^N, \boldsymbol{\eta})}{t + \xi_{3,\Omega}^N(\Omega^N, \boldsymbol{\eta}) x_3} \boldsymbol{E}^{(n+3)}\left(-\Omega^N, \boldsymbol{\eta}\right) ds ,$$

$$\partial_t^{(n+1)} \boldsymbol{\sigma}_j(t, \boldsymbol{x}) = -\frac{1}{2\pi} \sum_{n=4}^{6} H(t - t_N) \tag{7.3.10}$$
$$\times \oint_{|\eta|=1} \frac{(\Omega^N)^{n+1} \mathbf{C}_j(\Omega^N, \boldsymbol{\eta})}{t + \xi_{3,\Omega}^N(\Omega^N, \boldsymbol{\eta}) x_3} \mathbf{D}^N(\Omega^N, \boldsymbol{\eta}) \boldsymbol{E}^{(n+3)}(-\Omega^N, \boldsymbol{\eta}) ds ,$$

where new symbols mean:

$$\boldsymbol{E}^{(n+3)}(z, \boldsymbol{\eta}) = \frac{1}{2\pi \mathrm{i}} \partial_z^{(n+3)} \int_{-\infty}^{\infty} \frac{\boldsymbol{w}_0(p, \boldsymbol{\eta}) dp}{p - z} \quad (\operatorname{Im} z < 0) , \tag{7.3.11}$$

$$\boldsymbol{w}_0(p, \boldsymbol{\eta}) = \int_{-\infty}^{\infty} \int_{-\infty}^{\infty} \boldsymbol{w}(\boldsymbol{y}) \delta(p - \eta_\alpha y_\alpha) dy_1 \, dy_2 . \tag{7.3.12}$$

Finally, integrating (7.3.9, 10) once in t, obtain:

$$\partial_t^{(n+1)} \boldsymbol{u}(t, \boldsymbol{x}) \tag{7.3.13}$$
$$= -\frac{1}{2\pi} \sum_{N=4}^{6} \oint_{|\eta|=1} ds \int_{\Omega^N(t, \boldsymbol{x}, \boldsymbol{\eta})}^{\infty} \Omega^{n+1} \mathbf{D}^N(\Omega, \boldsymbol{\eta}) \boldsymbol{E}^{(n+3)}(-\Omega, \boldsymbol{\eta}) d\Omega ,$$

$$\partial_t^{(n)} \boldsymbol{\sigma}_j(t, \boldsymbol{x}) \tag{7.3.14}$$
$$= -\frac{1}{2\pi} \sum_{N=4}^{6} \oint_{|\eta|=1} ds \int_{\Omega^N(t, \boldsymbol{x}, \boldsymbol{\eta})}^{\infty} \Omega^{n} \mathbf{C}_j(\Omega, \boldsymbol{\eta}) \mathbf{D}^N(\Omega, \boldsymbol{\eta}) \boldsymbol{E}^{(n+3)}(-\Omega, \boldsymbol{\eta}) d\Omega$$

where $\operatorname{Im}\{\Omega\} > 0$, and the functions to be integrated behave as $O(\Omega^{-3})$ when $|\Omega| \to \infty$, since in this case

$$\mathbf{D}^N(\Omega, \boldsymbol{\eta}) = O(1), \ \mathbf{C}_j(\Omega, \boldsymbol{\eta}) = O(\Omega). \quad \boldsymbol{E}^{(n+3)}(-\Omega, \boldsymbol{\eta}) = O(\Omega^{-n-4}) .$$

7.3.2 Two-Dimensional Case

In two dimensions, the specified displacement is assumed to be constant over the line $\nu_\alpha x_\alpha = pt$ ($\boldsymbol{\nu} = \{\nu_1, \nu_2\}$), i.e.,

$$\boldsymbol{w}(t, x_\alpha) = t^{n+1} \boldsymbol{w}(p) \qquad (t > 0) . \tag{7.3.15}$$

Consequently, the vector $\boldsymbol{u}(t, \boldsymbol{x})$ depends only on t, pt, x_3 and is a homogeneous function of the $(n+1)$th order in these variables. Now, we hould treat $\boldsymbol{w}(t, x_\alpha)$ in the same way as $\boldsymbol{\tau}(t, x_\alpha)$ (the formulae (7.2.56, 59)) to convert equations (7.3.7, 8) into the following representation (analogous to (7.2.63, 64)) for the solution in the two-dimensional case:

$$\partial_t^{(n+2)} \boldsymbol{u}(t, \boldsymbol{x}) \tag{7.3.16}$$
$$= -2 \sum_{N=4}^{6} \operatorname{Im} \left\{ (\Omega^N)^{n+2} \frac{\mathbf{D}^N(\Omega^N, \boldsymbol{\nu})}{t + \xi_{3,\Omega}^N(\Omega^N, \boldsymbol{\nu}) x_3} \boldsymbol{J}^{(n+2)}(-\Omega^N, \boldsymbol{\nu}) \right\} H(t - t_N) ,$$

$$\partial_t^{(n+1)} \boldsymbol{\sigma}_j(t, \boldsymbol{x}) \tag{7.3.17}$$
$$= -2 \sum_{N=4}^{6} \operatorname{Im} \left\{ (\Omega^N)^{n+1} \frac{\mathbf{C}_j(\Omega^N, \boldsymbol{\nu})}{t + \xi_{3,\Omega}^N(\Omega^N, \boldsymbol{\nu}) x_3} \mathbf{D}^N(\Omega^N, \boldsymbol{\nu}) \boldsymbol{J}^{(n+2)}(-\Omega^N, \boldsymbol{\nu}) \right\}$$
$$\times H(t - t_N) ,$$

where

$$\Omega^N \equiv \Omega^N(\boldsymbol{\nu}), \quad \operatorname{Im} \Omega^N \geq 0 ,$$

$$\boldsymbol{J}^{(n+2)}(z, \boldsymbol{\nu}) = \frac{1}{2\pi \mathrm{i}} \partial_z^{(n+2)} \int_{-\infty}^{\infty} \frac{w(p) dp}{p - z} \qquad (\operatorname{Im} z < 0) . \tag{7.3.18}$$

By integrating (7.3.16, 17) once in t, find

$$\partial_t^{(n+1)} \boldsymbol{u}(t, \boldsymbol{x}) \tag{7.3.19}$$
$$= -2 \sum_{N=4}^{6} \operatorname{Im} \left\{ \int_{\Omega^N(t, \boldsymbol{x}, \boldsymbol{\nu})}^{\infty} \Omega^{n+1} \mathbf{D}^N(\Omega, \boldsymbol{\nu}) \boldsymbol{J}^{(n+2)}(-\Omega, \boldsymbol{\nu}) d\Omega \right\} H(t - t_N) ,$$

$$\partial_t^{(n)} \boldsymbol{\sigma}_j(t, \boldsymbol{x}) \tag{7.3.20}$$
$$= -2 \sum_{N=4}^{6} \operatorname{Im} \left\{ \int_{\Omega^N(t, \boldsymbol{x}, \boldsymbol{\nu})}^{\infty} \Omega^{n} \mathbf{C}_j(\Omega, \boldsymbol{\nu}) \mathbf{D}^N(\Omega, \boldsymbol{\nu}) \boldsymbol{J}^{(n+2)}(-\Omega, \boldsymbol{\nu}) d\Omega \right\}$$
$$\times H(t - t_N) ,$$

where $\operatorname{Im}\{\Omega\} > 0$, and the integrands behave as $\mathrm{O}(\Omega^{-2})$ when $|\Omega| \to \infty$, since in this case

$$\boldsymbol{J}^{(n+2)}(-\Omega, \boldsymbol{\nu}) = \mathrm{O}(\Omega^{-n-3}) .$$

7.4 Lamb's Problem

Let us exemplify the results of Sect. 7.2 by solution of the three- and two-dimensional Lamb problems.

7.4.1 Three-Dimensional Case

Let the concentrated impulse

$$\boldsymbol{\tau}(t, x_\alpha) = \boldsymbol{\tau}_0 \delta(t)\delta(x_1)\delta(x_2) \tag{7.4.1}$$

be applied to the boundary $x_3 = 0$ of the half-space $x_3 > 0$ at instance $t = 0$ (Fig. 7.3). Here, $\boldsymbol{\tau}_0$ is a constant vector. The displacement $\boldsymbol{u}(t, \boldsymbol{x})$ which arises in the medium subjected to this load may be written as

$$\boldsymbol{u}(t, \boldsymbol{x}) = \mathbf{U}(t, \boldsymbol{x})\boldsymbol{\tau}_0 , \tag{7.4.2}$$

where the matrix $\mathbf{U}(t, \boldsymbol{x})$ (the Green matrix) is such that the displacement vector

$$\boldsymbol{u}(t, \boldsymbol{x}) \equiv \mathbf{U}(t, \boldsymbol{x})\boldsymbol{e}_j = \{U_{ij}, U_{2j}, U_{3j}\}$$

is the solution to the problem (7.2.6–9) with the load $\boldsymbol{\tau}(t, x_\alpha) = \boldsymbol{e}_j\delta(t)\delta(x_1)\delta(x_2)$ (that is, $U_{ij}(t, \boldsymbol{x})$ is the ith component of the displacement caused by the concentrated impulse

$$\boldsymbol{\tau}(t, x_\alpha) = \boldsymbol{e}_j\delta(t)\delta(x_1)\delta(x_2)$$

applied to the boundary $x_3 = 0$ and aligned with the x_j-axis).

This problem cannot be solved immediately using the procedure of Sect. 7.2, since this procedure implicitly assumes that $n \geq -2$. However, $\mathbf{U}$ may be represented as

$$\mathbf{U}(t, \boldsymbol{x}) = \partial \mathbf{U}_1(t, \boldsymbol{x})/\partial t , \tag{7.4.3}$$

where the matrix $\mathbf{U}_1(t, \boldsymbol{x})$ is such that the vector $\boldsymbol{u}_j^1(t, \boldsymbol{x}) \equiv \mathbf{U}_1(t, \boldsymbol{x})\boldsymbol{e}_j$ becomes the solution to the problem (7.2.6–9) with the load

$$\boldsymbol{\tau}(t, x_\alpha) = \boldsymbol{e}_j H(t)\delta(x_1)\delta(x_2) .$$

The matrix $\mathbf{U}_1(t, \boldsymbol{x})$ can at once be found from equation (7.2.37) at $n = -2$ if we assume $\boldsymbol{u}(t, \boldsymbol{x}) \equiv \boldsymbol{u}_j^1(t, \boldsymbol{x})$ in (7.2.37) and take into account that

$$\begin{aligned}
\boldsymbol{\tau}^*(\Omega, \boldsymbol{\eta}) &\\
&= (2\pi)^{-3/2}\boldsymbol{e}_j \int_0^\infty dt \int_{-\infty}^\infty \int_{-\infty}^\infty \exp\{\mathrm{i}(\Omega t + \eta_\alpha x_\alpha)\}\, \delta(x_1)\delta(x_2) dx_1 dx_2 \\
&= \frac{\mathrm{i}}{(2\pi)^{3/2}\Omega}\boldsymbol{e}_j .
\end{aligned} \tag{7.4.4}$$

In view of (7.4.3), we have a result:

$$\mathbf{U}(t, \boldsymbol{x}) = \frac{\mathrm{i}}{(2\pi)^2}\frac{\partial}{\partial t}\sum_{N=4}^{6} H(t - t_N) \oint_{|\boldsymbol{\eta}|=1} \frac{\mathbf{B}^N(\Omega^N, \boldsymbol{\eta})ds}{\Omega^N\left[t + \xi_{3,\Omega}^N(\Omega^N, \boldsymbol{\eta})x_3\right]} . \tag{7.4.5}$$

In a somewhat differing form, this formula has been obtained by means of the Cagniard-deHoop method in [7.14].

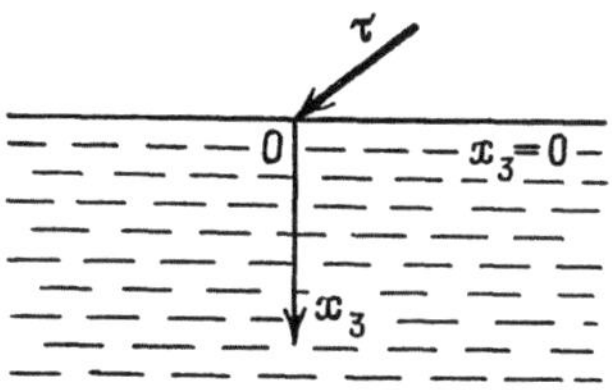

Fig. 7.3. Three-dimensional Lamb's problem

7.4.2 Two-Dimensional Case

Here, the load can be described by the formula,

$$\boldsymbol{\tau}(t, x_\alpha) = \boldsymbol{\tau}_0 \delta(t)\delta(\nu_\alpha x_\alpha) \,, \tag{7.4.6}$$

where $\boldsymbol{\nu}$ is the unit vector in the (x_1, x_2)-plane (Fig. 7.4). Again, represent $\boldsymbol{u}(t, \boldsymbol{x})$ in the form (7.4.2) where the vector $\mathbf{U}(t, \boldsymbol{x})\boldsymbol{e}_j$ is now a solution to the problem (7.2.6–9) for the load

$$\boldsymbol{\tau}(t, x_\alpha) = \boldsymbol{e}_j \delta(t)\delta(\nu_\alpha x_\alpha) \,.$$

The load (7.4.6) is homogeneous with respect to t and $\boldsymbol{x}$ (of order of $n = -2$), therefore, the solution in this case cannot be written immediately from (7.2.63), since this formula is valid for $n \geq -1$ only. However, $\mathbf{U}(t, \boldsymbol{x})$ may be expressed by means of (7.2.37) at $n = -2$ taking into account that, after (7.2.57, 58),

$$\boldsymbol{\tau}^*(\Omega, \boldsymbol{\eta}) = (2\pi)^{-1/2}\delta(\lambda_\alpha \eta_\alpha)\boldsymbol{\tau}_0 \,, \tag{7.4.7}$$

where $\boldsymbol{\lambda}$ is the unit vector (in the (x_1, x_2)-plane) perpendicular to the $\boldsymbol{\nu}$. Just as in deriving formula (7.2.63), account should be taken of the fact that only points $\boldsymbol{\eta} = \boldsymbol{\nu}$ and $\boldsymbol{\eta} = -\boldsymbol{\nu}$ are capable of making a contribution to the integral in (7.2.37), these contributions being complex conjugate values. After substituting (7.4.7) into (7.2.37), obtain the result:

$$\mathbf{U}(t, \boldsymbol{x}) = \frac{1}{\pi} \sum_{N=4}^{6} \mathrm{Re} \left\{ \frac{\mathbf{B}^N(\Omega^N, \boldsymbol{\nu})}{t + \xi^N_{3,\Omega}(\Omega^N, \boldsymbol{\nu})x_3} \right\} H(t - t_N) \,, \tag{7.4.8}$$

which agrees well with that of [7.14].

7.4.3 Isotropic Half-Space

In this case, the matrix $\mathbf{K}(\omega, \boldsymbol{\xi})$ takes the following form:

$$K_{ik}(\omega, \boldsymbol{\xi}) = (\mu\xi_j\xi_j - \rho\omega^2)\delta_{ik} + (\lambda + \mu)\xi_i\xi_k \,. \tag{7.4.9}$$

A simple analysis indicates that equation (7.2.16) has the roots

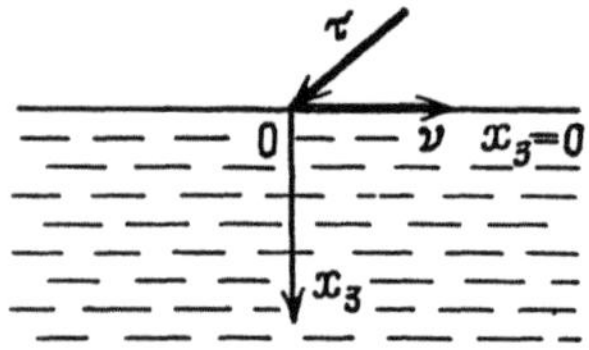

Fig. 7.4. Two-dimensional Lamb's problem

$$\begin{aligned} \xi_3^{\pm\alpha} &= \pm(\omega^2/\alpha^2 - |\boldsymbol{\xi}|^2)^{1/2} \qquad \text{(simple roots)}\,, \\ \xi_3^{\pm\beta} &= \pm(\omega^2/\beta^2 - |\boldsymbol{\xi}|^2)^{1/2} \qquad \text{(double roots)}\,, \end{aligned} \tag{7.4.10}$$

here, the upper or lower signs should be used simultaneously; α and β are the speeds of compressional and shear waves, respectively:

$$\alpha^2 \equiv c_1^2 = (\lambda + 2\mu)/\rho, \quad \beta^2 \equiv c_2^2 = \mu/\rho\,, \tag{7.4.11}$$

$|\boldsymbol{\xi}|^2 = \xi_1^2 + \xi_2^2$, as before. To choose the single-valued analytic branches of the radicals present in the expressions (7.4.10) (for the complex ω-plane) the cuts are drawn between the points $\pm\alpha|\boldsymbol{\xi}|$ (and $\pm\beta|\boldsymbol{\xi}|$), with branches taken such that $\mathrm{Im}\{(\xi_3^{-\alpha})\} < 0$ (and $\mathrm{Im}\{(\xi_3^{-\beta})\} < 0$) for $\mathrm{Im}\{\omega\} > 0$. In view of the degeneracy of the matrix $\mathbf{K}(\omega, \boldsymbol{\xi})$ (double roots $\xi_3^{\pm\beta}$) in the case of isotropy under consideration, one must change the derivation of the formulae obtained in Sect. 7.2 for the case of simple roots. First of all, note that equation (7.2.15) is satisfied by the vectors

$$\begin{aligned} \boldsymbol{v}^{\pm\alpha} &= \{\xi_1, \xi_2, \xi_3^{\pm\alpha}\}\,, \\ \boldsymbol{v}_1^{\pm\beta} &= \left\{\frac{\xi_1\xi_3^{\pm\beta}}{|\boldsymbol{\xi}|}, \frac{\xi_2\xi_3^{\pm\beta}}{|\boldsymbol{\xi}|}, -|\boldsymbol{\xi}|\right\}, \\ \boldsymbol{v}_2^{\pm\beta} &= \{-\xi_2, \xi_1, 0\}\,, \end{aligned} \tag{7.4.12}$$

where the first pair corresponds to the roots $\xi_3^{\pm\alpha}$, and two other pairs to double roots $\xi_3^{\pm\beta}$. This means that the general solution to (7.2.13) being limited when $x \to \infty$ ($\mathrm{Im}\{\omega\} > 0$) takes the following form

$$\begin{aligned} &\boldsymbol{u}^*(\omega, \xi_1, \xi_2, x_3) \\ &= \boldsymbol{v}^{-\alpha} b^\alpha \exp\{-\mathrm{i}\xi_3^{-\alpha} x_3\} + \left(\boldsymbol{v}_1^{-\beta} b_1^\beta + \boldsymbol{v}_2^{-\beta} b_2^\beta\right) \exp\left\{-\mathrm{i}\xi_3^{-\beta} x_3\right\}\,, \end{aligned} \tag{7.4.13}$$

where scalars $(b^\alpha, b_1^\beta, b_2^\beta)$ depend on (ω, ξ_1, ξ_2). In this case, (7.2.20) is replaced by equation (7.4.14):

$$\begin{bmatrix} -2\mathrm{i}\xi_1\xi_3^{-\alpha} & \frac{\mathrm{i}\xi_1}{|\boldsymbol{\xi}|}\left(2|\boldsymbol{\xi}|^2 - \frac{\omega^2}{\beta^2}\right) & \mathrm{i}\xi_2\xi_3^{-\beta} \\ -2\mathrm{i}\xi_2\xi_3^{-\alpha} & \frac{\mathrm{i}\xi_2}{|\boldsymbol{\xi}|}\left(2|\boldsymbol{\xi}|^2 - \frac{\omega^2}{\beta^2}\right) & \mathrm{i}\xi_1\xi_3^{-\beta} \\ \mathrm{i}\left(2|\boldsymbol{\xi}|^2 - \frac{\omega^2}{\beta^2}\right) & 2\mathrm{i}|\boldsymbol{\xi}|\xi_3^{-\beta} & 0 \end{bmatrix} \begin{bmatrix} b^\alpha \\ b_1^\beta \\ b_2^\beta \end{bmatrix} = \frac{\boldsymbol{\tau}^*(\omega, \xi_1, \xi_2)}{\mu}\,. \tag{7.4.14}$$

The solution to the latter is

$$\begin{bmatrix} b^\alpha \\ b_1^\beta \\ b_2^\beta \end{bmatrix} = \begin{bmatrix} -2\xi_1|\boldsymbol{\xi}|\xi_3^{-\beta} & -2\xi_2|\boldsymbol{\xi}|\xi_3^{-\beta} & |\boldsymbol{\xi}|\left(2|\boldsymbol{\xi}|^2 - \frac{\omega^2}{\beta^2}\right) \\ \xi_1\left(2|\boldsymbol{\xi}|^2 - \frac{\omega^2}{\beta^2}\right) & \xi_2\left(2|\boldsymbol{\xi}|^2 - \frac{\omega^2}{\beta^2}\right) & 2|\boldsymbol{\xi}|^2\xi_2^{-\alpha} \\ \frac{\xi_2 D(\omega,|\boldsymbol{\xi}|)}{|\boldsymbol{\xi}|\xi_3^{-\beta}} & \frac{-\xi_1 D(\omega,|\boldsymbol{\xi}|)}{|\boldsymbol{\xi}|\xi_3^{-\beta}} & 0 \end{bmatrix} \times \frac{\boldsymbol{\tau}^*(\omega,\xi_1,\xi_2)}{\mathrm{i}\mu|\boldsymbol{\xi}|D(\omega,|\boldsymbol{\xi}|)}, \tag{7.4.15}$$

the additional symbol D being

$$D(\omega,|\boldsymbol{\xi}|) = 4|\boldsymbol{\xi}|^2\left(\frac{\omega^2}{\alpha^2} - |\boldsymbol{\xi}|^2\right)^{1/2}\left(\frac{\omega^2}{\beta^2} - |\boldsymbol{\xi}|^2\right)^{1/2} + \left(2|\boldsymbol{\xi}|^2 - \frac{\omega^2}{\beta^2}\right)^2 . \tag{7.4.16}$$

The equation $D(\omega,|\boldsymbol{\xi}|) = 0$ is the Rayleigh equation, since, according to the expression (4.2.15), we may write $D(\omega,|\boldsymbol{\xi}|) = \omega^4\alpha^{-4}R(\alpha|\boldsymbol{\xi}|\omega^{-1})$. Substituting (7.4.15) into (7.4.13), find $\boldsymbol{u}^*$, which, in a brief form, can be written as

$$\boldsymbol{u}^*(\omega,\xi_1,\xi_2,x_3) = (2\pi)^{3/2}\mathbf{U}^*(\omega,\xi_1,\xi_2,x_3)\boldsymbol{\tau}^*(\omega,\xi_1,\xi_2) , \tag{7.4.17}$$

the matrix $\mathbf{U}^*$ being obtained by applying the Fourier transformation to the matrix $\mathbf{U}(t,\boldsymbol{x})$ introduced in (7.4.2). Because the complete expression of the matrix $\mathbf{U}^*$ is cumbersome, let us illustrate it by the element U_{33}^*:

$$\begin{aligned} U_{33}^*(\omega,\xi_1,\xi_2,x_3) &= \sum_{N=4}^{5} B_{33}^N(\omega,\xi_1,\xi_2)\exp\left\{-\mathrm{i}\xi_3^N x_3\right\} \\ &= -\frac{(\omega^2/\alpha^2 - |\boldsymbol{\xi}|^2)^{1/2}}{(2\pi)^{3/2}\mathrm{i}\mu D(\omega,|\boldsymbol{\xi}|)}\left[\left(2|\boldsymbol{\xi}| - \frac{\omega^2}{\beta^2}\right)\exp\left\{\mathrm{i}\left(\frac{\omega^2}{\alpha^2} - |\boldsymbol{\xi}|^2\right)^{1/2} x_3\right\}\right. \\ &\quad \left. -2|\boldsymbol{\xi}|^2\exp\left\{\mathrm{i}\left(\frac{\omega^2}{\beta^2} - |\boldsymbol{\xi}|^2\right)^{1/2} x_3\right\}\right] . \end{aligned} \tag{7.4.18}$$

Use (7.4.5) to invert (7.4.18) (with $N = 4,5$, since sheets with numbers $N = 5$ and $N = 6$ are identical; in addition, $\xi_3^5 = \xi_3^6 = -(\omega^2/\beta^2 - |\boldsymbol{\xi}|^2)^{1/2}$, $\xi_3^4 = -(\omega^2/\alpha^2 - |\boldsymbol{\xi}|^2)^{1/2}$):

$$\begin{aligned} U_{33}(t,\boldsymbol{x}) = &-\frac{1}{4\pi^2\mu}\frac{\partial}{\partial t}\oint_{|\eta|=1}\mathrm{Im}\left\{\frac{1}{\mathrm{i}\Omega_\alpha D(\Omega_\alpha,1)\left[t\left(\Omega_\alpha^2/\alpha^2 - 1\right)^{1/2} - \Omega_\alpha x_3\alpha^{-2}\right]}\right. \\ &\times\left[H\left(t - \frac{x_3}{\alpha}\right)\left(\frac{\Omega_\alpha^2}{\alpha^2} - 1\right)\left(2 - \frac{\Omega_\alpha^2}{\beta^2}\right) - 2\left(\frac{\Omega_\beta^2}{\alpha^2} - 1\right)^{1/2}\right. \\ &\times\left.\left.\left(\frac{\Omega_\beta^2}{\beta^2} - 1\right)^{1/2} H\left(t - \frac{x_3}{\beta}\right)\right]\right\}ds . \end{aligned} \tag{7.4.19}$$

Values of Ω_α and Ω_β, respectively, must be determined from equations

$$\Omega_\alpha t + \eta_\lambda x_\lambda - \left(\Omega_\alpha^2/\alpha^2 - 1\right)^{1/2} x_3 = 0\,,$$
$$\Omega_\beta t + \eta_\lambda x_\lambda - \left(\Omega_\beta^2/\beta^2 - 1\right)^{1/2} x_3 = 0\,. \tag{7.4.20}$$

If $x_3 = 0$, we can transform the expression (7.4.19) by replacing $x_1 = r\cos\varphi$, $x_2 = r\sin\varphi$, $\eta_1 = \cos\Psi$, $\eta_2 = \sin\Psi$ (whence $ds = d\Psi$) and denoting $\theta = -1/\Omega_\alpha = -1/\Omega_\beta$ to obtain

$$U_{33}(t, x_1, x_2, 0) = -\frac{1}{2\pi^2\mu c_2^2 r}\frac{\partial}{\partial t}\mathrm{Re}\int_{-\theta_0}^{\theta_0} \frac{\theta\sqrt{c_1^{-2} - \theta^2}\,d\theta}{S(\theta^2)\sqrt{\theta^2 - \theta_0^2}} \tag{7.4.21}$$

$$(\mathrm{Im}\,\theta \geq 0, \quad \theta_0 = t/r)\,.$$

This result may also be obtained from the rigth-hand side of the formula (3.8.26) by differentiating it with respect to t, then assuming $p = 1$ and changing the sign of this expression to the opposite one, since this formula has been found for $z \leq 0$. From the formula (3.8.27) it follows that the expression (7.4.21) can be written in elementary functions.

Proceeding analogously, one could find that the solution (7.4.18) for a two-dimensional case of isotropy becomes the expressions (3.4.41) (if the parameter p in the formulae (3.4.41) is set to -1 and if the sign at the term $u_y(x, 0, t)$ is changed to the opposite one, since these formulae are obtained for the half-plane $y \leq 0$).

To conclude this Section, let us write down the expression which will be useful later:

$$\mathbf{B}(\omega, \xi_1, \xi_2) \equiv \sum_{N=4}^{5} \mathbf{B}^N(\omega, \xi_1, \xi_2) = (2\pi)^{3\beta}\mathbf{U}^*(\omega, \xi_1, \xi_2, 0)$$
$$= \frac{1}{|\xi|}\begin{bmatrix} e + f\eta_2^2 & -f\eta_1\eta_2 & -ih\eta_1 \\ -f\eta_1\eta_2 & e + f\eta_1^2 & -ih\eta_2 \\ ih\eta_1 & ih\eta_2 & g \end{bmatrix}, \tag{7.4.22}$$

the added notation being

$$e = \frac{\Omega^2}{\mathrm{i}\mu D(\Omega, 1)\beta^2}\left(\frac{\Omega^2}{\beta^2} - 1\right)^{1/2},$$
$$f = \frac{-1}{\mathrm{i}\mu D(\Omega, 1)}\left[\frac{\Omega^2}{\beta^2}\left(\frac{\Omega^2}{\beta^2} - 1\right)^{1/2} - D(\Omega, 1)\left(\frac{\Omega^2}{\beta^2} - 1\right)^{-1/2}\right],$$
$$g = \frac{\Omega^2}{\mathrm{i}\mu D(\Omega, 1)\beta^2}\left(\frac{\Omega^2}{\alpha^2} - 1\right)^{1/2}, \tag{7.4.23}$$
$$h = -\frac{1}{\mu D(\Omega, 1)}\left[2 - \frac{\Omega^2}{\beta^2} + 2\left(\frac{\Omega^2}{\alpha^2} - 1\right)^{1/2}\left(\frac{\Omega^2}{\beta^2} - 1\right)^{1/2}\right],$$
$$D(\Omega, 1) = 4\left(\frac{\Omega^2}{\alpha^2} - 1\right)^{1/2}\left(\frac{\Omega^2}{\beta^2} - 1\right)^{1/2} + \left(2 - \frac{\Omega^2}{\beta^2}\right)^2.$$

The variables Ω and η_α are defined by the expressions employed in (7.2.27). It is worth noting that when Ω is real and $|\Omega| < \beta$, then (e, f, g, h) are real and the matrix $\mathbf{U}^*(\omega, \xi_1, \xi_2, 0)$ is hermitian: $U_{ij} = \bar{U}_{ij}$. In the formulae (7.4.23), imaginary parts of radicals $(\Omega^2/\alpha^2 - 1)^{1/2}$ and $(\Omega^2/\beta^2 - 1)^{1/2}$ for $\mathrm{Im}\{\Omega\} > 0$ should be considered positive, due to the adopted selection of branches of the functions $\xi_3^{-\alpha}$ and $\xi_3^{-\beta}$.

7.5 The Wedge-Shaped Punch Problem

Assume that a rigid symmetric punch with the semivertex angle α_0 is given a frictionless translational movement into an elastic half-space $x_3 > 0$ (Fig. 7.5) at the instance $t = 0$. Let the movement be "symmetric", the punch indentation velocity be directed along the axis and constant (v_0). The contact area $S(t)$ ($S(t)$: $x_3 = 0, |\nu_\lambda x_\lambda| \leq vt$) is, obviously, a strip (considered perpendicular to a certain constant vector $\boldsymbol{\nu} = \{\nu_1, \nu_2\}$) whose width increases at a constant speed v to be determined. At zero time ($t = 0$) the edge of the punch was in contact with the plane $x_3 = 0$ along the line $\nu_\lambda x_\lambda = 0$.

The problem must be solved at the following boundary conditions:

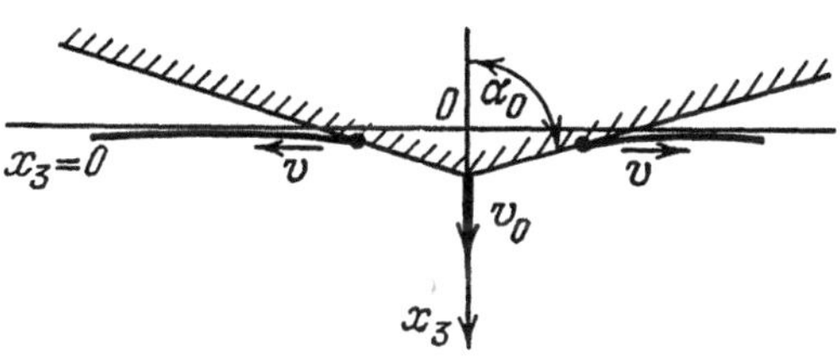

Fig. 7.5. Indentation of a half-space by a punch

$$
\begin{aligned}
&c_{\beta 3kl} u_{k,l}(t, \boldsymbol{x}) = 0 \qquad (x_3 = 0,\ \beta = 1, 2)\ , \\
&c_{33kl} u_{k,l}(t, \boldsymbol{x}) = 0 \qquad (x_3 = 0,\ \boldsymbol{x} \notin S(t))\ , \qquad (7.5.1)\\
&u_3(t, \boldsymbol{x}) = -\cot \alpha_0 |\nu_\lambda x_\lambda| + v_0 t \qquad (x_3 = 0,\ \boldsymbol{x} \in S(t))\ .
\end{aligned}
$$

Let us assume, as well, that the indentation speed, v_0, is such that the contact area boundary extension speed, v, is less than any speed c_j ($j = 1, 2, 3$) of wave propagation in the elastic anisotropic medium, as well as less than the Rayleigh surface-wave speed c_R. From equations (7.2.37, 40) at $x_3 = 0$ it follows that the Rayleigh wave, if any, is defined by those poles of the function $\mathbf{B}(\omega, \boldsymbol{\eta}) = \sum_{N=4}^{6} \mathbf{B}^N(\omega, \boldsymbol{\eta})$ which are on the axis $\mathrm{Im}\{\omega\} = 0$, i.e., we are interested in real zeros of the function $\det \mathbf{B}_0(\omega, \boldsymbol{\eta})$; at these points, the matrix $\mathbf{B}_0$ has the form

$$
\mathbf{B}_0(\omega, \boldsymbol{\eta}) = \sum_{N=4}^{6} C_3 \left(\boldsymbol{\xi}^N(\omega, \boldsymbol{\eta}) \right) \operatorname{adj} \mathbf{K} \left(\omega, \boldsymbol{\xi}^N(\omega, \boldsymbol{\eta}) \right) .
$$

As shown in [7.14, 25], the function $\det \mathbf{B}_0(\omega, \boldsymbol{\eta})$ can be zero at some real value $\omega = \Omega^0(\boldsymbol{\eta})$ with $|\Omega^0| < c_j(\eta_1, \eta_2, 0)$ for all speeds c_j of wave propagation ($j = 1, 2, 3$). Furthermore, the speeds and stresses in the vicinity of the contact boundaries ($\nu_\lambda x_\lambda = \pm vt$) are assumed not to be more singular than $\varepsilon^{-1/2}$; with this, the absence of an energy flux through the contact area edges is provided. (Note that, in the above stated problem, the symmetric punch is assumed to cause a symmetric distribution of the displacement component u_3 on the boundary $x_3 = 0$, therefore, the boundary conditions (7.5.1) assume the speeds of the contact strip boundaries to be equal to v at both sides. In the general case of anisotropy (21 independent constants c_{ijkl}), existence of such a solution symmetric with respect to u_3 is not obvious. However, we shall see below that the solution to the mathematical problem with these boundary conditions, (7.5.1), gives, indeed, a symmetric field of the displacement u_3 on the plane $x_3 = 0$. This result offers a way of determining the unknown speed v of the contact boundaries from conditions defined on the contact surface, thus validating the symmetric formulation adopted for the general case of anisotropy).

In the problem under consideration the value of n is zero. Substituting $x_3 = 0$ into the formulae (7.2.30, 63), taking into account the first of the conditions (7.5.1), and employing (7.2.63), obtain the following expression for the component $\ddot{u}_3(t, x_1, x_2, 0)$:

$$\begin{aligned}\ddot{u}_3(t, x_1, x_2, 0) &= -\frac{2}{t}\mathrm{Im}\left\{\Omega^2 B_{33}(\Omega, \boldsymbol{\nu}) G_3'(-\Omega, \boldsymbol{\nu})\right\} \\ &= -\frac{2}{t}\mathrm{Im}\left(-\frac{\nu_\lambda x_\lambda}{t}\right)^2 B_{33}\left(-\frac{\nu_\lambda x_\lambda}{t} + 0\mathrm{i}, \boldsymbol{\nu}\right) G_3'\left(\frac{\nu_\lambda x_\lambda}{t} - 0\mathrm{i}, \boldsymbol{\nu}\right)\Big\} ,\end{aligned} \tag{7.5.2}$$

where, according to (7.2.62) and the boundary condition, we have

$$G_3'(z, \boldsymbol{\nu}) = \frac{1}{2\pi \mathrm{i}} \frac{\partial}{\partial z} \int_{-v}^{v} \frac{\tau_3(p)dp}{p - z} \tag{7.5.3}$$

and the new symbol $B_{33}(\Omega, \boldsymbol{\nu})$ is an element of the matrix $\mathbf{B}(\Omega, \boldsymbol{\nu}) = \sum_{N=4}^{6} \mathbf{B}^N(\Omega, \boldsymbol{\nu})$.

The function $\boldsymbol{u}(t, x_1, x_2, 0)$ depends only on two variables $t, \nu_\lambda x_\lambda$ and is a homogeneous function of the first order with respect to these variables. So, let us designate

$$u_3(t, x_1, x_2, 0) = tw(p) \qquad (p = \nu_\lambda x_\lambda / t) .$$

With this, find

$$\dot{u}_3 = w - pw_p', \quad \ddot{u}_3 = p^2 w_{pp}''/t \tag{7.5.4}$$

so, from (7.5.2), obtain

$$p^2 w_{pp}'' = -2p^2 \mathrm{Im}\left\{B_{33}(-p + 0\mathrm{i}, \boldsymbol{\nu}) G_3'(p - 0\mathrm{i}, \boldsymbol{\nu})\right\} . \tag{7.5.5}$$

In the general case of anisotropy under consideration, the coefficient $B_{33}(\Omega, \eta)$ turns out to be real if Ω is real and $|\Omega| < c_j(\eta_1, \eta_2, 0)$ $(j = 1, 2, 3)$. Prove this statement. Comparing (7.2.23) to (7.4.2) and taking (7.4.4) into account, it follows that the matrix $(2\pi)^{-3/2}\mathbf{B}(\omega, \xi_\alpha)$ is a transform of the Green matrix $\mathbf{U}(t, \boldsymbol{x})$ $(x_3 = 0)$ in the Lamb's problem considered in Sect. 7.4. On the other hand, applying the Green–Volterra formula (2.6.12) to solutions of the two Lamb's problems, $\boldsymbol{u}$ and $\boldsymbol{u}^0$:

$$\boldsymbol{u} \equiv \boldsymbol{u}^{(k)}(t, \boldsymbol{x}, \boldsymbol{x}_0) = \{U_{1k}, U_{2k}, U_{3k}\} ,$$
$$\boldsymbol{u}^0 \equiv \boldsymbol{u}^{(l)}(t_0 - t, \boldsymbol{x}, \boldsymbol{y}_0) = \{U_{1l}, U_{2l}, U_{3l}\} ,$$

obtain

$$U_{lk}(t_0, \boldsymbol{y}_0, \boldsymbol{x}_0) = U_{kl}(t_0, \boldsymbol{x}_0, \boldsymbol{y}_0) . \tag{7.5.6}$$

Here, $U_{jm}(t, \boldsymbol{x}, \boldsymbol{x}_0)$ is the jth component of the displacement vector $\boldsymbol{U}^{(m)}(t, \boldsymbol{x}, \boldsymbol{x}_0)$ at the instance t at the point $\boldsymbol{x}$ on the boundary $x_3 = 0$ and is caused by the mth component of a concentrated impulse force applied at the point $\boldsymbol{x}_0$ on the boundary $x_3 = 0$ (note that the same result, (7.5.6), may be obtained as a consequence of the Betti theorem directly from the formula (2.6.16), if the points $\boldsymbol{x}_0$ and $\boldsymbol{x}_1$ tend to the boundary $x_3 = 0$). In each of the solutions $U_{lk}(t_0, y_0, x_0)$ and $U_{kl}(t_0, x_0, y_0)$ the coordinate system origin can be shifted to the force application point without changes to the solutions. After substitution of t for t_0 and $\boldsymbol{x}$ for $\boldsymbol{y}_0 - \boldsymbol{x}_0$, re-write (7.5.6) as

$$U_{lk}(t, \boldsymbol{x}, 0) = U_{kl}(t, -\boldsymbol{x}, 0) \qquad (x_3 = 0) \tag{7.5.7}$$

or, alternatively, dropping zero,

$$\mathbf{U}(t, \boldsymbol{x}) = \mathbf{U}^{\mathrm{T}}(t, -\boldsymbol{x}) \qquad (\boldsymbol{x} = \{x_1, x_2, 0\}) , \tag{7.5.8}$$

where $\mathbf{U}^{\mathrm{T}}$ identifies the transposition of a matrix $\mathbf{U}$. Taking (7.5.8) into account, the transforms of the functions present in (7.5.8) have the obvious property:

$$\mathbf{B}(\Omega, \boldsymbol{\eta}) = \mathbf{B}^{\mathrm{T}}(\Omega, -\boldsymbol{\eta}), \qquad \boldsymbol{\eta} = \{\eta_1, \eta_2\} \tag{7.5.9}$$

for any Ω. If Ω is real and the relationship $|\Omega| < c_j(\eta_1, \eta_2, 0)$ is valid for all speeds of wave propagation $(j = 1, 2, 3)$, then the equation,

$$\det K_{ik}(\Omega, \boldsymbol{\eta}, \xi_3) = 0 , \tag{7.5.10}$$

has only complex roots and, since this equation depends only on Ω^2, the relationship

$$\xi_3^N(\Omega + 0\mathrm{i}, \boldsymbol{\eta}) = \xi_3^M(-\Omega + 0\mathrm{i}, \boldsymbol{\eta}) \tag{7.5.11}$$

is valid. Note that (7.5.11) is valid for some M unnecessarily equal to N, however, being in the same range of numbers — (4, 5, 6) — in which N is, since ξ_3^M has a negative imaginary part.

Equations (7.2.24) and (7.5.11) suggest that

$$\mathbf{B}^N(\Omega+0\mathrm{i},\boldsymbol{\eta}) = \mathbf{B}^M(-\Omega+0\mathrm{i},\boldsymbol{\eta}) , \tag{7.5.12}$$

which implies

$$\mathbf{B}(\Omega+0\mathrm{i},\boldsymbol{\eta}) = \mathbf{B}(-\Omega+0\mathrm{i},\boldsymbol{\eta}) . \tag{7.5.13}$$

Thus, if $|\Omega| < c_j(\eta_1,\eta_2,0)$ $(j=1,2,3)$, the relations (7.5.13) and (7.5.9) give

$$\mathbf{B}(-\Omega+0\mathrm{i},-\boldsymbol{\eta}) = \mathbf{B}^{\mathrm{T}}(\Omega+0\mathrm{i},\boldsymbol{\eta}) . \tag{7.5.14}$$

For any Ω, equation (7.2.48) results in the relation

$$\mathbf{B}(-\overline{\Omega},-\boldsymbol{\eta}) = \overline{\mathbf{B}(\Omega,\boldsymbol{\eta})} , \tag{7.5.15}$$

so, from (7.5.14), we can find

$$\mathbf{B}(\Omega+0\mathrm{i},\boldsymbol{\eta}) = \overline{\mathbf{B}^{\mathrm{T}}(\Omega+0\mathrm{i},\boldsymbol{\eta})} , \tag{7.5.16}$$

i.e., the matrix $\mathbf{B}(\Omega+0\mathrm{i},\boldsymbol{\eta})$ is a hermitian one. In particular, $B_{33}(\Omega+0\mathrm{i},\boldsymbol{\eta})$ is a real even function in each of its arguments $\Omega,\boldsymbol{\eta}$ (this is obvious from (7.5.9, 14)). Of course, all these properties may be revealed by utilizing the explicit form of the functions (7.4.22, 23) in the case of isotropy.

Taking into account the fact that B_{33} is real, obtain from (7.5.5):

$$p^2 w''_{pp}(p) = -2B_{33}(-p+0\mathrm{i},\boldsymbol{\nu})\mathrm{Im}\left\{p^2 G'_3(p-0\mathrm{i},\boldsymbol{\nu})\right\} \quad (|p|<v) \tag{7.5.17}$$

(in order to avoid singularity at $p=0$ for the sake of convenience, we do not remove p^2 and deal with the function $p^2G'_3(p-0\mathrm{i},\boldsymbol{\nu})$).

Taking into account the fact that the function G'_3 is given in (7.5.3) explicitly, and considering the boundary conditions (7.5.1), we obtain the boundary value problem on determining the analytic function $z^2G'_3(z,\boldsymbol{\nu})$ in the half-plane $\mathrm{Im}\{z\}<0$, with the following boundary conditions on the axis $\mathrm{Im}\{z\}=0$:

$$\begin{aligned} &\mathrm{Im}\left[z^2G'_3(z,\boldsymbol{\nu})\right]=0 \qquad (|z|<v) ,\\ &\mathrm{Re}\left[z^2G'_3(z,\boldsymbol{\nu})\right]=0 \qquad (|z|>v) , \end{aligned} \tag{7.5.18}$$

and the condition at infinity:

$$z^2G'_3(z,\boldsymbol{\nu}) = O(1) \qquad (z\to\infty,\ \mathrm{Im}\,z<0) . \tag{7.5.19}$$

In addition, since no energy flux can exist at points $z=\pm v$, the function $z^2G'_3(z,\boldsymbol{\nu})$ is expected to possess singularities at these points, weaker than $\varepsilon^{-3/2}$.

It is due to the Keldysh–Sedov formula that the solution to this problem can be written immediately:

$$z^2G'_3(z,\boldsymbol{\nu}) = \frac{a+bz}{\mathrm{i}\sqrt{z^2-v^2}} , \tag{7.5.20}$$

here, a and b are real constants, and the branch of the radical $\sqrt{z^2 - v^2}$ is chosen so that $\sqrt{z^2 - v^2} \to z$ when $z \to \infty$, $\mathrm{Im}\{z\} < 0$ (a cut is drawn along the segment $[-v, v]$ of the real axis). Assuming the $\tau_3(p)$ to be integrable at $p = 0$, it appears from (7.5.3) that $z^2 G_3'(z, \boldsymbol{\nu}) \to 0$ as $z \to 0$, so $a = 0$ in the formula (7.5.20). Then, integrating with due regard for the condition (7.5.19), obtain from (7.5.20):

$$G_3(z, \boldsymbol{\nu}) = \frac{1}{2\pi \mathrm{i}} \int_{-v}^{v} \frac{\tau_3(p)dp}{p - z} = \frac{b}{2v} \ln \left\{ \frac{\sqrt{z^2 - v^2} + \mathrm{i}v}{\sqrt{z^2 - v^2} - \mathrm{i}v} \right\} . \tag{7.5.21}$$

The left- and right-hand sides of equation (7.5.21) obtained for $\mathrm{Im}\{z\} < 0$ are, obviously, analytically continuable into the upper half-plane through the boundary $\mathrm{Im}\{z\} = 0$ except for the cut $[-v, v]$. As a result (due to the uniqueness theorem), equation (7.5.21) will be satisfied everywhere in the complex z-plane except for the segment $[-v, v]$ of the real axis. So, the Sokhotskii–Plemelj formula (see, for example, [7.26]) yields:

$$\begin{aligned} \tau_3(p) &= G_3(p + 0\mathrm{i}, \boldsymbol{\nu}) - G_3(p - 0\mathrm{i}, \boldsymbol{\nu}) \\ &= \frac{2b}{v} \ln \left\{ \frac{v + \sqrt{v^2 - p^2}}{|p|} \right\} \qquad (|p| < v) . \end{aligned} \tag{7.5.22}$$

With (7.5.20) obtained, we can perform integration in (7.5.2) (once in t, from 0 to t) and employ the variable $\Omega = \nu_\lambda x_\lambda / t$. This results in

$$\dot{u}_3(t, \boldsymbol{x})|_{x_3=0} = 2b \,\mathrm{Re} \int_{\pm\frac{|\nu_\lambda x_\lambda|}{t}}^{\pm\infty} \frac{B_{33}(-\Omega + 0\mathrm{i}, \boldsymbol{\nu})}{\sqrt{(\Omega - 0\mathrm{i})^2 - v^2}} d\Omega , \tag{7.5.23}$$

where, if $\nu_\lambda x_\lambda > 0$, upper signs ("plus") should be taken in the integration limits (then the integration path goes along the positive part of the axis $\mathrm{Im}\{\Omega\} = 0$), whereas if $\nu_\lambda x_\lambda < 0$, lower signs ("minus") should be taken (which gives an integration contour passing along the negative part of the axis $\mathrm{Im}\{\Omega\} = 0$). If $|\nu_\lambda x_\lambda|/t < c_R$, these paths go above the poles $\Omega = \pm c_R$ of the integrand, contributions of these poles into (7.5.22) being equal to zero, since the function $B_{33,\Omega}(-\Omega + 0\mathrm{i}, \boldsymbol{\nu})$ is real at $|\Omega| = c_R$. With this, the function $\dot{u}_3(t, \boldsymbol{x})|_{x_3=0}$ turns out to be even with respect to the variable $y = \nu_\lambda x_\lambda$. Prove this statement. First, since $B_{33}(-\Omega + 0\mathrm{i}, \boldsymbol{\nu})$ is a real function for $|\Omega| < c_j$ ($j = 1, 2, 3$), the integrand does not make any contribution to (7.5.23) when $|\Omega| < v$. Secondly, from (7.5.9, 15) it follows that $\mathrm{Re}\{B_{33}(-\Omega + 0\mathrm{i}, \boldsymbol{\nu})\}$ is an even function of Ω, and $[(\Omega - 0\mathrm{i})^2 - v^2]^{1/2}$ is odd when $|\Omega| > v$, so the real part of the integrand in (7.5.23) is an even function of Ω if $|\Omega| > v$. Hence, real parts of the integrals taken over both paths ($-|\nu_\lambda x_\lambda|/t, -\infty$) and ($|\nu_\lambda x_\lambda|/t, \infty$) — are equal to each other. Thus, it is permissible to keep only the plus signs in the integration limits in (7.5.23):

$$\dot{u}_3(t, \boldsymbol{x})|_{x_3=0} = 2b \,\mathrm{Re} \int_{\frac{|\nu_\lambda x_\lambda|}{t}}^{\infty} \frac{B_{33}(-\Omega + 0\mathrm{i}, \boldsymbol{\nu})}{\sqrt{(\Omega - 0\mathrm{i})^2 - v^2}} d\Omega . \tag{7.5.24}$$

Integrating the expression (7.5.24) again in t and changing the integration sequence, obtain

$$u_3(t,\boldsymbol{x})|_{x_3=0} = 2b\,\mathrm{Re}\int\limits_{\frac{|\nu_\lambda x_\lambda|}{t}}^{\max c_j(\boldsymbol{\nu})}\left(t-\frac{|\nu_\lambda x_\lambda|}{\Omega}\right)\frac{B_{33}(-\Omega+0\mathrm{i},\boldsymbol{\nu})}{\sqrt{(\Omega-0\mathrm{i})^2-v^2}}d\Omega\,, \tag{7.5.25}$$

where the upper limit of integration is replaced by $\max c_j(\boldsymbol{\nu})$, i.e., by the maximum wave propagation speed in the $\boldsymbol{\nu}$-direction; this is caused by the fact that the $B_{33}(-\Omega+0\mathrm{i},\boldsymbol{\nu})$ is a purely imaginary function when $\Omega > \max c_j(\boldsymbol{\nu})$.

Equation (7.5.25) for $|\nu_\lambda x_\lambda| < vt$ can be written as

$$\begin{aligned} u_3(t,\boldsymbol{x})|_{x_3=0} &= -|\nu_\lambda x_\lambda|\cot\alpha_0 + v_0 t \\ &= 2b\,\mathrm{Re}\int\limits_{n}^{\max c_j(v)}\left(t-\frac{|\nu_\lambda x_\lambda|}{\Omega}\right)\frac{B_{33}(-\Omega+0\mathrm{i},\boldsymbol{\nu})d\Omega}{\sqrt{(\Omega-0\mathrm{i})^2-v^2}}\,. \end{aligned} \tag{7.5.26}$$

This relation makes it possible to determine the unknown constants b and v in terms of v_0 and $\cot\alpha_0$ (remember that we go above the pole $\Omega = c_R$). As a result, get from (7.5.26):

$$\begin{aligned} v_0 &= 2b\,\mathrm{Re}\int\limits_{v}^{\max c_j((\boldsymbol{\nu})}\frac{B_{33}(-\Omega+0\mathrm{i},\boldsymbol{\nu})}{\sqrt{(\Omega-0\mathrm{i})^2-v^2}}d\Omega\,, \\ b &= \frac{1}{2}\cot\alpha_0\,\mathrm{Re}\int\limits_{v}^{\max c_j(\boldsymbol{\nu})}\frac{B_{33}(-\Omega+0\mathrm{i},\boldsymbol{\nu})}{\Omega\sqrt{(\Omega-0\mathrm{i})^2-v^2}}d\Omega\,. \end{aligned} \tag{7.5.27}$$

After we take into account the property (7.5.13) for $B_{33}(\Omega+0\mathrm{i},\boldsymbol{\eta})$, the analyticity of the function $\sqrt{\Omega^2-v_2}$ outside the cut $|\Omega| < v$, and the abovestated evenness of these functions over the real axis, then the integral in the lower expression in (7.5.27) can be reduced to the following form:

$$\begin{aligned} \mathrm{Re}\int\limits_{v}^{\max c_j(\boldsymbol{\nu})}\frac{B_{33}(-\Omega+0\mathrm{i},\boldsymbol{\nu})}{\Omega\sqrt{(\Omega-0\mathrm{i})^2-v^2}}d\Omega &= \mathrm{Re}\int\limits_{v}^{\max c_j(\boldsymbol{\nu})}\frac{B_{33}(\Omega+0\mathrm{i},\boldsymbol{\nu})}{\Omega\sqrt{(\Omega+0\mathrm{i})^2-v^2}}d\Omega \\ &= \frac{1}{2}\mathrm{Re}\int\limits_{-\infty}^{\infty}\frac{B_{33}(\Omega+0\mathrm{i},\boldsymbol{\nu})d\Omega}{\Omega\sqrt{(\Omega+0\mathrm{i})^2-v^2}} + \frac{\pi}{2v}B_{33}(+0\mathrm{i},\boldsymbol{\nu})\,, \end{aligned} \tag{7.5.28}$$

where the integration path $(-\infty,\infty)$ goes above the integrand poles $\Omega = 0$ and $\Omega = \pm c_R$. Since $B_{33}(\Omega,\boldsymbol{\nu}) = \mathrm{O}(\Omega^{-1})$ when $\Omega \to \infty$ ($\mathrm{Im}\{\Omega\} > 0$), then, after deforming the integration path into a semicircle of infinitely large radius (in the region ($\mathrm{Im}\{\Omega\} > 0$), the residue theorem shows the first member of the sum in (7.5.28) to become zero. Consequently, the formulae (7.5.27) (with due attention to the property (7.5.13) and analyticity of the function $\sqrt{\Omega^2 - v^2}$ outside the cut $|\Omega| < v$) can be reduced finally to the form

$$v_0 = \frac{2v \cot \alpha_0}{\pi B_{33}(+0\mathrm{i}, \boldsymbol{\nu})} \mathrm{Re} \int\limits_v^{\max c_j(\boldsymbol{\nu})} \frac{B_{33}(\Omega + 0\mathrm{i}, \boldsymbol{\nu})}{\sqrt{(\Omega + 0\mathrm{i})^2 - v^2}} d\Omega \,, \qquad b = \pi \frac{v \cot \alpha}{B_{33}(+0\mathrm{i}, \boldsymbol{\nu})} \tag{7.5.29}$$

(poles of the integrand, $\Omega = \pm c_{\mathrm{R}}$, are underneath the integration path).

With this, the force $P(t)$ applied from the medium to the wedge-shaped punch is

$$\boldsymbol{P} = -\int\limits_{-vt}^{vt} \tau_3(t, x_1, x_2) d(\nu_\lambda x_\lambda) = -t \int\limits_{-v}^{v} \tau_3(p) dp = -2\pi bt \,. \tag{7.5.30}$$

In the particular case when the medium is isotropic, we get according to (7.4.22):

$$B_{33}(\Omega + 0\mathrm{i}, \boldsymbol{\nu}) = \frac{\Omega^2(\Omega^2/\alpha^2 - 1)^{1/2}}{\mathrm{i}\mu\beta^2 \left[4(\Omega^2/\alpha^2 - 1)^{1/2}(\Omega^2/\beta^2 - 1)^{1/2} + (2 - \Omega^2/\beta^2)^2\right]} \,. \tag{7.5.31}$$

If $\Omega \to +0\mathrm{i}$, find

$$B_{33}(+0\mathrm{i}, \boldsymbol{\nu}) = -\frac{\alpha^2}{2\mu(\alpha^2 - \beta^2)} \,. \tag{7.5.32}$$

Substitute the expressions (7.5.31, 32) into (7.5.29), change the integration path $(v, \max\{c_j(\boldsymbol{\nu})\})$ in (7.5.29) by the contour $(0, \infty)$ (this is feasible, since the additional parts of the integrals are purely imaginary), thereafter deform this contour into the positive imaginary semi-axis. Using a new variable $x = -1/\Omega\mathrm{i}$, obtain finally:

$$v_0 = \frac{4}{\pi} \cot \alpha_0 (\beta^{-2} - \alpha^{-2}) \times \int\limits_0^\infty \frac{\sqrt{\alpha^{-2} + x^2}\, dx}{\sqrt{v^{-2} + x^2} \left[(\beta^{-2} + 2x^2)^2 - 4x^2\sqrt{(\alpha^{-2} + x^2)(\beta^{-2} + x^2)}\right]} \,, \qquad b = -\frac{2\mu v(\alpha^2 - \beta^2) \cot \alpha_0}{\pi \alpha^2} \tag{7.5.33}$$

(radicals here are assumed to be arithmetic). These expressions, (7.5.22, 30, 33), are the same as in (3.6.16, 18); to prove this statement, replace the symbols p, α, β, α_0 by x/t, c_1, c_2, α, respectively.

It is worth noting that the presented solution to the wedge-shaped punch problem has been expounded in [7.18], whereas *Odintsev* [7.20] solved this problem by employing the Burridge technique.

7.6 Representing the Solutions for an Anisotropic Space in Terms of Displacement/Stress Discontinuities Across a Plane

Some particular cases (for example, discontinuity propagation problems, diffraction problems) require the solution to be found for the whole of the (x_1, x_2, x_3)-space when the boundary conditions are specified over the plane $x_3 = 0$. In considering such problems, general representation of a solution in terms of jumps in displacements and/or stresses over the plane $x_3 = 0$ can turn out to be useful. These representations will be derived in this Section on the basis of the above results from [7.18].

Consider two homogeneous anisotropic half-spaces occupying the region $x_3 > 0$ (elastic constants c^+_{ijkl}) and the region $x_3 < 0$ (elastic constants c^-_{ijkl}), respectively. We shall assume the general case of anisotropy, i.e., elastic properties of the upper (lower) half-space are described by no more than 21 independent elastic constants c^+_{ijkl} (c^-_{ijkl}). The solution $\boldsymbol{u}(t, \boldsymbol{x})$ to some dynamic problem in the (x_1, x_2, x_3)-space will be found decreasing when $|\boldsymbol{x}| \to \infty$, satisfying the zero initial conditions, and possessing the specified jumps in displacements $[\boldsymbol{u}]$ and/or stresses $[\boldsymbol{\sigma}_3]$ across the plane $x_3 = 0$:

$$[\boldsymbol{u}] = \boldsymbol{W}(t, x_\alpha), \qquad [\boldsymbol{\sigma}_3] = \boldsymbol{T}(t, x_\alpha) \qquad (x_3 = 0)\,. \tag{7.6.1}$$

Assume $\boldsymbol{W}(t, x_\alpha)$ and $\boldsymbol{T}(t, x_\alpha)$ to be homogeneous functions of $(n+1)$th and nth order, respectively, in the variables t, x_1, x_2 (in (7.6.1), we use $[f] = f|_{x_3=+0} - f|_{x_3=-0}$. $\boldsymbol{W}(t, x_\alpha) = \{W_1, W_2, W_3\}$, $\boldsymbol{T}(t, x_\alpha) = \{T_1, T_2, T_3\}$.

It is suggested that a solution to such a problem be written as a superposition of two solutions:

$$\boldsymbol{u} = \boldsymbol{u}^{(1)} + \boldsymbol{u}^{(2)}\,, \tag{7.6.2}$$

each satisfying the zero initial conditions and decreasing for $|\boldsymbol{x}| \to \infty$. Let us assume that, in going across the plane $x_3 = 0$, the first of the solutions satisfies conditions

$$\left[\boldsymbol{u}^{(1)}\right] = \boldsymbol{W}(t, x_\alpha), \quad \left[\boldsymbol{\sigma}_3^{(1)}\right] = 0 \qquad (x_3 = 0)\,, \tag{7.6.3}$$

whereas the second is such that

$$\left[\boldsymbol{u}^{(2)}\right] = 0, \quad \left[\boldsymbol{\sigma}_3^{(2)}\right] = \boldsymbol{T}(t, x_\alpha) \qquad (x_3 = 0)\,. \tag{7.6.4}$$

The first problem, with conditions (7.6.3), can be solved just in the same way as the first boundary value problem (see Sect. 7.2). Prove this. In view of (7.2.23), solutions for half-space may be represented by

$$\begin{aligned} & \boldsymbol{u}_+^*(\omega, \xi_\alpha, x_3) \\ & \quad = \sum_{N=4}^{6} \mathbf{B}^{+N}(\omega, \xi_\alpha)\boldsymbol{\sigma}_3^*(\omega, \xi_\alpha) \exp\left\{-\mathrm{i}\xi_3^N(\omega, \xi_\alpha)x_3\right\} \quad (x_3 > 0)\,, \\ & \boldsymbol{u}_-^*(\omega, \xi_\alpha, x_3) \\ & \quad = \sum_{N=1}^{3} \mathbf{B}^{-N}(\omega, \xi_\alpha)\boldsymbol{\sigma}_3^*(\omega, \xi_\alpha) \exp\left\{-\mathrm{i}\xi_3^N(\omega, \xi_\alpha)x_3\right\} \quad (x_3 < 0)\,, \end{aligned} \tag{7.6.5}$$

where $\boldsymbol{u}_\pm^*(\omega, \xi_\alpha, x_3)$ and $\boldsymbol{\sigma}_3^*(\omega, \xi_\alpha)$ are transforms of displacement and stress vectors, $\boldsymbol{u}^\pm(t, \boldsymbol{x})$ and $\boldsymbol{\sigma}_3(t, x_\alpha, 0)$, respectively; matrices $\mathbf{B}^{+N}(\omega, \xi_\alpha)$ and $\mathbf{B}^{-N}(\omega, \xi_\alpha)$ are derived from the right-hand side of (7.2.24) by means of replacing c_{ijkl} by c_{ijkl}^+ and c_{ijkl}^-, respectively; summation for the latter case (i.e., in the expression for $\mathbf{B}^{-N}(\omega, \xi_\alpha)$) being carried out for $N = 1, 2, 3$. Thus, according to the research in Sect. 7.2, the function ξ_3^N present in the lower formula of (7.6.5) has a positive imaginary part if $\mathrm{Im}\{\omega\} > 0$; therefore, $\boldsymbol{u}_-^*(\omega, \xi_\alpha, x_3) \to 0$ when $x_3 \to -\infty$. Apply now the Fourier transformation (7.2.11) to the conditions (7.6.3) and use (7.6.5) to find

$$\boldsymbol{W}^*(\omega, \xi_\alpha) = \mathbf{A}(\omega, \xi_\alpha)\boldsymbol{\sigma}_3^*(\omega, \xi_\alpha)\,, \tag{7.6.6}$$

where the matrix $\mathbf{A}$ is

$$\mathbf{A}(\omega, \xi_\alpha) = \sum_{N=4}^{6} \mathbf{B}^{+N}(\omega, \xi_\alpha) - \sum_{N=1}^{3} \mathbf{B}^{-N}(\omega, \xi_\alpha)\,. \tag{7.6.7}$$

After inverting equation (7.6.6), obtain

$$\boldsymbol{\sigma}_3^*(\omega, \xi_\alpha) = \mathbf{A}^{-1}(\omega, \xi_\alpha)\boldsymbol{W}^*(\omega, \xi_\alpha)\,. \tag{7.6.8}$$

So, according to (7.6.5), we find:

$$\begin{aligned} & \boldsymbol{u}_+^*(\omega, \xi_\alpha, x_3) \\ & \quad = \sum_{N=4}^{6} \mathbf{B}^{+N}(\omega, \xi_\alpha)\mathbf{A}^{-1}(\omega, \xi_\alpha)\boldsymbol{W}^*(\omega, \xi_\alpha) \exp\left\{-\mathrm{i}\xi_3^N(\omega, \xi_\alpha)x_3\right\}\,, \\ & \boldsymbol{u}_-^*(\omega, \xi_\alpha, x_3) \\ & \quad = \sum_{N=1}^{3} \mathbf{B}^{-N}(\omega, \xi_\alpha)\mathbf{A}^{-1}(\omega, \xi_\alpha)\boldsymbol{W}^*(\omega, \xi_\alpha) \exp\left\{-\mathrm{i}\xi_3^N(\omega, \xi_\alpha)x_3\right\}\,. \end{aligned} \tag{7.6.9}$$

Proceeding with (7.6.9) just as with (7.3.4), obtain for $x_3 > 0$:

$$\begin{aligned} \partial_t^{(n+2)}\boldsymbol{u}^+(t, \boldsymbol{x}) &= \frac{1}{2\pi\mathrm{i}} \sum_{N=4}^{6} H(t - t_N) \\ & \times \oint_{|\boldsymbol{\eta}|=1} (\Omega^N)^{n+2} B^{+N}(\Omega^N, \boldsymbol{\eta}) \frac{A^{-1}(\Omega^N, \boldsymbol{\eta})}{t + \xi_{3,\Omega}^N(\Omega^N, \boldsymbol{\eta})x_3} \boldsymbol{F}_1^{(n+3)}\left(-\Omega^N, \boldsymbol{\eta}\right) do \\ & (\mathrm{Im}\, \Omega^N \geq 0)\,, \end{aligned} \tag{7.6.10}$$

where the new function is defined as

$$F_1^{(n+3)}{}_j(z,\boldsymbol{\eta}) = \frac{1}{2\pi i}\partial_z^{(n+3)} \int_{-\infty}^{\infty} \frac{W_0(p,\boldsymbol{\eta})}{p-z}dp \qquad (\text{Im } z < 0) \tag{7.6.11}$$

and $W_0(p,\boldsymbol{\eta})$ is realted to $W(t, x_\alpha)$ by the Radon formula:

$$W_0(p,\boldsymbol{\eta}) = \int_{-\infty}^{\infty}\int_{-\infty}^{\infty} W(\boldsymbol{y})\delta(p-\eta_\alpha y_\alpha)dy_1 dy_2$$
$$\left(W(t,x_\alpha) = t^{n+1}W(\boldsymbol{y}), \boldsymbol{y} = \{y_1, y_2\}, t>0, y_\alpha = x_\alpha/t\right) . \tag{7.6.12}$$

Treating analogously, find for $x_3 < 0$:

$$\partial_t^{(n+2)}\boldsymbol{u}^-(t,\boldsymbol{x}) = \frac{1}{2\pi i}\sum_{N=1}^{3} H(t-t_N)$$
$$\times \oint_{|\eta|=1} \frac{(\Omega^N)^{n+2}\mathbf{B}^{-N}(\Omega^N,\boldsymbol{\eta})\mathbf{A}^{-1}(\Omega^N,\boldsymbol{\eta})}{t+\xi^N_{3,\Omega}(\Omega^N,\boldsymbol{\eta})x_3} F_1^{(n+3)}{}_\alpha(-\Omega^N,\boldsymbol{\eta})ds$$
$$(\text{Im } \Omega^N \geq 0) \tag{7.6.13}$$

(it is obvious that the expression (7.6.13) is derived from (7.6.10) by replacement of $(+N)$ by $(-N)$ and summation in N from 1 to 3). The function $\partial_t^{(n+1)}\boldsymbol{u}^\pm(t,\boldsymbol{x})$ may be derived from (7.6.10) and (7.6.13) just as in obtaining equation (7.3.13).

By analogy with (7.3.10), obtain for the stress vector $\boldsymbol{\sigma}_j^+(t,\boldsymbol{x})$ $(x_3 > 0)$:

$$\partial_t^{(n+1)}\boldsymbol{\sigma}_j^+(t,\boldsymbol{x}) = \frac{1}{2\pi i}\sum_{N=4}^{6} H(t-t_N) \oint_{|\eta|=1} (\Omega^N)^{n+1}\mathbf{C}_j^{+N}(\Omega^N,\boldsymbol{\eta})$$
$$\times \frac{\mathbf{B}^{+N}(\Omega^N,\boldsymbol{\eta})\mathbf{A}^{-1}(\Omega^N,\boldsymbol{\eta})}{t+\xi^N_{3,\Omega}(\Omega^N,\boldsymbol{\eta})x_3} F_1^{(n+3)}(-\Omega^N,\boldsymbol{\eta})ds , \tag{7.6.14}$$

new symbols being

$$\mathbf{C}_j^{+N}(\Omega^N,\boldsymbol{\eta}) = \mathbf{C}_j^+\left(\tilde{\boldsymbol{\xi}}^N\right), \quad \tilde{\boldsymbol{\xi}}^N = \{\eta_1, \eta_2, \xi_3^N(\Omega^N,\boldsymbol{\eta})\} . \tag{7.6.15}$$

An expression similar to (7.6.14) (with replacement of $(+N)$ by $(-N)$ and with summation in N from 1 to 3) will describe the function $\partial_t^{(n+1)}\boldsymbol{\sigma}_j^-(t,\boldsymbol{x})$.

Just as for (7.2.45, 46), one can prove that right-hand sides of the expressions (7.6.10, 13, 14) define real functions.

It is in a similar way that the second problem, with the conditions (7.6.4), may be solved. By treating this problem just as the second boundary value problem (see Sect. 7.3) and using the conditions (7.6.4), we find

$$\boldsymbol{u}_+^*(\omega,\xi_\alpha,x_3) \tag{7.6.16}$$
$$= (-\mathrm{i})\sum_{N=4}^{6} \mathbf{D}^{+N}(\omega,\xi_\alpha)\boldsymbol{u}^*(\omega,\xi_\alpha)\exp\left\{-\mathrm{i}\xi_3^N x_3\right\} \quad (x_3>0)\,,$$

$$\boldsymbol{u}_-^*(\omega,\xi_\alpha,x_3) \tag{7.6.17}$$
$$= (-\mathrm{i})\sum_{N=1}^{3} \mathbf{D}^{-N}(\omega,\xi_\alpha)\boldsymbol{u}^*(\omega,\xi_\alpha)\exp\left\{-\mathrm{i}\xi_3^N x_3\right\} \quad (x_3<0)\,,$$

$$\boldsymbol{\sigma}_{+j}^*(\omega,\xi_\alpha,x_3) \tag{7.6.18}$$
$$= -\sum_{N=4}^{6} \mathbf{C}_j^{+N}(\boldsymbol{\xi}^N)\mathbf{D}^{+N}(\omega,\xi_\alpha)\boldsymbol{u}^*(\omega,\xi_\alpha)\exp\left\{-\mathrm{i}\xi_3^N x_3\right\} \quad (x_3>0)\,,$$

$$\boldsymbol{\sigma}_{-j}^*(\omega,\xi_\alpha,x_3) \tag{7.6.19}$$
$$= -\sum_{N=1}^{3} \mathbf{C}_j^{-N}(\boldsymbol{\xi}^N)\mathbf{D}^{-N}(\omega,\xi_\alpha)\boldsymbol{u}^*(\omega,\xi_\alpha)\exp\left\{-\mathrm{i}\xi_3^N x_3\right\} \quad (x_3<0)\,,$$

where $\boldsymbol{u}^*(\omega,\xi_\alpha)$ is the transform of the displacement vector $\boldsymbol{u}(t,\boldsymbol{x})$ at $x_3=0$; matrices $\mathbf{D}^{+N}(\omega,\xi_\alpha)$ and $\mathbf{D}^{-N}(\omega,\xi_\alpha)$ should be determined from (7.3.6) after replacing c_{ijkl} by c_{ijkl}^+ and c_{ijkl}^-, respectively, summation in the formula (7.3.6) in the latter case (i.e., for $\mathbf{D}^{-N}$) being performed for $N=1,2,3$. By means of the conditions (7.6.4), find from (7.6.18, 19)

$$\boldsymbol{T}^*(\omega,\xi\alpha) = -\mathbf{Q}(\omega,\xi_\alpha)\boldsymbol{u}^*(\omega,\xi_\alpha)\,, \tag{7.6.20}$$

where the matrix $\mathbf{Q}$ is

$$\mathbf{Q}(\omega,\xi_\alpha) = \sum_{N=4}^{6}\mathbf{C}_3^{+N}(\boldsymbol{\xi}^N)\mathbf{D}^{+N}(\omega,\xi_\alpha) - \sum_{N=1}^{3}\mathbf{C}_3^{-N}(\boldsymbol{\xi}^N)\mathbf{D}^{-N}(\omega,\xi_\alpha)\,. \tag{7.6.21}$$

Consequently, obtain:

$$\boldsymbol{u}^*(\omega,\xi_\alpha) = -\mathbf{Q}^{-1}(\omega,\xi_\alpha)\boldsymbol{T}^*(\omega,\xi_\alpha)\,. \tag{7.6.22}$$

Substituting (7.6.22) for $\boldsymbol{u}^*$ in (7.6.16–19) yields

$$\boldsymbol{u}_{+}^{*}(\omega,\xi_\alpha,x_3)=\mathrm{i}\sum_{N=4}^{6}\mathbf{D}^{+N}(\omega,\xi_\alpha)\mathbf{Q}^{-1}(\omega,\xi_\alpha)\boldsymbol{T}^{*}(\omega,\xi_\alpha)\exp\left\{-\mathrm{i}\xi_3^N x_3\right\},$$

$$\boldsymbol{u}_{-}^{*}(\omega,\xi_\alpha,x_3)=\mathrm{i}\sum_{N=1}^{3}\mathbf{D}^{-N}(\omega,\xi_\alpha)\mathbf{Q}^{-1}(\omega,\xi_\alpha)\boldsymbol{T}^{*}(\omega,\xi_\alpha)\exp\left\{-\mathrm{i}\xi_3^N x_3\right\},$$

$$\boldsymbol{\sigma}_{+j}^{*}(\omega,\xi_\alpha,x_3)=\sum_{N=4}^{6}\mathbf{C}_j^{+N}(\boldsymbol{\xi}^N)\mathbf{D}^{+N}(\omega,\xi_\alpha)\mathbf{Q}^{-1}(\omega,\xi_\alpha)\boldsymbol{T}^{*}(\omega,\xi_\alpha)\exp\left\{-\mathrm{i}\xi_3^N x_3\right\},$$

$$\boldsymbol{\sigma}_{-j}^{*}(\omega,\xi_\alpha,x_3)=\sum_{N=1}^{3}\mathbf{C}_j^{-N}(\boldsymbol{\xi}^N)\mathbf{D}^{-N}(\omega,\xi_\alpha)\mathbf{Q}^{-1}(\omega,\xi_\alpha)\boldsymbol{T}^{*}(\omega,\xi_\alpha)\exp\left\{-\mathrm{i}\xi_3^N x_3\right\}.$$

Invert these expressions by analogy with (7.2.23, 39) to obtain

$$\partial_t^{(n+2)}\boldsymbol{u}_{+}(t,\boldsymbol{x})=\frac{1}{2\pi\mathrm{i}}\sum_{N=4}^{6}H(t-t_N)\times\oint_{|\boldsymbol{\eta}|=1}\frac{(\Omega^N)^{n+2}\mathbf{D}^{+N}(\Omega^N,\boldsymbol{\eta})\mathbf{Q}^{-1}(\Omega^N,\boldsymbol{\eta})}{t+\xi_{3,\Omega}^N(\Omega^N,\boldsymbol{\eta})x_3}\boldsymbol{F}_2^{(n+2)}(-\Omega^N,\boldsymbol{\eta})ds\,, \tag{7.6.23}$$

$$\partial_t^{(n+1)}\boldsymbol{\sigma}_j^{+}(t,\boldsymbol{x})=\frac{1}{2\pi\mathrm{i}}\sum_{N=4}^{6}H(t-t_N)\times\oint_{|\boldsymbol{\eta}|=1}\frac{(\Omega^N)^{n+1}\mathbf{C}_j^{+N}(\Omega^N,\boldsymbol{\eta})\mathbf{D}^{+N}(\Omega^N,\boldsymbol{\eta})\mathbf{Q}^{-1}(\Omega^N,\boldsymbol{\eta})}{t+\xi_{3,\Omega}^N(\Omega^N,\boldsymbol{\eta})x_3}\boldsymbol{F}_2^{(n+2)}(-\Omega^N,\boldsymbol{\eta})ds \tag{7.6.24}$$

$$(\mathrm{Im}\,\Omega^N\geq 0)\,,$$

where

$$\boldsymbol{F}_2^{(n+2)}(z,\boldsymbol{\eta})=\frac{1}{2\pi\mathrm{i}}\partial_z^{(n+2)}\int_{-\infty}^{\infty}\frac{\boldsymbol{T}_0(p,\boldsymbol{\eta})}{p-z}dp\qquad(\mathrm{Im}\,z<0) \tag{7.6.25}$$

and the vector $\boldsymbol{T}_0(p,\boldsymbol{\eta})$ is defined in terms of $\boldsymbol{T}(t,x_\alpha)$ by the Radon formula:

$$\boldsymbol{T}_0(p,\xi_\alpha)=\int_{-\infty}^{\infty}\int_{-\infty}^{\infty}\boldsymbol{T}(\boldsymbol{y})\delta(p-\xi_\alpha y_\alpha)dy_1dy_2 \tag{7.6.26}$$

$$(\boldsymbol{T}(t,x_\alpha)=t^n\boldsymbol{T}(\boldsymbol{y}))\,.$$

Analogous formulae for $\partial_t^{(n+2)}\boldsymbol{u}^{-}(t,\boldsymbol{x})$ and $\partial_t^{(n+1)}\boldsymbol{\sigma}^{-}(t,\boldsymbol{x})$ are obtainable from the formulae (7.6.23, 24), respectively, by replacing $(+N)$ by $(-N)$ with summation in N from 1 to 3.

Add now the right- and left-hand sides of expressions (7.6.10, 23), as well as those of the expressions (7.6.14, 24), to find the final form of the solution to the problem with conditions (7.6.1) ($x_3 > 0$):

$$\partial_t^{(n+2)} \boldsymbol{u}_+(t, \boldsymbol{x}) = \frac{1}{2\pi \mathrm{i}} \sum_{N=4}^{6} H(t - t_N) \oint_{|\boldsymbol{\eta}|=1} \frac{(\Omega^N)^{n+2} \boldsymbol{V} \, ds}{t + \xi_{3,\Omega}^N(\Omega^N, \boldsymbol{\eta}) x_3} , \tag{7.6.27}$$

$$\partial_t^{(n+1)} \boldsymbol{\sigma}_j^+(t, \boldsymbol{x}) = \frac{1}{2\pi \mathrm{i}} \sum_{N=4}^{6} H(t - t_N) \oint_{|\boldsymbol{\eta}|=1} \frac{(\Omega^N)^{n+1} \mathbf{C}_j^{+N}(\Omega^N, \boldsymbol{\eta}) \boldsymbol{V}}{t + \xi_{3,\Omega}^N(\Omega^N, \boldsymbol{\eta}) x_3} ds , \tag{7.6.28}$$

where

$$\begin{aligned} \boldsymbol{V} &= \mathbf{D}^{+N}(\Omega^N, \boldsymbol{\eta}) \mathbf{Q}^{-1}(\Omega^N, \boldsymbol{\eta}) \boldsymbol{F}_2^{(n+2)}(-\Omega^N, \boldsymbol{\eta}) \\ &\quad + \mathbf{B}^{+N}(\Omega^N, \boldsymbol{\eta}) \mathbf{A}^{-1}(\Omega^N, \boldsymbol{\eta}) \boldsymbol{F}_1^{(n+3)}(-\Omega^N, \boldsymbol{\eta}) , \\ &\mathrm{Im}\, \Omega^N \geq 0 . \end{aligned} \tag{7.6.29}$$

Analogous formulae for $x_3 < 0$ are derivable from (7.6.27–29) by means of replacement of $(+N)$ by $(-N)$ with summation in the range $N = 1, 2, 3$.

Formulae like these will also be obtained in the two dimensions if the specified functions in (7.6.1) satisfy conditions

$$\boldsymbol{T}(t, x_\alpha) = t^n \boldsymbol{T}(p), \quad \boldsymbol{W}(t, x_\alpha) = t^{n+1} \boldsymbol{W}(p), \quad p = \nu_\lambda x_\lambda / t .$$

These formulae may be based on superposition of solutions $\boldsymbol{u}^{(1)}$ and $\boldsymbol{u}^{(2)}$ to be obtained by analogy with the expressions (7.2.63, 64, 3.16, 17). Here are final results for the problem with conditions (7.6.1) in the two-dimensional case ($x_3 > 0$) :

$$\partial_t^{(n+2)} \boldsymbol{u}_+(t, \boldsymbol{x}) = -2 \sum_{N=4}^{6} \mathrm{Re} \left[\frac{(\Omega^N)^{n+2} \boldsymbol{Y}}{t + \xi_{3,\Omega}^N(\Omega^N, \boldsymbol{\nu}) x_3} \right] H(t - t_N) , \tag{7.6.30}$$

$$\partial_t^{(n+1)} \boldsymbol{\sigma}_j^+(t, \boldsymbol{x}) = -2 \sum_{N=4}^{6} \mathrm{Re} \left[\frac{(\Omega^N)^{n+1} \mathbf{C}_j^{+N}(\Omega^N, \boldsymbol{\nu}) \boldsymbol{Y}}{t + \xi_{3,\Omega}^N(\Omega^N, \boldsymbol{\nu}) x_3} \right] H(t - t_N) , \tag{7.6.31}$$

$$\mathrm{Im}\, \Omega^N \geq 0 ,$$

where supplementary symbols mean:

$$\begin{aligned} \boldsymbol{Y} &= \mathbf{B}^{+N}(\Omega^N, \boldsymbol{\nu}) \mathbf{A}^{-1}(\Omega^N, \boldsymbol{\nu}) \boldsymbol{G}_1^{(n+2)}(-\Omega^N, \boldsymbol{\nu}) \\ &\quad + \mathbf{D}^{+N}(\Omega^N, \boldsymbol{\nu}) \mathbf{Q}^{-1}(\Omega^N, \boldsymbol{\nu}) \boldsymbol{G}_2^{(n+1)}(-\Omega^N, \boldsymbol{\nu}) , \\ \boldsymbol{G}_1^{(n+2)}(z, \boldsymbol{\nu}) &= \frac{1}{2\pi \mathrm{i}} \partial_z^{(n+2)} \int_{-\infty}^{\infty} \frac{\boldsymbol{W}(p) dp}{p - z} , \\ \boldsymbol{G}_2^{(n+1)}(z, \boldsymbol{\nu}) &= \frac{1}{2\pi \mathrm{i}} \partial_z^{(n+1)} \int_{-\infty}^{\infty} \frac{\boldsymbol{T}(p) dp}{p - z} \quad (\mathrm{Im}\, z < 0) . \end{aligned} \tag{7.6.32}$$

For $x_3 < 0$, equations like (7.6.30–32) will be derived from the latter by substituting $(-N)$ for $(+N)$ with summation in N from 1 to 3.

It is necessary to mention that representation of solutions in terms of the jumps in displacements and stresses for an isotropic homogeneous medium has earlier been found by *Osaulenko* [7.21].

Let us now list briefly the problems where the formulae (7.6.27–29, 30–32) seem to be convenient.

1. Diffraction by welded rigid plates being in the plane $x_3 = 0$. In this case, the displacement vector is continuous across the plane $x_3 = 0$, thus requiring $\boldsymbol{W}(t, x_\alpha) \equiv 0$ in the conditions (7.6.1).

2. Diffraction by smooth rigid plates being in the plane $x_3 = 0$. Here, the shear stress $\boldsymbol{\sigma}_3$ and the normal component of the displacement vector $\boldsymbol{u}$ remain continuous in going across the plane $x_3 = 0$. Hence, one must assume

$$T_1(t, x_\alpha) \equiv T_2(t, x_\alpha) \equiv 0, \quad W_3(t, x_\alpha) \equiv 0$$

in (7.6.1).

3. The problems on propagation of cracks in the plane $x_3 = 0$. Since in this case the stress vector $\boldsymbol{\sigma}_3$ remains continuous in crossing the plane $x_3 = 0$, one shall adopt $\boldsymbol{T}(t, x_\alpha) \equiv 0$ in (7.6.1).

To exemplify the application of the representation obtained, let us consider below the elliptic crack propagation problem.

7.7 Expansion of an Elliptic Crack

Assume the infinite homogeneous elastic medium to be loaded so that the stress field $\sigma^0_{ij}(t, \boldsymbol{x})$ in the absence of a crack is a homogeneous function of the nth degree with respect to the variables t, $\boldsymbol{x}$. Assume that, at the moment $t = 0$, a crack (cut) appears at the coordinate system origin and expands in the plane $x_3 = 0$ as time goes on; at each time moment, t, the cut surface $S(t)$ (Fig. 7.6) has the shape of an ellipse

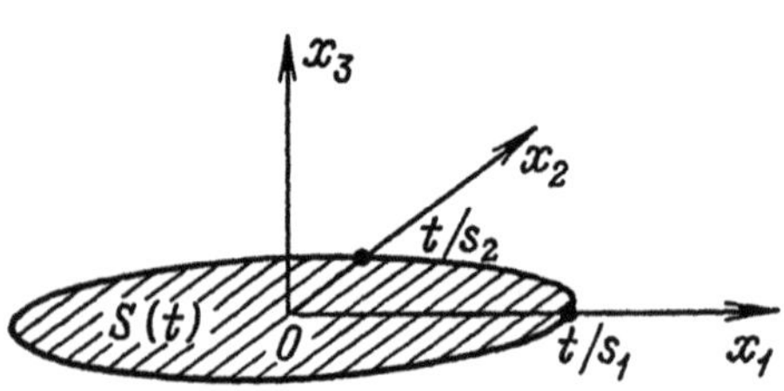

Fig. 7.6. An elliptic crack propagating from the point $x_1 = x_2 = x_3 = 0$ in the plane $x_3 = 0$

$$s_1^2 x_1^2 + s_2^2 x_2^2 \leq t^2 \, . \tag{7.7.1}$$

The crack "induces" additional displacements $u_i(t, \boldsymbol{x})$ and stresses $\sigma_{ij}(t, \boldsymbol{x})$, homogeneous function of the $(n+1)$th and nth degree, respectively, in the variables t, $\boldsymbol{x}$. These additional disturbances must satisfy (i) the equations of motion (7.2.3), (ii) zero initial conditions

$$\boldsymbol{u}(0, \boldsymbol{x}) = \dot{\boldsymbol{u}}(0, \boldsymbol{x}) = 0 \tag{7.7.2}$$

and (iii) the mixed boundary conditions

$$\begin{aligned} &\sigma_{i3}(t, \boldsymbol{x}) = c_{i3kl} u_{k,l}(t, \boldsymbol{x}) = -\tau_i(t, x_\alpha) \qquad (\boldsymbol{x} \in S(t), \; x_3 = \pm 0) \; , \\ &[\boldsymbol{u}(t, \boldsymbol{x})] = 0 \qquad (\boldsymbol{x} \notin S(t), \; x_3 = 0) \; , \\ &\boldsymbol{u}(t, \boldsymbol{x}) \to 0 \qquad (|\boldsymbol{x}| \to \infty) \\ &\left(\tau_i(t, x_\alpha) \equiv \sigma_{i3}^0(t, x_1, x_2, 0)\right) \; . \end{aligned} \tag{7.7.3}$$

Assume that (i) the boundary expansion speed is everywhere less than the Rayleigh speed, i.e., $s_\alpha^{-1} < c_{\mathrm{R}} (\alpha = 1, 2)$, and (ii) stresses at the cut edge are $\mathrm{O}(\varepsilon^{-1/2})$.

7.7.1 Solution to the Problem

In the problem under consideration the stress vector $\boldsymbol{\sigma}_3$ is continuous everywhere, whereas the displacement vector undergoes a jump across the plane $x_3 = 0$:

$$[\boldsymbol{u}(t, \boldsymbol{x})] = \boldsymbol{W}(t, x_\alpha) \, , \tag{7.7.4}$$

the jump being equal to zero if $\boldsymbol{x} \notin S(t)$.

Thus, the solution for the jump $\boldsymbol{W}(t, x_\alpha)$ is given by the formula (7.6.28) with $j = 3$, $x_3 = +0$, $\boldsymbol{F}_2 \equiv 0$. According to (7.2.30), Ω^N at $x_3 = 0$ does not depend upon $N (\Omega^N = \Omega = -\eta_\alpha x_\alpha / t)$, therefore, taking into account (7.2.24), obtain

$$\mathbf{A}^{-1}(\Omega, \boldsymbol{\eta}) \sum_{N=4}^{6} C_3^{+N}(\Omega^N, \boldsymbol{\eta}) \mathbf{B}^{+N}(\Omega^N, \boldsymbol{\eta}) = \mathrm{i} \mathbf{A}^{-1}(\Omega, \boldsymbol{\eta}) \, .$$

Consequently, the formula (7.6.28) at $j = 3$, $x_3 = +0$, $\boldsymbol{F}_2 \equiv 0$ takes the form as follows:

$$\begin{aligned} &\partial_t^{(n+1)} \boldsymbol{\sigma}_3^+(t, \boldsymbol{x}) \\ &\quad = \frac{1}{2\pi t} \oint\limits_{|\boldsymbol{\eta}|=1} (-\Omega_0)^{n+1} \mathbf{A}^{-1}(-\Omega_0 + 0\mathrm{i}, \boldsymbol{\eta}) \boldsymbol{F}_1^{(n+3)}(\Omega_0 - 0\mathrm{i}, \boldsymbol{\eta}) ds \\ &\quad = \frac{1}{2\pi t} \oint\limits_{|\boldsymbol{\eta}|=1} (-\Omega_0)^{n+1} \operatorname{Re} \left\{ \mathbf{A}^{-1}(-\Omega_0 + 0\mathrm{i}, \boldsymbol{\eta}) \boldsymbol{F}_1^{(n+3)}(\Omega_0 - 0\mathrm{i}, \boldsymbol{\eta}) \right\} ds \\ &(\Omega_0 \equiv \eta_\alpha x_\alpha / t) \, . \end{aligned} \tag{7.7.5}$$

Since $c^{+}_{ijkl} = c^{-}_{ijkl}$ and $\xi_3^N(\Omega+0\mathrm{i}, -\boldsymbol{\eta}) = -\xi_3^N(\Omega+0\mathrm{i}, \boldsymbol{\eta})$ for a certain N from the range $\{4, 5, 6\}$ (if N is in the range $\{1, 2, 3\}$), then

$$\sum_{N=1}^{3} \mathbf{B}^{-N}(\Omega+0\mathrm{i}, -\boldsymbol{\eta}) = -\sum_{N=4}^{6} \mathbf{B}^{+N}(\Omega+0\mathrm{i}, \boldsymbol{\eta}) = -\mathbf{B}(\Omega+0\mathrm{i}, \boldsymbol{\eta}) . \tag{7.7.6}$$

Next, from (7.6.7) it follows that

$$\mathbf{A}(\Omega+0\mathrm{i}, \boldsymbol{\eta}) = \mathbf{B}(\Omega+0\mathrm{i}, \boldsymbol{\eta}) = -\mathbf{B}(\Omega+0\mathrm{i}, -\boldsymbol{\eta}) . \tag{7.7.7}$$

With this, $\mathbf{A}(\Omega+0\mathrm{i}, \boldsymbol{\eta})$ is an even function of $\boldsymbol{\eta}$, and, from (7.5.9), the matrix $\mathbf{A}(\Omega+0\mathrm{i}, \boldsymbol{\eta})$ is symmetric. Consequently, according to (7.5.16), $\mathbf{A}(\Omega+0\mathrm{i}, \boldsymbol{\eta})$ is real when $|\Omega| < c_j(\eta_1, \eta_2, 0)$. Then, the matrix $\mathbf{A}^{-1}(\Omega+0\mathrm{i}, \boldsymbol{\eta}$ is also real when $|\Omega| < c_j(\eta_1, \eta_2, 0)$. After taking into account the boundary condition (7.7.3) and using the Sokhotskii–Plemelj formula

$$\boldsymbol{F}_1^{(n+3)}(z, \boldsymbol{\eta})_{z\to\Omega_0-\mathrm{i}0} = \partial_{\Omega_0}^{(n+3)} \left(-\frac{1}{2}\boldsymbol{W}_0(\Omega, \boldsymbol{\eta}) + \frac{1}{2\pi\mathrm{i}} \int_{-\infty}^{\infty} \frac{\boldsymbol{W}_0(p, \boldsymbol{\eta})dp}{p-\Omega_0} \right)$$

(where the integral shall be taken as the Cauchy principal value), equation (7.7.5) for $\boldsymbol{x} \in S(t)$ can be reduced finally to the form

$$\partial_t^{(n+1)}\boldsymbol{\tau}(t, x_\alpha) = \frac{1}{4\pi t} \oint_{|\boldsymbol{\eta}|=1} (-\Omega_0)^{n+1}\mathbf{A}^{-1}(-\Omega_0+0\mathrm{i}, \boldsymbol{\eta})\partial_{\Omega_0}^{(n+3)}\boldsymbol{W}_0(\Omega_0, \boldsymbol{\eta})ds , \tag{7.7.8}$$

where $\boldsymbol{\tau}(t, x_\alpha) = \{\tau_1, \tau_2, \tau_3\}$, $\boldsymbol{W}_0(\Omega_0, \boldsymbol{\eta})$ is the Radon transform of the function $\boldsymbol{W}(t, x_\alpha)$. Equation (7.7.8) is an integral equation for $\boldsymbol{W}_0$, thus, for $\boldsymbol{W}$, implicitly. It can be seen from (7.7.8), that if $\boldsymbol{W}_0(\Omega_0, \boldsymbol{\eta})$ is a polynomial of order $n+2$ in Ω_0, then the right-hand side of equation (7.7.8) becomes zero, thus providing $\partial_t^{(n+1)}\boldsymbol{\tau}(t, x_\alpha) \equiv 0$.

By analogy with derivation of (7.7.8), one could obtain the expression,

$$\partial_t^{(p)}\partial_{x_1}^{(q)}\partial_{x_2}^{(r)}\boldsymbol{\tau}(t, x_\alpha) = \frac{1}{4\pi t} \oint_{|\boldsymbol{\eta}|=1} (-\Omega_0)^p \eta_1^q \eta_2^r \mathbf{A}^{-1}(-\Omega_0+0\mathrm{i}, \boldsymbol{\eta})\partial_{\Omega_0}^{(n+3)}\boldsymbol{W}_0(\Omega_0, \boldsymbol{\eta})ds$$

for the case of $p+q+r = n+1 (p \geq 0, q \geq 0, r \geq 0)$ and for $\boldsymbol{x} \in S(t)$. This also means that if $\boldsymbol{W}_0(\Omega_0, \boldsymbol{\eta})$ is a polynomial of order $n+2$ in Ω_0, then

$$\partial_t^{(p)}\partial_{x_1}^{(q)}\partial_{x_2}^{(r)}\boldsymbol{\tau}(t, x_\alpha) \equiv 0 ,$$

i.e., all derivatives of the $(n+1)$th order are zero, therefore the function $\boldsymbol{\tau}(t, x_\alpha)$ is not only a homogeneous function of order n in t, x_1, x_2 but, definitely, a homogeneous polynomial of order n in these variables. Provided that $\boldsymbol{W}$ is

$$W(t,x_\alpha) = P_n(t,x_\alpha)q^{1/2}H(q) \qquad \left(q = t^2 - s_1^2x_1^2 - s_2^2x_2^2\right) , \tag{7.7.9}$$

(where $P_n(t,x_\alpha)$ represents a homogeneous polynomial of order n in t, x_1, x_2) the formula (7.6.12) shows $W_0(\Omega,\eta)$ to be a polynomial of order $n+2$ in Ω. Thus, the jump in displacements, equation (7.7.9), produces the stress $\tau(t,x_\alpha) = Q_n(t,x_\alpha)$ at cut sides, with $Q_n(t,x_\alpha)$ being a homogeneous polynomial of order n in t, x_1, x_2. In the problem under study the unknown coefficients of the polynomial P_n must be determined in terms of the coefficients of the prescribed polynomial Q_n by integrating equation (7.7.8) in time.

Consider a specific example, where $\tau(t,x_\alpha) \equiv \tau_0 = \text{const}$. In this case, n is equal to zero, which gives

$$W(t,x_\alpha) = U\left(t^2 - s_1^2x_1^2 - s_2^2x_2^2\right)^{1/2} H\left(t^2 - s_1^2x_1^2 - s_2^2x_2^2\right) . \qquad (U \equiv \text{const}). \tag{7.7.10}$$

Apply the Radon transformation to the function $W(t,x_\alpha)$:

$$\begin{aligned} W_0(\Omega_0,\eta) &= \int\limits_{1-s_1^2y_1^2-s_2^2y_2^2\geq 0}\int W(y)\delta(\Omega_0 - \eta_\alpha y_\alpha)dy_1dy_2 \\ &= \frac{\pi U}{2s_1s_2\omega_0}\left(1-\frac{\Omega_0^2}{\omega_0^2}\right)H\left(1-\frac{\Omega_0^2}{\omega_0^2}\right) , \end{aligned} \tag{7.7.11}$$

$$\omega_0 = \left(\frac{\eta_1^2}{s_1^2}+\frac{\eta_2^2}{s_2^2}\right)^{1/2} .$$

This yields:

$$F_1^{(3)}(z,\eta) = \frac{1}{2\pi \mathrm{i}}\partial_z^{(3)}\int\limits_{-\infty}^{\infty}\frac{W_0(p,\eta)dp}{p-z} = \frac{2U}{\mathrm{i}s_1s_2\left(\omega_0^2-z^2\right)^2} . \tag{7.7.12}$$

Consequently, equation (7.7.5) takes the form:

$$\begin{aligned} &\partial_1\sigma_3^+(t,x_\alpha,0) \\ &= -\frac{1}{\pi t s_1 s_2}\oint\limits_{|\eta|=1}\Omega_0\mathrm{Re}\left\{\mathsf{A}^{-1}(-\Omega_0+0\mathrm{i},\eta)U\frac{\mathrm{i}}{\left[\omega_0^2-(\Omega_0-0\mathrm{i})^2\right]^2}\right\}ds . \end{aligned} \tag{7.7.13}$$

Integrating this equation in t, we get

$$\begin{aligned} &\sigma_3(t,x_\alpha,0) \\ &= -\frac{1}{\pi s_1 s_2}\oint\limits_{|\eta|=1}\mathrm{Im}\left\{\int\limits_{\left|\frac{\eta_\alpha x_\alpha}{t}\right|}^{\infty}\frac{\mathsf{A}^{-1}(-\Omega_0+0\mathrm{i},\eta)Ud\Omega_0}{\left[\omega_0^2-(\Omega_0-0\mathrm{i})^2\right]^2}\right\}ds . \end{aligned} \tag{7.7.14}$$

The latter provides an equation for determining the unknown constant vector U at $x \in S(t)$:

$$\tau_0 = \frac{1}{\pi s_1 s_2} \oint\limits_{|\eta|=1} \mathrm{Im} \left\{ \int\limits_0^\infty \frac{\mathbf{A}^{-1}(-\Omega_0 + 0\mathrm{i}, \eta) \boldsymbol{U} d\Omega_0}{\left[\omega_0^2 - (\Omega_0 - 0\mathrm{i})^2\right]^2} \right\} ds \tag{7.7.15}$$

(the lower limit $|\eta_\alpha x_\alpha/t|$ of the integral in Ω_0 is replaced by 0, since the region $S(t)$ is characterized by the inequality $|\eta_\alpha x_\alpha/t| < c_R$, so, the added integral is real).

In order to find displacements of the upper and lower sides of the cut $S(t)$, one must use the formulae (7.6.10, 13) (at $n = 0$) with $x_3 = +0$ and $x_3 = -0$, respectively. Specifically, the formulae (7.5.15, 7.6, 7) with account for evenness in η of the functions $\boldsymbol{F}_1^{(3)}(-\Omega, \eta)$ and $\mathbf{A}^{-1}(\Omega, \eta)$ give

$$\partial_t^{(2)}(\boldsymbol{u}^+ + \boldsymbol{u}^-)\big|_{x_3=0}$$
$$= \frac{1}{\pi t} \int\limits_0^\pi \mathrm{Im} \left\{ \Omega^2 \left[B(\Omega, \eta) - B(\Omega, -\eta)\right] \mathbf{A}^{-1}(\Omega, \eta) \boldsymbol{F}_1^{(3)}(-\Omega, \eta) \right\} d\theta$$
$$\left(\Omega = -\eta_\alpha x_\alpha/t + 0\mathrm{i},\ \eta = \{\cos\theta, \sin\theta\}\right) . \tag{7.7.16}$$

In addition, we must use the fact that, for any $\boldsymbol{x}$ on $S(t)$, the function $\mathbf{A}^{-1}(\Omega, \eta)$ is real, whereas $\boldsymbol{F}_1^{(3)}(-\Omega, \eta)$ and the expression in square brackets in (7.7.16) are purely imaginary. This, when coupled with (7.7.16) and (7.7.4), allows us to find accelerations of sides of the cut:

$$\partial_t^{(2)} \boldsymbol{u}^+(t, x_\alpha, 0) = -\partial_t^{(2)} \boldsymbol{u}^-(t, x_\alpha, 0) = \partial_t^{(2)} \boldsymbol{W}(t, x_\alpha)/2 .$$

The results obtained above have been formulated earlier in [7.19] in a somewhat another form.

7.7.2 Isotropic Media

Let us consider the case of an isotropic medium. Here, the matrix $\mathbf{B}(\omega, \xi_\alpha) = |\xi|^{-1}\mathbf{B}(\Omega, \eta)$ is of the type (7.4.22) and the matrix $\mathbf{A}(\Omega + 0\mathrm{i}, \eta)$, according to (7.7.7), is two times the even part of $\mathbf{B}(\Omega + 0\mathrm{i}, \eta)$:

$$\mathbf{A}(\Omega + 0\mathrm{i}, \eta) = 2 \begin{bmatrix} e + f\eta_2^2 & -f\eta_1\eta_2 & 0 \\ -f\eta_1\eta_2 & e + f\eta_1^2 & 0 \\ 0 & 0 & g \end{bmatrix} , \tag{7.7.17}$$

where e, f, g are functions of Ω, real and even when Ω is subsonic, i.e., for $|\Omega| < \beta$. Hence, find

$$\mathbf{A}^{-1}(\Omega + 0\mathrm{i}, \eta) = \frac{1}{2} \begin{bmatrix} \frac{e+f\eta_1^2}{e(e+f)} & \frac{f\eta_1\eta_2}{e(e+f)} & 0 \\ \frac{f\eta_1\eta_2}{e(e+f)} & \frac{e+f\eta_2^2}{e(e+f)} & 0 \\ 0 & 0 & \frac{1}{g} \end{bmatrix} . \tag{7.7.18}$$

Introduce the new variable θ instead of η: $\eta_1 = \cos\theta$, $\eta_2 = \sin\theta$ $(-\pi < \theta \leq \pi)$; then, $ds = d\theta$. Taking into account equation (7.7.18) and the oddness of the function $\eta_1\eta_2$ in θ, reduce (7.7.15) to the form

$$\frac{U_1}{2\pi s_1 s_2}\int_{-\pi}^{\pi}\left[\operatorname{Im}\int_0^\infty \frac{(e+f\cos^2\theta)d\Omega_0}{e(e+f)\left[\omega_0^2-(\Omega_0-0\mathrm{i})^2\right]^2}\right]d\theta = \tau_1\,,$$
$$\frac{U_2}{2\pi s_1 s_2}\int_{-\pi}^{\pi}\left[\operatorname{Im}\int_0^\infty \frac{(e+f\sin^2\theta)d\Omega_0}{e(e+f)\left[\omega_0^2-(\Omega_0-0\mathrm{i})^2\right]^2}\right]d\theta = \tau_2\,, \tag{7.7.19}$$
$$\frac{U_3}{2\pi s_1 s_2}\int_{-\pi}^{\pi}\left[\operatorname{Im}\int_0^\infty \frac{d\Omega_0}{g\left[\omega_0^2-(\Omega_0-0\mathrm{i})^2\right]^2}\right]d\theta = \tau_3\,,$$
$$\left(\omega_0=\sqrt{\frac{\cos^2\theta}{s_1^2}+\frac{\sin^2\theta}{s_2^2}}\ \ \boldsymbol{\tau}_0=\{\tau_1,\,\tau_2,\,\tau_3\}\right)\,,$$

where the path of integration in Ω_0 (by analogy with the formula (7.7.15)) goes below the poles of the integrand, and the functions $(e,\ f,\ g)$ are given by the formulae (7.4.23) where we assume $\Omega = -\Omega_0 + 0\mathrm{i}$. From (7.7.19) it follows that the component U_k depends only on τ_k. Thus, from the formulae (7.7.19, 10) it follows that appearance of the kth component of the jump in the displacement $\boldsymbol{W}(t, x_\alpha)$ on sides of the cut $S(t)$ in an isotropic medium is due to action of the kth component of the stress vector $\boldsymbol{\tau}$ onto the cut.

For particular cases of loads $\boldsymbol{\tau}_0$, find from (7.7.4, 16, 19) at $x_3 = 0$ $(W_j \equiv W_j(t, x_\alpha))$:

$$\begin{aligned}
&u_1^+ = -u_1^- = W_1/2,\ u_2^+ = u_2^- = 0,\ u_3^+ = u_3^- \quad (\tau_2 = \tau_3 = 0)\,,\\
&u_1^+ = u_1^-,\ u_2^+ = u_2^-,\ u_3^+ = -u_3^- = W_3/2 \quad (\tau_1 = \tau_2 = 0)\,,\\
&u_1^+ = u_1^- = 0,\ u_2^+ = -u_2^- = W_2/2,\ u_3^+ = u_3^- \quad (\tau_1 = \tau_3 = 0)\,.
\end{aligned} \tag{7.7.20}$$

for a disk-shaped crack $(r < vt, r^2 = x_1^2 + x_2^2)$, obtain

$$\frac{U_m v^2}{2}\operatorname{Im}\int_0^\infty \frac{(2e+f)d\Omega_0}{e(e+f)\left[v^2-(\Omega_0-0\mathrm{i})^2\right]^2} = \tau_m \quad (m=1,2)\,,$$
$$U_3 v^2 \operatorname{Im}\int_0^\infty \frac{d\Omega_0}{g\left[v^2-(\Omega_0-0\mathrm{i}^2\right]^2} = \tau_3,\ v < c_R\,. \tag{7.7.21}$$

Deforming the integration path in (7.7.21) into the negative imaginary semi-axis, replacing the variable: $x = -1/\Omega_0^2$, and using (7.4.23) with $\Omega = -\Omega_0 + 0\mathrm{i}$, we can write the expressions (7.7.21) in the following form:

$$\tau_m = \frac{U_m \mu}{4v^2} \int_0^\infty \left\{ (\beta^{-2}+x)^{1/2} + \frac{4\beta^2 \left[(x+\beta^{-2}/2)^2 - x(\alpha^{-2}+x)^{1/2}(\beta^{-2}+x)^{1/2}\right]}{(\beta^{-2}+x)^{1/2}} \right\} \frac{dx}{(x+v^{-2})^2}$$

$(m = 1,2)$,

$$\tau_3 = \frac{2\mu\beta^2 U_3}{v^2} \int_0^\infty \frac{\left[(x+\beta^{-2}/2)^2 - x(\alpha^{-2}+x)^{1/2}(\beta^{-2}+x)^{1/2}\right] dx}{(\alpha^{-2}+x)^{1/2}(x+v^{-2})^2} , \quad (7.7.22)$$

the radicals present in (7.7.22) being assumed arithmetic. It can be seen that if $\tau_2 = \tau_3 = 0$ (the problem on a crack propagation under shear loading), then $U_2 = U_3 = 0$, so, for $x_3 = 0$ and $r < vt$, obtain from (7.7.20):

$$u_2^+ = u_2^- = 0, u_3^+ = u_3^-, \ u_1^+ = -u_1^- = \tfrac{1}{2} U_1 \sqrt{t^2 - r^2/v^2} , \quad (7.7.23)$$

U_1 is defined by the upper formula of (7.7.22). The result (7.7.23) is the same as the solution by *Kostrov* [7.27].

Proceeding analogously in the case $\tau_1 = \tau_2 = 0$ (the problem on a crack propagation under normal tension), obtain from (7.7.20) at $x_3 = 0, r < vt$:

$$u_1^+ = u_1^-, \ u_2^+ = u_2^-, \quad u_3^+ = -u_3^- = -\tfrac{1}{2} U_3 \sqrt{t^2 - r^2/v^2} , \quad (7.7.24)$$

with U_3 found from the lower formula of (7.7.22). In this case, the solution is the same as the solution obtained earlier in [7.28] (also, see the formula (3.9.17) of this monograph).

In conclusion, note that if the mass density of an isotropic medium ρ (present in expressions for the speeds α and β in the coefficients e, f, and g) in the formulae (7.7.19) tends to zero and use is made of symbols $a = t/s_1$, $b = t/s_2$, then a solution to the corresponding static problem for an elliptic crack with semi-axes a and b appears, see [7.29]. Specifically, if the tension τ_3 normal to the crack plane is applied at infinity, the displacements of the crack sides are as follows:

$$u_1^+ = u_1^-, u_2^+ = u_2^-, u_3^+ = -u_3^- = \frac{2\tau_3 b(1-\nu^2)}{E(k)E} \sqrt{1 - x_1^2/a^2 - x_2^2/b^2} . \quad (7.7.25)$$

Here, $k^2 = 1 - b^2/a^2$, $0 < b \le a$, ν is Poisson's ration, E is Young's modulus, $E(k)$ is the complete elliptic integral of the second kind.

7.8 Two-Dimensional Problems

In this section, we focus on examples of the use of the formulae derived in Sect. 7.6 for two dimensions.

7.8.1 A Strip-Shaped Crack

Assume the surface $S(t)$ of the cut to be a strip ($x_3 = 0$, $|\nu_\lambda x_\lambda| \leq vt$, $v < c_R$) (Fig. 7.7) and the stress vector $\boldsymbol{\tau}(t, x_\alpha)$ to be written as $t^n \boldsymbol{\tau}(p)$, with $\boldsymbol{\tau}(p)$ being defined for $|p| < v$ ($p = \nu_\lambda x_\lambda / t$). Additional disturbances must satisfy equations (7.2.3), the initial conditions (7.7.2), and the boundary conditions (7.7.3). The solution for the stresses on the plane $x_3 = 0$ can be described by the formula (7.6.31) with $j = 3$, $x_3 = +0$, $G_2 \equiv 0$ ($\Omega = -\nu_\lambda x_\lambda / t + 0\mathrm{i}$):

$$\partial_t^{(n+1)} \boldsymbol{\sigma}_3^+(t, x_\alpha, 0) = 2t^{-1} \mathrm{Im} \left[\Omega^{n+1} \mathbf{A}^{-1}(\Omega, \boldsymbol{\nu}) G_1^{(n+2)}(-\Omega, \boldsymbol{\nu}) \right] . \tag{7.8.1}$$

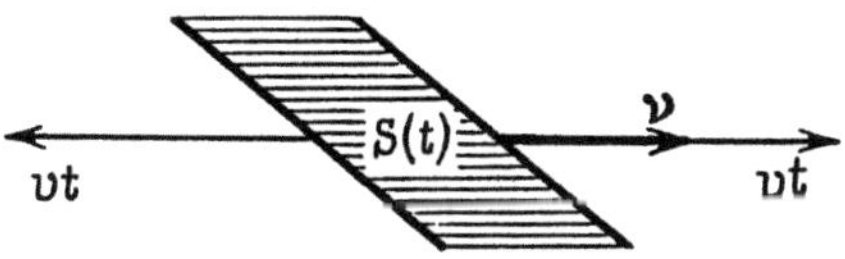

Fig. 7.7. Extension of a strip-shaped crack

After taking into account the fact that $\mathbf{A}^{-1}(\Omega, \boldsymbol{\nu})$ is real for $|\Omega| < c_j(\nu_1, \nu_2, 0)$, and using the Sokhotskii–Plemelj formula for $G_1^{(n+2)}(-\Omega, \boldsymbol{\nu})$, we can transform (7.8.1) for $\boldsymbol{x} \in S(t)$ into ($\widetilde{\boldsymbol{\tau}}(p) \equiv t\partial_t^{(n+1)} \boldsymbol{\tau}(t, x_\alpha)$)

$$\begin{aligned} \widetilde{\boldsymbol{\tau}}(p) = \mathrm{i}(-1)^{n+1} \mathbf{A}^{-1}(-p + 0\mathrm{i}, \boldsymbol{\nu}) \big[p^{n+1} \big(G_1^{(n+2)}(p - 0\mathrm{i}, \boldsymbol{\nu}) \\ + G_1^{(n+2)}(p + 0\mathrm{i}, \boldsymbol{\nu}) \big) \big] . \end{aligned} \tag{7.8.2}$$

By inverting equation (7.8.2), obtain

$$\begin{aligned} p^{(n+1)} G_1^{(n+2)}(p - 0\mathrm{i}, \boldsymbol{\nu}) + p^{n+1} G_1^{(n+2)}(p + 0\mathrm{i}, \boldsymbol{\nu}) \\ = \mathrm{i}(-1)^n \mathbf{A}(-p + 0\mathrm{i}, \boldsymbol{\nu}) \widetilde{\boldsymbol{\tau}}(p) . \end{aligned} \tag{7.8.3}$$

As a result, we have the problem where the function $z^{n+1} G_1^{(n+2)}(z, \boldsymbol{\nu})$ to be found must

(i) be analytic everywhere outside the cut $[-v, v]$ (on the real axis),
(ii) satisfy condition (7.8.3) over this cut,
(iii) behave as $O(z^{-2})$ when $|z| \to \infty$, and
(iv) behave as $O(z \pm v)^{-n-2+\varepsilon}$ when $z \to \pm v$, $\varepsilon > 0$ (because $[\boldsymbol{u}] \to 0$ if $|\nu_\lambda x_\lambda| \to vt$).

A solution to this problem may be found in [7.26]:

$$\begin{aligned} z^{n+1} G_1^{(n+2)}(z, \boldsymbol{\nu}) = \frac{(-1)^{n+1}}{\pi \sqrt{v^2 - z^2}} \int_{-v}^{v} \mathbf{A}(-p + 0\mathrm{i}, \boldsymbol{\nu}) \widetilde{\boldsymbol{\tau}}(p) \frac{\sqrt{v^2 - p^2} dp}{p - z} \\ + \boldsymbol{P}_{2n+1}(z)(v^2 - z^2)^{-n-3/2} \qquad \left(\sqrt{v^2 - z^2} \to v, \quad z \to -0\mathrm{i} \right) , \end{aligned} \tag{7.8.4}$$

where all $2n+2$ coefficients of the polynomial $\boldsymbol{P}_{2n+1}(z)$ must be chosen so that (I) the function $G_1^{(n+2)}(z,\boldsymbol{\nu})$ be limited when $z \to 0$ ($n+1$ equations) and (II) $n+1$ boundary conditions (on the cut $S(t)$) be satisfied which appear after $(n+1)$-fold integration of equation (7.8.1) in time. If $\boldsymbol{\tau}(t,x_\alpha)$ is a homogeneous polynomial of degree n in $(t,\nu_\lambda x_\lambda)$ then $\tilde{\boldsymbol{\tau}}(p) \equiv 0$ and $G_1^{(n+2)} = \boldsymbol{P}_n(z)(v^2-z^2)^{-n-3/2}$, whence

$$G_1(z,\boldsymbol{\nu}) = \boldsymbol{Q}_n(z)(v^2-z^2)^{1/2} + \boldsymbol{R}_{n+1}(z)\ , \tag{7.8.5}$$

where $\boldsymbol{Q}_n$ and $\boldsymbol{R}_{n+1}$ are polynomials of degree n and $n+1$, respectively; their coefficients should be determined from the requirement that $G_1(z,\boldsymbol{\nu})$ decreases when $z \to \infty$ (this gives $n+2$ equations) and from $n+1$ boundary conditions.

Using the Sokhotskii–Plemelj formula, find from (7.6.32)

$$\boldsymbol{W}(p) = -2\boldsymbol{Q}_n(p)(v^2-p^2)^{1/2}\ . \tag{7.8.6}$$

Thus, just as in the case of an elliptic cut, conclude that if the stress $\boldsymbol{\tau}(t,x_\alpha)$ specified on sides of the cut is a homogeneous polynomial $\boldsymbol{P}_n(t,\nu_\lambda x_\lambda)$ of degree n in $t,\nu_\lambda x_\lambda$ then the jump in displacements on the cut is

$$\boldsymbol{W}(t,x_\alpha) = \boldsymbol{Q}_n(t,\nu_\lambda x_\lambda)\left[v^2t^2-(\nu_\lambda x_\lambda)^2\right]^{1/2}\ , \tag{7.8.7}$$

where $\boldsymbol{Q}_n$ is a homogeneous polynomial of degree n in $t,\nu_\lambda x_\lambda$ and its $n+1$ coefficients must be determined from boundary conditions. Generalizing the abovesaid, note that in both this problem and the problem for the elliptic cut (considered in Sect. 7.7) the polynomial $\boldsymbol{P}_n$ and $\boldsymbol{Q}_n$ of degree n are permitted to be considered non-homogeneous since any such polynomial is a sum of homogeneous polynomials, so, the property proven above is valid after dropping the word "homogeneous" for $\boldsymbol{P}_n$ and $\boldsymbol{Q}_n$. In the particular case $\boldsymbol{\tau}(t,x_\alpha) = \boldsymbol{\tau}_0 \equiv$ const the n is zero, which results in $G_1(z,\boldsymbol{\nu}) = \boldsymbol{a}(v^2-z^2)^{1/2} + \boldsymbol{b}z + \boldsymbol{c}$. The function $G_1(z,\boldsymbol{\nu})$ must decrease when $|z| \to \infty$, therefore $\boldsymbol{c} = 0$, $\boldsymbol{b} = -\boldsymbol{a}\mathrm{i}$; using (7.8.6) for $\boldsymbol{x} \in S(t)$, we get

$$\begin{aligned} G_1(z,\boldsymbol{\nu}) &= \boldsymbol{a}\left(\sqrt{v^2-z^2} - \mathrm{i}z\right),\\ \boldsymbol{W}(t,x_\alpha) &= -2\boldsymbol{a}\left[v^2t^2-(\nu_\lambda x_\lambda)^2\right]^{1/2}\ . \end{aligned} \tag{7.8.8}$$

This yields: $\mathrm{Im}\{\boldsymbol{a}\} = 0$. After integrating equation (7.8.1) (at $n=0$) in t, thereafter substituting $G_1(z,\boldsymbol{\nu})$ from (7.8.8) and setting $\boldsymbol{\sigma}_3^+ = \boldsymbol{\tau}_0$, the obtained integral should be transformed by replacing the integration variable t by $p = \nu_\lambda x_\lambda / t$. We should take into account that the imaginary part of the integrand is an odd function in p and becomes zero if $|p| < v$. With this, obtain the relation

$$\boldsymbol{\tau}_0 = 2v^2\mathrm{Im}\int_0^\infty \frac{\mathsf{A}^{-1}(-p+0\mathrm{i},\boldsymbol{\nu})\boldsymbol{a}\,dp}{\left[v^2-(p-0\mathrm{i})^2\right]^{3/2}} \tag{7.8.9}$$

(where the radical in the denominator is assumed to be positive at $p=0$) from which the constant vector $\boldsymbol{a}$ is determined in terms of the specified vector $\boldsymbol{\tau}_0$.

Summing up equation (7.6.30) at $x_3 = +0$ and a similar equation at $x_3 = -0$, obtain, at $n = 0$ and $G_2 \equiv 0$,

$$\partial_t^{(2)}(\boldsymbol{u}^+ + \boldsymbol{u}^-)|_{x_3=0} = 2t^{-1}\mathrm{Re}\left\{\Omega^2\,[\mathbf{B}(\Omega,-\boldsymbol{\nu}) - \mathbf{B}(\Omega,\boldsymbol{\nu})] \times \mathbf{A}^{-1}(\Omega,\boldsymbol{\nu})G_1^{(2)}(-\Omega,\boldsymbol{\nu})\right\} \quad (\Omega = -\nu_\lambda x_\lambda/t + 0\mathrm{i})\,. \tag{7.8.10}$$

From this relation and using (7.7.4), one can find displacements on the plane $x_3 = 0$. In particular, taking into account that (i) the functions $\mathbf{A}^{-1}(\Omega,\boldsymbol{\nu})$ and $G_1^{(2)}(-\Omega,\boldsymbol{\nu})$ are real at points of the cut $S(t)$ and (ii) the expression in square brackets in (7.8.10) is purely imaginary (according to the formulae (7.5.13, 15)), obtain

$$\partial_t^{(2)}(\boldsymbol{u}^+ + \boldsymbol{u}^-)|_{x_3=0} = 0 \quad (|\nu_\lambda x_\lambda| < vt)\,.$$

This and equation (7.7.4) provide the formula for accelerations of sides of the cut $S(t)(|\nu_\lambda x_\lambda| < vt)$ to be found:

$$\partial_t^{(2)}\boldsymbol{u}^+(t,x_\alpha,0) = -\partial_t^{(2)}\boldsymbol{u}^-(t,x_\alpha,0) = \partial_t^{(2)}\boldsymbol{W}(t,x_\alpha)/2\,.$$

Let us assume $\boldsymbol{\nu} = \{1,0,0\}$ for an isotropic medium. Then, from (7.8.9),

$$\tau_1 = a_1\int_0^\infty \frac{R\,dx}{(x^2/\beta^2+1)^{1/2}}\,, \quad \tau_2 = -a_2 v^2\mu\int_0^\infty \frac{(x^2/\beta^2+1)^{1/2}dx}{(v^2+x^2)^{3/2}}\,,$$

$$\tau_3 = a_3\int_0^\infty \frac{R\,dx}{(x^2/\alpha^2+1)^{1/2}}\,, \quad \boldsymbol{a} = \{a_1,a_2,a_3\}\,, \tag{7.8.11}$$

$$R = \frac{4\left[(x^2/\alpha^2+1)(x^2/\beta^2+1)\right]^{1/2} - (2+x^2/\beta^2)^2}{x^2(v^2+x^2)^{3/2}}v^2\beta^2\mu\,.$$

Thus, the solution is obtained by superposition of the three independent solutions for each of the components τ_j. For different loads the following results — displacements in the plane $x_3 = 0$ — can be obtained from (7.7.4, 8.10) $\left(W_j \equiv -2a_j(v^2t^2 - x_1^2)^{1/2}H(v^2t^2 - x_1^2)\right)$:

$$\begin{aligned}
&u_1^+ = -u_1^- = W_1/2,\ u_2^+ = u_2^- = 0,\ u_3^+ = u_3 \quad (\tau_2 = \tau_3 = 0)\,,\\
&u_1^+ = u_1^-,\ u_2^+ = u_2^- = 0,\ u_3^+ = -u_3^- = W_3/2 \quad (\tau_1 = \tau_2 = 0)\,,\\
&u_1^+ = u_1^- = 0,\ u_2^+ = -u_2^- = W_2/2,\ u_3^+ = u_3^- = 0 \quad (\tau_1 = \tau_3 = 0)\,.
\end{aligned} \tag{7.8.12}$$

In these expressions the first, the second, and the third lines refer to the solutions of (i) the transverse shear plane crack problem, ii) the normal tensile plane crack problem, and (iii) the antiplane shear crack problem, respectively; these solutions may be found in [7.27, 30, 31].

Note that the case $\boldsymbol{\tau} = \boldsymbol{\tau}_0$ for an anisotropic medium was first considered by *Burridge* (see [7.19]) using a method different from the present one.

7.8.2 The Axisymmetric Case

If the medium under consideration is isotropic or transversely isotropic about the x_3-axis (5 independent constants c_{ijkl}), the matrix $\mathbf{B}(\omega, \xi_\alpha) = |\boldsymbol{\xi}|^{-1}\mathbf{B}(\Omega, \boldsymbol{\eta})$ is of the form (7.4.22) where $\{e, f, g, h\}$ are functions of Ω and can be described by the formulae (7.4.23) in the case of an isotropic medium. Then, according to (7.7.7), the matrix $\mathbf{A}(\Omega + 0\mathrm{i}, \boldsymbol{\eta})$ is two times the even part of the matrix $\mathbf{B}(\Omega + 0\mathrm{i}, \boldsymbol{\eta})$ and is present in (7.7.17) where $\{e, f, g\}$ are real even functions of Ω when $|\Omega| < c_j$ $(j = 1, 2, 3)$.

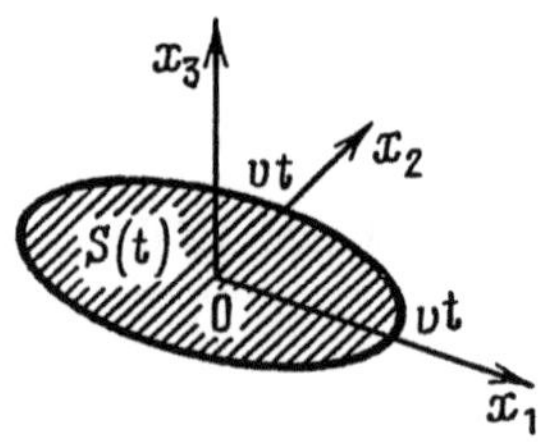

Fig. 7.8. Expansion of a circular crack

Assuming the cut $S(t)$ to be a circle ($x_3 = 0$, $r \le vt$, $v < c_R$, $r^2 = x_1^2 + x_2^2$) (Fig. 7.8), represent the vectors $t^{-n}\boldsymbol{\tau}(t, x_\alpha)$ and $t^{-n-1}\boldsymbol{W}(t, x_\alpha)$ by

$$t^{-n}\begin{bmatrix}\tau_1\\ \tau_2\\ \tau_3\end{bmatrix} = \tau_r(q)\begin{bmatrix}x_1/r\\ x_2/r\\ 0\end{bmatrix} + \tau_\theta(q)\begin{bmatrix}-x_2/r\\ x_1/r\\ 0\end{bmatrix} + \tau_3(q)\begin{bmatrix}0\\ 0\\ 1\end{bmatrix}, \tag{7.8.13}$$

$$t^{-n-1}\begin{bmatrix}W_1\\ W_2\\ W_3\end{bmatrix} = W_r(q)\begin{bmatrix}x_1/r\\ x_2/r\\ 0\end{bmatrix} + W_\theta(q)\begin{bmatrix}-x_2/r\\ x_1/r\\ 0\end{bmatrix} + W_3(q)\begin{bmatrix}0\\ 0\\ 1\end{bmatrix}, \tag{7.8.14}$$

where $q = r/t$, and $\tau_j(q)$ and $W_j(q)$ are projections of the vectors $t^{-n}\boldsymbol{\tau}(t, x_\alpha)$, $t^{-n-1}\boldsymbol{W}(t, x_\alpha)$, respectively, on the basis vector $\boldsymbol{e}_j$ $(j = r, \theta, 3)$. Applying the Radon transformation to the vector $t^{-n-1}\boldsymbol{W}$ yields a vector

$$\boldsymbol{W}_0(\Omega, \eta) = W_{0r}(\Omega)\begin{bmatrix}\eta_1\\ \eta_2\\ 0\end{bmatrix} + W_{0\theta}(\Omega)\begin{bmatrix}-\eta_2\\ \eta_1\\ 0\end{bmatrix} + W_{03}(\Omega)\begin{bmatrix}0\\ 0\\ 1\end{bmatrix}, \tag{7.8.15}$$

with $\{W_{0\lambda}, W_{03}\}$ and $\{W_\lambda, W_3\}$ $(\lambda = r, \theta)$ being interrelated by the formulae,

$$W_{0\lambda}(\Omega) = 2\Omega \int\limits_{|\Omega|}^{v} W_\lambda(q)\frac{dq}{\sqrt{q^2 - \Omega^2}},$$

$$W_\lambda(q) = -\frac{q}{\pi}\int_q^v \frac{d}{d\Omega}\left[\frac{W_{0\lambda}(\Omega)}{\Omega}\right]\frac{d\Omega}{\sqrt{\Omega^2-q^2}}\,, \tag{7.8.16}$$

$$W_{03}(\Omega) = 2\int_{|\Omega|}^v W_3(q)\frac{q\,dq}{\sqrt{q^2-\Omega^2}}\,,\quad W_3(q) = -\frac{1}{\pi}\int_q^v \frac{dW_{03}(\Omega)}{d\Omega}\frac{d\Omega}{\sqrt{\Omega^2-q^2}}$$

With this, obtain that

$$\mathbf{A}^{-1}(\Omega+0\mathrm{i},\boldsymbol{\eta})\boldsymbol{W}_0^{(n+3)}(-\Omega,\boldsymbol{\eta})$$
$$= \frac{W_{0r}^{(n+3)}(-\Omega)}{2e}\begin{bmatrix}\eta_1\\ \eta_2\\ 0\end{bmatrix} + \frac{W_{0\theta}^{(n+3)}(-\Omega)}{2(e+f)}\begin{bmatrix}-\eta_2\\ \eta_1\\ 0\end{bmatrix} + \frac{W_{03}^{(n+3)}(-\Omega)}{2g}\begin{bmatrix}0\\ 0\\ 1\end{bmatrix},$$

and the vector equation (7.7.8) (after introduction of a new integration variable $\Omega = \eta_\alpha x_\alpha/t$) reduces to an equation equivalent to the following three scalar equations:

$$\tau_r^{(n+1)}(q) = \frac{(-1)^{n+1}}{2\pi q}\int_0^q \frac{\Omega^{n+2}W_{0r}^{(n+3)}(\Omega)d\Omega}{e(q^2-\Omega^2)^{1/2}}\,,$$

$$\tau_\theta^{(n+1)}(q) = \frac{(-1)^{n+1}}{2\pi q}\int_0^q \frac{\Omega^{n+2}W_{0\theta}^{(n+3)}(\Omega)d\Omega}{(e+f)(q^2-\Omega^2)^{1/2}}\,, \tag{7.8.17}$$

$$\tau_3^{(n+1)}(q) = \frac{(-1)^{n+1}}{2\pi}\int_0^q \frac{\Omega^{n+1}W_{03}^{(n+3)}(\Omega)d\Omega}{g(q^2-\Omega^2)^{1/2}}\,.$$

Thus, a solution to the axisymmetric problem is superposition of the three independent solutions to problems for components of the vector $\boldsymbol{W}_0^{(n+3)}(\Omega,\boldsymbol{\eta})$ along the axes r,θ, and x_3. From (7.8.17) it follows that, applying the substitution $\Omega^2 = y$ to each of these equations, the Abel equation appears, and its solution is well known.

Specifically, if $n = 0$ and the components $\{\tau_r,\tau_\theta,\tau_3\}$ are constant, then invertion of equations (7.8.17) at $n = 0$ yields $W_{03}^{(3)}(\Omega) = W_{0r}^{(3)}(\Omega) = W_{0\theta}^{(3)}(\Omega) = 0$. Integrating these equations, we get

$$W_{0r}(\Omega) = \sum_{m=0}^{2} a_m\Omega^{2-m}\,,\quad W_{0\theta}(\Omega) = \sum_{m=0}^{2} b_m\Omega^{2-m}\,,$$
$$W_{03}(\Omega) = \sum_{m=0}^{2} c_m\Omega^{2-m}(0<\Omega<v);\ W_{0r} = W_{0\theta} = W_{03} = 0\,(\Omega > v)\,. \tag{7.8.18}$$

Accounting for the fact that the vector $\boldsymbol{W}(t,x_\alpha)/t$ is bounded for any x, find by the inversion formula (7.8.16): $a_2 = b_2 = c_1 = 0$. On the other hand, integrability of the stress for $r \to vt+0$ should be taken into account to obtain from the formula

(7.7.5): $a_1 = -a_0 v$, $b_1 = -b_0 v$, $c_2 = -c_0 v^2$. So, the formulae (7.8.18) take the form ($\Omega > 0$)

$$\begin{aligned} W_{0r}(\Omega) &= a_0 \Omega(\Omega - v)H(v - \Omega)\,, \\ W_{0\theta}(\Omega) &= b_0 \Omega(\Omega - v)H(v - \Omega)\,, \\ W_{03}(\Omega) &= c_0(\Omega^2 - v^2)H(v - \Omega)\,. \end{aligned} \tag{7.8.19}$$

Thereafter, use the formulae (7.8.16) for $q < v$ to derive

$$\begin{aligned} W_r(q) &= -q\pi^{-1} a_0 \ln\left[(v + \sqrt{v^2 - q^2})/q\right]\,, \\ W_\theta(q) &= -q\pi^{-1} b_0 \ln\left[(v + \sqrt{v^2 - q^2})/q\right]\,, \\ W_3(q) &= -2\pi^{-1} c_0 \sqrt{v^2 - q^2}\,. \end{aligned} \tag{7.8.20}$$

For determining the real constants $\{a_0, b_0, c_0\}$, we should employ equation (7.7.5) and the boundary condition $\boldsymbol{\sigma}_3^+(t, x_\alpha, 0) = \{-\tau_r, -\tau_\theta, -\tau_3\}$ for $\boldsymbol{x} \in S(t)$. For this purpose, form the vector $\boldsymbol{F}_1^{(3)}(z, \boldsymbol{\eta})$ ($\operatorname{Im}\{z\} < 0$) as in (7.6.11). Since $\boldsymbol{W}(-\Omega, -\boldsymbol{\eta}) = \boldsymbol{W}(\Omega, \boldsymbol{\eta})$ and the formula (7.8.15) is valid, it should be concluded that the functions $W_{0r}(\Omega)$ and $W_{0\theta}(\Omega)$ require an odd continuation into the region $\Omega < 0$), whereas $W_{03}(\Omega)$, an even continuation. Then, the formula (7.6.11) helps find the projections of $\boldsymbol{F}_1^{(3)}(z, \boldsymbol{\eta})$ on the axes r, θ, and x_3:

$$\begin{aligned} F_{1r}^{(3)}(z) &= \frac{2v^4 a_0 \mathrm{i}}{\pi z(v^2 - z^2)^2}\,, & F_{1\theta}^{(3)}(z) &= \frac{2v^4 b_0 \mathrm{i}}{\pi z(v^2 - z^2)^2}\,, \\ F_{13}^{(3)}(z) &= \frac{4v^3 c_0 \mathrm{i}}{\pi(v^2 - z^2)^2}\,, & &(\operatorname{Im} z < 0)\,. \end{aligned} \tag{7.8.21}$$

Consequently, obtain

$$\begin{aligned} &A^{-1}(\Omega, \boldsymbol{\eta})\boldsymbol{F}_1^{(3)}(-\Omega, \boldsymbol{\eta}) \\ &= \frac{F_{1r}^{(3)}(-\Omega)}{2e}\begin{bmatrix} \eta_1 \\ \eta_2 \\ 0 \end{bmatrix} + \frac{F_{1\theta}^{(3)}(-\Omega)}{2(e+f)}\begin{bmatrix} -\eta_2 \\ \eta_1 \\ 0 \end{bmatrix} + \frac{F_{13}^{(3)}(-\Omega)}{2g}\begin{bmatrix} 0 \\ 0 \\ 1 \end{bmatrix} \\ &(\Omega = -\eta_\alpha x_\alpha/t + 0\mathrm{i})\,. \end{aligned} \tag{7.8.22}$$

Substituting (7.5.16) into (7.7.5), introducing the new integration variable $\Omega_0 = \eta_\alpha x_\alpha/t$ instead of $\boldsymbol{\eta}$, thereafter integrating in t and transforming the integral obtained, find

$$-\tau_r = \frac{2v^4 a_0}{\pi^2}\operatorname{Im}\int_\varepsilon^\infty \frac{d\Omega_0}{e\left[v^2 - (\Omega_0 - 0\mathrm{i})^2\right]^2 \Omega_0}\,,$$

$$-\tau_\theta = \frac{2v^4 b_0}{\pi^2}\operatorname{Im}\int_\varepsilon^\infty \frac{d\Omega_0}{(e+f)\left[v^2 - (\Omega_0 - 0\mathrm{i})^2\right]^2 \Omega_0}\,,$$

$$-\tau_3 = \frac{2v^3 c_0}{\pi} \mathrm{Im} \int_0^\infty \frac{d\Omega_0}{g\left[v^2 - (\Omega_0 - 0\mathrm{i})^2\right]^2} \qquad (0 < \varepsilon < v)\,. \tag{7.8.23}$$

It follows from (7.8.20, 23) that each component of the jump in projection on the axes $\{r, \theta, x_3\}$ is defined by the corresponding component of the stress. Specifically, in the case of isotropy, when $\{e, f, g\}$ in (7.8.23) are given by the formulae (7.4.23) (at $\Omega = -\Omega_0 + 0\mathrm{i}$), the solution to the problem for τ_3 from the formula (7.8.23) is the same as the above one obtained earlier by using the lower formula (7.7.21). Note that the middle formula in (7.8.23) gives the solution to the problem of propagation of a circular cut under the load $-\tau_\theta$. In this case, only one displacement component u_θ is present; the solution shows that this component over the cut has the form

$$u_\theta^+ = -u_\theta^- = tW_\theta(q)/2 = r\tau_\theta(2\mu)^{-1} \ln\left[\left(v + \sqrt{v^2 - q^2}\right)/q\right]\,.$$

These expressions for displacements of sides of a cut are the same as those found in the elastostatics problem provided that the radius of a crack is taken equal to vt.

References

Chapter 1

1.1 L.I. Sedov: *A Course in Continuum Mechanics*, Vol. 1-4 (Wolters-Noordhoff, Groningen 1971-1972)
1.2 A.N. Guz', V.T. Golovchan: *Elastic Wave Diffraction in Multiple-Connected Bodies* (Naukova Dumka, Kiev 1972) [in Russian]
1.3 L.I. Slepyan: *Non-Stationary Elastic Waves* (Sudostroenie, Leningrad 1972) [in Russian]
1.4 J.D. Achenbach: *Wave Propagation in Elastic Solids* (North-Holland, Amsterdam 1973)
1.5 Y.H. Pao, C.C. Mow: *Diffraction of Elastic Waves and Dynamic Stress Concentrations* (Crane Russak, New York 1975)
1.6 C. Eringen, E.S. Suhubi: *Elastodynamics*, Vol. II: *Linear Theory* (Academic, New York 1975)
1.7 K.F. Graff: *Wave Motion in Elastic Solids* (Columbus State University, Ohio 1975)
1.8 V.M. Seymov: *Dynamic Contact Problems* (Naukova Dumka, Kiev 1976) [in Russian]
1.9 V.M. Babich, I.A. Molotkov: "Mathematical Methods in Elastic Body Theory", in *Mechanics of Deformable Bodies*, Vol. 10 (VINITI, Moscow 1977) [in Russian]
1.10 S.K. Datta: "Scattering of Elastic Waves", in *Mechanics Today*, Vol. 4 (Pergamon, Oxford 1978)
1.11 A.I. Guz', V.D. Kubenko, M.A. Cherevko: *Diffraction of Elastic Waves* (Naukova Dumka, Kiev 1978) [in Russian]
1.12 J. Miklowitz: *The Theory of Elastic Waves and Waveguides* (North-Holland, Amsterdam 1978)
1.13 G.I. Petrashen': "Fundamentals of Mathematical Theory for Elastic Wave Propagation", in *Aspects of Seismic Wave Propagation Theory*, Vol. XVIII (Nauka, Leningrad 1978) [in Russian]
1.14 I.I. Vorovich, V.A. Babeshko: *Mixed Elastodynamics Problems for Non-Classical Domains* (Nauka, Moscow 1979) [in Russian]
1.15 J.A. Hudson: *The Excitation and Propagation of Elastic Waves* (Cambridge University Press, Cambridge 1980)
1.16 G.I. Petrashen': *Propagation of Waves in Anisotropic Elastic Continua* (Nauka, Leningrad 1980) [in Russian]
1.17 K. Aki, P. Richards: *Quantitative Seismology – Theory and Methods* (W.H. Freeman, San Francisco 1980)
1.18 A. Ben-Menahem, S.J. Singh: *Seismic Waves and Sources* (Springer, Berlin Heidelberg New York 1981)
1.19 V.T. Grinchenko, V.V. Meleshko: *Harmonic Vibrations and Waves in Elastic Bodies* (Naukova Dumka, Kiev 1980) [in Russian]
1.20 J.D. Achenbach, A.K. Gautesen, H. McMaken: *Ray Methods for Waves in Elastic Solids* (Pitman, Boston 1982)
1.21 G.I. Petrashen', L.A. Molotkov, P.V. Krauklis: *Waves in Layered Homogeneous Isotropic Elastic Continua* (Nauka, Leningrad 1982) [in Russian]
1.22 Y.H. Pao: Proc. ASME, J. Appl. Mech. **50**, 4 (b), 1152 (1983)
1.23 V.Z. Parton, P.I. Perlin: *Mathematical Methods of the Theory of Elasticity*, Vol. 1,2 (Mir, Moscow 1984)
1.24 G.I. Petrashen', L.A. Molotkov, P.V. Krauklis: *Waves in Layered Homogeneous Isotropic Elastic Continua* II (Nauka, Leningrad 1985) [in Russian]
1.25 J. Miklowitz: "Modern Corner, Edge, and Crack Problems in Linear Elastodynamics Involving Transient Waves", in *Advances in Appl. Mech.*, Vol. 25 (Orlando ed. 1987)

1.26 V.A. Dobrushkin: *Boundary-Value Problems in Elastodynamics for Wedge-Shaped Domains* (Nauka i Tekhnika, Minsk 1988) [in Russian]
1.27 V.M. Babich, V.S. Buldyrev: *Asymptotic Methods for Problems on Diffraction of Short Waves* (Nauka, Moscow 1972) [in Russian]
1.28 J.R. Willis: Phil. Trans. Roy. Soc. London, Ser. A **274**, 1240, 435 (1973)

Chapter 2

2.1 L.I. Sedov: *A Course in Continuum Mechanics*, Vol. 1-4 (Wolters-Noordhoff, Groningen 1971-1972)
2.2 A.E.H. Love: *The Mathematical Theory of Elasticity* (Dover, New York 1944)
2.3 W. Nowacki: *The Theory of Elasticity* (Mir, Moscow 1970) [in Russian]
2.4 B.V. Kostrov, S. Das: *Principles of Earthquake Source Mechanics* (Cambridge University Press, Cambridge 1988)
2.5 G.I. Petrashen': "Fundamentals of Mathematical Theory for Elastic Waves Propagation", in *Aspects of Seismic Wave Propagation Theory*, Vol. XVIII (Nauka, Leningrad 1978) [in Russian]
2.6 S.L. Sobolev: "Some Aspects of the Vibration Propagation Theory", in *F. Frank, R. Mises: Differential and Integral Equations in Physics*, II (ONTI, Leningrad 1937) [in Russian]
2.7 R.J. Knops, L.E. Payne: *Uniqueness in Linear Elasticity* (Springer, Berlin Heidelberg New York 1971)
2.8 G.S. Brockway: Arch. Rat. Mech. Analys. **48**, 213 (1972)
2.9 V.I. Smirnov: *A Course of Higher Mathematics*, Vol. 4 (Pergamon, Oxford 1964)
2.10 R. Courant: *Partial Differential Equations* (Interscience, New York 1962)
2.11 G.I. Petrashen': *Propagation of Waves in Anisotropic Elastic Continua* (Nauka, Leningrad 1980) [in Russian]
2.12 M.J.P. Musgrave: *Crystal Acoustics* (Holden-Day, San Francisco 1970)
2.13 L.I. Slepyan, A.L. Fishkov "Mixed Two-Dimensional Problem with Non-Uniform Motion of the Point Separating the Boundary Conditions", in *Current Problems of Continuum Mechanics*, issue 13 (LGU, Leningrad 1980) [in Russian]
2.14 P.M. Morse, H. Feshbach: *Methods of Theoretical Physics*, (McGraw-Hill, New York 1953) p.1
2.15 G. Muentz: *Integral Equations*, Vol. 1 (GTTI, Leningrad 1934) [in Russian]
2.16 P.M. Morse, H. Feshbach: *Methods of Theoretical Physics*, (McGraw-Hill, New York 1953) p. 2
2.17 M. Iacovache: "O Extindere a Metodei lui Galerkin Pentru Sistemul Ecuatiilor Elasticitatii", in *Bull. Acad. Sci. RPR*, Ser. A **1** (1949)
2.18 E. Sternberg, R.A. Eubanks: Quart. Appl. Math. **15**, 2, 149 (1957)
2.19 V.D. Kupradze: "Dynamical Problems in Elasticity", in *Progress in Solid Mechanics*, Vol. 3, I.N. Sneddon, R. Hill (eds.) (North Holland, Amsterdam 1963)
2.20 V.D. Kupradze, T.G. Gegeliya, M.O. Basheleyshvili, T.V. Burchuladze: *Three-Dimensional Problems of the Mathematical Theory of Elasticity and Thermoelasticity* (Nauka, Moscow 1976) [in Russian]
2.21 Y.H. Pao, C.C. Mow: *Diffraction of Elastic Waves and Dynamic Stress Concentrations* (Crane Russak, New York 1973)
2.22 A.I. Guz', V.D. Kubenko, M.A. Cherevko: *Diffraction of Elastic Waves* (Naukova Dumka, Kiev 1978) [in Russian]
2.23 J. Miklowitz: *The Theory of Elastic Waves and Waveguides* (North-Holland, Amsterdam 1978)
2.24 N.I. Muskhelishvili: *Singular Integral Equations* (Noordhoff, Groningen 1953)
2.25 N.I. Muskhelishvili: *Some Basic Problems of the Mathematical Theory of Elasticity* (Noordhoff, Groningen 1953)
2.26 L.A. Galin: *Contact Problems in the Theory of Elasticity* (Department of Mathematics of North Carolina State University, Raleigh 1961)
2.27 I.N. Sneddon: Rend. Circ. Mat. Palermo, Ser. 2, **1**, 1, 57 (1952)
2.28 J.R.M. Radok: Quart. Appl. Math. **XIV**, 3, 289 (1956)

2.29 I.N. Sneddon, D.S. Berry: *The Classical Theory of Elasticity*, Handbuch der Physik, Vol. 6 (Springer, Berlin Göttingen Heidelberg 1958)
2.30 R.V. Goldstein: Inzh. Zhurn. **5**, 93 (1966) [in Russian]
2.31 G.I. Barenblatt, R.V. Goldstein: Inter. J. Fract. Mech. **8**, 427 (1972)
2.32 G.P. Cherepanov: *Mechanics of Brittle Fracture* (McGraw-Hill, New York 1979)
2.33 A.V. Zwyagin: Vestnik MGU, Ser. 1, **3**, 60 (1979) [in Russian]
2.34 I.V. Simonov: Izv. Akad. Nauk SSSR, Mekh. Tverd. Tela **6**, 90 (1980) [in Russian]
2.35 I.V. Simonov: J. Appl. Math. Mech. (PMM) **48**, 1, 74 (1984)

Chapter 3

3.1 V.I. Smirnov, S.L. Sobolev: Trudy Inst. Seism. Akad. Nauk SSSR **20**, 37 (1932) [in Russian]; V.I. Smirnov, S.L. Sobolev: C. R. Acad. Sci. Paris **8**, 1437 (1932); V.I. Smirnov, S.L. Sobolev: C. R. Acad. Sci. Paris **8**, 1797 (1932); V.I. Smirnov, S.L. Sobolev: Trudy Inst. Seism. Akad. Nauk SSSR **29**, 43 (1933) [in Russian]
3.2 V.I. Smirnov: *A Course of Higher Mathematics*, Vol. 3, part 2 (Pergamon, Oxford 1964)
3.3 E.A. Naryshkina: Trudy Inst. Seism. Acad. Nauk SSSR **45**, 71 (1934) [in Russian]
3.4 V.A. Sveklo: Dokl. Akad. Nauk SSSR **95**, 4, 737 (1954) [in Russian]
3.5 I.N. Vekua: Trudy Tbilisskogo Geofiz. Inst. **2**, 23 (1937) [in Russian]
3.6 E.A. Naryshkina: Trudy Inst. Seism. Akad. Nauk SSSR **90**, 11 (1940) [in Russian]
3.7 M.G. Gogoladze: Trudy Inst. Seism. Akad. Nauk SSSR **125**, 34 (1947) [in Russian]
3.8 L.P. Zaytsev, N.V. Zvolinsky: Izv. Akad. Nauk SSSR, Ser. Geofiz. **5**, 40 (1951) [in Russian]
3.9 N.V. Zvolinski: Izv. Akad. Nauk SSSR, Ser. Geofiz. **5**, 40 (1951) [in Russian]
3.10 N.V. Zvolinski: Izv. Akad. Nauk SSSR, Ser. Geofiz. **1**, 3 (1958) [in Russian]
3.11 N.V. Zvolinski: Izv. Akad. Nauk SSSR, Ser. Geofiz. **2**, 165 (1958) [in Russian]
3.12 S.L. Sobolev: Trudy Inst. Seism. Akad. Nauk SSSR **41**, 23 (1934) [in Russian]
3.13 M.M. Fridman: Dokl. Akad. Nauk SSSR **60**, 7, 1145 (1948) [in Russian]
3.14 M.M. Fridman: Dokl. Akad. Nauk SSSR **66**, 1, 21 (1949) [in Russian]
3.15 A.F. Filippov: Prikl. Mat. Mekh. **20**, 688 (1956) [in Russian]
3.16 A.F. Filippov: Appl. Math. Mech. (PMM) **23**, 989 (1959)
3.17 N.I. Yerugin: Dokl. Akad. Nauk SSSR **42**, 9, 385 (1944) [in Russian]
3.18 V.A. Sveklo: Dokl. Akad. Nauk SSSR **59**, 5, 871 (1948) [in Russian]
3.19 V.A. Sveklo: Dokl. Akad. Nauk SSSR **59**, 6, 1077 (1948) [in Russian]
3.20 V.A. Sveklo: Uch. Zap. Leningr. Gos. Univ., Ser. Math. Nauk **17**, 28 (1949) [in Russian]
3.21 V.A. Sveklo: J. Appl. Math. Mech. (PMM) **25**, 1324 (1961)
3.22 S.A. Chaplygin: *On Gas Jets. Collected Works*, Vol. 2 (Izd. Akad. Nauk SSSR, Moscow 1933) [in Russian]
3.23 A. Busemann: Schriften Deut. Akad. Luftfahrforschung **7B**, 105 (1943)
3.24 G.N. Ward: *Linearized Theory Of Steady High-Speed Flow* (Cambridge University Press, London 1955)
3.25 V.B. Poruchikov: Fluid Dyn. **9**, 293 (1974)
3.26 A.Ya. Sagomonyan, V.B. Poruchikov: *Three-Dimensional Problems on Transient Motion of Compressible Fluid* (MGU, Moscow 1970) [in Russian]
3.27 J.W. Miles: Appl. Math. **18**, 1, 37 (1960)
3.28 A.W. Maue: ZAMM **34**, 1 (1954)
3.29 B.V. Kostrov: Izv. Akad. Nauk SSSR, Otd. Tekh. Nauk, Mekh. Mashinostr. **4**, 54 (1964) [in Russian]
3.30 B.V. Kostrov: Appl. Math. Mech. (PMM) **28**, 793 (1964)
3.31 B.V. Kostrov: J. Appl. Math. Mech. (PMM) **28**, 1077 (1964)
3.32 J.W. Craggs: Int. J. Eng. Sci. **4**, 2, 113 (1966)
3.33 E.F. Afanas'ev, G.P. Cherepanov: J. Appl. Math. Mech. (PMM) **37**, 584 (1973)
3.34 G.P. Cherepanov, E.F. Afanas'ev: Int. J. Eng. Sci. **12**, 665 (1974)
3.35 E.F. Afanas'ev; G.P. Cherepanov: Sov. Phys. - Dokl. **15**, 185 (1970)

3.36 L.V. Nikitin, V.N. Odintsev: Arch. Mech. Stosow. **25**, 2, 351 (1973)
3.37 L.V. Nikitin, V.N. Odintsev: "Axisymmetric Self-Similar Dynamic Problem on Crack Propagation in Elastic Continuum", in *Propagation of Elastic and Elastoplastic Waves*: Proc. of the 5th All-Union Symposium (Nauka, Alma-Ata 1973) [in Russian]
3.38 A.R. Robinson, J.C. Thompson: Z. Angew. Math. Mech. **54**, 3, 139 (1974)
3.39 A.R. Robinson, J.C. Thompson: Proc. Camb. Phil. Soc. **76**, 1, 369 (1974)
3.40 L.M. Brock: Int. J. Eng. Sci. **14**, 181 (1976)
3.41 L.M. Brock: Int. J. Eng. Sci. **15**, 147 (1977)
3.42 L.M. Brock: Int. J. Eng. Sci. **17**, 365 (1979)
3.43 L.M. Brock: Int. J. Eng. Sci. **19**, 81 (1981)
3.44 L.P. Zaytsev, L.M. Flitman: Izv. Akad. Nauk SSSR, Fiz. Zemli **11**, 13 (1965) [in Russian]
3.45 J.D. Achenbach, L.M. Brock: J. Elasticity **1**, 1, 51 (1971)
3.46 E.F. Afanas'ev: "Some Homogeneous Solutions in Elastodynamics", in *Continuum Mechanics and Related Problems of Calculus* (Nauka, Moscow 1972) [in Russian]
3.47 E.F. Afanas'ev: Dokl. Akad. Nauk SSSR **210**, 3, 555 (1973) [in Russian]
3.48 V.B. Poruchikov: Appl. Math. Mech. (PMM) **37**, 74 (1973)
3.49 F.R. Norwood: Int. J. Solids Struct **9**, 7, 789 (1973)
3.50 J.R. Willis: Phil. Trans. Roy. Soc. London, Ser. A **274**, 1240, 435 (1973)
3.51 N.I. Muskhelishvili: *Singular Integral Equations* (Noordhoff, Groningen 1953)
3.52 R. Courant: *Partial Differential Equations* (Interscience, New York 1962)
3.53 V.S. Vladimirov: *Equations of Mathematical Physics* (Mir, Moscow 1984)
3.54 N.I. Muskhelishvili: *Some Basic Problems of the Mathematical Theory of Elasticity* (Noordhoff, Groningen 1953)
3.55 V.B. Poruchikov: *Methods of the Dynamic Theory of Elasticity* (Nauka, Moscow 1986) [in Russian]
3.56 V.S. Vladimirov: *Generalized Functions in Mathematical Physics* (Mir, Moscow 1979)
3.57 J.W. Craggs: Proc. Cambr. Phil. Soc. **56**, 3, 269 (1960)
3.58 R.V. Goldstein: Prikl. Mat. Mekh. **29**, 3, 516 (1965) [in Russian]
3.59 H. Lamb: Phil. Trans. Roy. Soc. **A203**, 1 (1904)
3.60 P.M. Morse, H. Feshbach: *Methods of Theoretical Physics* (McGraw-Hill, New York 1953) p. 1
3.61 K.V. Broberg: Arkiv. Fys. **18**, 2, 159 (1960)
3.62 I. Sneddon: *Fourier Transforms* (McGraw-Hill, New York 1951)
3.63 R.J. Bedding, J.R. Willis: J. Elasticity **3**, 4, 289 (1973)
3.64 B.V. Kostrov, L.V. Nikitin: Arch. Mech. Stosow. **6**, 22, 750 (1970)
3.65 G.V. Logvinovich: *Hydrodynamics of Free Boundary Flows* (NASA-TTF-658, Washington 1972)
3.66 V.I. Smirnov, S.L. Sobolev: Trudy Inst. Seism. Akad. Nauk SSSR **29**, 43 (1933) [in Russian]
3.67 L.I. Slepyan: *Non-Stationary Elastic Waves* (Sudostroenie, Leningrad 1972) [in Russian]
3.68 I.S. Gradshteyn, I.M. Ryzhik: *Tables of Integrals, Series and Products* (Academic, New York 1965)
3.69 P.G. Richards: Bull. Seism. Soc. Amer. **69**, 4, 947 (1979)
3.70 A.I. Lur'e: *Three-Dimensional Problems of the Theory of Elasticity* (Interscience, New York 1964)
3.71 E.I. Shemyakin, V.L. Faynshmidt: Uch. Zap. Leningr. Gos. Univ., Ser. Mat. Nauk **148**, 28, 148 (1954) [in Russian]
3.72 G.M.L. Gladwell: *Contact Problems in the Classical Theory of Elasticity* (Sijthoff and Nordhoff, Germantown 1980)
3.73 R. Burridge: J. Phys. Earth, **16** (special issue), 169 (1968)
3.74 R. Burridge, J.R. Willis: Proc. Cambr. Phil. Soc. 66, **2**, 443 (1969)

Chapter 4

4.1 H. Lamb: Phil. Trans. Roy. Soc. **A203**, 1 (1904)
4.2 G.I. Petrashen', G.I. Marchuk, K.I. Ogurtsov: Uch. Zap. Leningr. Gos. Univ., Ser. Mat. Nauk **35**, 21, 71 (1950) [in Russian]

4.3 G.L. Pekeris: Proc. Nat. Acad. Sci. **41**, 469 (1955)
4.4 G.L. Pekeris: Proc. Nat. Acad. Sci. **41**, 629 (1955)
4.5 J. Miklowitz: *Elastic Wave Propagation* (Spartan Books, Washington 1966)
4.6 L. Cagniard: *Reflection and Diffraction of Progressive Seismic Waves* (McGraw-Hill, New York 1962)
4.7 A.T. De Hoop: Appl. Sci. Res., Sect. B, **8**, 4, 349 (1960)
4.8 A.W. Mane: Z. Angew. Math. Mech. **33**, 1/2, 1 (1953)
4.9 L.M. Flitman: J. Appl. Math. Mech. (PMM) **23**, 997 (1959)
4.10 L.M. Flitman: Appl. Math. Mech. (PMM) **26**, 1582 (1962)
4.11 B.V. Kostrov: Appl. Math. Mech. (PMM) **28**, 113 (1964)
4.12 B.V. Kostrov: Appl. Math. Mech. (PMM) **30**, 1241 (1966)
4.13 R. Skalak: J. Appl. Mech. **24**, 1, 59 (1957)
4.14 C. Atkinson: Int. J. Eng. Sci. **6**, 1, 27 (1968)
4.15 C. Atkinson, D. Innes: Int. J. Frac. Mech. **5**, 245 (1969)
4.16 D. Webb, C. Atkinson: Int. J. Eng. Sci. **7**, 525 (1969)
4.17 R. Folk, G. Fox, C.A. Shook, C.W. Curtis: J. Acoust. Soc. Amer. **30**, 6, 552 (1958)
4.18 L.B. Freund: J. Mech. Phys. Solids **20**, 3, 141 (1972)
4.19 B.V. Kostrov: J. Appl. Math. Mech. (PMM) **38**, 511 (1974)
4.20 B.V. Kostrov: "Dynamic Crack Propagation with Variable Rate", in *Mechanica Zniszczenia. Teoria i Zastosowania* (Wydawn. Polskiej Acad. Nauk, Warszawa 1976)
4.21 L.I. Slepyan: *Mechanics of Cracks* (Sudostroenie, Leningrad 1981) [in Russian]
4.22 J.R. Willis: Phil. Trans. Roy. Soc. London, Ser. A **274**, 1240, 435 (1973)
4.23 V.B. Poruchikov: J. Appl. Math. Mech. (PMM) **40**, 847 (1976)
4.24 V.B. Poruchikov: Inzh. Zhurn. Mekh. Tverd. Tela **6**, 126 (1968) [in Russian]
4.25 L.I. Slepyan: *Non-Stationary Elastic Waves* (Sudostroenie, Leningrad 1972) [in Russian]
4.26 V.M. Babich, I.A. Molotkov: "Mathematical Methods in Elastic Body Theory", in *Mechanics of Deformable Bodies*, Vol. 10 (VINITI, Moscow 1977) [in Russian]
4.27 J. Miklowitz: *The Theory of Elastic Waves and Waveguides* (North-Holland, Amsterdam 1978)
4.28 Y.H. Pao: Proc. ASME J. Appl. Mech. **50**, 4(b), 1152 (1983)
4.29 J. Miklowitz: "Modern Corner, Edge, and Crack Problems in Linear Elastodynamics Involving Transient Waves", in *Advances in Appl. Mech.* Vol. 25 (Orlando ed. 1987)
4.30 V.A. Dobrushkin: *Boundary-Value Problems in Elastodynamics for Wedge-Shaped Domains* (Nauka i Tekhnika, Minsk 1988) [in Russian]
4.31 N.V. Zvolinski, M.I. Reitman, G.S. Shapiro: "Dynamics of Deformable Bodies", in *Mechanics in the USSR for 50 Years*, Vol. 3 (Nauka, Moscow 1972) [in Russian]
4.32 B. van der Pol, H. Bremmer: *Operational Calculus Based on the Two-Sided Laplace Integral* (Cambridge 1950)
4.33 I.N. Sneddon: *The Use of Integral Transforms* (McGraw-Hill, New York 1972)
4.34 G.I. Petrashen', L.A. Molotkov, P.V. Krauklis: *Waves in Layered Homogeneous Isotropic Elastic Continua* (Nauka, Leningrad 1982) [in Russian]
4.35 B. Noble: *Methods Based on the Wiener Hopf Technique for the Solution of Partial Differential Equations* (Pergamon, London 1958)
4.36 A.A. Kharkevich: *Non-Steady Wave Phenomena* (Gostekhizdat, Moscow 1950) [in Russian]
4.37 Y.N. Mnev, A.K. Pertsev: *Hydroelasticity of Shells* (Wright-Patterson Air Force Base, Ohio, FTD-MT-24-119-71 1971)
4.38 G.N. Watson: *A Treatise on the Theory of Bessel Functions* (University Press, Cambridge 1945)
4.39 H. Hönl, A.W. Maue, K. Westpfahl: "Theorie der Beugung", in *Handbuch der Physik*, Band XXV/I (Springer, Berlin Göttingen Heidelberg 1961)
4.40 H.M. Nussenzveig: Ann. Phys. **34**, 1, 23 (1965)
4.41 J.D. Achenbach, A.K. Gautesen, H. McMaken: *Ray Methods for Waves in Elastic Solids* (Pitman, Boston 1982)
4.42 V.M. Babich, V.S. Buldyrev: *Asymptotic Methods for Problems on Diffraction of Short Waves* (Nauka, Moscow 1972) [in Russian]

4.43 Y.H. Pao, C.C. Mow: *Diffraction of Elastic Waves and Dynamic Stress Concentrations* (Crane Russak, New York 1973)

4.44 F.G. Friedlander: *Sound Pulses* (University Press, Cambridge 1958)

4.45 A.N. Kovshov, I.V. Simonov: Izv. Akad. Nauk SSSR, Mekh. Tverd. Tela **5**, 155 (1967) [in Russian]

4.46 V.B. Poruchikov: "Expansion in Terms of Legendre Differential Equation Eigenfunctions and its Application to Diffraction Problems", in *Interaction of Acoustic/Shock Waves with Elastic Structures* (MGU, Moscow 1981) [in Russian]

4.47 H. Bateman, A. Erdelyi: *Higher Transcendental Functions*, Vol. 1 (McGraw-Hill, New York 1953)

4.48 J.L. Griffith: J. Proc. Roy. Soc., New South Wales **90**, part 4, 157 (1957)

4.49 I.S. Gradshteyn, I.M. Ryzhik: *Tables of Integrals, Series and Products* (Academic, New York 1965)

4.50 V.B. Poruchikov, T.M. Prismakova: Moscow University Mechanics Bulletin **42**, 2, 23 (1987)

4.51 S.L. Sobolev: Trudy Inst. Seism. Akad. Nauk SSSR **41**, 23 (1934) [in Russian]

4.52 V.B. Poruchikov: J. Appl. Math. Mech. (PMM) **32**, 312 (1968)

4.53 L. Ting: Quart. Appl. Math. **18**, 1, 89 (1960)

4.54 H. Bateman, A. Erdlyi: *Higher Transcendental Functions*, Vol. 2 (McGraw-Hill, New York 1953)

4.55 P.C. Waterman: J. Math. Phys. **42**, 323 (1963)

4.56 E.W. Hobson: *The Theory of Spherical and Ellipsoidal Harmonics* (University Press, Cambridge 1931)

4.57 V.B. Poruchikov: Fluid Dyn. **11**, 345 (1976)

4.58 A.F. Filippov: J. Appl. Math. Mech. (PMM) **28**, 372 (1964)

4.59 V.B. Poruchikov: Mech. Solids (USSR) **11**, 2, 135 (1976)

4.60 L. Pochhammer: J. Reine Angew. Math. **81**, 4, 324 (1876)

4.61 H. Kolsky: *Stress Waves in Solids* (Oxford 1953)

4.62 A. Erdelyi: *Asymptotic Expansions* (Dover Publications, Inc. 1956)

4.63 L.I. Sedov: *A Course in Continuum Mechanics*, Vol. 1-4 (Wolters-Noordhof, Groningen 1971-1972)

4.64 V.Z. Parton, P.I. Perlin: *Mathematical Methods of the Theory of Elasticity*, Vol. 1, 2 (Mir, Moscow 1984)

4.65 V.N. Kukudzhanov: "Numerical Solution of Multidimensional Problems on Stress Waves in Solids", in *Communications on Applied Mathematics*, issue 6 (Vopr. Tekh. Akad. Nauk SSSR, Moscow 1976) [in Russian]

4.66 G.P. DeVault, C.W. Curtis: J. Acoust. Soc. Amer. **34**, 421 (1962)

4.67 U.K. Nigul: Appl. Math. Mech. (PMM) **27**, 1602 (1963/1964)

4.68 M.A. Malkov: Sov. Phys.- Dokl. **8**, 209 (1963)

4.69 M.A. Malkov: J. Appl. Math. Mech. **32**, 472 (1968)

4.70 W.R. Spillers, A. Callegari: Int. J. Mech. Sci. **11**, 845 (1969)

4.71 A.G. Bagdoyev, A.I. Martirosyan: Dokl. Akad. Nauk SSSR **226**, 3, 537 (1976) [in Russian]

4.72 R. Folk, A. Nerczynski: J. Acoust. Soc. Amer. **80**, 4, 1103 (1986)

4.73 E.I. Grigolyuk, I.T. Selezov: "Non-classical Theories on Vibrations of Bars, Plates and Shells", in *Results in Science and Technology*, Mechanics of Solids Ser., Vol. 5 (VINITI, Moscow 1973) [in Russian]

Chapter 5

5.1 L. Knopoff: "Elastic Wave Propagation in a Wedge", in *Wave Propagation in Solids*, J. Miklowitz (ed.) (American Society of Mechanical Engineers, New York 1969)

5.2 J. Miklowitz: "Modern Corner, Edge, and Crack Problems in Linear Elastodynamics Involving Transient Waves", in *Advances in Appl. Mech.*, Vol. 25 (Orlando ed. 1987)

5.3 V.A. Dobrushkin: *Boundary-Value Problems in Elastodynamics for Wedge-Shaped Domains* (Nauka i Tekhnika, Minsk 1988) [in Russian]

5.4 B.V. Kostrov: Prikl. Mat. Mekh. **30**, 1, 198 (1966) [in Russian]

5.5 V.A. Sveklo, V.A. Siukiiajnen: Sov. Phys.- Dokl. **119**, 451 (1958)
5.6 S.M. Kapustyanskii: J. Appl. Math. Mech. (PMM) **40**, 172 (1976)
5.7 S.H. Zemell: SIAM J. Appl. Math. **29**, 4, 582 (1975)
5.8 V.B. Poruchikov: Mechanics of Solids (Mekh. Tverd. Tela) **11**, 5, 121 (1976)
5.9 V.B. Poruchikov: "Some Unsteady Problems on Mechanics of Compressible Medium"; D. Sc. Thesis, Moscow University (1976)
5.10 S.A. Thau, A. Umek: Trans. ASME, Ser. E, J. Appl. Mech. **41**, 697 (1974)
5.11 M. Dravinski, S.A. Thau: Trans. ASME, Ser. E, J. Appl. Mech. **43**, 291 (1976)
5.12 M.Sh. Israilov: Sov. Phys. - Dokl. **25**, 537 (1980)
5.13 V.B. Poruchikov: J. Appl. Math. Mech. (PMM) **40**, 847 (1976)
5.14 V.B. Poruchikov: Sov. Phys. - Dokl. **26**, 626 (1981)
5.15 M.Sh. Israilov: Sov. Phys. - Dokl. **24**, 679 (1979)
5.16 S.Kh. Sharipova: Mech. Solids (USSR) **17**, 4, 115 (1982)
5.17 V.B. Poruchikov: MM **42**, 908 (1978) [in Russian]
5.18 V.A. Borovikov: *Diffraction by Polygons and Polyhedrons* (Nauka, Moscow 1966) [in Russian]
5.19 A.F. Filippov: J. Appl. Math. Mech. (PMM) **28**, 372 (1964)
5.20 A. Sommerfeld: *Vorlesungen über theoretische Physik*, Band II: Mechanik der deformierbaren Medien (Leipzig 1949)
5.21 L.M. Flitman: Izv. Akad. Nauk SSSR, Ser. Geofiz. **1**, 105 (1958)
5.22 F.G. Friedlander: *Sound Pulses* (University Press, Cambridge 1958)
5.23 V.B. Poruchikov: *Methods of the Dynamic Theory of Elasticity* (Nauka, Moscow 1986) [in Russian]
5.24 H. Bateman, A. Erdelyi: *Higher Transcendental Functions*, Vol. 1 (McGraw-Hill, New York 1953)

Chapter 6

6.1 N. Wiener, E. Hopf: Sitzungsber. Preuss. Akad. Wiss. 696 (1931)
6.2 V.A. Fok: Dokl. Akad. Nauk SSSR **36**, 4-5, 147 (1942) [in Russian]
6.2 V.A. Fok: Dokl. Akad. Nauk SSSR **36**, 4-5, 147 (1942) [in Russian]
6.3 V.A. Fok: Matem. Sbornik **14**, 1-2, 3 (1944) [(in Russian]
6.4 P.M. Morse, H. Feshbach: *Methods of Theoretical Physics*, (McGraw-Hill, New York 1953) p. 1
6.5 B. Noble: *Methods Based on the Wiener-Hopf Technique for the Solution of Partial Differential Equations* (Pergamon, London 1958)
6.6 A.W. Maue: Z. Angew. Math. Mech. **33**, 1/2, 1 (1953)
6.7 A.W. Maue: Z. Angew. Math. Mech. **34**, 1 (1954)
6.8 L.M. Flitman: J. Appl. Math. Mech. (PMM) **23**, 997 (1959)
6.9 L.M. Flitman: J. Appl. Math. Mech. (PMM) **26**, 1582 (1962)
6.10 A. Sommerfeld: *Vorlesungen über theoretische Physik*, Band II: Mechanik der deformierbaren Medien (Leipzig 1949)
6.11 L.M. Flitman: Izv. Akad. Nauk SSSR, Ser. Geofiz. **1**, 105 (1958)
6.12 A.Ya. Sagomonyan, V.B. Poruchikov: *Three-Dimensional Problems on Transient Motion of Compressible Fluid* (MGU, Moscow 1970) [in Russian]
6.13 L.M. Flitman: Prikl. Mat. Mekh. **22**, 6, 829 (1958) [in Russian]
6.14 V.M. Seymov: *Dynamic Contact Problems* (Naukova Dumka, Kiev 1976) [in Russian]
6.15 L.M. Flitman: J. Appl. Math. Mech. (PMM) **26**, 1582 (1962)
6.16 L.I. Slepyan, Yu.S. Yakovlev: *Integral Transforms in Non-Stationary Problems of Mechanics* (Sudostroenie, Leningrad 1980) [in Russian]
6.17 M.C. Gutzwiller: Phil. Trans. Roy. Soc., London, Ser. A **255**, 153 (1962)
6.18 V.A. Afanas'ev: Izv. Akad. Nauk SSSR, Ser. Fiz. Zemli **1**, 38 (1969) [in Russian]
6.19 B.V. Kostrov: J. Appl. Math. Mech. (PMM) **30**, 1241 (1966)
6.20 B.V. Kostrov: J. Appl. Math. Mech. (PMM) **38**, 511 (1974)
6.21 B.V. Kostrov: "Dynamic Crack Propagation with Variable Rate", in *Mechanica Zniszczenia. Teoria i Zastosowania* (Wydawn. Polskiej Acad. Nauk, Warszawa 1976)

6.22 L.I. Slepyan, A.L. Fishkov: "Mixed Two-Dimensional Problem with Non-Uniform Motion of the Point Separating the Boundary Conditions", in *Current Problems of Continuum Mechanics*, issue 13 (LGU, Leningrad 1980) [in Russian]
6.23 L.I. Slepyan: *Mechanics of Cracks* (Sudostroenie, Leningrad 1981) [in Russian]
6.24 V.A. Saraykin, L.I. Slepyan: Izv. Akad. Nauk SSSR, Mekh. Tverd. Tela **4**, 54 (1979) [in Russian]
6.25 V.S. Vladimirov: *Equations of Mathematical Physics* (Mir, Moscow 1984)
6.26 V.S. Vladimirov: *Generalised Functions in Mathematical Physics* (Mir, Moscow 1979)
6.27 B.V. Kostrov, L.V. Nikitin, L.M. Flitman: Izv. Akad. Nauk SSSR, Mekh. Tverd. Tela **3**, 112 (1969) [in Russian]
6.28 Ye.N. Sher: PMTF **3**, 175 (1969) [in Russian]

Chapter 7

7.1 F.I. Fedorov: *Theory of Elastic Waves in Crystals* (Plenum, New York 1968)
7.2 M.J.P. Musgrave: *Crystal Acoustics* (Holden-Day, San Francisco 1970)
7.3 G.I. Petrashen': *Propagation of Waves in Anisotropic Elastic Continua* (Nauka, Leningrad 1980) [in Russian]
7.4 E.A. Kraut: Rev. Geophys. **1**, 401 (1963)
7.5 S. Champin: Wave Motion **3**, 343 (1981)
7.6 V.A. Sveklo: Dokl. Akad. Nauk SSSR **59**, 5, 871 (1948) [in Russian]
7.7 V.A. Sveklo: J. Appl. Math. Mech. (PMM) **26**, 1354 (1962)
7.8 V.A. Sveklo: Uch. Zap. Leningr. Gos. Univ., Ser. Mat. Nauk **17**, 28 (1949) [in Russian]
7.9 I.O. Osipov: "On a Plane-Strain Problem with a Point Source Inside an Anisotropic Solid", in *Propagation of Elastic and Elastic/Plastic Waves* (FAN, Tashkent 1969) [in Russian]
7.10 G.F. Miller, M.J.P. Musgrave: Proc. Roy. Soc. A **236**, 352 (1956)
7.11 M.J.P. Musgrave: Proc. Camb. Phil. Soc. **53**, 897 (1957)
7.12 I.O. Osipov: J. Appl. Math. Mech. (PMM) **36**, 874 (1972)
7.13 V.A. Sveklo: J. Appl. Math. Mech. (PMM) **25**, 1324 (1961)
7.14 R. Burridge: Quart. J. Mech. Appl. Math. **24**, 1, 81 (1971)
7.15 V.A. Saraykin: Phys. Technol. Development Min. Resources **3**, 52 (1974) [in Russian]
7.16 V.A. Saraykin: Phys. Technol. Development Min. Resources **4**, 65 (1974) [in Russian]
7.17 C. Atkinson: Int. J. Eng. Sci. **3**, 1, 77 (1965)
7.18 J.R. Willis: Phil. Trans. Roy. Soc. London, Ser.A **274**, 1240, 435 (1973)
7.19 R. Burridge, J.R. Willis: Proc. Cambr. Phil. Soc. **66**, **2**, 443 (1969)
7.20 V.N. Odintsev: "Some Three-Dimensional Self-Similar Elastodynamic Problems"; Ph.D. Thesis, Moscow Phys.-Techn. Institute (1973) [in Russian]
7.21 V.I. Osaulenko: "Mixed Elastodynamic Problems for Domains with Circular Line Separating Boundary Conditions and Geophysical Applications of These Problems"; Ph.D. Thesis, Moscow Institute of Physics of the Earth (1982) [in Russian]
7.22 V.S. Vladimirov: *Generalized Functions in Mathematical Physics* (Mir, Moscow 1979)
7.23 L. Hörmander: *Linear Partial Differential Operators* (Springer, Berlin Göttingen Heidelberg 1963)
7.24 R. Ludwig: Comm. Pure Appl. Math. **19**, 1, 49 (1966)
7.25 R. Burridge: Quart. J. Mech. Appl. Math. **23**, 2, 217 (1970)
7.26 N.I. Muskhelishvili: *Singular Integral Equations* (Nordhoff, Groningen 1953)
7.27 B.V. Kostrov: J. Appl. Math. Mech. (PMM) **28**, 1077 (1964)
7.28 B.V. Kostrov: J. Appl. Math. Mech. (PMM) **28**, 1, 793 (1964)
7.29 V.V. Panasyuk: *A Limiting Equilibrium State of Brittle Solids with Cracks* (Naukova Dumka, Kiev 1968) [in Russian]
7.30 K.B. Broberg: Arkiv Fys. **18**, 2, 159 (1960)
7.31 J.D. Achenbach, L.M. Brock: J. Elasticity **1**, 1, 51 (1971)

Subject Index

Springer-Verlag and the Environment

We at Springer-Verlag firmly believe that an international science publisher has a special obligation to the environment, and our corporate policies consistently reflect this conviction.

We also expect our business partners – paper mills, printers, packaging manufacturers, etc. – to commit themselves to using environmentally friendly materials and production processes.

The paper in this book is made from low- or no-chlorine pulp and is acid free, in conformance with international standards for paper permanency.

MIX
Papier aus verantwortungsvollen Quellen
Paper from responsible sources
FSC® C105338

If you have any concerns about our products,
you can contact us on
ProductSafety@springernature.com

In case Publisher is established outside the EU,
the EU authorized representative is:
Springer Nature Customer Service Center GmbH
Europaplatz 3, 69115 Heidelberg, Germany

Printed by Libri Plureos GmbH
in Hamburg, Germany